BODY TENSOR FIELDS IN CONTINUUM MECHANICS

BODY TENSOR FIELDS IN CONTINUUM MECHANICS

With Applications to Polymer Rheology

ARTHUR S. LODGE
UNIVERSITY OF WISCONSIN
MADISON, WISCONSIN

ACADEMIC PRESS New York San Francisco London 1974
A Subsidiary of Harcourt Brace Jovanovich, Publishers

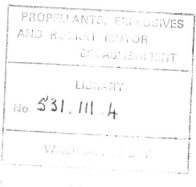

ACADEMIC PRESS, INC.
111 Fifth Avenue, New York, New York 10003

United Kingdom Edition published by
ACADEMIC PRESS, INC. (LONDON) LTD.
24/28 Oval Road, London NW1

Library of Congress Cataloging in Publication Data

Lodge, A S

Body tensor fields in continuum mechanics, with applications to polymer rheology.

Bibliography: p.

1. Continuum mechanics. 2. Calculus of tensors. 3. Viscoelasticity. 4. Polymers and polymerization. I. Title.

QA808.2.L6 531 73-18968
ISBN 0-12-454950-0

AMS(MOS) 1970 Subject Classifications: 73B05, 73B20, 73G15, 76A10

PRINTED IN THE UNITED STATES OF AMERICA

CONTENTS

3 The Kinematics of Shear Flow and Shear-Free Flow

4 Cartesian Vector and Tensor Fields

5 Relative Tensors, Field Transfer, and the Body Stress Tensor Field

6 Constitutive Equations for Viscoelastic Materials

7 Reduced Constitutive Equations for Shear Flow and Shear-Free Flow

8 Covariant Differentiation and the Stress Equations of Motion

9 Stress Measurements in Unidirectional Shear Flow: Theory

10 Constitutive Predictions and Experimental Data

11 Relations between Body- and Space-Tensor Formalisms

PREFACE

"When *I* use a word," Humpty Dumpty said, in a rather scornful tone,
"It means just what I choose it to mean—neither more nor less."

Lewis Carroll, 1872

The main object of this book is to define body tensor fields and to show how they can be used to advantage in continuum mechanics, which has hitherto been treated with space tensor fields. General tensor analysis is developed from first principles, using a novel approach that also lays the foundations for other applications, e.g., to differential geometry and relativity theory. The applications given here lie in the field of polymer rheology, treated on the macroscopic level, in which relations between stress and finite-strain histories are of central interest.

For our purposes, it is essential to abandon a widely used terminology that confuses a tensor field with its component functions. Just as a function is logically distinct from its value at a point, so a tensor field is logically distinct from its representative matrix in some coordinate system. In fact, we shall define a tensor field associated with a given geometric manifold as a function of special type whose domain is a set of coordinate systems for the manifold and whose range is a set of matrix functions of position on the manifold. Body tensor fields are obtained by taking the geometric manifold to be the continuous deforming material under investigation in continuum mechanics. Space tensor fields are obtained by taking the geometric manifold to be the space of places through which the body moves, and are *Cartesian* or *general* according as the space coordinate systems (used in the definition) are, or are not, restricted to be rectangular Cartesian. Body tensor fields are *general* in this sense.

This method of defining tensor fields, kindly suggested by Dr. H. B. Shutrick and Professor C.-W. de Boor, involves only the familiar notions "function" and "coordinate system" and is less abstract than the alternative, coordinate-free method used by Bishop and Goldberg (1968) and the alternative method involving abstraction from equivalence classes (F. and R. Nevanlinna, 1959; Lodge, 1964, pp. 306, 307, 310; 1972). Moreover, coordinate systems used in the definition are, of course, useful in subsequent applications to actual problems. For the most part, continuum mechanics can be treated by introducing a few second-rank tensor fields, in contrast to relativity theory in which fourth-rank tensor fields play a central role; we shall not follow Penrose (1968) who uses a notation that enables one to recognize the rank and kind of a given tensor field. Our readers are asked to remember the meaning of the few symbols used here to denote tensor fields in our treatment of continuum mechanics: $\boldsymbol{\pi}(\mathrm{P}, t)$ and $\boldsymbol{\gamma}(\mathrm{P}, t)$ for the contravariant body stress and covariant body metric tensor fields evaluated at particle P and time t; $\boldsymbol{p}(\mathrm{Q}, t)$ and $\boldsymbol{g}(\mathrm{Q})$ for the contravariant space stress and covariant space metric tensor fields evaluated at place Q and time t; and $\mathbf{p}(\mathrm{Q}, t)$ and $\mathbf{U}$ for the Cartesian space stress and unit tensor fields.

In applications to continuum mechanics, body tensor fields, general space tensor fields, and Cartesian space tensor fields can be regarded as three different but interchangeable tools; all results derived with one tool can be rederived using either of the others. In the classical theories of hydrodynamics and small-strain elasticity, where the constitutive equations are given, there is little, if any, point in using body tensor fields. In polymer rheology, however, where the determination of constitutive equations applicable to a given rubberlike solid or polymeric, viscoelastic, liquid is one of the main tasks of current research, we claim certain advantages of convenience for the use of body tensor fields, and we believe that it is worthwhile to be able to use both space and body tensor fields as the occasion demands. This book aims to make this possible.

In comparison with space tensor fields, body tensor fields give a simpler description of stress, strain, and their rheologically useful time derivatives and time integrals. In particular, $\boldsymbol{\pi}(\mathrm{P}, t)$ gives a one-state description of stress;

$\gamma(P, t) - \gamma(P, t')$ gives a two-state, additive description (antisymmetric in t and t') of strain (whether small or large); both tensors are unaltered by any superposed rigid body motion (relative to axes fixed in space); and body tensor equations in which π and γ are the only tensor variables are automatically admissible as possible constitutive equations for elastic or viscoelastic materials. There are no space tensor fields for which all statements corresponding to these are valid. Using body tensor fields, manipulations of constitutive equations involving time derivatives and time integrals are simplified, and so also are the proofs of certain general results (e.g., the superposition theorem 3.3(25) for shear flows, and the low-frequency relations (1)–(5) of Section 7.2); one is also free to choose, according to context, whether or not to introduce a "reference state." On the other hand, it could be argued that the stress equation of motion expressed in body tensors is more complicated than the corresponding Cartesian tensor equation, due to the term involving the divergence of stress; but the Cartesian tensor equation when referred to an arbitrary curvilinear space coordinate system is equally complicated, and the body tensor equation can always be referred to a body coordinate system, specially chosen for a given problem, which is rectangular Cartesian, or at least orthogonal, at the instant t at which the stress equation of motion has to be invoked.

The body tensor formalism is very closely related to the convected component formalism used by Oldroyd (1950a) to give a fundamental and very general discussion of the formulation of constitutive equations (Section 11.1). The relation between body tensors and convected components of space tensors was stated without proof (Lodge, 1951) and with proof (Lodge, 1972). The advantages of using body tensor fields were summarized (Lodge, 1954, 1964), definitions were given (Lodge, 1964, 1972), and a notation for the body tensors themselves was introduced (Lodge, 1972). The relation to the Lagrangian formulation of hydrodynamics is discussed in Section 11.4, and to the "embedded vector" formalism of Lodge (1964) in Section 11.2.

This book is a development of Chapter 12 of "Elastic Liquids" (Lodge, 1964), and is based on a course of 45 lectures given at the graduate level to students of engineering and mathematics having a knowledge of the elements of matrices, determinants, and elementary vector analysis. The usual knowledge of differential and integral calculus is assumed. Much of the important bookwork is set as problems for the reader; complete solutions are given at the end of the book. Guides for readers having various goals are offered in the following tabulation.

Background	Present interests	Suggested first reading
Elementary vector analysis	General tensor analysis and continuum mechanics (excluding polymer rheology)	Chapters 1, 2, 4, 5, 8; Sections 6.1, 6.3, 6.9 (omit Sections 1.4, 2.8, 5.4, 5.7, 5.8, and 8.8)
Cartesian tensors and continuum mechanics	Body tensor fields in continuum mechanics	Sections 2.1–2.7, 3.1–3.4, 4.1–4.5, 5.4, 5.6, 6.1, 8.3, 8.5, 8.10, 11.3
Elementary vector analysis	General and Cartesian tensor analysis	Chapters 1, 2, 4, 5, 8 (omit Sections 1.4, 5.6–5.8, 8.5, 8.8)
Cartesian tensors and polymer rheology	Polymer rheology	Chapters 3, 6, 7, 9, 10; Sections 2.3–2.7, 5.4–5.8
Polymer rheology	New results in polymer rheology	Sections 3.2(15), 3.2(20), 3.3(25), 3.3(37), 3.7(10), 3.8(14); 5.8, 6.1; 7.1(11), 7.4(5), 9.2(9), 10.1(4)(i), 10.2(16), 10.4(4)

ACKNOWLEDGMENTS

Much of the work in this book was performed from 1948 to 1960 at the British Rayon Research Association. It was further developed at the University of Manchester Institute of Science and Technology and at the University of Wisconsin—Madison. I owe a great deal to many discussions with present and former colleagues and students: Professors R. B. Bird, C.-W. de Boor, M. W. Johnson, Jr., J. Barkley Rosser, K. Walters, and J. Kramer; Drs. A. Kaye, H. B. Shutrick, and R. C. Armstrong; and Mr. D. Segalman. I am particularly indebted to Dr. R. C. Armstrong, who read much of the first draft, and to Professor K. Walters, who read all the final draft and prepared Appendix F. I acknowledge gratefully research support received from the Petroleum Research Fund of the American Chemical Society (Unrestricted Grant PRF 1758-C to R. B. Bird), the National Science Foundation (Grant GK-15611 to A.S.L.), and the United States Army (Contract No. DA-31-124-ARO-D-462 to the Mathematics Research Center, University of Wisconsin—Madison).

LIST OF NOTATION

$g = [g_{rc}]$	3×3 matrix with element g_{rc} in row r and column c
$\tilde{g} = [g_{cr}]$	transpose of g
$\lvert g \rvert$	determinant of g
$u = [u_r]$	3×1 column matrix having element u_r in row r
$\tilde{u} = [u_c]$	3×1 row matrix having element u_c in column c
$u_i v^i$	$u_1 v^1 + u_2 v^2 + u_3 v^3$
$u_{\underset{\sim}{i}} v^{\underset{\sim}{i}}$	$u_1 v^1, u_2 v^2, u_3 v^3$ (when $i = 1, 2, 3$)
$\boldsymbol{u} \cdot \boldsymbol{v}$	scalar product of vectors $\boldsymbol{u}$ and $\boldsymbol{v}$
$\boldsymbol{uv}$	outer (or direct) product of $\boldsymbol{u}$ and $\boldsymbol{v}$
$\boldsymbol{u} \times \boldsymbol{u}'$	vector product of $\boldsymbol{u}$ and $\boldsymbol{u}'$
$\{P\}$	set of elements of which P denotes a typical member
$B : P \rightarrow \xi$	body coordinate system B; particle P has coordinates $\xi = [\xi^1, \xi^2, \xi^3]$
$S : Q \rightarrow x$	space coordinate system S; place Q has coordinates $x = [x^1, x^2, x^3]$
$C : Q \rightarrow y$	rectangular Cartesian space coordinate system; $y = [y^1, y^2, y^3]$
$\delta_{ij}, \delta^{ij}, \delta^i_j$	1 if $i = j$; 0 if $i \neq j$
$I = [\delta_{rc}]$	3×3 unit matrix

$\tilde{\delta}_k$	$[\delta_{rk}]$ ($k = 1, 2, 3$); e.g., $\tilde{\delta}_1 = [1, 0, 0]$
D/Dt	hydrodynamic derivative, applied to a function of x^i and t
$\mathscr{D}/\mathscr{D}t$	convected derivative, applied to a function of x^i and t
diag$[a,b,c]$	$\begin{bmatrix} a & 0 & 0 \\ 0 & b & 0 \\ 0 & 0 & c \end{bmatrix}$
$\mathscr{R}$	set of all real numbers
A_3	set of all ordered triples of real numbers
$A_{3\times 1}$	set of all real 3×1 column matrices
$A_{m\times n}$	set of all real $m \times n$ matrices
$\lVert \boldsymbol{u} \rVert$	magnitude of $\boldsymbol{u}$

Type of tensor field:	Body	General space	Cartesian space
Type used:	Bold Greek	Bold Italic	Bold sans serif (roman)
Unit tensor	$\boldsymbol{\delta}$	$\boldsymbol{I}$	U
Base vectors	$\boldsymbol{\beta}_i(B, \mathrm{P})$, $\boldsymbol{\beta}^i(B, \mathrm{P})$	$\boldsymbol{b}_i(S, \mathrm{Q})$, $\boldsymbol{b}^i(S, \mathrm{Q})$	$\mathsf{b}_i(S, \mathrm{Q})$, $\mathsf{b}^i(S, \mathrm{Q})$
Metric tensor	$\boldsymbol{\gamma}(\mathrm{P}, t)$	$\boldsymbol{g}(\mathrm{Q})$	U
Stress tensor	$\boldsymbol{\pi}(\mathrm{P}, t)$	$\boldsymbol{p}(\mathrm{Q}, t)$	$\mathsf{p}(\mathrm{Q}, t)$
Extra-stress tensor	$\boldsymbol{\Pi}(\mathrm{P}, t)$	$\boldsymbol{P}(\mathrm{Q}, t)$	$\mathsf{P}(\mathrm{Q}, t)$

$\Lambda_{\mathrm{P}}(B, \bar{B})$	$[\partial\bar{\xi}^r/\partial\xi^c]_{\mathrm{P}}$ (transformation matrix for body coordinate systems B, $\bar{B}$).
$L_{\mathrm{Q}}(S, \bar{S})$	$[\partial\bar{x}^r/\partial x^c]_{\mathrm{Q}}$ (transformation matrix for space coordinate systems S, $\bar{S}$)
$L(C, \bar{C})$	$[\partial\bar{y}^r/\partial y^c]$ (transformation matrix for rectangular Cartesian coordinate systems C, $\bar{C}$)

1 SOME MATHEMATICAL PREREQUISITES

1.1 Primitive concepts

Our aim is to build a mathematical formalism capable of describing the behavior and properties of a continuous geometric manifold. One develops formalisms in mathematics by choosing certain *primitive concepts* (or *undefined elements*) which are to be accepted without further logical analysis. One then assigns properties to these primitive concepts by means of *axioms* (or unproved propositions), and then deduces various theorems. The primitive concepts may invite deeper logical analysis, but such analysis would be part of a different activity. The primitive concepts need not necessarily be the smallest number which could be used; in fact, we shall choose them to enable us to get on with the development of the required formalism as quickly as possible. It is necessary, however, to list the primitive concepts to be used (Robison, 1969, p. 2). Our list is as follows.

(1) **Primitive concepts** Set of distinguishable elements; point, particle, place; order, relation, correspondence; rigid body, mass, force, time.

We shall also use the standard formalisms of analysis, calculus, and geometry. "Set" is synonymous with "collection, aggregate, class." "Correspondence" is synonymous with "mapping, function."

We use the standard notation $\{x\}$ for a set of elements of which x is a member. When it is necessary, we include the condition for membership after a vertical line, and we use the symbol $\in$ to mean "is a member of the set . . ."; thus, for example, if N denotes the set of integers, then $\{2n \mid n \in N\}$ denotes the set of even numbers.

Let $X = \{x\}$ and $Y = \{y\}$ denote sets. A *function* f from X into Y, written $f: X \to Y$, is a rule which assigns to *each* element $x \in X$ a unique element $y \in Y$, written $y = fx$ or $y = f(x)$ or $f: x \to y$. For example, if $X = Y = \mathscr{R}$ (the set of all real numbers), we could have $y = \sin x$, in which case $f \equiv$ sine. Other words used for function are mapping, map, transformation, correspondence, and operator.

The most familiar use of the term function is that in which $X = Y = \mathscr{R}$. It is essential for our purposes to realize that the idea "function" is in no way restricted to such applications and can be applied to any sets of distinguishable elements. Thus, for example, we shall define a coordinate system $B: \mathrm{P} \to \xi$ as a correspondence or function which assigns to each particle P of a body $\mathscr{P} = \{\mathrm{P}\}$ a unique ordered set ξ of three real numbers ξ^1, ξ^2, ξ^3. Thus we may write $B: \mathscr{P} \to A_3$, where A_3 denotes the arithmetic three-space, the set of all ordered sets of three real numbers. The reader will notice that we have defined a coordinate system in terms of the primitive concepts: particle, correspondence, order. One might feel uneasy about defining things in terms of undefined elements (or primitive concepts), but such a procedure is in fact valid and unavoidable at the outset; one simply gets used to it.

For any function $f: X \to Y$, with $y = fx$, the set X is called the *domain* of f (i.e., it is the set of all values of x for which fx is defined), and the set $\{fx \mid x \in X\}$ is called the *range* of f (i.e., it is the set of all values of $y = fx$). If $fX = Y$, i.e., if Y is composed entirely of elements fx, we say that f maps X *onto* Y; otherwise, f maps X *into* Y. If for every $y \in fX$ there is only one x such that $y = fx$, then the function or correspondence f is said to be *one-to-one*, and the inverse f^{-1} exists which maps fX onto X. In the example given above, a coordinate system $B: \mathscr{P} \to A_3$ has a subset of A_3 as range and the body $\mathscr{P}$ as domain, and the correspondence B is one-to-one.

An integral $y = \int_a^b f(t')\,dt'$ is an example of an important type of mapping called a *functional*: Such an integral assigns a unique real number y to each integrable real function f with domain (a, b). More generally, a functional $\mathscr{F}: F \to \mathscr{R}$, where F denotes the set of all functions $f: \mathscr{R} \to \mathscr{R}$ [usually subject to some restriction, such as integrability, and having some specified domain, (a, b), say] is a rule which assigns to each *function* $f \in F$ a unique number

$y = \mathscr{F}f$. When it is necessary to make explicit the domain of f, we write $y = \mathscr{F}\left\{ \overset{b}{\underset{a}{f(t')}} \right\}$. We shall need more elaborate functionals in which $f(t')$ is replaced by a 3×3 matrix function of t' and the value y is a 3×3 matrix.

A set $X = \{x\}$ is called a *group* if there exists a rule which assigns to each ordered pair x, x' of elements in X an element, written xx', in X with the following properties: if $x'' \in X$, then $(xx')x'' = x(x'x'')$; there exists a unit element $1 \in X$ such that $1x = x1 = x$ for every $x \in X$; there exists an inverse element $x^{-1} \in X$ such that $xx^{-1} = x^{-1}x = 1$ for every $x \in X$. The notation $(xx')x''$ means: form xx' first, and then take the element $(xx')x''$ which corresponds to element xx' and x'', in that order.

1.2 Linear spaces

(1) **Definition** A set $X = \{\boldsymbol{x}\}$ of elements $\boldsymbol{x}$, $\boldsymbol{y}$, $\boldsymbol{z}$, ... is called a *linear space* if operations of addition and multiplication by numbers a, b, ... are defined with the following properties.

(i) *Addition* To any $\boldsymbol{x} \in X$ and any $\boldsymbol{y} \in X$ there corresponds a unique element, written $\boldsymbol{x} + \boldsymbol{y}$, in X, such that

$$\boldsymbol{x} + \boldsymbol{y} = \boldsymbol{y} + \boldsymbol{x} \qquad \text{(addition is commutative)},$$
$$(\boldsymbol{x} + \boldsymbol{y}) + \boldsymbol{z} = \boldsymbol{x} + (\boldsymbol{y} + \boldsymbol{z}) \qquad \text{(addition is associative)},$$
$$\boldsymbol{x} + \boldsymbol{0} = \boldsymbol{x} \qquad \text{(a unique zero element } \boldsymbol{0} \in X \text{ exists)},$$
$$\boldsymbol{x} + (-\boldsymbol{x}) = \boldsymbol{0} \qquad \text{(each } \boldsymbol{x} \text{ has a unique inverse, } -\boldsymbol{x}).$$

(ii) *Multiplication by a number* To any element $\boldsymbol{x}$ and any number a there corresponds a unique element, written $a\boldsymbol{x}$ or $\boldsymbol{x}a$, which is also in the set $\{\boldsymbol{x}\}$ and has the following properties:

$$a(\boldsymbol{x} + \boldsymbol{y}) = a\boldsymbol{x} + a\boldsymbol{y};$$
$$a(b\boldsymbol{x}) = (ab)\boldsymbol{x};$$
$$1\boldsymbol{x} = \boldsymbol{x};$$
$$(-1)\boldsymbol{x} = -\boldsymbol{x};$$
$$(a + b)\boldsymbol{x} = a\boldsymbol{x} + b\boldsymbol{x}.$$

The linear space $\{\boldsymbol{x}\}$ is said to be a *complex space* if a, b, ... range through all complex numbers. For most of the linear spaces used in this book, the numbers a, b, ... are real, and the linear spaces are said to be *real*.

A set of elements $\boldsymbol{y}_1, \boldsymbol{y}_2, \ldots, \boldsymbol{y}_n$ in a linear space $\{\boldsymbol{x}\}$ are said to be *linearly independent* if the only solution to the equation

(2) $$a_i \boldsymbol{y}_i = \boldsymbol{0}$$

is $a_i = 0$ ($i = 1, 2, \ldots, n$). If one or more of the coefficients a_i is not zero, the elements are said to be *linearly dependent.*

Here, and elsewhere, we use the *summation convention,* according to which a repeated letter index in a term is taken to imply summation of that term over the values of the index assigned from the context. In Eq. (2), the index i takes values 1, 2, ..., n. In the other chapters of this book, the summation usually runs over the values 1, 2, and 3. A repeated letter index is called a dummy index and can be replaced by any other letter wherever it occurs in any given term. Thus, for example, $a_i y_i = a_j y_j = \sum_{k=1}^{n} a_k y_k$, in the present context. *When we wish to suspend the summation convention, we place a caret over a repeated letter index*: Thus $a_{\hat{i}} y_{\hat{i}}$ means $a_i y_i$ (not summed). The main precaution to observe in using the summation convention is to use different repeated letter indexes if brackets are removed in a term which involves two or more separate summations; thus, for example,

$$\left(\sum_{i=1}^{n} a_i b_i\right)\left(\sum_{i=1}^{n} c_i y_i\right) = a_i b_i c_j y_j \neq a_i b_i c_i y_i .$$

A repeated index implies summation whether it occurs on one factor or on two factors; thus, for example, A_{ii} means $A_{11} + A_{22} + \cdots + A_{\hat{n}\hat{n}}$.

A linear space is said to be of *finite dimension n* if the greatest number of linearly independent elements in the space is n (a positive integer). In such a space, which is also said to be n dimensional, any set $\boldsymbol{e}_1, \boldsymbol{e}_2, \ldots, \boldsymbol{e}_n$ of n linearly independent elements is said to form a *basis* for the space, because every element $\boldsymbol{x}$ can be expressed as a linear combination of the basis elements, i.e., one can find numbers x_i, for any given $\boldsymbol{x}$, such that

$$\boldsymbol{x} = x_i \boldsymbol{e}_i . \tag{3}$$

(4) **Problem** Prove the converse of (3), i.e., if $\boldsymbol{e}_i$ ($i = 1, 2, \ldots, n$) are linearly independent elements such that every element $\boldsymbol{x}$ in the space can be expressed in the form (3), then the space is n dimensional.

(5) **Definition** A real linear space $\{\boldsymbol{x}\}$ is said to be *normed* if to every pair of elements $\boldsymbol{x}$ and $\boldsymbol{y}$ in the space there corresponds a unique real number, written $\langle \boldsymbol{x}, \boldsymbol{y} \rangle$, with the following properties:

$$\begin{aligned}
&\langle \boldsymbol{x}, \boldsymbol{y} \rangle = \langle \boldsymbol{y}, \boldsymbol{x} \rangle; \\
&\langle a\boldsymbol{x}_1 + b\boldsymbol{x}_2, \boldsymbol{y} \rangle = a\langle \boldsymbol{x}_1, \boldsymbol{y} \rangle + b\langle \boldsymbol{x}_2, \boldsymbol{y} \rangle \qquad \text{for all numbers} \quad a, b; \\
&\langle \boldsymbol{x}, \boldsymbol{x} \rangle > 0 \qquad \text{whenever} \quad \boldsymbol{x} \neq \boldsymbol{0}; \\
&\langle \boldsymbol{0}, \boldsymbol{0} \rangle = 0.
\end{aligned}$$

The number $\langle \boldsymbol{x}, \boldsymbol{y} \rangle$ is called the *inner product* of $\boldsymbol{x}$ and $\boldsymbol{y}$; the number $\|\boldsymbol{x}\| = \sqrt{\langle \boldsymbol{x}, \boldsymbol{x} \rangle}$ is called the *norm* of $\boldsymbol{x}$.

(6) **Problem** Prove that $\|\boldsymbol{x}+\boldsymbol{y}\| \leqslant \|\boldsymbol{x}\| + \|\boldsymbol{y}\|$ and that the equality sign is valid only if $\boldsymbol{x}$ and $\boldsymbol{y}$ are linearly dependent. This is called the triangle inequality.

(7) **Definition** Elements $\boldsymbol{x}$ and $\boldsymbol{y}$ in a normed linear space are *orthogonal* if $\langle \boldsymbol{x}, \boldsymbol{y} \rangle = 0$. A *basis* $\boldsymbol{e}_i$ is *orthogonal* if all its members are orthogonal to one another, and *orthonormal* if, in addition, all its members are of unit norm. Thus we may write

$$\langle \boldsymbol{e}_i, \boldsymbol{e}_j \rangle = \delta_{ij} \qquad (\boldsymbol{e}_i \quad \text{orthonormal}).$$

The Kronecker symbol δ_{ij} has the value 1 if i and j are equal, and has the value 0 if i and j are unequal. In other contexts, we shall write δ^{ij} and δ^i_j with the same meaning as δ_{ij}.

(8) **Definition** *An operator* (or transformation) $\boldsymbol{T}$ is a correspondence between elements of linear spaces $\{\boldsymbol{x}\}$ and $\{\boldsymbol{X}\}$ such that to any given element $\boldsymbol{x}$ of $\{\boldsymbol{x}\}$ there corresponds a unique element $\boldsymbol{X}$ of $\{\boldsymbol{X}\}$; we write this in the form

$$\boldsymbol{T}\colon \quad \boldsymbol{x} \to \boldsymbol{X} = \boldsymbol{T}\boldsymbol{x}.$$

The spaces $\{\boldsymbol{x}\}$ and $\{\boldsymbol{X}\}$ may be the same or different. If they are the same, an operator $\boldsymbol{T}$ is said to be *linear* if (i) it is *additive* in the sense that $\boldsymbol{T}(\boldsymbol{x}_1 + \boldsymbol{x}_2) = \boldsymbol{T}\boldsymbol{x}_1 + \boldsymbol{T}\boldsymbol{x}_2$, and (ii) it is *continuous* in the sense that, for any sequence $\boldsymbol{x}_1, \boldsymbol{x}_2, \ldots$ of elements in $\{\boldsymbol{x}\}$ which possesses a limit $\boldsymbol{x}$ also in $\{\boldsymbol{x}\}$, we have $\lim_{n\to\infty} \boldsymbol{T}\boldsymbol{x}_n = \boldsymbol{T}\boldsymbol{x}$.

(9) **Definition** The Kronecker (or direct, or tensor) product $\{\boldsymbol{x}\} \otimes \{\boldsymbol{y}\}$ of two linear spaces $\{\boldsymbol{x}\}$ and $\{\boldsymbol{y}\}$ is a linear space $\{\boldsymbol{z}\}$ constructed from *ordered* pairs $(\boldsymbol{x}, \boldsymbol{y})$ of elements (one from $\{\boldsymbol{x}\}$ and one from $\{\boldsymbol{y}\}$) and their linear combinations according to the following rules:

(i) $(a\boldsymbol{x}_1 + b\boldsymbol{x}_2, \boldsymbol{y}) = a(\boldsymbol{x}_1, \boldsymbol{y}) + b(\boldsymbol{x}_2, \boldsymbol{y})$;
(ii) $(\boldsymbol{x}, a\boldsymbol{y}_1 + b\boldsymbol{y}_2) = a(\boldsymbol{x}, \boldsymbol{y}_1) + b(\boldsymbol{x}, \boldsymbol{y}_2)$;
(iii) if $\boldsymbol{e}_i$ $(i = 1, 2, \ldots, n)$ is a basis for $\{\boldsymbol{x}\}$ and $\boldsymbol{f}_k$ $(k = 1, 2, \ldots, m)$ is a basis for $\{\boldsymbol{y}\}$, then the set $(\boldsymbol{e}_i, \boldsymbol{f}_k)$ is a basis for $\{\boldsymbol{z}\}$.

It can be shown (Eilenberg and Steenrod, 1952) that (iii) follows from (i) and (ii). The spaces $\{\boldsymbol{x}\}$ and $\{\boldsymbol{y}\}$ can be the same or different. When they are the same, an element in the product space is usually written $\boldsymbol{x}_1\boldsymbol{x}_2$ [instead of $(\boldsymbol{x}_1, \boldsymbol{x}_2)$] and is called a *dyadic product*.

We complete this brief review of linear spaces by giving some examples.

(10) The *arithmetic n-space* A_n is a linear space whose elements are $n \times 1$ matrices $[y_1, y_2, \ldots, y_n] = \boldsymbol{y}$, say, where y_i can assume all real values. It is easy to verify that the matrix rules of addition and multiplication by a number, namely $a\boldsymbol{y} + b\boldsymbol{y}' = [ay_c + by'_c]$, satisfy the rules (i) and (ii) of (1), with the zero matrix $\boldsymbol{0} = [0, 0, \ldots, 0]$ playing the part of the zero element in the space A_n, and that the space is n dimensional; one basis is the following:

$$\boldsymbol{e}_1 = [1, 0, 0, \ldots, 0], \qquad \boldsymbol{e}_2 = [0, 1, 0, \ldots, 0], \ldots, \qquad \boldsymbol{e}_n = [0, 0, \ldots, 0, 1].$$

Every real $n \times n$ matrix $\boldsymbol{T} = [T_{ij}]$ is linear operator in A_n when used according to the matrix multiplication rule

$$\boldsymbol{T}\colon \quad \boldsymbol{x} \to \boldsymbol{y} = \boldsymbol{T}\boldsymbol{x}, \qquad \text{where} \quad y_i = T_{ij} x_j.$$

The direct product $A_m \otimes A_n$ is a space whose elements are $m \times n$ matrices: $(\boldsymbol{x}, \boldsymbol{y}) = \tilde{\boldsymbol{x}}\boldsymbol{y} = [x_r y_c]$, where $\boldsymbol{x}$ is in A_m and $\boldsymbol{y}$ is in A_n, and a superior tilde denotes the transpose. A norm for A_n can be defined in various ways: The most familiar is $\langle \boldsymbol{y}, \boldsymbol{y}' \rangle = y_i y'_i$; a more general inner product is given by the equation

$$\langle \boldsymbol{y}, \boldsymbol{y}' \rangle = g_{ij} y_i y'_j, \tag{11}$$

where g_{ij} are coefficients of a positive-definite quadratic form.

(12) **Displacement vectors** In our three-dimensional Euclidean space, as defined in elementary vector analysis, the displacement vectors form a three-dimensional linear space. If Q_i are places (points) in the Euclidean space, the addition of vectors is defined by the usual triangle law $(\overrightarrow{Q_1Q_2} + \overrightarrow{Q_2Q_3} = \overrightarrow{Q_1Q_3})$, and multiplication by a number is defined by the equation $a\overrightarrow{Q_1Q_2} = \overrightarrow{Q_1Q_3}$, where Q_1, Q_2, and Q_3 are collinear and $aQ_1Q_2 = Q_1Q_3$. It is easy to verify that the rules (i) and (ii) of (1) are obeyed, and that the usual scalar product $\overrightarrow{Q_1Q_2} \cdot \overrightarrow{Q_1Q_3} = (Q_1Q_2)(Q_1Q_3)\cos\theta$ (where θ is the angle between Q_1Q_2 and Q_1Q_3) satisfies the conditions (5) for an inner product.

(13) **Function space** A simple example of a linear space of infinite dimension is the function space whose elements are continuously differentiable functions $f(t)$, $g(t)$, ... of a real variable t in some interval $(a \leqslant t \leqslant b)$. The sum of elements $f(t)$ and $g(t)$ is just the sum $f(t) + g(t)$ of the functions, and the product of $f(t)$ by a number x is just the function $xf(t)$. Fourier's theorem shows that the set of functions $\exp(\pm 2\pi nit)$ $(n = 0, 1, 2, \ldots)$ form a basis for this space, whose dimension is therefore countably infinite. The definition $\langle f, g \rangle = \int_a^b f(t)g(t)\,dt$ gives one possible inner product, and a suitably restricted

function $T(s, t)$ of two real variables s, t is a linear operator in the space when used according to the equation

$$(14) \qquad T\colon\ f \to g = Tf, \qquad \text{where} \quad g(s) = \int_a^b T(s, t)f(t)\, dt.$$

(15) **Nomenclature: Algebraic vectors and geometric vectors** The elements of a linear space are often called *vectors* in the mathematical literature. When necessary, we shall call them *algebraic vectors* in order to distinguish them from what we shall later define as *geometric vectors*. Similarly, an operator $\boldsymbol{T}$ in a linear space is often called a tensor (of second rank or second order); we shall call it an *algebraic tensor* to distinguish it from *geometric tensors*, defined later. Geometric vectors and tensors are (in our nomenclature) always associated (in certain specific ways) with one or another geometric manifold (e.g., the space of places in which we live, or the body manifold consisting of the point particles of some continuous deforming body), but algebraic vectors and tensors may, or may not, be so associated. A geometric vector or tensor is always an element of some linear space and hence a geometric vector is always an algebraic vector. The converse is not true, however; for example, the elements of the arithmetic n-space A_n are not geometric vectors; they can be put in one-to-one correspondence with geometric vectors, as we shall frequently show when we choose a particular coordinate system for a geometric manifold: The *components* of a geometric vector in a given coordinate system are given by an $n \times 1$ matrix with real elements.

1.3 Matrices and determinants

In this section, we summarize those properties of matrices and determinants that we use in later chapters. It is sufficient for our purposes to consider real, $n \times n$ matrices, with $n = 3$ in most cases. For the most part, we state the results and leave the proofs as exercises for the reader. Proofs can be found in many standard texts on matrices, but in some cases our restriction to real 3×3 matrices enables us to give (pp. 279–282) elementary, unsophisticated proofs, which, we hope, will save time for those whose main concern is with the applications which we make later in this book.

In this section, letter suffixes assume the values $1, 2, \ldots, n$, where n is any positive integer. An $n \times n$ matrix $A = [A_{rc}]$ is said to be of *order* n. We deal with matrix equations in which each matrix is a square matrix of order n. If x and y are numbers, the sum of matrices and product by a number is defined by the equation $xA + yB = [xA_{rc} + yB_{rc}]$, where $B = [B_{rc}]$. The product of square matrices is defined by the equation $AB = [A_{ri} B_{ic}]$. The zero matrix 0 has every element equal to zero; the unit matrix $I = [\delta_{rc}]$. A diagonal matrix $[x_{\hat{r}} \delta_{\hat{r}c}]$ has elements off the main diagonal equal to zero, and is written in the form $[x_{\hat{r}} \delta_{\hat{r}c}] = \text{diag}[x_i]$ or $\text{diag}[x_1, x_2, \ldots, x_n]$. We use the following notation and definitions:

$$
\begin{aligned}
|A| &= \det A_{ij} \qquad \text{(the determinant of } A\text{);} \\
\tilde{A} &= [A_{cr}] \qquad \text{(the transpose of } A\text{);} \\
\text{if } \tilde{A} &= A, \quad A \text{ is symmetric;} \\
\text{if } \tilde{A} &= -A, \quad A \text{ is antisymmetric (or skew symmetric),} \\
\text{Tr } A &= A_{ii} \qquad \text{(the trace of } A\text{);} \\
\text{if } |A| &\neq 0, \quad A \text{ is nonsingular.}
\end{aligned}
$$

When A is nonsingular, there is a unique inverse (or reciprocal) matrix, written A^{-1}, such that $AA^{-1} = I$; it has the property that $A^{-1}A = I$, and is given by the equation

$$A^{-1} = |A|^{-1}[a^{cr}], \tag{1}$$

where a^{rc} is the cofactor of A_{rc} in $|A|$ ($[a^{cr}]$ is often called the adjoint of A, but we avoid this usage because it differs from the usage in the term self-adjoint). The inverse of a product of matrices is the product of inverse matrices with order reversed. The transpose of a product of matrices is the product of transposed matrices with order reversed. Transpose and inverse operations commute. The value of the trace of a product of matrices is unaltered by cyclic permutation of the factors.

A symmetric matrix A is said to be *positive definite* if it is the coefficient matrix of a positive-definite quadratic form, i.e., if $\tilde{x}Ax = x_i A_{ij} x_j \geqslant 0$ for all real values of the variables x_i, and the equality sign holds only if all x_i are zero.

(2) **Problem** Prove that the following is a necessary and sufficient set of conditions for the 3×3 matrix A to be positive definite:

$$A_{11} > 0; \qquad a^{33} \equiv \begin{vmatrix} A_{11} & A_{12} \\ A_{12} & A_{22} \end{vmatrix} > 0; \qquad |A| > 0.$$

(For an $n \times n$ matrix, there is a similar set of conditions.) Alternative sets of conditions can evidently be obtained by starting with A_{22}, or with A_{33}, instead of A_{11}. It follows, in particular, that a positive-definite matrix possesses an inverse. Since a positive-definite matrix is usually used in connection with a quadratic form, there is no real loss of generality in restricting the matrix to be symmetric; in future, we shall therefore take the term positive-definite matrix to imply that the matrix is symmetric.

(3) **Problem** If A is an $n \times n$ matrix, a^{rc} is the cofactor of A_{rc} in $|A|$, and α_{rc} is the cofactor of a^{rc} in $|a| = \det a^{rc}$, prove that

$$\text{(i)}\quad |a| = |A|^{n-1} \qquad \text{and} \qquad \text{(ii)}\quad [\alpha_{rc}] = |A|^{n-2}A.$$

(4) **Problem** If A is a positive-definite 3×3 matrix, prove that A^{-1} is positive definite. (A similar result holds for an $n \times n$ matrix.)

A square matrix R is said to be *orthogonal* if

(5) $$\tilde{R} = R^{-1}.$$

It follows that

(6) $$\tilde{R}R = R\tilde{R} = I$$

(with finite matrices, a left inverse is necessarily also a right inverse), and hence that

(7) $$|R| = \pm 1.$$

If $|R| = +1$, R is said to be a *proper* orthogonal matrix.

(8) **Problem** Show that any 3×3 orthogonal matrix R can be written in the form

$$R = \begin{bmatrix} l_1 & l_2 & l_3 \\ m_1 & m_2 & m_3 \\ n_1 & n_2 & n_3 \end{bmatrix},$$

where (l_1, m_1, n_1), (l_2, m_2, n_2), and (l_3, m_3, n_3) denote actual direction cosines of three mutually perpendicular directions in space. Show also that every such matrix is orthogonal, and that (l_1, l_2, l_3), (m_1, m_2, m_3), and (n_1, n_2, n_3) are also actual direction cosines of three mutually orthogonal directions in space.

The roots λ_i $(i = 1, 2, \ldots, n)$ in λ of the *characteristic equation*

(9) $$|A - \lambda I| = 0$$

for any $n \times n$ matrix A are called *latent roots*, *characteristic values*, or *eigenvalues*. A column matrix $x^{(1)} = [x_r^{(1)}]$ which satisfies the equation

(10) $$(A - \lambda_1 I)x^{(1)} = 0$$

is called an *eigenvector* of A belonging to the eigenvalue λ_1. It is clear that (9), with $\lambda = \lambda_1$, is a necessary and sufficient condition that (10) should have a nonzero solution for $x^{(1)}$; the solution is undetermined to the extent of an arbitrary numerical factor. Similar statements hold for eigenvectors belonging to the other eigenvalues.

When A is a 3×3 matrix, the characteristic equation is a cubic

(11) $$|A - \lambda I| \equiv -\lambda^3 + A_{\mathrm{I}}\lambda^2 - A_{\mathrm{II}}\lambda + A_{\mathrm{III}} = 0,$$

where

(12) $$A_{\mathrm{I}} = A_{11} + A_{22} + A_{33} = \lambda_1 + \lambda_2 + \lambda_3;$$

(13) $$A_{\mathrm{II}} = A_{22}A_{33} + A_{33}A_{11} + A_{11}A_{22} - A_{23}A_{32} - A_{31}A_{13} - A_{12}A_{21} = \lambda_2\lambda_3 + \lambda_3\lambda_1 + \lambda_1\lambda_3;$$

(14) $$A_{\mathrm{III}} = |A| = \lambda_1\lambda_2\lambda_3.$$

(15) **Problem** Prove that the eigenvalues and eigenvectors for a real symmetric matrix are all real.

(16) **Problem** If two eigenvalues λ_1 and λ_2 of a real, symmetric matrix A are distinct, prove that the two corresponding eigenvectors are orthogonal in the sense that $\tilde{x}^{(1)}x^{(2)} = 0$.

(17) **Problem** Given any real, symmetric 3×3 matrix A, show that there exists an orthogonal matrix R such that

$$\tilde{R}AR = \mathrm{diag}[\lambda_1, \lambda_2, \lambda_3],$$

where λ_1, λ_2, and λ_3 are the eigenvalues of A. [Consider separately the cases (i) $\lambda_1, \lambda_2, \lambda_3$ unequal, (ii) $\lambda_1 = \lambda_2 \neq \lambda_3$, and (iii) $\lambda_1 = \lambda_2 = \lambda_3$.]

(18) **Problem** If A and R are $n \times n$ matrices with A positive definite and R orthogonal, prove that $\tilde{R}AR$ is positive definite.

The matrix A^n, where n is any positive integer, is defined by the n-factored product $AAA \cdots A$, for any square matrix A; if A is nonsingular, A^{-n} is defined as $(A^{-1})^n$. Integral powers of matrices obey the usual laws for indices. Fractional powers can also be defined, but even a simple matrix such as the 2×2 unit matrix has an unlimited number of square roots (Frazer *et al.*, 1952, p. 38). For our present purposes, all we require is a positive-definite square root of a positive-definite 3×3 matrix, for use in the polar decomposition theorem (23). For this purpose, we define a square root $A^{1/2}$ as follows.

Since A is positive definite, it is symmetric, and therefore (17) applies with $\lambda_1, \lambda_2, \lambda_3$ positive. We may therefore define a real, symmetric matrix $A^{1/2}$ by the equation

(19) $$\tilde{R}A^{1/2}R = \mathrm{diag}[\lambda_1^{1/2}, \lambda_2^{1/2}, \lambda_3^{1/2}],$$

and it is easy to verify [using (6)] that $A^{1/2}A^{1/2} = A$. We interpret $x^{1/2}$ as the positive root throughout, unless the contrary is explicitly stated. From (18), it follows that $A^{1/2}$ is positive definite. It is clear that other fractional powers of A can be defined in a similar way, but we shall not need them. The definition can also be extended to an $n \times n$ matrix.

(20) **Problem** Prove that $(A^{-1})^{1/2} = (A^{1/2})^{-1}$.

(21) **Problem** If A and B are positive-definite 3×3 matrices which have the same eigenvectors, prove that $AB = BA$ and that $(AB)^{1/2} = A^{1/2}B^{1/2} = B^{1/2}A^{1/2}$.

(22) **Problem** If F is a nonsingular square matrix, prove that $F\tilde{F}$ and $\tilde{F}F$ are positive-definite matrices.

(23) **Theorem** Every nonsingular 3×3 matrix F can be expressed in the form $F = RS$, where R is orthogonal and S is symmetric.

Proof F is nonsingular and therefore $\tilde{F}F$ is positive definite, by (22). Hence, by (19), a symmetric square root matrix $(\tilde{F}F)^{1/2} = S$, say, can be defined which, being positive definite, is nonsingular. Hence S^{-1} exists, and $S^{-2} = (\tilde{F}F)^{-1} = F^{-1}\tilde{F}^{-1}$. Writing $FS^{-1} = R$, we have $R\tilde{R} = FS^{-1}\tilde{S}^{-1}\tilde{F} = FS^{-2}\tilde{F} = I$. Thus R is orthogonal, S is symmetric, and $F = RS$, which proves the theorem. A similar result is valid with the order of the orthogonal and the symmetric matrices changed, but the matrices are then different from R and S.

(24) **Cayley–Hamilton theorem** Let A be a real 3×3 matrix, not necessarily symmetric, and let (11) be its characteristic equation. Then A satisfies "its own characteristic equation," namely,

$$-A^3 + A_{\mathrm{I}}A^2 - A_{\mathrm{II}}A + A_{\mathrm{III}}I = 0,$$

which is obtained from (11) by putting A^n in place of λ^n, with $A^0 = I$.

Proof Let $B = [B_{rc}]$, where B_{cr} is the cofactor of the element in row r and column c in $|A - \lambda I|$; then we can write $B = B_0 + B_1\lambda + B_2\lambda^2$, where B_0, B_1, and B_2 are 3×3 matrices, independent of λ. From (1), for any value of λ, we have

$$(A - \lambda I)(B_0 + B_1\lambda + B_2\lambda^2) = |A - \lambda I|I = (-\lambda^3 + A_{\mathrm{I}}\lambda^2 - A_{\mathrm{II}}\lambda + A_{\mathrm{III}})I,$$

from (11). Since λ is arbitrary, we can equate matrix coefficients of λ^n in this equation (because this is equivalent to equating the r, c elements in the matrix coefficients of λ^n, for r, $c = 1, 2, 3$); the results are $0 = B_2 - I$; $AB_2 = B_1 + A_{\mathrm{I}}I$; $AB_1 = B_0 - A_{\mathrm{II}}I$; and $AB_0 = A_{\mathrm{III}}I$. When these equations are multiplied from the left by A^3, A^2, A, and I, respectively, it is found that the terms in the B_i cancel and that the resulting equation is just that stated in the theorem (24).

It is convenient for later use to treat determinants by means of the completely antisymmetric quantities $e_{ijk} = e^{ijk}$ ($i, j, k = 1, 2, 3$) defined as follows:

$$(25) \quad e_{ijk} = e^{ijk} = \begin{Bmatrix} 1 \\ -1 \\ 0 \end{Bmatrix} \quad \text{according as} \quad \begin{Bmatrix} (ijk) = (123),\ (231),\ \text{or}\ (312); \\ (ijk) = (132),\ (213),\ \text{or}\ (321); \\ \text{any two of}\ \ i, j, k\ \ \text{are equal.} \end{Bmatrix}.$$

We then have the following results [see, e.g., McConnell (1957)] for the determinant $|A|$ of a 3×3 matrix $A = [A_{rc}]$, with a^{rc} being used to denote the cofactor of A_{rc} in $|A|$:

$$(26) \quad |A| = e^{ijk} A_{1i} A_{2j} A_{3k} = e^{ijk} A_{i1} A_{j2} A_{k3};$$

$$(27) \quad |A| e_{rst} = e^{ijk} A_{ri} A_{sj} A_{tk};$$

$$(28) \quad a^{1i} = e^{ijk} A_{2j} A_{3k};$$

$$(29) \quad a^{ri} = \tfrac{1}{2} e^{rst} e^{ijk} A_{sj} A_{tk};$$

$$(30) \quad a^{ri} A_{si} = |A| \delta^r_s;$$

$$(31) \quad \partial |A| / \partial A_{ri} = a^{ri} \qquad \text{for} \quad A_{ri}\ (r, i = 1, 2, 3) \quad \text{independent};$$

$$(32) \quad d|A|/dt = a^{ri}\, dA_{ri}/dt \qquad \text{for} \quad A_{ri} \quad \text{functions of} \quad t;$$

$$(33) \quad d \log_e |A| / dt = \operatorname{Tr}(A^{-1}\, dA/dt) \qquad \text{for} \quad A \quad \text{nonsingular}; \quad dA/dt \equiv [dA_{rc}/dt].$$

(34) **Problem** Prove that the sum of two positive-definite, $n \times n$ matrices is positive definite.

(35) **Problem** If A is an arbitrary, positive-definite matrix and M is a given square matrix such that $A = MA\tilde{M}$, prove that $M = cI$, for some number c.

(36) **Problem** Prove that the eigenvalues of a 3×3 positive-definite matrix are all positive.

(37) **Problem** Prove that $\frac{1}{2} e^{ijk} e_{irs} = \delta^j_r \delta^k_s - \delta^j_s \delta^k_r$.

1.4 Functionals

In describing materials with memory, we shall make use of *hereditary functionals*, of which a simple example is the definite integral

$$(1) \qquad \int_a^t K(t - t') x(t')\, dt' \equiv \mathscr{F}(x), \quad \text{say.}$$

Here, $K(t - t')$, the kernel or memory function, can be regarded as a given function (determined by the properties of the given material) and $x(t')$ as a variable function, which can range through all functions in some given set (e.g., the set of all real functions for which $\int_a^t x^2(t')\, dt'$ exists). The value of the integral is a real number (dependent on t). The symbol $\mathscr{F}$ denotes a *functional*, which is thus *a mapping from the set of functions* $\{x\}$ *onto the set of real numbers*. The particular example (1) is evidently a *linear* functional. For our applications, it is essential to deal also with nonlinear functionals. The value of a *functional* $\mathscr{F}(x)$ depends on the values of the function $x(t')$ at *all* points t' in the range (a, t) of definition; in contrast, the value of a *function* $f(x(t'))$ of a function $x(t')$ depends only on the value of x at *one* point t'. In the mathematical literature, the interval of definition (a, t) is usually specified separately; in the rheological literature, it is often convenient to specify the interval by notations of the type $\overset{t}{\underset{a}{\mathscr{F}}}(x)$, in which t usually denotes a current time, and $a = -\infty$. In one important example, $\sigma(t) = \overset{t}{\underset{-\infty}{\not{f}}}(\dot{s})$, considered in Sections 7.1 and 7.2, $\sigma(t)$ denotes the value at time t of the shear stress at a particle P in a viscoelastic liquid which has undergone unidirectional shear flow throughout the interval $(-\infty, t)$ with a time-dependent shear rate $\dot{s}(t')$. In (1), the functional $\mathscr{F}$ is *hereditary* (because t and t' in the kernel occur only as $t - t'$) in the sense that it is invariant under a translation along the time axis, provided that the lower limit a is such that its replacement by $a - t_0$ has negligible effect on the integral's value: If we consider two functions x_1 and x_2 such that $x_1(t') = x_2(t' + t_0)$ for some constant t_0, then the values $y_i(t) = \overset{t}{\underset{a}{\mathscr{F}}}(x_i)$ $(i = 1, 2)$, with $\mathscr{F}$ defined by (1), are related by the equation $y_1(t) = y_2(t + t_0)$. In practice, the memory function $K(\tau)$ often decreases exponentially as τ increases, so that one can put a equal to some large negative number in (1), if one wants to avoid possible formal complications which might arise with $a = -\infty$. With this understanding, we shall nevertheless, for brevity, usually write $-\infty$ for a in the functionals used here.

We shall consider functionals which are differentiable (Fréchet differentiable) in the sense that, *for any function* $h(t')$ in our given set, there shall exist a function $k(t - t')$ in the set such that

(2)
$$\mathscr{F}(x + h) - \mathscr{F}(x) = \int_{-\infty}^{t} k(t - t')h(t')\, dt' + \rho,$$

where k can depend on x *but not on* h and ρ denotes a term which can be neglected, in comparison with the first term on the right-hand side, as $h \to 0$; for example, one might write $\rho = O\left(\int_{-\infty}^{t} h^2(t')\, dt'\right)$, and interpret the statement $h \to 0$ as meaning $\int_{-\infty}^{t} h^2(t')\, dt' \to 0$. The function k [which occurs in (2) with argument $t - t'$, instead of t and t' separately, because $\mathscr{F}$ is taken to be hereditary] is called the *Fréchet derivative* of $\mathscr{F}$ at x, and evidently satisfies the equation

$$\text{(3)} \qquad \int_{-\infty}^{t} k(t - t')h(t')\,dt' = \frac{d}{du}\,\mathscr{F}(x + uh)\bigg|_{u=0},$$

where u is a real variable. The right-hand side of this equation involves an operation which we may loosely describe as "differentiating along the line in function space joining the points x and h"; an important feature of a functional which is Fréchet differentiable at x is that the Fréchet derivative shall be independent of the direction of the line along which one differentiates in function space. An example arising from the equality of derivatives taken along different directions in function space is treated in Section 7.2. We shall also need to consider functionals, twice differentiable at x, such that a function $m(u, v) = m(v, u)$ exists which can depend on x but not on h, and (for all h)

$$\text{(4)} \qquad \mathscr{F}(x + h) - \mathscr{F}(x) = \int_{-\infty}^{t} k(t - t')h(t')\,dt' + \int_{-\infty}^{t}\int_{-\infty}^{t} m(t - t', t - t'')h(t')h(t'')\,dt'\,dt'' + \rho.$$

In Section 6.7, we use series expansions which are formally similar to (2) and (4) but involve tensor-valued (instead of scalar-valued) functions in place of k, m, h, and ρ. According to the usage in the mathematical literature, the values of a functional are *scalars* (whether the argument is tensor- or scalar-valued) and many of the results of functional analysis depend on this definition. If one sticks to this definition of the term functional, then the phrase "tensor-valued functional" (found in some rheological literature) is inadmissible and can be replaced (we suggest) by the phrase *tensor-valued function of a tensor history*. Thus a functional of x can also be called a function of the history of x, the word *history* being appropriate when t in $x(t)$ denotes time.

For reasons of brevity, we give no more than the above heuristic summary of those aspects of functional analysis that we shall make use of in our applications. A useful summary of certain aspects of functional analysis is given by Noble and Sewell (1972, Appendix), who also refer to various mathematical texts on the subject. Coleman and Noll (1960) have given a more detailed treatment with rheological applications in mind, and, in particular, introduce an "influence function" (or convergence factor) in the appropriate integrands in order to deal with difficulties which arise at the lower limit $(-\infty)$ of integration; it is true, of course, that integrals of the type $\int_{-\infty}^{t} h^2(t')\,dt'$ diverge in various important applications (e.g., when

$h = \dot{s}$, a constant shear rate), but the kernels k and m in (4) can usually be safely assumed (in rheological applications to viscoelastic liquids) to decrease exponentially as their arguments increase, and so we can reasonably hope to be able to use expansions of the type (4) without having to go into detail about the definition and tending to zero of the remainder term ρ. This is probably safe enough when the function $h(t')$ is algebraic or trigonometric (as in steady and oscillatory shear flow applications), but genuine difficulties arise when $h(t')$ is itself exponential (as in steady elongational flow applications); in the latter case, it is safer to investigate the behavior of a given functional rather than to attempt to use expansion theorems for some general class of functionals. In any case, we believe that the value of expansions of the type (4) in rheological calculations is rather limited, and that it would be of doubtful value to seek (for example) necessary and sufficient conditions for the validity of such expansions for general classes of functionals.

A fundamental theorem [the Riesz–Fréchet representation theorem; Riesz and Sz.-Nagy (1955, p. 61)] states that a linear functional $\mathscr{F}(x)$ is always expressible as a single integral of the form

$$\mathscr{F}(x) = \int_a^t g(t')x(t')\,dt', \tag{5}$$

where g is uniquely determined by $\mathscr{F}$, when x is any complex-valued function such that $\int_a^t x(t')x^*(t')\,dt'$ exists (in the Lebesgue sense).

2 THE BODY METRIC TENSOR FIELD

2.1 Introduction

In order to describe mathematically the rheological properties of a viscoelastic liquid or the deformation of an elastic solid, it is reasonable and convenient to treat them as continuous bodies which fill a region of space at each instant. The appropriate theories are thus field theories in the sense that we have to deal with various dependent variables which describe the properties of the materials and which are functions of time and of position variables. As with other field theories in physics (e.g., electromagnetic theory, relativity theory), it is convenient to introduce coordinate systems to label the points of the continuous manifolds involved; one must use the coordinate systems in such a way that the field theory equations express the intrinsic properties of the materials in a manner which contains no unwanted dependence on the choice of coordinate system, since this is essentially arbitrary.

It is well known that the natural tool for such a situation is general tensor analysis (also called the absolute differential calculus or Ricci calculus). Tensor equations automatically possess the desired invariance properties with respect to arbitrary coordinate transformations. In our applications, we deal with two distinct geometric manifolds—the space manifold and the body manifold—and *we claim that it is possible to define two distinct kinds of tensor field: space tensor fields and body tensor fields.* Space tensor fields can be further subdivided into two mutually exclusive classes: general space tensor fields and Cartesian space tensor fields. Body tensor fields are necessarily general tensor fields; Cartesian body tensor fields can be defined throughout a nonzero time interval only in the degenerate case in which the body moves rigidly.

The aim of this chapter is to define body tensor fields and general space tensor fields and to derive, from the definitions, some of the simpler operations which such fields admit. We use boldface symbols to stand for general space tensors and body tensors (italic Latin boldface type for the former, and Greek boldface for the latter). We use italic type (not boldface) for the components (or matrices of components) which represent the tensors in specified coordinate systems.

Most published treatments of general tensor analysis (as applied to general relativity theory, electromagnetic theory, relativistic quantum field theories of elementary particles, and the geometry of differentiable manifolds) use the *components* of tensors, and neither define nor use the general tensors themselves. Exceptions include mathematical texts by Nevanlinna and Nevanlinna (1959), Lang (1962), Bishop and Goldberg (1968), and Warner (1971). Apart from our own work (Lodge, 1951, 1954, 1964, 1972) and possibly also two recent papers by Noll (1972, 1973), general tensor fields (as distinct from their components) have not been used in the literature on rheology and continuum mechanics, as far as we are aware. We claim that working with the tensors themselves (instead of working with their components exclusively) leads not only to certain added convenience in manipulation but also to enhanced understanding of certain of the concepts involved in the formalism.

The definition of general tensors and the derivation of their admissible operations from the definition unavoidably involve a greater abstract content at the early stages of the development than is perhaps common in many engineering texts and courses. We have tried to sweeten this abstract pill by introducing a few of the more important general vectors and tensors as we require them in describing the motion of a deforming body, thereby furnishing a clear motivation for the more abstract definitions. We tend to treat important particular cases first, in the belief that, once the reader has grasped the point and the direction of travel, the subsequent task of generalization will be the more readily accomplished.

In particular, we define the body metric tensor field $\boldsymbol{\gamma}(\mathrm{P}, t)$ and the space metric tensor field $\boldsymbol{g}(\mathrm{Q})$, and we define base vectors, angles, and arbitrary first- and second-rank general tensors. We develop sufficient formalism to enable us, in Chapter 3, to describe the kinematics of shear flow in sufficient generality to include all the cases which have so far arisen in actual experiments with shear flow. Chapter 3 serves both as a clarifying example of the general formalism of the present chapter and as a simple treatment of shear flow, which is of fundamental importance in experimental investigations of the rheological properties of polymeric liquids. Our method of defining tensor fields makes use of the familiar concepts "function" and "coordinate system"; methods used in the texts cited above require topological concepts and the concept of an equivalence class which are less familiar to many engineering students.

2.2 Body and space manifolds and coordinate systems

(1) **Definitions** A *geometric manifold* is an infinite set of undefined elements called *points*, to which certain properties are assigned. The *body manifold* $\mathscr{P} = \{P\}$ is an infinite set of undefined elements called *particles*; the *space manifold* $\mathscr{Q} = \{Q\}$ is an infinite set of undefined elements called *places*. A *coordinate system* for a manifold is a one-to-one correspondence between points of the manifold and ordered sets of n real numbers, called *coordinates*, such that to each point there corresponds one, and only one, set of coordinates; if there exists a coordinate system such that, for every set of values of the coordinates in n intervals, there exists one, and only one point, then the manifold is said to be n-dimensional. We consider the case in which both the space manifold and the body manifold are three-dimensional, and we use the following notation for coordinate systems:

(2) Body coordinate system $B : P \to \xi$; space coordinate system $S : Q \to x$.

Here, ξ denotes the ordered set of three real numbers (ξ^1, ξ^2, ξ^3) that are the coordinates of particle P in the body coordinate system B; x denotes the ordered set of three real numbers (x^1, x^2, x^3) that are the coordinates of the place Q in the space coordinate system S.

The preceding is an abbreviated notation for the statement, "The body coordinate system B is a mapping $\mathscr{P} \to A_3$ in which an arbitrary particle P has coordinates (ξ^1, ξ^2, ξ^3); the space coordinate system S is a mapping $\mathscr{Q} \to A_3$ in which an arbitrary place Q has coordinates (x^1, x^2, x^3)." We do *not* use the conventional further abbreviations, "the body coordinate system ξ^i, the space coordinate system x^i," because we find it essential to retain different symbols (e.g., B and ξ^i) for the coordinate system and for the coordinates of an arbitrary point.

A *coordinate surface* is a two-dimensional submanifold containing all points which have the same value of some one coordinate; a *coordinate line* is a one-dimensional submanifold containing all points for which all but one coordinate have fixed values. A *body coordinate surface* $\xi^1 = c^1$, say, where c^1 is a constant, is a material surface composed of all particles having the same value c^1 for coordinate ξ^1. A *body coordinate line* (the intersection of two coordinate surfaces), say the ξ^3-line, is a material line composed of all particles for which coordinate ξ^1 has the same value and coordinate ξ^2 has the same value. A body coordinate system is a three-parameter family of material surfaces; through each particle there passes one, and only one, surface of each family. No two coordinate surfaces of the same family can intersect; if they did, then a particle on the intersection would have two distinct values for one coordinate, which contradicts the definition that a coordinate system is a one-to-one correspondence between points and ordered sets of real numbers.

Coordinates are real numbers; they may or may not have physical dimensions (length) attached to them. For a body coordinate system, deformation of the body means that distances between particles or along material lines change, and so it is convenient to use body coordinates which have no dimensions of length. *A body coordinate system is thus a purely numerical method of labeling unambiguously the particles of the body*, regardless of the motion of the body in space; a particle P always has the coordinates ξ^i. A body coordinate system is thus, in a sense, "fixed" in the body, although the body may be deforming. A space coordinate system, similarly, is "fixed" in space. The familiar phrase "a moving coordinate system" means "a one-parameter family of space coordinate systems," the time t being the parameter.

(3) **Definition** A *configuration* t of the body is a one-to-one correspondence $\mathscr{P} \leftrightarrow \mathscr{Q}$ between particles of the body and places in a region of space. For given coordinate systems B and S, a configuration t generates a one-to-one correspondence between the coordinate sets ξ and x (restricted to appropriate ranges) which we write in the form

$$x^i = f^i(\xi^1, \xi^2, \xi^3, t) \quad \text{or} \quad x^i = f^i(\xi, t) \quad (i = 1, 2, 3). \tag{4}$$

These equations have an inverse set, which we write in the form

$$\xi^i = \varphi^i(x^1, x^2, x^3, t) \quad \text{or} \quad \xi^i = \varphi^i(x, t) \quad (i = 1, 2, 3). \tag{5}$$

We shall assume that, for any given space coordinate system S, one can choose a body coordinate system B such that the functions f^i and φ^i are sufficiently differentiable with respect to ξ^i and x^i, respectively. By sufficiently differentiable, we mean that all partial derivatives up to order m exist and are continuous, where m is a number (at least three) which we do not need to specify: We use a convention that when a derivative is used in the text, that derivative is presumed to exist and to be continuous, unless the contrary is explicitly stated. Since the partial derivatives exist and the correspondence $x \leftrightarrow \xi$ is one-to-one, it follows that f^i and φ^i are one valued and have non-zero Jacobians:

$$(6) \qquad |\partial f^i/\partial \xi^j| \neq 0, \qquad |\partial \varphi^i/\partial x^j| \neq 0.$$

Since we are assuming that the body fills a region of space, the assumption that f^i and φ^i are sufficiently differentiable, for some one configuration t, involves no practical loss of generality. We shall, however, maintain the assumption for different configurations, $t = t_1$ and $t = t_2$, say, which means that the deformation $t_1 \to t_2$ (i.e., the change of configuration) is continuous in the sense that neighboring particles remain neighboring: i.e., if P and P_1 are two particles whose body coordinate differences $d\xi^i$ are arbitrarily small, then they occupy places whose space coordinate differences, dx_1^i and dx_2^i say, in configurations t_1 and t_2 are also arbitrarily small. These differences are the values of

$$(7) \qquad dx^i = \frac{\partial f^i(\xi, t)}{\partial \xi^j}\, \partial \xi^j$$

obtained for $t = t_1$ and $t = t_2$. By *neighboring particles*, we simply mean particles which occupy places whose coordinate differences are arbitrarily small, but not all zero. By confining our attention to continuous deformations, we are excluding phenomena such as fracture and self-diffusion.

A *flow* is a one-parameter family of configurations; it may be described by (4) with t assuming values in some interval (t_1, t_2). In this context, t is usually taken to denote the time at which a given configuration occurs. With the possible exception of a finite number of instants, the functions f^i (and φ^i) will also be assumed to be sufficiently differentiable with respect to time t; the exceptions are of importance in the theory, when we consider instantaneous deformations described by a simple discontinuity in f^i at some value of t. Although such discontinuities do not presumably actually occur in any experiment, they represent a useful idealization of very rapid deformations, which do occur. It is easy, however, to give them special treatment when they arise, and so we will take f^i to be sufficiently differentiable in t unless the contrary is explicitly stated. We can thus, in particular, define the velocity components

$$(8) \qquad v^i = \frac{\partial f^i(\xi, t)}{\partial t}, \qquad v^i(x, t) = \left.\frac{\partial f^i(\xi, t)}{\partial t}\right|_{\xi = \varphi(x, t)}.$$

These are not what we commonly think of as velocity components, however, because they may or may not involve any reference to distance, depending on the choice of the space coordinate system S. We return to this point later, after we have defined the metric tensor. The v^i simply represent the rate of change of place coordinates for a given particle. We could (in the interests of symmetry between the body manifold and the space manifold) also consider the derivatives $\partial\varphi^i(x, t)/\partial t$, which describe the rate of change of particle coordinates at a given place, but we have no use for them.

(9) **Definition** A *metric* for a geometric manifold is a correspondence between unordered pairs of *neighboring* points and positive real numbers *ds*. The number *ds* is called the *separation* of the points or the *distance* between them. The separation is independent of the order of the points and is zero if, and only if, the points coincide. To each pair of neighboring points corresponds one, and only one, value *ds*. A manifold is *Riemannian* if $(ds)^2$ is expressible as a quadratic form in the coordinate differences.

(10) We shall *assume that the space manifold is Riemannian with a metric which is independent of time*, and that the process of assigning distance is such that, at any time t, *the value of ds for any two neighboring* particles is equal to the distance between the places which they then occupy. It then follows that the body manifold is Riemannian with a metric which depends on time, as the following argument shows.

Let Q and Q_1 be any two neighboring places in space; let x^i and $x^i + dx^i$ denote their coordinates in S. Then their separation ds is given by an equation of the form

(11) $$(ds)^2 = g_{ij}(x)\, dx^i\, dx^j \qquad \text{(all } t),$$

where the functions $g_{ij}(x)$ are called *components of the space metric tensor field.* For any x, the quadratic form (11) is positive definite, and so the coefficients g_{ij} satisfy 1.3(2). On substituting for dx^i from (7), we obtain the equation

(12) $$(ds)^2 = \gamma_{mn}(\xi, t)\, d\xi^m\, d\xi^n,$$

where

(13) $$\gamma_{mn}(\xi, t) = \left[g_{ij}(x)\frac{\partial f^i}{\partial \xi^m}\frac{\partial f^j}{\partial \xi^n}\right]_{x=f(\xi,t)}.$$

(12) gives the separation at time t of particles P and P_1 whose coordinates in B are ξ^i and $\xi^i + d\xi^i$. We shall call the functions $\gamma_{ij}(\xi, t)$ *components of the body metric tensor field at time t.* Equation (12) shows that the body manifold is Riemannian.

In the theory of general relativity, one usually considers a four-dimensional manifold, called space–time, in which our variables x^i and t are taken to be coordinates and which has a metric which depends on x^i and t and is not positive definite. Our assumption for space is thus appropriate in the non-relativistic approximation. Oldroyd (1970) has treated continuum mechanics without this restriction.

It follows that the quadratic form (12) is positive definite, and hence, from 1.3(2), that

(14) $$\gamma_{11} > 0; \qquad \begin{vmatrix} \gamma_{11} & \gamma_{12} \\ \gamma_{12} & \gamma_{22} \end{vmatrix} > 0; \qquad |\gamma_{ij}| > 0.$$

Without loss of generality, we can take

(15) $$g_{ij} = g_{ji}, \qquad \gamma_{ij} = \gamma_{ji}.$$

The body metric components γ_{ij} play a fundamental role in this book. As functions of ξ^i and t, they give a complete description of changes of separation with time for all pairs of neighboring particles in the body. A motion is said to be *rigid* if ds is independent of t for every pair of particles; clearly we have

(16) $$\gamma_{ij}(\xi, t) \quad \text{independent of } t \text{ for a rigid motion.}$$

(17) **Definition** *Strain* means change of metric with configuration.

If ds and ds' denote the values at times t and t', respectively, of the separation of P and P_1, then it follows from (12) (and the fact that body coordinates for given particles are constant) that

(18) $$(ds)^2 - (ds')^2 = [\gamma_{ij}(\xi, t) - \gamma_{ij}(\xi, t')]\, d\xi^i\, d\xi^j$$

(square brackets here do not denote a matrix), and, on differentiating (12), that

(19) $$\frac{d^n}{dt^n}(ds)^2 = \frac{\partial^n \gamma_{ij}(\xi, t)}{\partial t^n}\, d\xi^i\, d\xi^j.$$

The quantities

(20) $$\Delta\gamma_{ij}(\xi, t, t') = \gamma_{ij}(\xi, t) - \gamma_{ij}(\xi, t')$$

are called *strain components*. The quantities $\partial^n \gamma_{ij}(\xi, t)/\partial t^n$ are called *nth-strain-rate components*.

(21) **Problem** Let $\sigma(\xi) = c$ and $\sigma(\xi) = c + dc$ be two neighboring members of a one-parameter family of material surfaces. Let P (of coordinates ξ^i) be any given particle on one of the surfaces, and let P_1 (of coordinates $\xi^i + d\xi^i$) be a variable particle on the other. By finding dh, the least value of the separation ds of P and P_1 as P_1 varies (or otherwise), show that

$$\left(\frac{dc}{dh}\right)^2 = \gamma^{ij}(\xi, t)\frac{\partial\sigma}{\partial\xi^i}\frac{\partial\sigma}{\partial\xi^j},$$

where

(22) $$[\gamma^{ij}] = [\gamma_{ij}]^{-1}.$$

Comparing this result with (12), we see that γ^{ij} is simply related to the separation of material surfaces and that γ_{ij} is simply related to the separation of particles. The reciprocal matrix $[\gamma_{ij}]^{-1}$ certainly exists, because $[\gamma_{ij}]$ is nonsingular in view of the positive-definite condition (14).

The postulated differentiability of the functions f^i in (4) implies that the set of functions considered includes the identity transformation $x^i = \xi^i$.

(23) **Definition** Coordinate systems B and S are *congruent in a state* t if $x^i = \xi^i$ at t. We shall write this in the form $B \overset{t}{\equiv} S$.

It is clear that, to any given space coordinate system S and flow, there corresponds a unique body coordinate system B which is congruent to S at a given time t. The congruent coordinate systems have coordinate surfaces and coordinate lines which coincide at time t and also have the same parameter values attached to coincident coordinate surfaces. Coordinate systems which are congruent at one instant will not, in general, be congruent at another instant.

2.3 The definition of contravariant vector fields

In Section 2.2, we defined quantities of interest in connection with the flow or deformation of a continuous material: v^i describes the velocity relative to space, $d\xi^i$ describes a material direction as defined by two neighboring particles, and $\gamma_{ij}(\xi, t)$ describes the metric (or shape) as a function of time t. The values of all these quantities, however, depend not only on the properties described (as we wish) but also on the choice of coordinate system (S for v^i, and B for $d\xi^i$ and γ_{ij}), which is arbitrary. Our aim now is *to define quantities which describe the required properties but which do not depend on any choice of coordinate system.*

We consider the infinite set $\mathscr{B} = \{B\}$ consisting of all body coordinate systems that are related to one another by one-valued, sufficiently differentiable "coordinate transformation" functions which possess a unique inverse. If $B: \mathrm{P} \to \xi$ and $\bar{B}: \mathrm{P} \to \bar{\xi}$ denote any two body coordinate systems, then the correspondences $\mathrm{P} \to \xi$ and $\mathrm{P} \to \bar{\xi}$, both of which are one-to-one by definition, generate a correspondence $\xi \to \bar{\xi}$ which must also be one-to-one and can be written in the form

(1) $$\bar{\xi}^i = \bar{\xi}^i(\xi^1, \xi^2, \xi^3), \qquad \text{with inverse} \quad \xi^i = \xi^i(\bar{\xi}^1, \bar{\xi}^2, \bar{\xi}^3) \qquad (i = 1, 2, 3).$$

From the definition of body coordinate systems, the functions (1) are necessarily independent of time. We are assuming that at least one body coordinate system, B say, exists; this assumption assigns a property to the body manifold. It then follows that an unlimited number of body coordinate systems exist; we achieve sufficient generality for our applications if we consider the set $\mathscr{B}$ whose members are obtained from B by all possible transformations of the type (1) for which all partial derivatives of the functions $\bar{\xi}^i(\xi)$ exist and are continuous at least up to the second order. From now on, the phrase all body coordinate systems will mean all body coordinate systems in the set $\mathscr{B}$. It follows, in particular, that the Jacobians are nonzero:

(2) $$|\Lambda_P(B, \bar{B})| \neq 0, \qquad |\Lambda_P^{-1}(B, \bar{B})| \neq 0,$$

where $\Lambda_P(B, \bar{B})$ is the *coordinate transformation matrix* at P and is given by the equations

(3) $$\Lambda_P(B, \bar{B}) = [\partial\bar{\xi}^r/\partial\xi^c]_P, \qquad \tilde{\Lambda}_P(B, \bar{B}) = [\partial\bar{\xi}^c/\partial\xi^r]_P,$$
$$\Lambda_P^{-1}(B, \bar{B}) = [\partial\xi^r/\partial\bar{\xi}^c]_P.$$

The notation $[\;]_P$ means that $[\;]$ is to be evaluated at P.

The fundamental property underlying tensor analysis and the definition of general tensors is expressed by the equations

(4) $$\frac{\partial\xi^r}{\partial\bar{\xi}^k}\frac{\partial\bar{\xi}^k}{\partial\bar{\bar{\xi}}^c} = \frac{\partial\xi^r}{\partial\bar{\bar{\xi}}^c} \qquad (r, c = 1, 2, 3),$$

where $\bar{\bar{B}}$: $P \to \bar{\bar{\xi}}$ denotes any third body coordinate system. In the present matrix notation, (4) becomes

(5) $$\Lambda_P^{-1}(B, \bar{B})\Lambda_P^{-1}(\bar{B}, \bar{\bar{B}}) = \Lambda_P^{-1}(B, \bar{\bar{B}}).$$

In this sense, we can say that the coordinate transformation matrices at particle P form a group, for (5) expresses the transitive property, and the inverse $\Lambda_P(\bar{B}, B) = \Lambda_P^{-1}(B, \bar{B})$ and the identity $\Lambda_P(B, B) = I$ are evidently members of the set $\{\Lambda_P(B, \bar{B}) \mid B, \bar{B} \in \mathscr{B}\}$. On taking the reciprocal and the transpose of (5), we obtain the similar equation

(6) $$\tilde{\Lambda}_P(B, \bar{B})\tilde{\Lambda}_P(\bar{B}, \bar{\bar{B}}) = \tilde{\Lambda}_P(B, \bar{\bar{B}}).$$

(7) **Problem** Prove that $\tilde{\Lambda}_P^2(B, \bar{B})\tilde{\Lambda}_P^2(\bar{B}, \bar{\bar{B}}) \neq \tilde{\Lambda}_P^2(B, \bar{\bar{B}})$, for all B, $\bar{B}$, and $\bar{\bar{B}}$.

To motivate the definition of a contravariant vector, we consider again the pair of neighboring particles P and P_1 whose coordinate differences are given by the following 3×1 (single-column) matrices:

(8) $$d\xi \;\text{ in }\; B; \qquad d\bar{\xi} = \Lambda_P(B, \bar{B})\, d\xi \;\text{ in }\; \bar{B};$$
$$d\bar{\bar{\xi}} = \Lambda_P(B, \bar{\bar{B}})\, d\xi \;\text{ in }\; \bar{\bar{B}}; \ldots .$$

For a given pair of particles, there is thus one, and only one, coordinate matrix associated with each body coordinate system in $\mathscr{B}$. In other words, *a given ordered pair of neighboring particles,* P *and* P_1, *generates a correspondence between the set* $\mathscr{B}$ *of all body coordinate systems and the set* $A_{3\times 1}$ *of all* 3×1 *matrices*. For this correspondence, or function, we introduce a symbol $\boldsymbol{d\xi}$, which can therefore be defined by the equation

$$(9)\qquad \boldsymbol{d\xi}\, B = d\xi = \Lambda_{\mathrm{P}}^{-1}(B, \bar{B})\, d\bar{\xi} \qquad (\text{for all} \quad B \quad \text{in} \quad \mathscr{B}),$$

if we choose an arbitrary coordinate system $\bar{B}$, associate any given infinitesimal column matrix $d\bar{\xi}$ with $\bar{B}$, and then allow B in (9) to run through all possible body coordinate systems. Then $\boldsymbol{d\xi}$, thus defined, is a particular example of what we call a contravariant body vector at P. In this definition, the important feature is not the value of any one representative matrix (such as $d\bar{\xi}$), but the relation (9) between representative matrices in different coordinate systems. Accordingly, we extend the definition by using an arbitrary 3×1 column matrix $\bar{\theta}$ in place of $d\bar{\xi}$.

(10) **Definition** A *contravariant body vector* $\boldsymbol{\theta}(\mathrm{P})$ at a particle P is a function $\boldsymbol{\theta}(\mathrm{P}) : \mathscr{B} \to A_{3\times 1}$ defined as follows. Choose any body coordinate system, $\bar{B}$ say, and any 3×1 real column matrix $\bar{\theta}$. Then the body vector $\boldsymbol{\theta}(\mathrm{P})$ associates with any other body coordinate system, B say, a 3×1 column matrix θ, where

$$(11)\qquad \boldsymbol{\theta}(\mathrm{P})B = \theta = \Lambda_{\mathrm{P}}^{-1}(B, \bar{B})\bar{\theta} \qquad (\text{for all} \quad B \quad \text{in} \quad \mathscr{B}).$$

(12) **Theorem** $\boldsymbol{\theta}(\mathrm{P})$ is independent of the choice $(\bar{B}, \bar{\theta})$ in the sense that, if we replace $(\bar{B}, \bar{\theta})$ by $(\bar{\bar{B}}, \bar{\bar{\theta}})$, where $\bar{\bar{B}}$ is any other body coordinate system and $\bar{\bar{\theta}}$ is the associated matrix given by the equation

$$(13)\qquad \bar{\bar{\theta}} = \Lambda_{\mathrm{P}}^{-1}(\bar{\bar{B}}, \bar{B})\bar{\theta} \qquad \left[\text{or} \quad \bar{\bar{\theta}}^r(\bar{\bar{\xi}}) = \left(\frac{\partial \bar{\bar{\xi}}^r}{\partial \bar{\xi}^i}\right)_{\mathrm{P}} \bar{\theta}^i(\bar{\xi})\right],$$

[obtained from (11) by putting $B = \bar{\bar{B}}$ and $\theta = \bar{\bar{\theta}}$], then we obtain the same vector $\boldsymbol{\theta}(\mathrm{P})$.

Proof Suppose, if possible, that the stated procedure gives another vector, $\boldsymbol{\theta}'(\mathrm{P})$ say; then, for all B in $\mathscr{B}$, we have

$$\begin{aligned} \boldsymbol{\theta}'(\mathrm{P})B = \theta' &= \Lambda_{\mathrm{P}}^{-1}(B, \bar{\bar{B}})\bar{\bar{\theta}} && \text{by definition (10)} \\ &= \Lambda_{\mathrm{P}}^{-1}(B, \bar{\bar{B}})\Lambda_{\mathrm{P}}^{-1}(\bar{\bar{B}}, \bar{B})\bar{\theta} && \text{by (13)} \\ &= \Lambda_{\mathrm{P}}^{-1}(B, \bar{B})\bar{\theta} && \text{by (5)} \\ &= \boldsymbol{\theta}(\mathrm{P})B && \text{by (11).} \end{aligned}$$

It follows that $\boldsymbol{\theta}'(\mathrm{P}) = \boldsymbol{\theta}(\mathrm{P})$, which proves the theorem.

Among all possible functions which map $\mathscr{B}$ into $A_{3\times 1}$, a contravariant vector $\boldsymbol{\theta}(\mathrm{P})$ is a function of one special form which is completely determined as soon as the representative matrix, $\bar{\theta}$ say, in any one coordinate system, $\bar{B}$ say, is given.

(14) We may represent this symbolically by a graph [Fig. 2.3(14)] which is meant to represent the variation of θ with B. One axis is meant to represent the set $\mathscr{B}$ of body coordinate systems; the other axis is meant to represent the

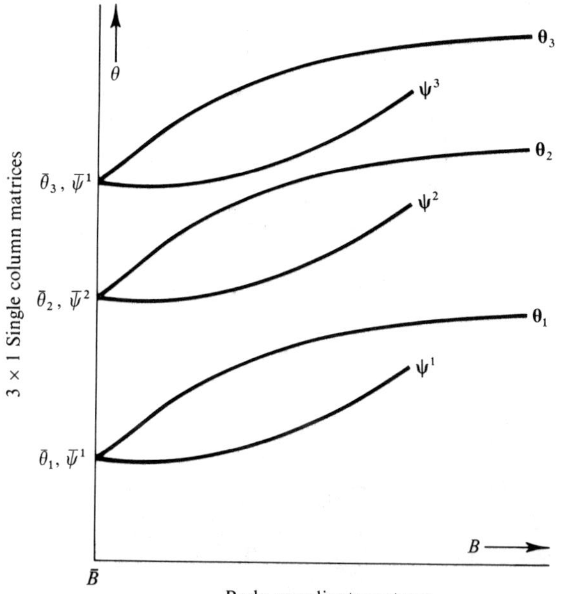

Figure 2.3(14) Schematic representation of contravariant body vectors $\boldsymbol{\theta}_i$ and covariant body vectors $\boldsymbol{\psi}^i$ at a particle P, which have the same representative matrices $\bar{\theta}_i = \bar{\psi}^i$ in a body coordinate system $\bar{B}$; the representative matrices in other coordinate systems are, in general, unequal.

set $A_{3\times 1}$ of 3×1 single-column real matrices. If we specify the representative matrix, $\bar{\theta}_1$ say, for coordinate system $\bar{B}$, then the contravariant body vector at P, $\boldsymbol{\theta}_1(\mathrm{P})$ say, that has $\bar{\theta}_1$ as its "$\bar{B}$-representative" is completely determined, and may be represented symbolically by a graph which passes through the point $(\bar{B}, \bar{\theta}_1)$. Any other matrix, $\bar{\theta}_2$ say, associated with $\bar{B}$, will give rise to another contravariant body vector at P, $\boldsymbol{\theta}_2(\mathrm{P})$ say, *which is represented by a graph of similar form* passing through the point $(\bar{B}, \bar{\theta}_2)$. The form of the graph in this diagram is of no significance, of course, because we are using one axis to represent a space of three dimensions (the θ axis) and the other to represent a space of infinite dimensions (the B axis). The definition (10) shows, however, that to each B there corresponds one, and only one, θ, for any given $\boldsymbol{\theta}(\mathrm{P})$. The converse is not true, as the following problem shows.

(15) **Problem** Given any contravariant body vector $\boldsymbol{\theta}(\mathrm{P})$ at P and any body coordinate system B, show that there exists a second body coordinate system $\bar{B}$ such that $\Lambda_{\mathrm{P}}(B, \bar{B}) \neq I$ and $\theta = \bar{\theta}$, where θ and $\bar{\theta}$ are the B and $\bar{B}$ representatives of $\boldsymbol{\theta}(\mathrm{P})$, respectively.

Theorem (12) shows that, in an important sense, *all body coordinate systems are equivalent insofar as the definition of any contravariant body vector at* P *is concerned.* Whichever pair $(\bar{B}, \bar{\theta})$ we start with, we get the same vector $\boldsymbol{\theta}(\mathrm{P})$. Hence *contravariant body vectors constitute one type of variable which is suitable for describing properties of a continuous body in a manner which is (automatically) independent of any particular choice of body coordinate system from the set* $\mathscr{B}$. The force of this statement may perhaps be felt better when one considers the following problem, which furnishes a counterexample.

(16) **Problem** Repeat the definition (10), using $\Lambda_{\mathrm{P}}^{-2}$ instead of $\Lambda_{\mathrm{P}}^{-1}$, thereby defining a different function, $\boldsymbol{\varphi}(\mathrm{P}) : \mathscr{B} \to A_{3\times 1}$, where $\boldsymbol{\varphi}(\mathrm{P})B = \Lambda_{\mathrm{P}}^{-2}(B, \bar{B})\bar{\theta}$. Prove that $\boldsymbol{\varphi}(\mathrm{P})$ depends on $\bar{B}$ and that the analog of Theorem (12) [with $\Lambda_{\mathrm{P}}^{-1}$ in (13) replaced by $\Lambda_{\mathrm{P}}^{-2}$] is false.

This counterexample emphasizes the important role played by the contravariant vector transformation law (13) and the transformation matrix transitive property (5); it is the latter which is responsible for the equivalence of coordinate systems, and the former which, being linear and homogeneous in $\bar{\theta}$ and $\bar{\bar{\theta}}$, gives rise to the basic operations (addition, multiplication by a number) which can be performed with vectors $\boldsymbol{\theta}(\mathrm{P})$ at P.

(17) **Theorem** If, for any given particle P and any given body coordinate system $\bar{B}$, we allow the matrix $\bar{\theta}$ to assume all possible values in $A_{3\times 1}$, then the contravariant body vectors $\{\boldsymbol{\theta}(\mathrm{P}) \,|\, \bar{\theta} \in A_3\}$ defined by (11) form a three-dimensional linear space in which the operations of addition and multiplication by numbers $a_1, a_2, \ldots$ are defined by the equation

$$(a_1\boldsymbol{\theta}_1 + a_2\boldsymbol{\theta}_2)B = a_1\theta_1 + a_2\theta_2 = \Lambda_{\mathrm{P}}^{-1}(B, \bar{B})(a_1\bar{\theta}_1 + a_2\bar{\theta}_2), \tag{18}$$

and the zero vector $\mathbf{0}$ is defined by the equation

$$\mathbf{0}B = 0 \qquad \text{for all} \quad B. \tag{19}$$

Proof Equation (18) (valid for all B) has the same form as (11) and therefore defines a contravariant body vector $a_1\boldsymbol{\theta}_1 + a_2\boldsymbol{\theta}_2$ at P, where $\boldsymbol{\theta}_1$ and $\boldsymbol{\theta}_2$ are any two contravariant body vectors at P. It is easy to verify, using (18), that the laws (i) and (ii) of 1.2(1) are satisfied, because they are satisfied for single-column matrices. Hence the vectors form a linear space, as stated. The dimensionality follows from the fact that $A_{3\times 1}$ is three-dimensional and $\{\boldsymbol{\theta}(\text{P})\}$ is isomorphic to $A_{3\times 1}$ [see also (23)].

(20) **Definition** The *tangent base vectors* $\boldsymbol{\beta}_i(B, \text{P})$ at particle P for a body coordinate system B are three contravariant body vectors at P whose representative matrices in B are δ_i, where

(21) $$\tilde{\delta}_1 = [1, 0, 0], \qquad \tilde{\delta}_2 = [0, 1, 0], \qquad \tilde{\delta}_3 = [0, 0, 1].$$

Thus (in B only) we have

(22) $$\boldsymbol{\beta}_i(B, \text{P})B = \delta_i \quad (i = 1, 2, 3) \qquad \text{or} \qquad \beta^j_{(i)} = \delta^j_i,$$

where $\beta^j_{(i)}$ denote the components (i.e., the elements of the representative matrix) of the vector $\boldsymbol{\beta}_i$ in B.

We use a superscript, as in θ^i, to label components of a *contravariant* vector; thus $\boldsymbol{\theta}$ has components θ^i in B, where $\theta = [\theta^r]$ is the representative matrix in B. We use a subscript to distinguish one contravariant vector from another: thus $\boldsymbol{\theta}_1$ and $\boldsymbol{\theta}_2$ are contravariant vectors, and their components are written $\theta^i_{(1)}$ and $\theta^i_{(2)}$, respectively, with parentheses on the subscripts to emphasize the fact that they are labels for different vectors, not different components of the same vector.

(23) **Theorem** The tangent base vectors $\boldsymbol{\beta}_i(B, \text{P})$ $(i = 1, 2, 3)$ for any given body coordinate system B form a basis for all contravariant body vectors at P; if $\theta^i(\xi)$ are the components in B of any given contravariant body vector $\boldsymbol{\theta}(\text{P})$ at P, then

(24) $$\boldsymbol{\theta}(\text{P}) = \theta^i(\xi)\boldsymbol{\beta}_i(B, \text{P}).$$

Proof Both sides of (24) are contravariant body vectors at P, by (17), and therefore, if the representative matrices of both sides in any one coordinate system are equal, the validity of the vector equation (24) is established. For the representative matrices in B, we have

$$\begin{aligned}(\boldsymbol{\theta}(\text{P}) - \theta^i(\xi)\boldsymbol{\beta}_i(B, \text{P}))B &= \theta - \theta^i\,\delta_i \qquad \text{by (11) and (21)},\\ &= \theta - \theta = 0,\end{aligned}$$

since

(25) $$\theta = \theta^i\,\delta_i$$

for any 3×1 single-column matrix $\theta = [\theta^r]$. Hence (24) is proved, and this proves the theorem, because $\boldsymbol{\theta}$ is any contravariant body vector at P. Since there are three tangent base vectors, we have obtained a proof that $\{\boldsymbol{\theta}(\mathrm{P})\}$ is three-dimensional, provided we show that the $\boldsymbol{\beta}_i$ are linearly independent. This proof is immediate, because, if $\boldsymbol{\theta} = \mathbf{0}$ in (24), then $\theta = 0$ and thus $\theta^i = 0$ $(i = 1, 2, 3)$.

The term *tangent* base vector for $\boldsymbol{\beta}_i(B, \mathrm{P})$ is justified because, if P_1 is a neighboring particle to P and lies on the ξ^1 curve of B that passes through P, then P_1 has coordinates in B of the form $(\xi^1 + d\xi^1_{(1)}, \xi^2, \xi^3)$, where (ξ^1, ξ^2, ξ^3) are the coordinates of P. As $d\xi^1_{(1)}$ tends to zero, the line PP_1 defines an infinitesimal body vector $\boldsymbol{d\xi}_1$, tangent to the ξ^1 curve of the coordinate system B, given by the equation

(26) $$\boldsymbol{d\xi}_1 = d\xi^1_{(1)}\,\boldsymbol{\beta}_1(B, \mathrm{P}),$$

which is a particular case of (24) with $\boldsymbol{\theta} = \boldsymbol{d\xi}_1$, $\theta^1 = d\xi^1_{(1)}$, and $\theta^2 = \theta^3 = 0$. Similarly, $\boldsymbol{\beta}_2$ and $\boldsymbol{\beta}_3$ are associated with tangents to the ξ^2 curve and the ξ^3 curve at P, respectively.

For any two neighboring particles P and P_1, the coordinate difference matrices $d\xi$, $d\bar{\xi}$, ... in different coordinate systems satisfy the contravariant body vector transformation law (9) and are therefore the representatives of a contravariant body vector $\boldsymbol{d\xi}$ at P given by the equation

(27) $$\boldsymbol{d\xi} = d\xi^i\boldsymbol{\beta}_i(B, \mathrm{P}).$$

In a continuous material moving and deforming in any manner in space, the body vector $\boldsymbol{d\xi}$ *is independent of time*, contains no reference to the separation of particles P and P_1, yet defines unambiguously their relative position in the material in the sense that, given P and $\boldsymbol{d\xi}$ at P, a second particle P_1 is defined uniquely.

(28) Figure 2.3(28) illustrates two configurations of a deforming body to emphasize the rather unfamiliar fact that $\boldsymbol{d\xi}$ is independent of time.

For the body vector $\boldsymbol{d\xi}$, we use a boldface type for $\boldsymbol{d}$ as well as for $\boldsymbol{\xi}$ to emphasize the fact that $\boldsymbol{d\xi}$ is a single entity [defined by (27)] and is not the difference of two vectors $\boldsymbol{\xi}$: The *coordinate matrices* ξ, $\bar{\xi}$, ... *for a particle do not satisfy the transformation law for a contravariant vector* and so we cannot define a body vector $\boldsymbol{\xi}$ whose components equal the coordinates of P. The d in $d\xi^i$ in (27), on the other hand, does denote a difference of coordinates. To use a symbol $d\boldsymbol{\xi}$ (instead of $\boldsymbol{d\xi}$) might therefore be misleading.

In the example of $\boldsymbol{d\xi}$, we have a graphic illustration of the fact that *a contravariant body vector is an entity different from every Cartesian vector of elementary vector analysis*, for a given Cartesian vector has a unique magnitude and direction (in space); *the body vector* $\boldsymbol{d\xi}$ has a unique direction (in the material) but *does not have a unique magnitude*. Its magnitude, as usually

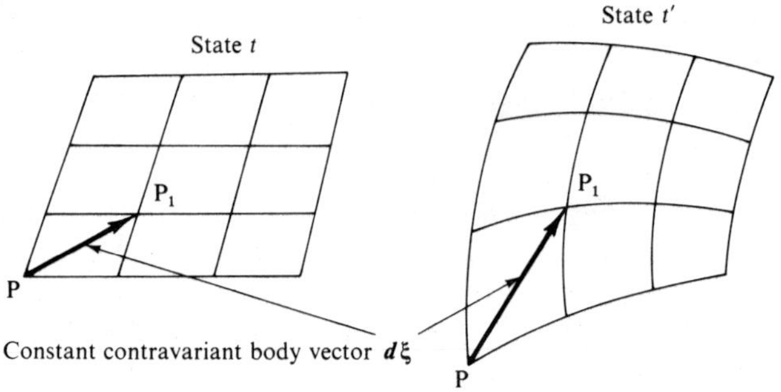

Figure 2.3(28) Coordinate curves of a single body coordinate system in two states t and t' of a deforming body. Although its length and orientation in space are different at t and at t', the contravariant body vector $d\xi = \overrightarrow{PP_1}$ (for any two neighboring particles P and P_1) is constant.

defined, depends on the body metric and this, in general, varies with time. The concepts *magnitude* and *direction* in a geometric manifold are plainly distinct; body vector and tensor analysis separates them and is to this extent simpler and more basic than Cartesian tensor analysis, which does not. On the other hand, Cartesian tensor analysis is simpler because Cartesian vectors at different places can be added, whereas *body vectors at different particles cannot be added.*

The contravariant vectors $\{\boldsymbol{\theta}(P)\}$ defined here are called *body* vectors because each is a mapping of *body* coordinate systems on to $A_{3\times 1}$ and each is associated with a *particle* P; a *field* of such vectors means a set of vectors $\{\boldsymbol{\theta}(P) \mid P \in \mathscr{P}\}$ which contains one vector for each particle. It is clear that the method of defining contravariant body vector fields can also be used to define contravariant vector fields for any geometric manifold and for the space manifold in particular. *We thereby define contravariant space vector fields, which are distinct from contravariant body vector fields.* It will suffice to state the definition and results briefly for space fields because we have given a detailed discussion for body fields and the reader need only consider a similar discussion in which the terms particle, body manifold, body coordinate system, and body vector are replaced by the terms place, space manifold, space coordinate system, and space vector.

(29) **Definition** Let $\bar{v}$ be any 3×1 column matrix associated with any given space coordinate system $\bar{S}$ and place Q. A *contravariant space vector* $\boldsymbol{v}(Q)$ at Q is a function $\boldsymbol{v}(Q): \mathscr{S} \to A_{3\times 1}$ which associates with any space coordinate system S a 3×1 column matrix v, where

(30) $$\boldsymbol{v}(Q)S = v = L_Q^{-1}(S, \bar{S})\bar{v},$$

(31) $$L_Q(S, \bar{S}) = [\partial \bar{x}^r/\partial x^c]_Q, \quad \tilde{L}_Q(S, \bar{S}) = [\partial \bar{x}^c/\partial x^r]_Q, \quad L_Q^{-1}(S, \bar{S}) = [\partial x^r/\partial \bar{x}^c].$$

$\mathscr{S} \equiv \{S\}$ denotes the set of all space coordinate systems which are related to one another by one-valued, sufficiently differentiable transformation functions possessing a unique inverse. The coordinate transformation matrix L_Q satisfies the equations

$$L_Q^{-1}(S, \bar{S})L_Q^{-1}(\bar{S}, \bar{\bar{S}}) = L_Q^{-1}(S, \bar{\bar{S}}), \tag{32}$$

$$\tilde{L}_Q(S, \bar{S})\tilde{L}_Q(\bar{S}, \bar{\bar{S}}) = \tilde{L}_Q(S, \bar{\bar{S}}). \tag{33}$$

The space vector $\boldsymbol{v}(Q)$ so defined is independent of the choice $(\bar{S}, \bar{v})$ in the sense that any other choice $(\bar{\bar{S}}, \bar{\bar{v}})$ would give the same vector $\boldsymbol{v}(Q)$ on the understanding that $\bar{v}$ and $\bar{\bar{v}}$ are related by the space contravariant vector transformation law

$$\bar{\bar{v}} = L_Q^{-1}(\bar{\bar{S}}, \bar{S})\bar{v}. \tag{34}$$

The *tangent base vectors for a space coordinate system S at place* Q are defined as the set of three contravariant space vectors $\boldsymbol{b}_i(S, Q)$ at Q whose representative matrices in S are δ_i, i.e.,

$$\boldsymbol{b}_i(S, Q)S = \delta_i \qquad (i = 1, 2, 3). \tag{35}$$

These vectors form a basis for contravariant space vectors at Q; for any such vector $\boldsymbol{v}(Q)$, we have

$$\boldsymbol{v}(Q) = v^i(x)\boldsymbol{b}_i(S, Q), \tag{36}$$

where $[v^r(x)]$ is the matrix representing $\boldsymbol{v}(Q)$ in S. The set $\{\boldsymbol{v}(Q) \mid \bar{v} \in A_{3\times 1}\}$ of all contravariant space vectors at Q [defined by (30)] is a linear space in which addition and multiplication by numbers are defined by the equation

$$(a_1\boldsymbol{v}_1 + a_2\boldsymbol{v}_2)S = a_1 v_1 + a_2 v_2 = L_Q^{-1}(S, \bar{S})(a_1\bar{v}_1 + a_2\bar{v}_2), \tag{37}$$

where $\boldsymbol{v}_1$ and $\boldsymbol{v}_2$ are any two vectors at Q and $\bar{v}_1$ and $\bar{v}_2$ are their representative matrices in $\bar{S}$.

An important example of a contravariant space vector at Q is the displacement vector

$$\boldsymbol{dx} = dx^i\boldsymbol{b}_i(S, Q), \tag{38}$$

which is determined by any two neighboring places Q and Q_1; dx^i denotes their coordinate differences in S. If Q and Q_1 are the places occupied by a particle P at times t and $t + dt$, respectively, it follows that the velocity

$$\boldsymbol{v}(Q, t) = v^i(x, t)\boldsymbol{b}_i(S, Q) \qquad (v^i = dx^i/dt \quad \text{at constant} \quad P) \tag{39}$$

is a contravariant space vector at Q. The *velocity field* for a body is the set of all velocity vectors at time t, there being one vector for each place in a region occupied by the body.

(40) **Problem** Prove that the velocity components $v^i(x, t)$, as defined by 2.2(8) and similar equations with S replaced by $\bar{S}$, satisfy the transformation law (30) for a contravariant space vector.

Finally, we note that *in order to specify unambiguously a geometric vector field* (as defined previously), *we require the following information*:

(41) (a) the manifold (e.g., $\mathscr{P}$ or $\mathscr{Q}$) over which it is defined;
(b) the set (e.g., $\mathscr{B}$ or $\mathscr{S}$) of coordinate systems *to be used in the definition* (this includes a knowledge of the "allowable" coordinate transformation matrices, e.g., Λ_P or L_Q);
(c) the transformation law (e.g., $\theta = \Lambda^{-1}\bar{\theta}$ or $v = L^{-1}\bar{v}$);
(d) the values of the component functions [e.g., $\theta^i(\xi)$ or $v^i(x)$] in any one specified coordinate system.

A similar list of information is required for other kinds of vectors and for tensors, as we shall presently see. We emphasize the words "to be used in the definition" in (b) because in Chapter 4 we restrict the set of coordinate systems [used in the context similar to that of $\mathscr{B}$ in (11)] to be rectangular Cartesian and thereby obtain a different kind of vector, called a *Cartesian* vector, which has some different properties because of this restriction on the "allowable" coordinate systems; the Cartesian vectors, however, once defined in this way (or in other, more common but equivalent ways), can then be *used* in conjunction with curvilinear coordinate systems. It is the nature of the group of allowable coordinate transformations *used in the definition* of vectors that determines whether the vectors are general or Cartesian and also determines what operations can be performed with them.

In the remainder of this chapter, we continue with general vectors and define covariant vectors [by using (6) instead of (5)] and tensors (by forming direct products of vector spaces already defined). The procedure will be relatively brief because it involves only straightforward extensions of the analysis already given in detail for contravariant vector fields. Some additional operations and definitions will be postponed to Chapter 5, after we have given the application to shear flow in Chapter 3.

The wording in (41a) admits two different interpretations. In our usage, a body field is defined over the body manifold in the sense that it is the choice of manifold which governs the choice of coordinate systems used in the definition of the field [e.g., $\mathscr{B}$ in (11)]. An alternative usage (more common than ours) would have it that a field defined over the body manifold is any set of objects (e.g., space vectors or body vectors) associated with points of the manifold, with one object to one point.

2.4 The definition of covariant vector fields

We have seen that the group property 2.3(5) possessed by transformation matrices $\Lambda_P^{-1}(B, \bar{B})$, ... plays a fundamental role in the definition 2.3(10) of contravariant body vectors and leads to the essential property 2.3(12), which ensures that the vector has a significance independent of the choice of any particular coordinate system from the set $\mathscr{B}$. But 2.3(6) shows that the matrices $\tilde{\Lambda}_P(B, \bar{B})$, ... also possess the same group property; it follows that we can define another kind of vector (called a covariant vector) if we simply replace Λ_P^{-1} in the transformation law (11) by $\tilde{\Lambda}_P$.

(1) **Definition** Let $\bar{\psi}$ be any 3×1 column matrix associated with any given body coordinate system $\bar{B}$ and particle P. A *covariant body vector* $\boldsymbol{\psi}(\mathrm{P})$ at P is a function $\boldsymbol{\psi}(\mathrm{P}): \mathscr{B} \to A_{3\times 1}$ which associates with any body coordinate system B a 3×1 column matrix ψ where, for all B in $\mathscr{B}$,

(2) $$\boldsymbol{\psi}(\mathrm{P})B = \psi = \tilde{\Lambda}_P(B, \bar{B})\bar{\psi} \qquad [\psi_r(\xi) = (\partial\bar{\xi}^i/\partial\xi^r)_P \bar{\psi}_i(\bar{\xi})].$$

(3) **Problem** Prove that $\boldsymbol{\psi}(\mathrm{P})$ is independent of the choice of $(\bar{B}, \bar{\psi})$ in the sense that, if in (2) the pair $(\bar{B}, \bar{\psi})$ is replaced by $(\bar{\bar{B}}, \bar{\bar{\psi}})$, where $\bar{\bar{\psi}} = \tilde{\Lambda}_P(\bar{\bar{B}}, \bar{B})\bar{\psi}$, the same vector $\boldsymbol{\psi}(\mathrm{P})$ is obtained.

(4) **Problem** Prove that, if $\bar{\psi}$ in (2) assumes all possible values in $A_{3\times 1}$, the vectors $\{\boldsymbol{\psi}(\mathrm{P}) \mid \bar{\psi} \in A_{3\times 1}\}$ are elements of a linear space in which addition and multiplication by numbers are defined by the equation

(5) $$(a_1\boldsymbol{\psi}^1 + a_2\boldsymbol{\psi}^2)B = a_1\psi^1 + a_2\psi^2 = \tilde{\Lambda}_P(B, \bar{B})(a_1\bar{\psi}^1 + a_2\bar{\psi}^2)$$

and the zero vector is given by 2.3(19).

(6) **Definition** The *normal base vectors* $\boldsymbol{\beta}^i(B, \mathrm{P})$ for any given body coordinate system B and particle P are three covariant body vectors at P whose representative matrices in B are $\delta^i = \delta_i$, i.e.,

(7) $$\boldsymbol{\beta}^i(B, \mathrm{P})B = \delta^i \quad (i = 1, 2, 3) \qquad \text{or} \qquad \beta_j^{(i)} = \delta_j^i \quad (\text{in } B \text{ only}).$$

(8) **Problem** Prove that any covariant body vector $\boldsymbol{\psi}(\mathrm{P})$ at P can be written in the form

$$\boldsymbol{\psi}(\mathrm{P}) = \psi_i(\xi)\boldsymbol{\beta}^i(B, \mathrm{P}),$$

where $[\psi_r(\xi)] = \psi$ is the representative matrix of $\boldsymbol{\psi}(\mathrm{P})$ in B.

(9) **Problem** Let $\boldsymbol{\theta}(\mathrm{P})$ be any given contravariant body vector at P, and let $\bar{\theta}$ be its representative matrix in some coordinate system $\bar{B}$. Let $\boldsymbol{\psi}(\mathrm{P})$ be

that covariant body vector at P that has the same representative matrix $\bar{\psi} = \bar{\theta}$ in $\bar{B}$. Prove that, unless $\bar{\psi} = \bar{\theta} = 0$, there is another coordinate system B in which the representative matrices, ψ and θ, of $\boldsymbol{\psi}$(P) and $\boldsymbol{\theta}$(P) are unequal [see Fig. 2.3(14)].

This result shows that (with the possible exception of the zero vector) covariant vectors and contravariant vectors are different geometric objects belonging to different linear spaces. In particular, *we cannot add a covariant vector at* P *and a contravariant vector at* P.

(10) **Definition** A *body scalar field* is a mapping $s : \mathscr{P} \to \mathscr{R}$ which assigns to each particle P a real number s(P). The *transformation law for a body scalar field* is thus of the form

(11) $$s(\mathrm{P}) = \sigma(\xi) = \bar{\sigma}(\bar{\xi}),$$

where $\sigma(\xi)$ and $\bar{\sigma}(\bar{\xi})$ are the component functions representing the scalar field in any two body coordinate systems B and $\bar{B}$, respectively.

(12) **Problem** Prove that, for any given body scalar field s(P), there exists a covariant body vector field (called grad s) whose representative matrix in an arbitrary body coordinate system B is $[\partial\sigma/\partial\xi^r]$, where $\sigma = \sigma(\xi)$ is the component function (assumed differentiable) of s(P) in B.

We are here using the shortened notation s(P) for a scalar field instead of the full notation $\{s(\mathrm{P}) \mid \mathrm{P} \in \mathscr{P}\}$. We shall often use a similar short notation for other fields where no confusion is likely to ensue.

(13) **Theorem** Given any contravariant body vector field $\boldsymbol{\theta}$(P) and any covariant body vector field $\boldsymbol{\psi}$(P), there exists a scalar field $\boldsymbol{\theta}(\mathrm{P}) \cdot \boldsymbol{\psi}(\mathrm{P})$, called the *scalar product* of $\boldsymbol{\theta}$ and $\boldsymbol{\psi}$, defined by the equation

(14) $$\boldsymbol{\theta}(\mathrm{P}) \cdot \boldsymbol{\psi}(\mathrm{P}) = \theta^i(\xi)\psi_i(\xi) = \tilde{\theta}\psi,$$

where $\theta = [\theta^r(\xi)]$ and $\psi = [\psi_r(\xi)]$ are the representative matrices of θ and ψ in an arbitrary coordinate system B. The process of equating a superscript and a subscript letter in a term and carrying out the implied summation is called *contraction*, provided that one index labels components of a contravariant vector and the other labels components of a covariant vector at the same point.

Proof If $\bar{\theta}$ and $\bar{\psi}$ are the representative matrices of $\boldsymbol{\theta}$(P) and $\boldsymbol{\psi}$(P) in any other body coordinate system $\bar{B}$, we have $\tilde{\bar{\theta}}\bar{\psi} = \tilde{\theta}\psi$, from 2.3(11) and (2); the process of contraction thus assigns a unique number (equal to $\tilde{\theta}\psi$) to each particle P, independent of B, which proves the result. Moreover, since $\tilde{\theta}\psi = \tilde{\psi}\theta$, the same number is obtained if one changes the order of the factors in the scalar product. Since $\boldsymbol{\theta}$ and $\boldsymbol{\psi}$ are elements in different linear spaces, the scalar product is not an inner product as defined in 1.2(5). We call $\boldsymbol{\theta} \cdot \boldsymbol{\psi}$ a scalar product because it is a scalar field.

(15) **Definition** Given any material surface whose equation in an arbitrary body coordinate system B is $\sigma(\xi) = c$, where c is a constant. A *normal* to the surface at P is a covariant body vector $\mathbf{v}(\mathrm{P})$ at P whose component matrix in B is $\lambda[\partial\sigma/\partial\xi^r]$ where λ is any scalar. From the analysis used in proving (12), it follows that the component matrix in any other body coordinate system $\bar{B}$ is $\lambda[\partial\bar{\sigma}/\partial\bar{\xi}^r]$. The justification for calling $\mathbf{v}$ a normal lies in the equation

(16) $$\mathbf{v}(\mathrm{P}) \cdot \boldsymbol{d\xi} = 0,$$

which is valid when $\boldsymbol{d\xi}$ is the contravariant vector determined by any two neighboring particles P and P_1 in the given surface. Thus $\boldsymbol{d\xi}$ is a tangent to the surface at P if the scalar product of $\boldsymbol{d\xi}$ with the normal $\mathbf{v}$ is zero. We shall presently choose λ to give $\mathbf{v}$ unit magnitude.

(17) **Problem** Prove that the "normal base vector" $\boldsymbol{\beta}^i(B, \mathrm{P})$, defined in (6), is normal at P to the ξ^i surface of the coordinate system B.

Covariant space vector fields, and the definition of the scalar product for space vector fields, can be treated in a similar manner.

(18) **Definition** Let $\bar{u}$ be any given 3×1 column matrix associated with a given place Q and a given space coordinate system $\bar{S}$. A *covariant space vector* $\boldsymbol{u}(\mathrm{Q})$ at Q is a function $\boldsymbol{u}(\mathrm{Q}) : \mathscr{S} \to A_{3\times 1}$ which associates with any given space coordinate system S a 3×1 column matrix u, where

(19) $$\boldsymbol{u}(\mathrm{Q})S = u = \tilde{L}_{\mathrm{Q}}(S, \bar{S})\bar{u}.$$

The vector $\boldsymbol{u}(\mathrm{Q})$ is independent of the choice of $(\bar{S}, \bar{u})$ in the usual sense. The set $\{\boldsymbol{u}(\mathrm{Q}) \mid \bar{u} \in A_{3\times 1}\}$ is a linear space in which addition and multiplication by numbers is defined by the equation

(20) $$(a_1\boldsymbol{u}_1 + a_2\boldsymbol{u}_2)S = a_1u_1 + a_2u_2 = \tilde{L}_{\mathrm{Q}}(S, \bar{S})(a_1\bar{u}_1 + a_2\bar{u}_2).$$

The normal base vectors $\boldsymbol{b}^i(S, \mathrm{Q})$ at Q for a space coordinate system S are covariant space vectors at Q whose representative matrices in S are δ^i, i.e.,

(21) $$\boldsymbol{b}^i(S, \mathrm{Q})S = \delta^i \qquad \text{or} \qquad b_j^{(i)} = \delta_j^i \qquad (\text{in } S \text{ only}).$$

Any covariant space vector $\boldsymbol{u}(\mathrm{Q})$ at Q can be written in the form

(22) $$\boldsymbol{u}(\mathrm{Q}) = u_i(x)\boldsymbol{b}^i(S, \mathrm{Q}),$$

where $[u_r(x)] = u$ is the representative matrix of $\boldsymbol{u}(\mathrm{Q})$ in S. The scalar product of $\boldsymbol{u}(\mathrm{Q})$ and $\boldsymbol{v}(\mathrm{Q})$ is defined by the equation

$$\boldsymbol{u}(\mathrm{Q}) \cdot \boldsymbol{v}(\mathrm{Q}) = u_i(x)v^i(x), \tag{23}$$

where $\boldsymbol{v}(\mathrm{Q})$ is a contravariant space vector at Q having representative matrix $[v^i(x)]$ in S. A space scalar field s is a mapping $s: \mathscr{Q} \to \mathscr{R}$ which assigns to each place Q a unique real number $s(\mathrm{Q})$; the transformation law is of the form

$$s(\mathrm{Q}) = \bar{s}(\bar{x}) = \bar{\bar{s}}(\bar{\bar{x}}), \tag{24}$$

where $\bar{s}(\bar{x})$ and $\bar{\bar{s}}(\bar{\bar{x}})$ are the component functions of the scalar field in coordinate systems $\bar{S}$ and $\bar{\bar{S}}$, respectively.

A normal $\boldsymbol{n}(\mathrm{Q})$ at Q to a surface $s(x) = c$ in space is a covariant space vector at Q, having a representative matrix $\lambda[\partial s(x)/\partial x^r]$ in S, where λ is a scalar; $\boldsymbol{dx}$ is a tangent to the surface at Q if

$$\boldsymbol{n}(\mathrm{Q}) \cdot \boldsymbol{dx} = 0. \tag{25}$$

(26) **Problem** Prove the following results:

(i) $\boldsymbol{\beta}_i(B, \mathrm{P}) \cdot \boldsymbol{\beta}^j(B, \mathrm{P}) = \delta_i^j$; $\quad \mathbf{b}_i(S, \mathrm{Q}) \cdot \mathbf{b}^j(S, \mathrm{Q}) = \delta_i^j$.
(ii) $\theta^i(\xi) = \boldsymbol{\theta}(\mathrm{P}) \cdot \boldsymbol{\beta}^i(B, \mathrm{P})$ [from 2.3(24)];
(iii) $\psi_i(\xi) = \boldsymbol{\psi}(\mathrm{P}) \cdot \boldsymbol{\beta}_i(B, \mathrm{P})$ [from 2.4(8)];
(iv) $v^i(x) = \boldsymbol{v}(\mathrm{Q}) \cdot \boldsymbol{b}^i(S, \mathrm{Q})$ [from 2.3(36)];
(v) $u_i(x) = \boldsymbol{u}(\mathrm{Q}) \cdot \boldsymbol{b}_i(S, \mathrm{Q})$ [from 2.4(22)].

By virtue of the relations (i), the sets of vectors $\boldsymbol{\beta}_i$ and $\boldsymbol{\beta}^i$ are said to be *reciprocal* to one another. The sets $\boldsymbol{b}_i$ and $\boldsymbol{b}^i$ are reciprocal to one another. It is easy to show that any given set of base vectors has a unique reciprocal set.

(27) **Problem** Show that any given set of three linearly independent vectors has a unique reciprocal set.

(28) **Problem** Show that, for any two coordinate systems $B: \mathrm{P} \to \xi$ and $\bar{B}: \mathrm{P} \to \bar{\xi}$, the base vectors are related by the equations

(i) $\boldsymbol{\beta}_i(\bar{B}, \mathrm{P}) = (\partial \xi^k / \partial \bar{\xi}^i)\boldsymbol{\beta}_k(B, \mathrm{P})$;
(ii) $\boldsymbol{\beta}^i(\bar{B}, \mathrm{P}) = (\partial \bar{\xi}^i / \partial \xi^k)\boldsymbol{\beta}^k(B, \mathrm{P})$.

2.5 The definition of second-rank tensor fields

Vectors are also called *first-rank tensors*; in any coordinate system, a vector is represented by components, having a single dummy index, which, for convenience, we have taken to be elements of a 3×1 single-column matrix. A second-rank tensor is represented by components having two dummy indexes, which we shall take to be elements of a 3×3 (square) matrix. An nth-rank tensor has components, with n dummy indexes, which form an ordered array. In this sense, a scalar can be regarded as a zeroth-rank tensor. *Tensors of second and higher rank can conveniently be defined as elements of those linear spaces that are formed by taking the direct (or tensor) products of the sets of contravariant and covariant vectors at the same point of a geometric manifold*; the direct product of linear spaces has been defined in 1.2(9), and all that remains is to specify how the processes of addition and multiplication by numbers in the product space have to be performed. Since there are (for a given manifold and a given point) two distinct vector spaces, there are four distinct kinds of second-rank tensor: contravariant, covariant, left-covariant mixed, and right-covariant mixed. In our applications, second-rank tensors play the major role. The method of defining tensors and the derivation of their properties is for the most part sufficiently exemplified by second-rank tensors; the extension to higher rank tensors is fairly obvious.

(1) **Definition** Let $\boldsymbol{\theta}_1$ and $\boldsymbol{\theta}_2$ denote any two elements in the set $\{\boldsymbol{\theta}\} = \{\boldsymbol{\theta}(\mathrm{P}) \,|\, \bar{\theta} \in A_3\}$ of all contravariant body vectors at P, so that

$$\boldsymbol{\theta}_i B = \theta_i = \Lambda_{\mathrm{P}}^{-1}(B, \bar{B})\bar{\theta}_i \qquad (i = 1, 2),$$

where θ_i and $\bar{\theta}_i$ are the 3×1 column representative matrices of $\boldsymbol{\theta}_i$ in B and $\bar{B}$, respectively. Form the direct product $\{\boldsymbol{\theta}\} \otimes \{\boldsymbol{\theta}\} \equiv \{\boldsymbol{\Theta}\}$ in which each element $\boldsymbol{\Theta}$ maps $\mathscr{B}$ onto $A_{3\times 3}$, the set of all 3×3 real matrices, and the element $(\boldsymbol{\theta}_1, \boldsymbol{\theta}_2)$ is defined by the equation

$$(2) \qquad (\boldsymbol{\theta}_1, \boldsymbol{\theta}_2)B = \theta_1\tilde{\theta}_2 = \Lambda_{\mathrm{P}}^{-1}(B, \bar{B})\bar{\theta}_1\tilde{\bar{\theta}}_2\tilde{\Lambda}_{\mathrm{P}}^{-1}(B, \bar{B}) \qquad \text{for all } B \text{ in } \mathscr{B}.$$

Any element $\boldsymbol{\Theta}(\mathrm{P})$ in the product space $\{\boldsymbol{\theta}\} \otimes \{\boldsymbol{\theta}\}$ is called a *contravariant second-rank body tensor* at P and is represented in an arbitrary body coordinate system B by a square matrix $\Theta = [\Theta^{rc}(\xi)]$, where, for all B in $\mathscr{B}$,

$$(3) \qquad \boldsymbol{\Theta}(\mathrm{P})B = \Theta = \Lambda_{\mathrm{P}}^{-1}(B, \bar{B})\bar{\Theta}\tilde{\Lambda}_{\mathrm{P}}^{-1}(B, \bar{B}) \qquad \left(\Theta^{rc}(\xi) = \left(\frac{\partial \xi^r}{\partial \bar{\xi}^i}\frac{\partial \xi^c}{\partial \bar{\xi}^j}\right)_{\mathrm{P}} \bar{\Theta}^{ij}(\bar{\xi})\right)$$

and $\bar{\Theta}$ is the representative matrix in $\bar{B}$. An element $(\boldsymbol{\theta}_1, \boldsymbol{\theta}_2)$ will usually be written $\boldsymbol{\theta}_1\boldsymbol{\theta}_2$ and called the *direct* or *outer product* of $\boldsymbol{\theta}_1$ and $\boldsymbol{\theta}_2$. Alternatively, $\boldsymbol{\Theta}$ can be defined by (3), for given $\bar{\Theta}$ and $\bar{B}$.

(4) **Problem** For any given pair $(\bar{B}, \bar{\Theta})$, define a tensor $\boldsymbol{\Theta}(\mathrm{P})$ by (3); similarly, define a tensor $\boldsymbol{\Theta}'(\mathrm{P})$ by replacing $(\bar{B}, \bar{\theta})$ in (3) by $(\bar{\bar{B}}, \bar{\bar{\theta}})$, where $\bar{\bar{B}}$ is any other body coordinate system and

$$(5) \qquad \bar{\bar{\Theta}} = \Lambda_{\mathrm{P}}^{-1}(\bar{\bar{B}}, \bar{B})\bar{\Theta}\tilde{\Lambda}_{\mathrm{P}}^{-1}(\bar{\bar{B}}, \bar{B}).$$

Prove that $\mathbf{\Theta}(\mathrm{P}) = \mathbf{\Theta}'(\mathrm{P})$, thus showing that the tensor $\mathbf{\Theta}(\mathrm{P})$ is independent of the choice of coordinate system $\bar{B}$.

(6) **Problem** Without using (iii) of 1.2(9), prove that any contravariant second-rank tensor $\mathbf{\Theta}(\mathrm{P})$ at P can be written in the form

$$\mathbf{\Theta}(\mathrm{P}) = \Theta^{ij}(\xi)\boldsymbol{\beta}_i(B, \mathrm{P})\boldsymbol{\beta}_j(B, \mathrm{P}),$$

where $[\Theta^{rc}(\xi)]$ is the representative matrix of $\mathbf{\Theta}(\mathrm{P})$ in B. It follows that the nine tensors $\boldsymbol{\beta}_i\,\boldsymbol{\beta}_j$ form a basis for contravariant second-rank tensors at P. Prove that the *tangent base tensors* $\boldsymbol{\beta}_i\,\boldsymbol{\beta}_j$ are linearly independent, thus showing that the space $\{\mathbf{\Theta}(\mathrm{P})\}$ is nine-dimensional (as expected, because it is isomorphic to $A_{3\times 3}$).

The other kinds of second-rank tensor may be defined in similar ways. The results are as follows. We include for comparison the contravariant tensors just defined. $\{\boldsymbol{\theta}\}$ and $\{\boldsymbol{\psi}\}$ are respectively the sets of all contravariant and covariant body vectors at P.

(7) *Second-rank body tensors at* P *are elements of direct product spaces*:

(i)	$\mathbf{\Theta} \in \{\boldsymbol{\theta}\} \otimes \{\boldsymbol{\theta}\}$	(contravariant),
(ii)	$\mathbf{\Psi} \in \{\boldsymbol{\psi}\} \otimes \{\boldsymbol{\psi}\}$	(covariant),
(iii)	$\boldsymbol{\mu} \in \{\boldsymbol{\theta}\} \otimes \{\boldsymbol{\psi}\}$	(right covariant mixed),
(iv)	$\tilde{\boldsymbol{\mu}} \in \{\boldsymbol{\psi}\} \otimes \{\boldsymbol{\theta}\}$	(left covariant mixed).

(8) *Second-rank body tensors at* P *map* $\mathscr{B}$ *onto* $A_{3\times 3}$ *according to the rules*

(i)	$\mathbf{\Theta}(\mathrm{P})B = [\Theta^{rc}(\xi)] = \Lambda_{\mathrm{P}}^{-1}(B, \bar{B})\bar{\Theta}\tilde{\Lambda}_{\mathrm{P}}^{-1}(B, \bar{B})$	(contravariant),
(ii)	$\mathbf{\Psi}(\mathrm{P})B = [\Psi_{rc}(\xi)] = \tilde{\Lambda}_{\mathrm{P}}(B, \bar{B})\bar{\Psi}\Lambda_{\mathrm{P}}(B, \bar{B})$	(covariant),
(iii)	$\boldsymbol{\mu}(\mathrm{P})B = [\mu^r_c(\xi)] = \Lambda_{\mathrm{P}}^{-1}(B, \bar{B})\bar{\mu}\Lambda_{\mathrm{P}}(B, \bar{B})$	(right covariant mixed),
(iv)	$\tilde{\boldsymbol{\mu}}(\mathrm{P})B = [\mu^c_r(\xi)] = \tilde{\Lambda}_{\mathrm{P}}(B, \bar{B})\tilde{\bar{\mu}}\tilde{\Lambda}_{\mathrm{P}}^{-1}(B, \bar{B})$	(left covariant mixed).

(9) *Second-rank body tensors at* P *can be expressed in terms of outer products of base vectors at* P *for an arbitrary body coordinate system B as follows*:

(i)	$\mathbf{\Theta}(\mathrm{P}) = \Theta^{ij}(\xi)\boldsymbol{\beta}_i(B, \mathrm{P})\boldsymbol{\beta}_j(B, \mathrm{P})$	(contravariant),
(ii)	$\mathbf{\Psi}(\mathrm{P}) = \Psi_{ij}(\xi)\boldsymbol{\beta}^i(B, \mathrm{P})\boldsymbol{\beta}^j(B, \mathrm{P})$	(covariant),
(iii)	$\boldsymbol{\mu}(\mathrm{P}) = \mu^i_j(\xi)\boldsymbol{\beta}_i(B, \mathrm{P})\boldsymbol{\beta}^j(B, \mathrm{P})$	(right covariant mixed),
(iv)	$\tilde{\boldsymbol{\mu}}(\mathrm{P}) = \mu^j_i(\xi)\boldsymbol{\beta}^i(B, \mathrm{P})\boldsymbol{\beta}_j(B, \mathrm{P})$	(left covariant mixed).

In (8), B is an arbitrary body coordinate system, and $\bar{\Theta}$, $\bar{\Psi}$, and $\bar{\mu}$ are square matrices associated with any other given coordinate system $\bar{B}$. It is easy to verify that the tensors $\mathbf{\Theta}$, $\mathbf{\Psi}$, $\boldsymbol{\mu}$, and $\tilde{\boldsymbol{\mu}}$ are independent of the choice of $\bar{B}$. The right-hand equations in (8) represent the transformation laws for the various tensors.

We may use (8) to form new tensors by replacing the $\bar{B}$-representative matrices by their transposes. In cases (i) and (ii), the tensors so obtained are independent of the choice of the coordinate system $\bar{B}$ in which the representative matrices were transposed, because the contravariant and covariant transformation laws are unchanged in form when their transposes are taken. The mixed-tensor transformation laws, however, go into one another when their transposes are taken. Hence we have the following result for second-rank body tensors at any particle P.

(10) **Theorem** To any given contravariant tensor $\mathbf{\Theta}$ there corresponds a unique contravariant tensor $\tilde{\mathbf{\Theta}}$, called the *transpose* of $\mathbf{\Theta}$; their representative matrices in any B are transposes of one another. A similar statement is true for covariant tensors. To any given right covariant tensor $\boldsymbol{\mu}$ there corresponds a unique left covariant tensor $\tilde{\boldsymbol{\mu}}$ whose representative matrix in any B is the transpose of that of $\boldsymbol{\mu}$; a similar statement with left and right interchanged is true. In each case, the transpose of the transpose equals the original tensor. The transpose of an outer product of vectors equals the same vectors in reverse order.

A contravariant (or covariant) tensor is said to be *symmetric* if it equals its transpose and *antisymmetric* (or skew symmetric) if it equals its transpose times minus one.

(11) **Problem** Prove that any given contravariant (or covariant) tensor of second rank can be expressed as a sum of a symmetric tensor and an antisymmetric tensor; prove also that this decomposition is unique.

Symmetry in this sense *is not a property of mixed tensors* because a mixed tensor and its transpose belong in different spaces and cannot be added.

Each tensor space evidently contains a *zero tensor* $\mathbf{0}$ whose representative matrix in every coordinate system is the zero 3×3 matrix. This follows from the fact that all the transformation laws in (8) are linear and homogeneous in the representative matrices.

Choose an arbitrary but definite body coordinate system $\bar{B}$, and consider the tensors defined by (8) with representative matrices in $\bar{B}$ each equal to the 3×3 unit matrix I. In cases (i) and (ii), it is obvious that in other coordinate systems the representative matrices will not be equal to I, and therefore the contravariant and covariant tensors so defined will depend on the choice of $\bar{B}$. In cases (iii) and (iv), however, it is evident that the representative matrices in every coordinate system equal I, and hence we can define a right covariant mixed unit tensor $\boldsymbol{\delta}$ and a left covariant mixed unit tensor $\tilde{\boldsymbol{\delta}}$ such that $\boldsymbol{\delta}\bar{B} = \tilde{\boldsymbol{\delta}}\bar{B} = I$. The unit tensors $\boldsymbol{\delta}$ and $\tilde{\boldsymbol{\delta}}$ do not depend on the choice of $\bar{B}$ in the definition. Moreover, $\boldsymbol{\delta}$ and $\tilde{\boldsymbol{\delta}}$ are tensors at P whose components δ_j^i have the same values at every particle P. The zero tensors have the same feature. *In general tensor analysis, there is no covariant or contravariant unit tensor.*

(12) **Problem** Prove that

$$\text{(i)} \quad \boldsymbol{\delta} = \boldsymbol{\beta}_i(B, \text{P})\boldsymbol{\beta}^i(B, \text{P}); \qquad \tilde{\boldsymbol{\delta}} = \boldsymbol{\beta}^i(B, \text{P})\boldsymbol{\beta}_i(B, \text{P});$$
$$\text{(ii)} \quad \boldsymbol{\delta} = \boldsymbol{\theta}_i(\text{P})\boldsymbol{\psi}^i(\text{P}); \qquad \tilde{\boldsymbol{\delta}} = \boldsymbol{\psi}^i(\text{P})\boldsymbol{\theta}_i(\text{P});$$

where $\boldsymbol{\theta}_1$, $\boldsymbol{\theta}_2$, and $\boldsymbol{\theta}_3$ denote any three linearly independent contravariant body vectors at P, and $\boldsymbol{\psi}^1$, $\boldsymbol{\psi}^2$, $\boldsymbol{\psi}^3$ is the set of covariant body vectors reciprocal to them, i.e., such that

$$\text{(iii)} \quad \boldsymbol{\theta}_i \cdot \boldsymbol{\psi}^j = \delta_i^j .$$

On taking the reciprocals (i.e., the matrix inverses) of the transformation laws in (8), it is evident that the contravariant law goes over into the covariant law, and vice versa, but the mixed laws go over into themselves. This enables us to define the *reciprocal* of a covariant tensor as being that contravariant tensor whose representative matrix in any coordinate system B is the reciprocal of the representative matrix of the given tensor; the reciprocal tensor so defined is independent of the choice of B. A similar statement with covariant and contravariant interchanged is true. It is necessary that the representative matrices be nonsingular (i.e., have nonzero determinants). On taking the determinants of the transformation laws (8), it is clear that if any one representative matrix is nonsingular, then every representative matrix will be nonsingular; we may therefore speak of a *nonsingular second-rank tensor*. Such tensors possess reciprocals. The reciprocal of a right covariant mixed tensor is a right covariant mixed tensor; the reciprocal of a left covariant mixed tensor is a left covariant mixed tensor. Thus *the second-rank mixed tensors alone possess the useful feature that a tensor and its reciprocal belong to the same linear space*; this is one reason why it is more convenient, when defining various functions of tensors, to use mixed tensors in the first instance (Section 5.7). The reciprocal of any second-rank tensor $\boldsymbol{\Phi}$ is written $\boldsymbol{\Phi}^{-1}$.

(13) **Problem** Prove that the operations of taking the transpose and the reciprocal of a second-rank tensor (of any kind) commute.

(14) The effects of taking the transpose and the reciprocal of various kinds of tensor are summarized in Table 2.5(14).

Table 2.5(14) Transpose $\tilde{\mathbf{\Phi}}$ and Reciprocal $\mathbf{\Phi}^{-1}$ of a Second-Rank Tensor $\mathbf{\Phi}$

$\mathbf{\Phi}$	$\mathbf{\Phi}^{-1}$	$\tilde{\mathbf{\Phi}}$	$\tilde{\mathbf{\Phi}}^{-1}$
Contravariant	Covariant	Contravariant	Covariant
Covariant	Contravariant	Covariant	Contravariant
Right-covariant mixed	Right-covariant mixed	Left-covariant mixed	Left-covariant mixed
Left-covariant mixed	Left-covariant mixed	Right-covariant mixed	Right-covariant mixed

(15) **Contraction** The process of contraction defined in 2.4(13) and denoted by a dot extends at once to second-rank tensors. Some of the main uses are illustrated as follows:

(i)	$\operatorname{tr} \boldsymbol{\mu} = \operatorname{tr} \mu = \mu_i^i$	(scalar),
(ii)	$\mathbf{\Theta} \cdot \boldsymbol{\psi} B = \Theta \psi = [\Theta^{ri} \psi_i]$	(contravariant vector),
(iii)	$\mathbf{\Psi} \cdot \boldsymbol{\theta} B = \Psi \theta = [\Psi_{ri} \theta^i]$	(covariant vector),
(iv)	$\mathbf{\Theta} \cdot \mathbf{\Psi} B = \Theta \Psi = [\Theta^{ri} \Psi_{ic}]$	(right covariant mixed tensor),
(v)	$\mathbf{\Psi} \cdot \mathbf{\Theta} B = \Psi \Theta = [\Psi_{ri} \Theta^{ic}]$	(left covariant mixed tensor),
(vi)	$\boldsymbol{\mu} \cdot \boldsymbol{\mu}' B = \mu \mu' = [\mu_i^r \mu_c'^i]$	(right covariant mixed tensor),
(vii)	$\mathbf{\Theta} : \mathbf{\Psi} = \operatorname{tr} \Theta \Psi = \Theta^{ij} \Psi_{ji}$	(scalar),
(viii)	$\boldsymbol{\psi} \cdot \mathbf{\Theta} B = \tilde{\Theta} \psi = [\psi_i \Theta^{ir}]$	(contravariant vector),
(ix)	$\boldsymbol{\theta} \cdot \mathbf{\Psi} B = \tilde{\Psi} \theta = [\theta^i \Psi_{ir}]$	(covariant vector).

Where a single contraction is used, it is the component indexes "adjacent to the dot" which are put equal and summed. Where more than one contraction is used in the same term, it may be necessary to state explicitly which component indexes are to be put equal and summed; thus, for example, $\mathbf{\Theta} : \mathbf{\Psi}$ could mean either $\Theta^{ij} \Psi_{ij}$ or $\Theta^{ij} \Psi_{ji}$ [as chosen in (vii)]. This ambiguity is a disadvantage of the tensor notation (not present when the components are used) which is not, however, very serious when one deals with first- and second-rank tensors only. The convention (vii) is obtained if one forms $\mathbf{\Theta} : \mathbf{\Psi}$ by contracting $\mathbf{\Theta} \cdot \mathbf{\Psi}$.

(16) **Problem** Prove that the quantities given by contraction in the list (15) are quantities of the kinds stated on the right.

(17) **Problem** If $\mathbf{\Phi}$ denotes any given nonsingular second-rank tensor, either covariant or right covariant mixed, and $\mathbf{\Phi}^{-1}$ is its reciprocal as defined above, prove that

(i) $\mathbf{\Phi} \cdot \mathbf{\Phi}^{-1} = \tilde{\boldsymbol{\delta}}, \qquad \mathbf{\Phi}^{-1} \cdot \mathbf{\Phi} = \boldsymbol{\delta},$

and that $\boldsymbol{\Phi}^{-1}$ is the only tensor which satisfies these equations, i.e., that the reciprocal is a unique inverse. Prove also that

(ii) $\tilde{\boldsymbol{\delta}} \cdot \boldsymbol{\Phi} = \boldsymbol{\Phi} \cdot \boldsymbol{\delta} = \boldsymbol{\Phi}$.

There are similar results and definitions for space tensors of second rank which can all be obtained from the above body tensor results by making the obvious changes. It will suffice to state the results which correspond to (8), (9), and (12).

(18) *Second-rank space tensors* $\boldsymbol{p}$, $\boldsymbol{q}$, $\boldsymbol{m}$, *and* $\tilde{\boldsymbol{m}}$ *at any place* Q *map* $\mathscr{S}$ *on to* $A_{3\times 3}$ *according to the rules*

(i) $\boldsymbol{p}(\mathrm{Q})S = [p^{rc}(x)] = L_{\mathrm{Q}}^{-1}(S, \bar{S})\bar{p}\tilde{L}_{\mathrm{Q}}^{-1}(S, \bar{S})$ (contravariant),
(ii) $\boldsymbol{q}(\mathrm{Q})S = [q_{rc}(x)] = \tilde{L}_{\mathrm{Q}}(S, \bar{S})\bar{q}L_{\mathrm{Q}}(S, \bar{S})$ (covariant),
(iii) $\boldsymbol{m}(\mathrm{Q})S = [m^r_c(x)] = L_{\mathrm{Q}}^{-1}(S, \bar{S})\bar{m}L_{\mathrm{Q}}(S, \bar{S})$ (right covariant mixed)
(iv) $\tilde{\boldsymbol{m}}(\mathrm{Q})S = [m_r^c(x)] = \tilde{L}_{\mathrm{Q}}(S, \bar{S})\bar{\tilde{m}}\tilde{L}_{\mathrm{Q}}^{-1}(S, \bar{S})$ (left covariant mixed).

These tensors can be expressed in terms of outer products of space base vectors at Q for a space coordinate system S, as follows:

(a) $\boldsymbol{p}(\mathrm{Q}) = p^{ij}(x)\boldsymbol{b}_i(S, \mathrm{Q})\boldsymbol{b}_j(S, \mathrm{Q})$ (contravariant),
(b) $\boldsymbol{q}(\mathrm{Q}) = q_{ij}(x)\boldsymbol{b}^i(S, \mathrm{Q})\boldsymbol{b}^j(S, \mathrm{Q})$ (covariant),
(c) $\boldsymbol{m}(\mathrm{Q}) = m^i_j(x)\boldsymbol{b}_i(S, \mathrm{Q})\boldsymbol{b}^j(S, \mathrm{Q})$ (right covariant mixed),
(d) $\tilde{\boldsymbol{m}}(\mathrm{Q}) = m_i^j(x)\boldsymbol{b}^i(S, \mathrm{Q})\boldsymbol{b}_j(S, \mathrm{Q})$ (left covariant mixed).

Further, the right covariant unit space tensor $\boldsymbol{I}$ at Q can be written in the form

(19) $$\boldsymbol{I} = \boldsymbol{b}_i(S, \mathrm{Q})\boldsymbol{b}^i(S, \mathrm{Q}); \qquad \tilde{\boldsymbol{I}} = \boldsymbol{b}^i(S, \mathrm{Q})\boldsymbol{b}_i(S, \mathrm{Q}),$$

and has the same representative matrix I in every space coordinate system.

2.6 Metric tensor fields for the body and space manifolds

By hypothesis, the separation ds of any two neighboring places Q and Q_1 in the space manifold is given by an equation of the form 2.2(11), which can be written in matrix form as follows:

(1) $$(ds)^2 = \widetilde{dx}\, g\, dx,$$

where $dx = [dx^r]$ and $g = [g_{rc}(x)]$. For any other space coordinate system $\bar{S}: \mathrm{Q} \to \bar{x}$, we have

(2) $$d\bar{x} = L_{\mathrm{Q}}(S, \bar{S})\, dx,$$

in the notation of 2.3(31). Hence $\widetilde{dx}\, g\, dx = \widetilde{d\bar{x}}\, \tilde{L}_{\mathrm{Q}}^{-1}(S, \bar{S}) g L_{\mathrm{Q}}^{-1}(S, \bar{S})\, d\bar{x} = \widetilde{d\bar{x}}\, \bar{g}\, d\bar{x}$, where

(3) $$\bar{g} = \tilde{L}_{\mathrm{Q}}^{-1}(S, \bar{S}) g L_{\mathrm{Q}}^{-1}(S, \bar{S}).$$

Thus the matrices g and $\bar{g}$, which are associated with coordinate systems S and $\bar{S}$, respectively, satisfy the transformation law (ii) of 2.5(18), and therefore represent a covariant second-rank space tensor $\boldsymbol{g}(\mathrm{Q})$ at Q which is defined by the equation

$$(4) \qquad \boldsymbol{g}(\mathrm{Q})S = g = \tilde{L}_{\mathrm{Q}}(S, \bar{S})\bar{g}L_{\mathrm{Q}}(S, \bar{S}) \qquad \text{for all} \quad S \quad \text{in} \quad \mathscr{S}.$$

$\boldsymbol{g}(\mathrm{Q})$ is called the *space metric tensor*, and has the following properties:

(5) $\tilde{\boldsymbol{g}}(\mathrm{Q}) = \boldsymbol{g}(\mathrm{Q})$, i.e., $\boldsymbol{g}$ is symmetric [from 2.2(15) and 2.5(10)];

(6) $(ds)^2 = \boldsymbol{dx} \cdot \boldsymbol{g}(\mathrm{Q}) \cdot \boldsymbol{dx}$ [from 2.5(15)(iii) and (1)];
$\quad = (\boldsymbol{dx}\,\boldsymbol{dx}) : \boldsymbol{g}(\mathrm{Q})$;

(7) $\boldsymbol{g}(\mathrm{Q})$ is a positive-definite tensor, and possesses a reciprocal $\boldsymbol{g}^{-1}(\mathrm{Q})$ which is a contravariant, positive-definite tensor.

We say that a second-rank tensor (covariant or contravariant) is *positive definite* if the associated quadratic form [e.g., (6)] is positive definite; the corresponding quadratic form [e.g., (1)] in an arbitrary coordinate system is necessarily positive definite, and hence the tensor is nonsingular and therefore possesses a reciprocal. We also have

$$(8) \qquad \boldsymbol{g}(\mathrm{Q}) = g_{ij}(x)\boldsymbol{b}^i(S, \mathrm{Q})\boldsymbol{b}^j(S, \mathrm{Q}), \qquad \boldsymbol{g}^{-1}(\mathrm{Q}) = g^{ij}(x)\boldsymbol{b}_i(S, \mathrm{Q})\boldsymbol{b}_j(S, \mathrm{Q}),$$

from 2.5(18a,b), where $[g^{ij}(x)] = g^{-1}$, and

$$(9) \qquad \boldsymbol{g}(\mathrm{Q}) \cdot \boldsymbol{g}^{-1}(\mathrm{Q}) = \tilde{\boldsymbol{I}}, \qquad \boldsymbol{g}^{-1}(\mathrm{Q}) \cdot \boldsymbol{g}(\mathrm{Q}) = \boldsymbol{I} \qquad \text{[from 2.5(19)]}.$$

A similar procedure for the body manifold leads to the definition

$$(10) \qquad \boldsymbol{\gamma}(\mathrm{P}, t)B = \gamma = \tilde{\Lambda}_{\mathrm{P}}(B, \bar{B})\bar{\gamma}\Lambda_{\mathrm{P}}(B, \bar{B}) \qquad \text{for all} \quad B \quad \text{in} \quad \mathscr{B},$$

of the *covariant body metric tensor* $\boldsymbol{\gamma}(\mathrm{P}, t)$ at particle P and time t, where

$$(11) \qquad \gamma = [\gamma_{rc}(\xi, t)] \qquad \text{and} \qquad \bar{\gamma} = [\bar{\gamma}_{rc}(\bar{\xi}, t)]$$

denote the representative matrices in B and $\bar{B}$, respectively, and 2.2(12) takes the form

$$(12) \qquad (ds)^2 = \widetilde{d\xi}\,\gamma\,d\xi = \widetilde{d\bar{\xi}}\,\bar{\gamma}\,d\bar{\xi}.$$

The body metric tensor has the following properties, which correspond to properties of the space metric tensor:

(13) $\tilde{\boldsymbol{\gamma}}(\mathrm{P}, t) = \boldsymbol{\gamma}(\mathrm{P}, t)$, i.e., $\boldsymbol{\gamma}$ is symmetric;

(14) $(ds)_t^2 = \boldsymbol{d\xi} \cdot \boldsymbol{\gamma}(\mathrm{P}, t) \cdot \boldsymbol{d\xi} = (\boldsymbol{d\xi}\,\boldsymbol{d\xi}) : \boldsymbol{\gamma}(\mathrm{P}, t)$;

(15) $\boldsymbol{\gamma}(\mathrm{P}, t)$ is a positive-definite tensor and possesses a reciprocal $\boldsymbol{\gamma}^{-1}(\mathrm{P}, t)$ which is a contravariant, positive-definite, body tensor;

(16) $\boldsymbol{\gamma}(\mathrm{P}, t) = \gamma_{ij}(\xi, t)\boldsymbol{\beta}^i(B, \mathrm{P})\boldsymbol{\beta}^j(B, \mathrm{P})$;

(17) $\boldsymbol{\gamma}^{-1}(\mathrm{P}, t) = \gamma^{ij}(\xi, t)\boldsymbol{\beta}_i(B, \mathrm{P})\boldsymbol{\beta}_j(B, \mathrm{P})$, where $[\gamma^{rc}(\xi, t)] = \gamma^{-1}$;

(18) $\boldsymbol{\gamma}(\mathrm{P}, t) \cdot \boldsymbol{\gamma}^{-1}(\mathrm{P}, t) = \tilde{\boldsymbol{\delta}}$; $\boldsymbol{\gamma}^{-1}(\mathrm{P}, t) \cdot \boldsymbol{\gamma}(\mathrm{P}, t) = \boldsymbol{\delta}$;

(19) $(dc/dh)_t^2 = (\text{grad } \sigma) \cdot \boldsymbol{\gamma}^{-1}(\mathrm{P}, t) \cdot (\text{grad } \sigma)$ [from 2.2(21)],

where dh denotes the separation at particle P at time t of material surfaces $\sigma(\xi) = c$ and $\sigma(\xi) = c + dc$, and grad σ is a covariant body vector at Q whose components in B are $\partial\sigma(\xi)/\partial\xi^i$ [see 2.4(12)].

The body metric tensor field $\boldsymbol{\gamma}(\mathrm{P}, t)$ *varies with time* (unless the motion is rigid); *the space metric tensor field* $\boldsymbol{g}(\mathrm{Q})$ *is independent of time.*

Considering an arbitrary pair of neighboring particles P and P_1 during any continuous flow of the material, the separation is given by (14) with $\boldsymbol{d\xi}$ independent of time; hence

$$\frac{d^n}{dt^n}(ds)_t^2 = (\boldsymbol{d\xi}\, \boldsymbol{d\xi}) : \frac{\partial^n}{\partial t^n}\boldsymbol{\gamma}(\mathrm{P}, t), \tag{20}$$

and

$$(ds)^2 - (ds')^2 = (\boldsymbol{d\xi}\, \boldsymbol{d\xi}) : \Delta\boldsymbol{\gamma}(\mathrm{P}, t, t'), \tag{21}$$

where

$$\Delta\boldsymbol{\gamma}(\mathrm{P}, t, t') \equiv \boldsymbol{\gamma}(\mathrm{P}, t) - \boldsymbol{\gamma}(\mathrm{P}, t') \tag{22}$$

is a covariant body tensor at P, which evidently describes the strain (i.e., the change of metric) for any two states t and t'. The covariant *strain-rate body tensor* $\dot{\boldsymbol{\gamma}}$ [which occurs in (20) when $n = 1$] is defined as a limit

$$\dot{\boldsymbol{\gamma}} = \frac{\partial}{\partial t}\boldsymbol{\gamma}(\mathrm{P}, t) = \lim_{t' \to t} \frac{\Delta\boldsymbol{\gamma}(\mathrm{P}, t, t')}{(t - t')}, \tag{23}$$

and the nth-strain-rate tensor [which occurs in (20)] is defined by successive applications of this process.

Equations (20) and (21) involve differences of body tensors at a given particle P; these differences are themselves body tensors (of the same kind) at P because body tensors of given kind at any given particle form a linear space. In general, a given particle P will occupy different places Q and Q′ at different times t and t'; the space tensors of given kind at Q and at Q′ form two disjoint linear spaces and cannot be added or compared by any elementary process. In describing the rheological properties of an arbitrary infinitesimal material element, it is necessary to relate various variables evaluated at a given particle P during a time interval in which P moves through various places. It is clear that body tensor fields are more convenient than general space tensor fields for this purpose.

By differentiating (19), we obtain the equation

$$\frac{d^n}{dt^n}\left(\frac{dc}{dh}\right)^2 = (\text{grad } \sigma) \cdot \frac{\partial^n}{\partial t^n} \gamma^{-1}(\mathrm{P}, t) \cdot (\text{grad } \sigma), \tag{24}$$

since grad σ, a body vector normal to the material surface $\sigma(\xi) = c$, is independent of time. This equation gives an interpretation to the contravariant nth strain-rate body tensor, which may be compared with the corresponding interpretation given by (20) for the covariant nth strain-rate body tensor. It is clear that:

(25) *The covariant body strain tensor field* $\Delta\gamma(\mathrm{P}, t, t')$ defined in (22) *has the following properties*:

- (i) it is independent of any particular choice of coordinate system;
- (ii) it contains no reference to any configuration other than t and t';
- (iii) it describes unambiguously the strain $t \leftrightarrow t'$;
- (iv) it is unchanged by any rigid displacement of the body from t or t';
- (v) it is antisymmetric in t and t' but otherwise treats t and t' on an equal footing;
- (vi) $\Delta\gamma(\mathrm{P}, t, t') + \Delta\gamma(\mathrm{P}, t', t'') = \Delta\gamma(\mathrm{P}, t, t'')$ (i.e., $\Delta\gamma$ is additive).

We claim that these are the simplest properties which one can reasonably require of a mathematical description of strain. As far as we are aware, *there is no space tensor field* (general or Cartesian) *which has all these properties*. It is clear that the same properties are also possessed by the contravariant body tensor field:

$$\Delta\gamma^{-1}(\mathrm{P}, t, t') = \gamma^{-1}(\mathrm{P}, t) - \gamma^{-1}(\mathrm{P}, t'). \tag{26}$$

Since the body base vectors $\boldsymbol{\beta}_i$ and $\boldsymbol{\beta}^i$ for a given body coordinate system $B: \mathrm{P} \to \xi$ are independent of time (or configuration), it follows from (16) and (17) that

$$\overset{(n)}{\gamma} = \frac{\partial^n}{\partial t^n} \gamma(\mathrm{P}, t) = \frac{\partial^n}{\partial t^n} \gamma_{ij}(\xi, t)\boldsymbol{\beta}^i(B, \mathrm{P})\boldsymbol{\beta}^j(B, \mathrm{P}), \tag{27}$$

$$\Delta\gamma(\mathrm{P}, t, t') = (\gamma_{ij}(\xi, t) - \gamma_{ij}(\xi, t'))\boldsymbol{\beta}^i(B, \mathrm{P})\boldsymbol{\beta}^j(B, \mathrm{P}), \tag{28}$$

$$\frac{\partial^n}{\partial t^n} \gamma^{-1}(\mathrm{P}, t) = \frac{\partial^n}{\partial t^n} \gamma^{ij}(\xi, t)\boldsymbol{\beta}_i(B, \mathrm{P})\boldsymbol{\beta}_j(B, \mathrm{P}), \tag{29}$$

$$\Delta\gamma^{-1}(\mathrm{P}, t, t') = (\gamma^{ij}(\xi, t) - \gamma^{ij}(\xi, t'))\boldsymbol{\beta}_i(B, \mathrm{P})\boldsymbol{\beta}_j(B, \mathrm{P}). \tag{30}$$

In applications to rheology, physics, and engineering, it is usual to assign "units" of length to the distance ds between two neighboring points when dealing with the space manifold or the body manifold. In the process of assigning a value ds to a given pair of points, one usually uses (or contemplates using) a rigid body [listed as a primitive concept in 1.1(1)] as a "measuring rod" or "length gauge" and adopts a procedure which enables one to compare the separation of points of the manifold with points of the measuring rod (a second body manifold). One uses labels like cm or in. (centimeters or inches) for an agreed pair of particles on the measuring rod, and then expresses the separation of the points of the manifold in terms of these "units," e.g., $ds = 10^{-4}$ cm. We thereby assign units of length to ds. For the mathematical formalism, it is simpler in the first instance to regard coordinates as pure numbers (as defined above) without dimensions of length; it then follows that the metric tensors $\boldsymbol{g}$ and γ have dimensions $(\text{length})^2$. We shall not need to consider the question of changing from one length gauge to another for our applications; the associated question of gauge invariance is, however, important in electromagnetic and other theories (Weyl, 1922).

(31) **Problem** Prove that

$$\text{(i)}\quad \partial\gamma^{-1}/\partial t = -\gamma^{-1}\cdot\dot{\gamma}\cdot\gamma^{-1},$$
$$\text{(ii)}\quad \partial^2\gamma^{-1}/\partial t^2 = -\gamma^{-1}\cdot\ddot{\gamma}\cdot\gamma^{-1} + 2(\gamma^{-1}\cdot\dot{\gamma})^2\cdot\gamma^{-1},$$

where

$$\text{(iii)}\quad \gamma = \gamma(\mathrm{P}, t), \qquad \dot{\gamma} = \partial\gamma(\mathrm{P}, t)/\partial t, \qquad \text{and} \qquad \ddot{\gamma} = \partial^2\gamma(\mathrm{P}, t)/\partial t^2,$$

and $\boldsymbol{\mu}^2 = \boldsymbol{\mu}\cdot\boldsymbol{\mu}$, where $\boldsymbol{\mu}$ is a mixed tensor. Similarly, we define $\boldsymbol{\mu}^n = \boldsymbol{\mu}\cdot\boldsymbol{\mu}\cdot\boldsymbol{\mu}\cdot\cdots\cdot\boldsymbol{\mu}$ (n factors). In (ii), $\gamma^{-1}\cdot\dot{\gamma}$ is a mixed tensor (right covariant). There are results similar to these with γ and γ^{-1} interchanged.

(32) **Problem** Prove that, for any positive integer m,

$$\text{(i)}\quad \overset{(m)}{\gamma}\cdot\gamma^{-1} + m\overset{(m-1)}{\gamma}\cdot\dot{\gamma}^{-1} + \frac{m(m-1)}{1\cdot 2}\overset{(m-2)}{\gamma}\cdot\ddot{\gamma}^{-1} + \cdots + \gamma\cdot\overset{(m)}{\gamma^{-1}} = \mathbf{0},$$

where

$$\text{(ii)}\quad \dot{\gamma}^{-1} = \partial\gamma^{-1}(\mathrm{P}, t)/\partial t, \ldots, \qquad \text{and} \qquad \overset{(m)}{\gamma^{-1}} = \partial^m\gamma^{-1}(\mathrm{P}, t)/\partial t^m.$$

In (ii), the reciprocal of γ is taken first, and the time derivatives are taken second. For the reverse order, we use the notation $(\dot{\gamma})^{-1}$, etc.

(33) **Problem** Let $\gamma B = \gamma$, where B is any body coordinate system. If $\partial|\gamma|/\partial t = 0$, show that $\gamma^{-1}:\ddot{\gamma} = (\gamma^{-1}\cdot\dot{\gamma}\cdot\gamma^{-1}):\dot{\gamma}$. [The Cartesian space tensor version of this result, namely, $\mathrm{Tr}\,\mathbf{A}^2 = \mathrm{Tr}(\mathbf{A}^1)^2$ for flow at constant volume, is due to Rivlin (1962).]

2.7 Magnitudes and angles

We call ds, given by 2.6(14), the *magnitude* of the vector $\boldsymbol{d\xi}$. Although the body vector $\boldsymbol{d\xi}$ is independent of time, its magnitude ds varies with time when the body metric tensor γ varies with time. Since the metric tensor plays an important role in all discussions of the properties of the body manifold, it is convenient to use it to define magnitudes (or norms) for all body vectors, as follows. (Any other positive-definite, second-rank covariant tensor could also be used.)

(1) **Definition** The *magnitudes* $\|\boldsymbol{\theta}\|$ and $\|\boldsymbol{\psi}\|$ for any contravariant body vector $\boldsymbol{\theta}$ and covariant body vector $\boldsymbol{\psi}$ are given by the equations

$$\text{(i)} \quad \|\boldsymbol{\theta}\| = \{\boldsymbol{\theta}(\mathrm{P}) \cdot \boldsymbol{\gamma}(\mathrm{P}, t) \cdot \boldsymbol{\theta}(\mathrm{P})\}^{1/2} = \{\theta^i(\xi)\gamma_{ij}(\xi, t)\theta^j(\xi)\}^{1/2};$$
$$\text{(ii)} \quad \|\boldsymbol{\psi}\| = \{\boldsymbol{\psi}(\mathrm{P}) \cdot \boldsymbol{\gamma}^{-1}(\mathrm{P}, t) \cdot \boldsymbol{\psi}(\mathrm{P})\}^{1/2} = \{\psi_i(\xi)\gamma^{ij}(\xi, t)\psi_j(\xi)\}^{1/2}.$$

The *magnitudes* $\|\boldsymbol{v}\|$ and $\|\boldsymbol{u}\|$ for any contravariant space vector $\boldsymbol{v}$ and covariant space vector $\boldsymbol{u}$ are given by the equations

$$\text{(iii)} \quad \|\boldsymbol{v}\| = \{\boldsymbol{v}(\mathrm{Q}) \cdot \boldsymbol{g}(\mathrm{Q}) \cdot \boldsymbol{v}(\mathrm{Q})\}^{1/2} = \{v^i(x)g_{ij}(x)v^j(x)\}^{1/2};$$
$$\text{(iv)} \quad \|\boldsymbol{u}\| = \{\boldsymbol{u}(\mathrm{Q}) \cdot \boldsymbol{g}^{-1}(\mathrm{Q}) \cdot \boldsymbol{u}(\mathrm{Q})\}^{1/2} = \{u_i(x)g^{ij}(x)u_j(x)\}^{1/2}.$$

The magnitude of a vector is a scalar whose value depends on the metric tensor as well as on the vector. For a vector function of time, the magnitude will vary with time; for body vectors, the metric tensor introduces an additional time dependence not present with space vectors. The magnitude of a vector is a particular application of the scalar product of two vectors of the same kind, defined as follows:

(2) **Definition** $\boldsymbol{\theta}_1 \cdot \boldsymbol{\gamma} \cdot \boldsymbol{\theta}_2$ and $\boldsymbol{\psi}^1 \cdot \boldsymbol{\gamma}^{-1} \cdot \boldsymbol{\psi}^2$ are called *scalar products* of contravariant body vectors $\boldsymbol{\theta}_1$ and $\boldsymbol{\theta}_2$ and covariant body vectors $\boldsymbol{\psi}^1$ and $\boldsymbol{\psi}^2$, all quantities being evaluated at the same particle P. The quantities $\boldsymbol{v}_1 \cdot \boldsymbol{g} \cdot \boldsymbol{v}_2$ and $\boldsymbol{u}^1 \cdot \boldsymbol{g}^{-1} \cdot \boldsymbol{u}^2$ are called *scalar products* of contravariant space vectors $\boldsymbol{v}_1$ and $\boldsymbol{v}_2$ and covariant space vectors $\boldsymbol{u}^1$ and $\boldsymbol{u}^2$, all quantities being evaluated at the same place Q.

(3) **Problem** Prove that the scalar products in (2) satisfy the conditions 1.2(5) for an inner product. Thus *vectors of a given kind at a given point form a normed linear space.*

(4) **Problem** Prove the following results for the magnitudes of base vectors for arbitrary coordinate systems B and S:

$$\text{(i)} \quad \|\boldsymbol{\beta}_k(B, \mathrm{P})\| = \{\gamma_{\bar{k}\bar{k}}(\xi, t)\}^{1/2}, \qquad \|\boldsymbol{\beta}^k(B, \mathrm{P})\| = \{\gamma^{\bar{k}\bar{k}}(\xi, t)\}^{1/2};$$
$$\text{(ii)} \quad \|\boldsymbol{b}_k(S, \mathrm{Q})\| = \{g_{\bar{k}\bar{k}}(x)\}^{1/2}, \qquad \|\boldsymbol{b}^k(S, \mathrm{Q})\| = \{g^{\bar{k}\bar{k}}(x)\}^{1/2}.$$

(5) **Definition** A vector with unit magnitude is called a *unit vector*.

(6) **Problem** Prove that $\mathbf{v}(\mathrm{P}, t)$, the unit normal to a material surface whose equation in an arbitrary body coordinate system $B:\mathrm{P} \to \xi$ is $\sigma(\xi) = c$, has components in B given by

$$\text{(i)} \quad v_k(\xi, t) = \pm \frac{\partial \sigma(\xi)}{\partial \xi^k} \left\{ \gamma_{ij}(\xi, t) \frac{\partial \sigma(\xi)}{\partial \xi^i} \frac{\partial \sigma(\xi)}{\partial \xi^j} \right\}^{-1/2}$$

Prove that $\boldsymbol{n}(\mathrm{Q})$, the unit normal to a surface whose equation in an arbitrary space coordinate system $S:\mathrm{Q} \to x$ is $s(x) = c$, has components in S given by

$$\text{(ii)} \quad n_k(x) = \pm \frac{\partial s(x)}{\partial x^k} \left\{ g_{ij}(x) \frac{\partial s(x)}{\partial x^i} \frac{\partial s(x)}{\partial x^j} \right\}^{-1/2}$$

(7) **Definition** The *angle* χ between two vectors of the same kind at the same point is given by the following equations:

(i) $\|\boldsymbol{\theta}_1\| \; \|\boldsymbol{\theta}_2\| \cos \chi = \boldsymbol{\theta}_1(\mathrm{P}) \cdot \boldsymbol{\gamma}(\mathrm{P}, t) \cdot \boldsymbol{\theta}_2(\mathrm{P})$ (contravariant body vectors),

(ii) $\|\boldsymbol{\psi}^1\| \; \|\boldsymbol{\psi}^2\| \cos \chi = \boldsymbol{\psi}^1(\mathrm{P}) \cdot \boldsymbol{\gamma}^{-1}(\mathrm{P}, t) \cdot \boldsymbol{\psi}^2(\mathrm{P})$ (covariant body vectors),

(iii) $\|\boldsymbol{v}_1\| \; \|\boldsymbol{v}_2\| \cos \chi = \boldsymbol{v}_1(\mathrm{Q}) \cdot \boldsymbol{g}(\mathrm{Q}) \cdot \boldsymbol{v}_2(\mathrm{Q})$ (contravariant space vectors),

(iv) $\|\boldsymbol{u}^1\| \; \|\boldsymbol{u}^2\| \cos \chi = \boldsymbol{u}^1(\mathrm{Q}) \cdot \boldsymbol{g}^{-1}(\mathrm{Q}) \cdot \boldsymbol{u}^2(\mathrm{Q})$ (covariant space vectors).

If $\chi = \pi/2$ or $3\pi/2$ ($\cos \chi = 0$), the vectors are said to be *orthogonal. Surfaces* are said to be *orthogonal* at a point of intersection if their normals at that point are orthogonal. A *basis* is said to be *orthogonal* if every pair of its vectors is orthogonal, and *orthonormal* if it is an orthogonal basis of unit vectors; this terminology agrees with that of 1.2(7). *Lines* are *orthogonal* at a point of intersection if their tangents are orthogonal.

(8) **Definition** A *coordinate system* for a manifold is *orthogonal* if, for every point in the manifold, the coordinate surfaces are all orthogonal to one another at that point.

(9) **Problem** Prove that the following conditions are necessary and sufficient for coordinate systems $B:\mathrm{P} \to \xi$ and $S:\mathrm{Q} \to x$ to be orthogonal:

$$\gamma^{ij}(\xi, t) = 0 \quad (i \neq j) \quad \text{for} \quad B; \qquad g^{ij}(x) = 0 \quad (i \neq j) \quad \text{for} \quad S.$$

For an orthogonal space coordinate system S, it is often convenient to use the notation $g^{-1} = \text{diag}[h_1^{-2}, h_2^{-2}, h_3^{-2}]$, $g = \text{diag}[h_1{}^2, h_2{}^2, h_3{}^2]$, or

(10) $$g^{ij} = (h_i)^{-2}\,\delta^{ij}, \qquad g_{ij} = h_i^2\,\delta_{ij} \qquad (S \text{ orthogonal}).$$

The equations (7) are formally similar to the equation $Q_1Q_2 \cdot Q_1Q_3 \cos\chi = \overrightarrow{Q_1Q_2} \cdot \overrightarrow{Q_1Q_3}$ which is used in elementary vector analysis to define the scalar product $\overrightarrow{Q_1Q_2} \cdot \overrightarrow{Q_1Q_3}$ of displacement vectors $\overrightarrow{Q_1Q_2}$ and $\overrightarrow{Q_1Q_3}$, where Q_1, Q_2, and Q_3 are any three places and χ is the angle between Q_1Q_2 and Q_1Q_3. The logical status of our equations (7) is different: They are used to define angle after the scalar product has been defined.

(11) **Definition** The *angle between a contravariant body vector* $\boldsymbol{\theta}$ *and a covariant body vector* $\boldsymbol{\psi}$ is the angle between the contravariant body vectors $\boldsymbol{\theta}$ and $\boldsymbol{\gamma}^{-1} \cdot \boldsymbol{\psi}$. The *angle between a contravariant space vector* $\boldsymbol{v}$ *and a covariant space vector* $\boldsymbol{u}$ is the angle between the contravariant space vectors $\boldsymbol{v}$ and $\boldsymbol{g}^{-1} \cdot \boldsymbol{u}$. It follows from 2.4(16) that the normal to a surface is orthogonal to every local tangent.

(12) **Problem** Prove that the angle between $\boldsymbol{\theta}$ and $\boldsymbol{\gamma}^{-1} \cdot \boldsymbol{\psi}$ is equal to the angle between $\boldsymbol{\psi}$ and $\boldsymbol{\gamma} \cdot \boldsymbol{\theta}$.

(13) **Problem** If χ denotes the angle between any two contravariant space vectors $\boldsymbol{v}_1(Q)$ and $\boldsymbol{v}_2(Q)$, show that

$$(\|\boldsymbol{v}_1\|\,\|\boldsymbol{v}_2\| \sin\chi)^2 = (g_{ri}\,g_{sj} - g_{rs}\,g_{ij})v_1^i v_2^j\,v_1^r v_2^s,$$

where the components on the right-hand side are referred to an arbitrary space coordinate system S.

(14) **Problem** Show that the condition for orthogonality at time t of two material surfaces whose equations in B are $\sigma(\xi) = c$ and $\sigma'(\xi) = c'$ is

$$(\text{grad}\,\sigma) \cdot \boldsymbol{\gamma}^{-1}(P, t) \cdot (\text{grad}\,\sigma') = 0, \qquad \text{or} \qquad \gamma^{ij}(\xi, t)\frac{\partial\sigma}{\partial\xi^i}\frac{\partial\sigma}{\partial\xi^j} = 0.$$

Because of the time dependence of the body metric tensor, it is clear that, for a general flow of a material, any two material surfaces which are orthogonal at one instant will not be orthogonal at other instants. There is, however, an important class of flows (called *shear free*, and discussed in Chapter 3) for which there does exist a body coordinate system that is always orthogonal.

(15) **Definition** A coordinate system $S: Q \to x$ is *rectangular Cartesian* if $g_{ij}(x) = \delta_{ij}$ for all values of x, and is *Cartesian* if $\partial g_{ij}(x)/\partial x^k = 0$ for all values of x. In a general Riemannian manifold, a Cartesian coordinate system does

not exist. The condition on a manifold for a Cartesian (and therefore also a rectangular Cartesian) coordinate system to exist is given in 8.7(5); such a manifold is called *Euclidean*.

The space manifold (in the nonrelativistic approximation) is assumed to be Euclidean. We shall show that it follows that the body manifold is also Euclidean; but it is clear that a body coordinate system that is rectangular Cartesian at one instant will not be rectangular Cartesian at other instants unless the body is moving rigidly so that the $\gamma_{ij}(\xi, t)$ are independent of t.

(16) **Problem** Let $\boldsymbol{\alpha}_i(\mathrm{P}, t)$ $(i = 1, 2, 3)$ be a set of contravariant body vectors at P which are orthonormal at time t. Prove the results:

(i) $\boldsymbol{\alpha}_i(\mathrm{P}, t)$ are linearly independent;
(ii) the set $\boldsymbol{\alpha}^i(\mathrm{P}, t) = \boldsymbol{\gamma}(\mathrm{P}, t) \cdot \boldsymbol{\alpha}_i(\mathrm{P}, t)$ is reciprocal to the set $\boldsymbol{\alpha}_i(\mathrm{P}, t)$;
(iii) $\boldsymbol{\gamma}(\mathrm{P}, t) = \boldsymbol{\alpha}^i(\mathrm{P}, t)\boldsymbol{\alpha}^i(\mathrm{P}, t)$; $\quad \boldsymbol{\gamma}^{-1}(\mathrm{P}, t) = \boldsymbol{\alpha}_i(\mathrm{P}, t)\boldsymbol{\alpha}_i(\mathrm{P}, t)$;
(iv) $\tilde{\boldsymbol{\delta}} = \boldsymbol{\alpha}^i(\mathrm{P}, t)\boldsymbol{\alpha}_i(\mathrm{P}, t)$; $\quad \boldsymbol{\delta} = \boldsymbol{\alpha}_i(\mathrm{P}, t)\boldsymbol{\alpha}^i(\mathrm{P}, t)$.

Since orthonormal vectors $\boldsymbol{\alpha}_i(\mathrm{P}, t)$ form a basis for contravariant body vectors at P, by (16)(i), it follows that their outer products $\boldsymbol{\alpha}_i\boldsymbol{\alpha}_j$ form a basis for contravariant, second-rank tensors at P, by 1.2(9)(iii) and 2.5(7). For any such tensor $\boldsymbol{\Theta}(\mathrm{P})$, we have

$$\begin{aligned} \boldsymbol{\Theta} &= \boldsymbol{\delta} \cdot \boldsymbol{\Theta} \cdot \tilde{\boldsymbol{\delta}} && \text{[by 2.5(17)(ii)]} \\ &= (\boldsymbol{\alpha}_i\boldsymbol{\alpha}^i) \cdot \boldsymbol{\Theta} \cdot (\boldsymbol{\alpha}^j\boldsymbol{\alpha}_j) && \text{[by (16)(iv)];} \end{aligned}$$

hence

(17) $$\boldsymbol{\Theta} = \widehat{\Theta^{ij}}\boldsymbol{\alpha}_i\boldsymbol{\alpha}_j, \qquad \text{where} \quad \widehat{\Theta^{ij}} = \boldsymbol{\alpha}^i \cdot \boldsymbol{\Theta} \cdot \boldsymbol{\alpha}^j.$$

$\widehat{\Theta^{ij}}$ are called the *physical components* of $\boldsymbol{\Theta}$ for the orthonormal basis $\boldsymbol{\alpha}_i$. We have not in fact used the orthonormal property of the basis $\boldsymbol{\alpha}_i$; since (16)(iv) is valid for any basis and its reciprocal, so also is (17), but most of the applications are made when the basis is orthonormal. There are similar results for other kinds of tensor. Suppose we have a body coordinate system B which is instantaneously orthogonal at time t. Then the base vectors $\boldsymbol{\beta}_i(B, \mathrm{P})$ are orthogonal at t and so are the base vectors $\boldsymbol{\beta}^i(B, \mathrm{P})$. From these vectors, we can construct orthonormal sets by normalization, i.e., by dividing them by their magnitudes at time t, which are given by (4). The physical components of any tensor for these bases are called *physical components for the coordinate system B at time t*, and are related to the (ordinary) components for B by the following equations:

(18) $$\widehat{\Theta^{ij}}(\xi, t) = \{\gamma^{\hat{\imath}\hat{\imath}}(\xi, t)\gamma^{\hat{\jmath}\hat{\jmath}}(\xi, t)\}^{-1/2}\Theta^{\hat{\imath}\hat{\jmath}}(\xi) \qquad \text{(contravariant);}$$

(19) $\widehat{\Psi}_{ij}(\xi, t) = \{\gamma_{ii}(\xi, t)\gamma_{jj}(\xi, t)\}^{-1/2}\Psi_{ij}(\xi)$ (covariant);

(20) $\widehat{\mu}_j^i(\xi, t) = \{\gamma^{ii}(\xi, t)\gamma_{jj}(\xi, t)\}^{-1/2}\mu_j^i(\xi)$ (mixed);

(21) $\widehat{\theta}^i(\xi, t) = \{\gamma^{ii}(\xi, t)\}^{-1/2}\theta^i(\xi)$ (contravariant vector);

(22) $\widehat{\psi}_i(\xi, t) = \{\gamma_{ii}(\xi, t)\}^{-1/2}\theta_i(\xi)$ (covariant vector).

There are similar formulas for physical components of space vectors and tensors for an orthogonal space coordinate system, but in this case, the process of forming physical components introduces no time dependence because the $g_{ij}(x)$ are independent of time.

(23) **Problem** Let $\pi(P, t)$ be a second-rank contravariant body tensor, and write $\gamma(P, t) \cdot \pi(P, t) \cdot \gamma(P, t) = \pi^0(P, t)$ (a covariant tensor). Taking physical components for any body coordinate system orthogonal at time t, prove the results:

(i) $\widehat{\gamma}_{ij}(\xi, t) = \widehat{\gamma}^{ij}(\xi, t) = \delta_{ij}$;

(ii) $\widehat{\pi}^{ij}(\xi, t) = \widehat{\pi}^{(0)}_{ij}(\xi, t)$;

(iii) $\widehat{\dot{\pi}}^{ij} = \widehat{\dot{\pi}}^{(0)}_{ij} - \widehat{\dot{\gamma}}_{ik}\widehat{\pi}^{(0)}_{kj} - \widehat{\pi}^{(0)}_{ik}\widehat{\dot{\gamma}}_{kj}$;

a superior dot denotes $\partial/\partial t$ at constant ξ.

2.8 Principal axes and principal values

In this section, we consider body vectors and tensors at an arbitrary but definite particle P.

If μ denotes any given right-covariant, mixed, second-rank tensor and θ denotes any contravariant vector, then $\theta' \equiv \mu \cdot \theta$ is also a contravariant vector which, in general, is not a scalar multiple of θ. We can, however, always choose θ (for given μ) so that θ' is a scalar multiple of θ.

(1) **Definition** θ is a *contravariant principal axis* (or right eigenvector) of a right covariant, mixed, second-rank tensor μ belonging to the *principal value* (or eigenvalue) λ if $\mu \cdot \theta = \lambda\theta$ and $\theta \neq 0$. ψ is a *covariant principal axis* (or left eigenvector) belonging to the principal value λ if $\psi \cdot \mu = \lambda\psi$, $\psi \neq 0$, ψ being a covariant vector. λ is a scalar.

(2) **Theorem** If $\mu = \Theta \cdot \Psi$, where Θ and Ψ are *symmetric*, second-rank tensors (contravariant and covariant, respectively), one of which is positive

definite, then $\boldsymbol{\mu}$ has three linearly independent contravariant principal axes $\boldsymbol{\theta}_i$ and three linearly independent covariant principal axes $\boldsymbol{\psi}^i$ ($i = 1, 2, 3$) which can be chosen so as to satisfy the equations

(i) $\boldsymbol{\theta}_i \cdot \boldsymbol{\psi}^j = \delta^j_i$,

i.e., to be reciprocal sets of vectors. Further, $\boldsymbol{\mu}$ can be expressed in the form

(ii) $\boldsymbol{\mu} = \lambda_i \boldsymbol{\theta}_i \boldsymbol{\psi}^i$,

where λ_i is the principal value belonging to $\boldsymbol{\theta}_i$ and $\boldsymbol{\psi}^i$.

The conditions of the theorem are satisfied in our subsequent applications, because we will take $\boldsymbol{\Psi} = \boldsymbol{\gamma}$ or $\boldsymbol{\Theta} = \boldsymbol{\gamma}^{-1}$; both these tensors are positive definite. The conditions are sufficient to ensure that the principal values and principal axes are all real.

Proof $\boldsymbol{\mu} \cdot \boldsymbol{\theta} - \lambda\boldsymbol{\theta} = (\boldsymbol{\mu} - \lambda\boldsymbol{\delta}) \cdot \boldsymbol{\theta}$ is a contravariant vector. For any B, we have $(\boldsymbol{\mu} - \lambda\boldsymbol{\delta}) \cdot \boldsymbol{\theta}B = (\mu - \lambda I)\theta$, say, in terms of matrix representatives in B. Hence, in order that a principal axis $\boldsymbol{\theta}$ shall exist, it is necessary and sufficient that a nonzero matrix θ exist such that

$$(\mu - \lambda I)\theta = 0. \tag{3}$$

The condition for this is

$$|\mu - \lambda I| = 0, \tag{4}$$

which is a cubic in λ. Suppose that $\boldsymbol{\Psi}$ is positive definite; then $\boldsymbol{\Psi}^{-1}$ exists, and (4) is equivalent to

$$|\Theta - \lambda\Psi^{-1}| = 0. \tag{5}$$

Let λ be any root of this equation. Then there exists a nonzero column matrix φ, say, such that

$$\Theta\varphi = \lambda\Psi^{-1}\varphi. \tag{6}$$

Multiply from the left by the conjugate transpose $\tilde{\varphi}^*$, and write $\varphi = \chi + iv$, where χ and v are real column matrices. We obtain the result

$$\tilde{\chi}\Theta\chi + \tilde{v}\Theta v = \lambda\{\tilde{\chi}\Psi^{-1}\chi + \tilde{v}\Psi^{-1}v\},$$

because the terms in i cancel due to the symmetry of Θ and Ψ. The coefficient of λ is positive (and real) because $\boldsymbol{\Psi}^{-1}$ is positive definite. The left-hand side is real. Hence λ is real, and therefore a real matrix solution φ exists which is nonzero. Hence $\theta = \Psi^{-1}\varphi$ is a real, nonzero matrix satisfying (3).

For $i = 1, 2, 3$, let θ_i be a matrix solution of (3) corresponding to the root $\lambda = \lambda_i$ of (4). We suppose that the three roots are all different. Writing $\varphi_i = \Psi\theta_i$, we have

$$\tilde{\varphi}_2\Theta\varphi_1 = \lambda_1\tilde{\varphi}_2\Psi^{-1}\varphi_1, \tag{7}$$

and

(8) $$\tilde{\varphi}_1 \Theta \varphi_2 = \lambda_2 \tilde{\varphi}_1 \Psi^{-1} \varphi_2 .$$

Since Θ and Ψ are symmetric and $\lambda_1 \neq \lambda_2$, it follows that

(9) $$\tilde{\varphi}_i \Psi^{-1} \varphi_j = 0 \qquad \text{when} \quad i \neq j.$$

It follows that φ_i are linearly independent, because the equation $a_i \varphi_i = 0$ (where the a_i are any numbers) implies that $a_1 \tilde{\varphi}_1 \Psi^{-1} \varphi_1 = 0$, by (9), and this implies that $a_1 = 0$, since Ψ is positive definite; similarly, $a_2 = a_3 = 0$. We thus see that the θ_i are linearly independent, for $a_i \theta_i = \Psi^{-1}(a_i \varphi_i) = 0$ if and only if all the a_i are zero.

Similarly, to each λ_i there corresponds a column matrix ψ^i such that

(10) $$\tilde{\psi}^{\hat{\imath}}(\mu - \lambda_{\hat{\imath}} I) = 0 \qquad \text{and hence} \qquad \tilde{\psi}^{\hat{\imath}} \Theta = \lambda_{\hat{\imath}} \tilde{\psi}^{\hat{\imath}} \Psi^{-1}.$$

A similar argument shows that the ψ^i are linearly independent, and that

(11) $$\tilde{\psi}^i \Psi^{-1} \psi^j = 0 \qquad \text{when} \quad i \neq j.$$

Moreover, on comparing (10) with the transpose of (6), we see that we can take $\psi^i = x_{\hat{\imath}} \varphi_{\hat{\imath}}$, where the x_i are numbers at our disposal. Hence $\tilde{\psi}^i \theta_j = x_{\hat{\imath}} \tilde{\varphi}_{\hat{\imath}} \Psi^{-1} \varphi_j = x_{\hat{\imath}} \delta_{\hat{\imath} j}$, and we can clearly choose the x_i so that

(12) $$\tilde{\psi}^i \theta_j = \delta^i_j.$$

Finally, we may, for $i = 1, 2, 3$, define a contravariant vector $\boldsymbol{\theta}_i$ whose representative matrix in B is θ_i, and a covariant vector $\boldsymbol{\psi}^i$ whose representative matrix in B is ψ^i. Then $a_i \boldsymbol{\theta}_i = \mathbf{0}$ if and only if $a_i \theta_i = 0$, so the vectors $\boldsymbol{\theta}_i$ are linearly independent; similarly, the vectors $\boldsymbol{\psi}^i$ are linearly independent. Further, $\boldsymbol{\psi}^i \cdot \boldsymbol{\theta}_j = \tilde{\psi}^i \theta_j = \delta^i_j$, by (12), which proves (2)(i). It is clear that a proof similar to the preceding can be given if $\boldsymbol{\Theta}$ instead of $\boldsymbol{\Psi}$ is positive definite. To save space, we do not give the proof when two or three of the principal values are equal; proofs for these cases can be given which are similar to other eigenvalue proofs with degeneracy [e.g., as in 1.3(17), cases (ii) and (iii)].

It remains only to prove (2)(ii). We have shown that the $\boldsymbol{\theta}_i$ form a basis for contravariant vectors at P and that the $\boldsymbol{\psi}^i$ form a basis for covariant vectors at P. It follows that the products $\boldsymbol{\theta}_i \boldsymbol{\psi}^j$ form a basis for right-covariant, second-rank, mixed tensors at P, and hence that $\boldsymbol{\mu} = x^i_j \boldsymbol{\theta}_i \boldsymbol{\psi}^j$ for some numbers x^i_j. Hence

$$\lambda_{\hat{k}} \boldsymbol{\psi}^{\hat{k}} = \boldsymbol{\psi}^k \cdot \boldsymbol{\mu} = x^i_j \boldsymbol{\psi}^k \cdot \boldsymbol{\theta}_i \boldsymbol{\psi}^j = x^i_j \delta^k_i \boldsymbol{\psi}^j = x^k_j \boldsymbol{\psi}^j,$$

from (2)(i). Since the $\boldsymbol{\psi}^j$ are linearly independent, it follows that $x^k_j = \lambda_{\hat{k}} \delta^{\hat{k}}_j$, which proves the required result (2)(ii).

We note that 2.7(16)(iv) is a particular case of (2)(ii) in which $\boldsymbol{\mu} = \boldsymbol{\delta}$. Theorem (2) has been formulated for a right-covariant tensor $\boldsymbol{\mu}$; it can obviously be reformulated for a left-covariant tensor $\tilde{\boldsymbol{\mu}}$: (i) is still valid, where

(13) $$\tilde{\boldsymbol{\mu}} \cdot \boldsymbol{\psi}_i = \lambda_i \boldsymbol{\psi}_i \qquad \text{and} \qquad \boldsymbol{\theta}_i \cdot \tilde{\boldsymbol{\mu}} = \lambda_i \boldsymbol{\theta}_i,$$

and the transpose of (2)(ii) is

(14) $$\tilde{\boldsymbol{\mu}} = \lambda_i \boldsymbol{\psi}^i \boldsymbol{\theta}_i \qquad (\tilde{\boldsymbol{\mu}} \quad \text{left covariant}).$$

(15) **Problem** If $\boldsymbol{\theta}_i$ and $\boldsymbol{\psi}^i$ are eigenvectors of a right-covariant mixed tensor $\boldsymbol{\mu}$ belonging to eigenvalues λ_i and normalized according to (2)(i), prove that

$$\boldsymbol{\mu}^n = \lambda_i^n \boldsymbol{\theta}_i \boldsymbol{\psi}^i,$$

where n is any positive integer, and n can be any negative integer provided that $\lambda_i \neq 0$ $(i = 1, 2, 3)$.

(16) **Problem** If the three eigenvalues of a mixed tensor $\boldsymbol{\mu}$ are all equal to λ_1, prove that $\boldsymbol{\mu} = \lambda_1 \boldsymbol{\delta}$ and that every vector is an eigenvector.

Although we have not used the fact in our proof of Theorem (2), it is worth noting that if λ is a characteristic value of the representative matrix μ for a given mixed tensor referred to any coordinate system B, then λ is also a characteristic value of the representative matrix $\bar{\mu}$ in any other coordinate system $\bar{B}$; this follows from the transformation law 2.5(8)(iii), which gives

(17) $$\bar{\mu} - \lambda I = \Lambda(\mu - \lambda I)\Lambda^{-1}, \qquad \text{where} \quad \Lambda = \Lambda_P(B, \bar{B}),$$

and hence $|\bar{\mu} - \lambda I| = |\mu - \lambda I| = 0$, since λ is a characteristic value of μ, by hypothesis.

(18) **Definition** The principal values $\dot{s}_i$ $(i = 1, 2, 3)$ of the right-covariant tensor $\boldsymbol{\gamma}^{-1} \cdot \dot{\boldsymbol{\gamma}}$ are called *principal strain rates*, and the principal axes are called *principal strain-rate axes*.

(19) **Problem** If $\boldsymbol{\Psi}$ is any symmetric covariant tensor, show that $\boldsymbol{\gamma}^{-1} \cdot \boldsymbol{\Psi}$ has a $\boldsymbol{\gamma}$-*orthonormal* set of principal axis vectors $\boldsymbol{\theta}_i$, satisfying the equations

(i) $(\boldsymbol{\Psi} - \lambda_i \boldsymbol{\gamma}) \cdot \boldsymbol{\theta}_i = \mathbf{0}$,
(ii) $\boldsymbol{\theta}_i \cdot \boldsymbol{\gamma} \cdot \boldsymbol{\theta}_j = \delta_{ij}$.

If $\boldsymbol{\Theta}$ is any symmetric contravariant tensor, show that $\boldsymbol{\Theta} \cdot \boldsymbol{\gamma}$ has a $\boldsymbol{\gamma}$-*orthonormal* set of principal axis vectors $\boldsymbol{\psi}^i$ satisfying the equations

(iii) $\boldsymbol{\psi}^i \cdot (\boldsymbol{\Theta} - \lambda_i \boldsymbol{\gamma}^{-1}) = \mathbf{0}$,
(iv) $\boldsymbol{\psi}^i \cdot \boldsymbol{\gamma}^{-1} \cdot \boldsymbol{\psi}^j = \delta^{ij}$.

In such cases, we say that λ_i and $\boldsymbol{\theta}_i$ are *principal values and principal axes of* $\boldsymbol{\Psi}$, and that λ_i and $\boldsymbol{\psi}^i$ are *principal values and principal axes of* $\boldsymbol{\Theta}$, both sets being at time t when $\boldsymbol{\gamma} = \boldsymbol{\gamma}(\mathrm{P}, t)$. Prove also that

(20) $$\boldsymbol{\gamma}^{-1} \cdot \boldsymbol{\Psi} \cdot \boldsymbol{\gamma}^{-1} = \lambda_i \boldsymbol{\theta}_i \boldsymbol{\theta}_i, \qquad \boldsymbol{\gamma} \cdot \boldsymbol{\Theta} \cdot \boldsymbol{\gamma} = \lambda_i \boldsymbol{\psi}^i \boldsymbol{\psi}^i.$$

(21) **Definition** The principal axes $\boldsymbol{\theta}_i$ of $\boldsymbol{\gamma}^{-1} \cdot \boldsymbol{\gamma}'$, where $\boldsymbol{\gamma} = \boldsymbol{\gamma}(\mathrm{P}, t)$ and $\boldsymbol{\gamma}' = \boldsymbol{\gamma}(\mathrm{P}, t')$, are called (contravariant) *principal axes of strain* for the states t and t'. The *principal elongation ratios* $e_i(t, t')$ are the positive square roots of the principal values of $\boldsymbol{\gamma}^{-1} \cdot \boldsymbol{\gamma}'$.

(22) **Problem** Show that the contravariant principal axes of strain $\boldsymbol{\theta}_i$ are tangential at P to three material lines through P which are orthogonal in state t and orthogonal in state t'.

(23) **Problem** [converse of (20)] If a covariant second-rank tensor $\boldsymbol{\Psi}$ can be written in the form (20), where the $\boldsymbol{\theta}_i$ satisfy (19)(ii), prove that $\boldsymbol{\theta}_i$ are the principal axes of $\boldsymbol{\Psi}$ belonging to the principal values λ_i $(i = 1, 2, 3)$. If $\boldsymbol{\Theta}$ is a contravariant, second-rank tensor which can be written in the form (20) where $\boldsymbol{\psi}^i$ satisfy (19)(iv), prove that $\boldsymbol{\psi}^i$ are principal axes of $\boldsymbol{\Theta}$ belonging to the principal values λ_i.

(24) **Problem** Prove the following relation between the principal strain rates $\dot{s}_i$ and the principal elongation ratios e_i:

$$\dot{s}_i = \lim_{t' \to t} \frac{e_i^2(t, t') - 1}{t' - t}.$$

[From (18) and the convention in (19), we can say that the $\dot{s}_i$ are the principal values of $\dot{\boldsymbol{\gamma}}$.]

(25) **Problem** Let χ be the angle between two material line element vectors $\boldsymbol{d\xi}_1$ and $\boldsymbol{d\xi}_2$ which, at time t, are instantaneously orthogonal and coincident with principal (contravariant) axes of $\dot{\boldsymbol{\gamma}}$. Prove that $\dot{\chi} = 0$ instantaneously at time t. [This result is suggested by (22) and (24) and shows that the contravariant principal axes of $\dot{\boldsymbol{\gamma}}$ are material lines which instantaneously are changing in length but not in mutual inclination.]

(26) **Definition** Let P be any given particle; the set of all neighboring particles $\{\mathrm{P}_1\}$ such that P and P_1 determine a body vector $\boldsymbol{d\xi}$ which satisfies the equation

(i) $\boldsymbol{d\xi} \cdot \boldsymbol{\gamma}(\mathrm{P}, t') \cdot \boldsymbol{d\xi} = (ds')^2 \qquad (ds' = \text{const})$

is the surface of a sphere (center P) in state t' and of a *strain ellipsoid* in any other state t. If instead $d\boldsymbol{\xi}$ satisfies the equation

$$\text{(ii)}\quad d\boldsymbol{\xi}\cdot\dot{\boldsymbol{\gamma}}(\text{P}, t)\cdot d\boldsymbol{\xi} = \varepsilon^2 \qquad (\varepsilon = \text{const}),$$

the particles at time t form the surface of a *strain-rate quadric* which is not necessarily an ellipsoid.

(27) **Problem** Prove that $d\boldsymbol{\xi}_i$ $(i = 1, 2, 3)$, the main semiaxes of the strain ellipsoid (26)(i) at time t, coincide with the principal axes of strain $t' \to t$ and (regarded as material lines throughout) change in length by factors $e_i(t, t')$ in the deformation $t' \to t$.

In the preceding treatment of principal axes and principal values, we have started with *mixed* tensors (rather than with covariant or contravariant tensors) because mixed second-rank tensors map (by contraction) each of the two vector spaces (contravariant and covariant) onto themselves. The situation is then identical to that which occurs when one considers a linear operator for an algebraic vector space, and the eigenvalue and eigenvector properties and definitions for this are well known. On the other hand, a second-rank *covariant* tensor maps (again by contraction) the space of contravariant vectors onto the space of covariant vectors; we have, in effect, used the covariant metric tensor to map the latter space back onto the former. Any positive-definite covariant tensor could have been used for this purpose, but in practice the metric tensor is usually the one of interest.

It is evident that the preceding definitions of principal axes and principal values can all be reformulated in an entirely similar manner for space tensors instead of body tensors, or for any Riemannian manifold of any finite dimension.

(28) **Problem** Show that the principal strain rates are the roots in $\dot{s}$ of

$$|\dot{\boldsymbol{\gamma}} - \dot{s}\boldsymbol{\gamma}| = 0.$$

3 THE KINEMATICS OF SHEAR FLOW AND SHEAR-FREE FLOW

3.1 Introduction

The simplest example of a shear flow is *rectilinear* shear flow, which is usually defined by the statement that, when referred to a rectangular Cartesian space coordinate system $C: \mathrm{Q} \to y$, the velocity components $v^i = v^i(y, t)$ can be written in the form

(1) $$v^1 = \dot{s}(t)y^2, \qquad v^2 = v^3 = 0.$$

Here, $\dot{s}$ is the *velocity gradient* or *shear rate* and is independent of time if the shear flow is steady. The material planes $y^2 = \text{const}$ move rigidly at constant separation and are called *shearing planes.* They slide past one another in the direction of material lines (called *lines of shear*, or *shear lines*) which, in this case, are parallel to the y^1 axis. The flow is at constant volume.

(2) The aim of this chapter is to generalize the definition and treatment of shear flow, retaining the essential feature of a family of material surfaces which move without stretching, in order to encompass all the types of shear flow [listed in Table 3.1(2)] which are of practical importance. It is necessary

Table 3.1(2) Shear Flows Used in Measurements with Viscoelastic Liquids

Apparatus geometry	Cause of flow	Type	Selected references[a]
1. Tube, circular cross section	Pressure gradient	Unidirectional	Eisenschitz *et al.* (1929), Rabinowitsch (1929), Gaskins and Philippoff (1959), Metzner *et al.* (1961); Middleman and Gavis (1961), Harris (1961, 1963, 1965), Sakiadis (1962), Davies and Walters (1972, 1973a), Wales and Philippoff (1973).
2. Annulus, circular cross section	Pressure gradient	Unidirectional	Hayes and Tanner (1965), Huppler (1965).
3. Tube, slit cross section	Pressure gradient	Unidirectional	Janeschitz-Kriegl (1965), Davies and Walters (1972, 1973a), Han and Kim (1973), Higashitani (1973), Wales and Philippoff (1973).
4. Coaxial circular cylinders	Relative rotation	Unidirectional	Russell (1946), Krieger and Maron (1954), Padden and DeWitt (1954), Kaye and Saunders (1964), Bancroft and Kaye (1970), Osaki *et al.* (1963), Broadbent and Lodge (1971).
5. Coaxial circular cylinders	Rotation and pressure gradient	Unidirectional if steady (helical flow)	Rivlin (1956), Dirckes and Schowalter (1966).
6. Coaxial circular cylinders	Steady rotation and oscillatory axial motion	Not unidirectional; helical flow	Simmons (1966, 1968), Tanner and Simmons (1967).
7. Parallel circular coaxial plates	Relative rotation	Unidirectional	Russell (1946), Greensmith and Rivlin (1953), Kotaka *et al.* (1959), Markovitz and Brown (1964), Adams and Lodge (1964), Kaye *et al.* (1968), Pritchard (1971).

Table 3.1(2) *(continued)*

Apparatus geometry	Cause of flow	Type	Selected references[a]
8. Cone and touching plate	Relative rotation	Unidirectional	Russell (1946), Weissenberg (1947), Roberts (1952, 1954), Markovitz and Brown (1964), Adams and Lodge (1964), Kaye *et al.* (1968), Ginn and Metzner (1969), Williams (1965), Pritchard (1971), Miller and Christiansen (1972), Meissner (1972), Olabisi and Williams (1972), van Es (1972), Christiansen and Leppard (1974), Brindley and Broadbent (1973).
9. Cone and plate, separated	Relative rotation	Unidirectional	Jackson and Kaye (1966), Marsh and Pearson (1968).
10. Sphere and touching plate	Relative rotation	Unidirectional	Jackson (1967).
11. Inclined channel, semicircular cross section	Gravity	Unidirectional	Wineman and Pipkin (1966), Tanner (1970), Tanner and Kuo (1972).
12. Flow past slit mouth	Pressure gradient	Unidirectional	Tanner and Pipkin (1969), Kearsley (1970), Higashitani and Pritchard (1972), Higashitani (1973).
13. Noncoaxial, parallel, circular plates	Rotation at same speed about parallel axes	Not unidirectional (orthogonal rheometer)	Maxwell and Chartoff (1965), Blyler and Kurtz (1967), Maxwell (1967), Bird and Harris (1968), Huilgol (1969), Abbott and Walters (1970), Macosko and Starita (1971).
14. Concentric hemispheres	Rotation at same speed about inclined axes	Not unidirectional (balance rheometer)	Kaelble (1969), Walters (1970b).

[a] References covering several shear flows: Garner *et al.* (1950), Lodge (1964), Oldroyd (1965), Truesdell and Noll (1965), Coleman *et al.* (1966), Pipkin and Tanner (1973).

to allow the shearing surfaces and lines of shear to be curved, to allow the shear rate to vary throughout the material, and to allow the shear lines and shear rate to vary with time. In this chapter, we give the kinematics (i.e., the description of strain, strain rate, etc.) and in Chapter 9 we apply the results obtained here to the problem of determining the stress distribution in viscoelastic materials in shear flow. We give the kinematics here as an immediate and important application of the body tensor formalism developed in Chapter 2.

3.2 Shear flow

(1) **Definition** A material surface moves *isometrically* if it moves without stretching, i.e., if the separation of every pair of *neighboring* particles is independent of time.

In particular, a rigidly moving surface moves isometrically. Most of the types of shear flow in Table 3.1(2) have shearing surfaces which move rigidly; however, type 12 does not. Using body tensors, it is as easy to treat isometric surfaces as it is to treat rigid surfaces. An example of an isometric surface moving nonrigidly is also furnished by a thin sheet of paper which flexes as it moves; this example shows the importance of including the word *neighboring* in the definition (1), because the separation of nonneighboring particles changes as two distant parts of the paper are brought closer together by flexing. The separation of neighboring particles in the sheet does not change, provided that we exclude "sharp" folds; in our applications, these are excluded by our differentiability assumptions for the functions 2.2(4) describing continuous flows.

(2) **Definition** A flow is a *shear flow* if (i) each material element moves with constant volume and (ii) there exists a one-parameter family of isometric material surfaces which can be used as one family of coordinate surfaces, $\xi^2 = \text{const}$, say, of some body coordinate system $B: \mathrm{P} \to \xi$. The isometric surfaces are called *shearing surfaces*, and B will be called a *shear flow body coordinate system.*

This definition and the following analysis is sufficiently general to include all shear flows listed in Table 3.1(2) *without restriction on the time dependence* of the various parameters which characterize the flows. Since we have defined body vector and tensor fields which have a significance which is independent of any particular choice of body coordinate system from the infinite set $\mathscr{B}$, we are at liberty to choose a particular body coordinate system so as to simplify the analysis of a given flow problem and to use, if we wish, vectors and tensors associated with the coordinate lines and curves of this coordinate system. For shear flow, it is natural to use a body coordinate system constructed from the shearing surfaces $\xi^2 = \text{const}$, and it is a further convenience to choose the coordinate system [see (21)] so that the ξ^1 and ξ^3 coordinate lines are orthogonal at any instant; they will then be orthogonal at every instant, because they lie in shearing surfaces $\xi^2 = \text{const}$ which move isometrically.

(3) **Definition** A body coordinate system $B:\mathrm{P} \to \xi$ is a *shear-flow body coordinate system* if one family, $\xi^2 = \text{const}$ say, of coordinate surfaces are shearing surfaces and the ξ^1 and ξ^3 coordinate lines are orthogonal.

We shall use a shear-flow body coordinate system $B:\mathrm{P} \to \xi$ throughout this chapter (*except in Section 3.9*). The following results for the components $\gamma_{ij}(\xi, t)$ of the covariant body metric tensor field, referred to B, are readily obtained:

(4) $\dot{\gamma}_{11} = \dot{\gamma}_{13} = \dot{\gamma}_{31} = \dot{\gamma}_{33} = 0$ ($\xi^2 = \text{const}$ moves isometrically);

(5) $\gamma_{13} = \gamma_{31} = 0$ (ξ^1 and ξ^3 curves orthogonal);

(6) $\partial|\gamma|/\partial t = 0$ [constant volume; see 5.5(4)];

(7) $|\gamma| = \gamma_{11}\gamma_{22}\gamma_{33} - \gamma_{33}\gamma_{12}^2 - \gamma_{11}\gamma_{32}^2$ [from (5)];

(8) $$\gamma^{23} = -\frac{\gamma_{11}\gamma_{23}}{|\gamma|}, \qquad \gamma^{21} = -\frac{\gamma_{33}\gamma_{21}}{|\gamma|}, \qquad \gamma^{22} = \frac{\gamma_{11}\gamma_{33}}{|\gamma|}$$ [from 1.3(1), (5)];

(9) $\dot{\gamma}^{22} = 0$ [from (5), (6), and (8)];

(10) $\dot{\gamma}_{22} = 2(\gamma_{12}\dot{\gamma}_{12}/\gamma_{11} + \gamma_{32}\dot{\gamma}_{32}/\gamma_{33})$ [from (4), (6), and (7)].

Equation (4) follows from the fact that for any two neighboring particles in a shearing surface, $d\xi^2 = 0$, and hence

$$(ds)^2 = \gamma_{11}(d\xi^1)^2 + 2\gamma_{13}\, d\xi^1\, d\xi^3 + \gamma_{33}(d\xi^3)^2,$$

which must be independent of time for all values of $d\xi^1$ and $d\xi^3$. Equation (5) follows from the fact that in B, a tangent to a ξ^1 curve has components $(d\xi^1, 0, 0)$, and a tangent to a ξ^3 curve has components $(0, 0, d\xi^3)$; the orthogonality condition $\chi = \pi/2$ in 2.7(70)(i) for contravariant vectors reduces to $\gamma_{13}\, d\xi^1\, d\xi^3 = 0$. Equation (6) follows from 5.5(4) in Chapter 5, where volume is defined. The remaining equations follow from the definition of a determinant and from the fact that $[\gamma^{rc}] = [\gamma_{rc}]^{-1}$.

From (9) and 2.2(21), it follows at once that *the separation of neighboring shearing surfaces*, $\xi^2 = c$ *and* $\xi^2 = c + dc$, *is independent of time*; this result is otherwise obvious, because the volume of each material element is constant and the area of each material element in a shearing surface is constant because the shearing surfaces move isometrically.

For an arbitrary continuous flow, at most six of the metric tensor components $\gamma_{ij}(\xi, t)$ are arbitrary functions of time, because of the symmetry condition $\gamma_{ij} = \gamma_{ji}$. (We shall see in Chapter 5 that at most three are arbitrary functions of the ξ^i, but we are here concerned with the time dependence.) Equations (4) and (6) represent four additional restrictions on the time dependence, and hence, *in shear flow, at most two of the functions* $\gamma_{ij}(\xi, t)$ *can be regarded as arbitrary functions of time*. Equation (4) shows that in a shear-flow body coordinate system, the only time-dependent γ_{ij} are γ_{12} $(=\gamma_{21})$, γ_{32} $(=\gamma_{23})$, and γ_{22}; Eq. (7) gives γ_{22} in terms of γ_{12}, γ_{32}, and constants; Eq. (10) expresses $\dot{\gamma}_{22}$ in terms of $\dot{\gamma}_{12}$, $\dot{\gamma}_{32}$, and constants. Similar equations can be written down for the components γ^{ij} of the contravariant metric tensor, but we shall not require them. Instead of regarding γ_{12} and γ_{32} as the two basic functions of time in shear flow, we can regard the shear rate $\dot{s}$ and the angle ζ which a line of shear makes with a ξ^1 curve as the basic functions of time (at each particle). We now define these quantities.

(11) **Definitions** In shear flow, the *shear rate* $\dot{s}(\mathrm{P}, t)$ is a nonzero principal strain rate at particle P and time t; a *shear line* (or *line of shear*) at time t is a line on a shear surface which is everywhere tangential to the local "shear direction" at time t. Let perpendiculars drawn to a shear surface at times t, t' from a given neighboring particle P_0 meet the surface in particles P, P′; as $t' \to t$, the material line element PP′ defines the *shear direction* at P and t.

The principal strain rates are defined by 2.8(18), and are therefore, from 2.8(19)(i), the roots in λ of the equation

(12) $$|\dot{\gamma}_{ij} - \lambda\gamma_{ij}| = 0.$$

Using (4), (5), and (10), it is a straightforward matter to show that, for shear flow, the principal strain rates are 0, $\dot{s}$, and $-\dot{s}$, where

(13) $$\dot{s} = \{(\gamma_{11}\dot{\gamma}_{23}^2 + \gamma_{33}\dot{\gamma}_{21}^2)/|\gamma|\}^{1/2}.$$

The shear rate can be positive or negative, and is given by (13).

(14) See Fig. 3.2(14).

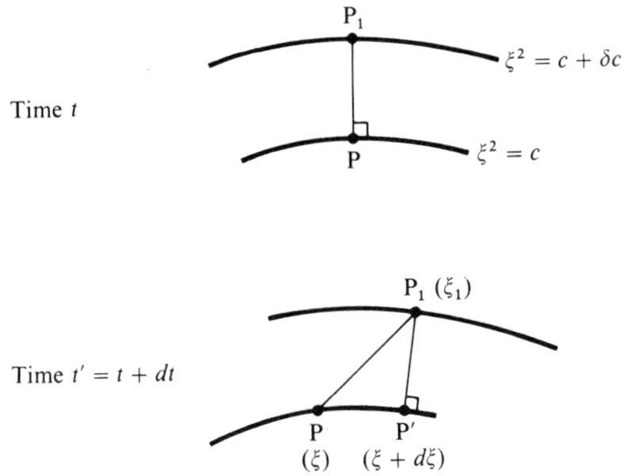

Figure 3.2(14) Definition of a shear line on a shear surface $\xi^2 = c$: As c and dt tend to zero, PP′ defines the direction of the shear line which passes through P at time t.

(15) **Theorem** Referred to a shear-flow body coordinate system, the *differential equations of the lines of shear* on a shearing surface $\xi^2 = c$ at time t are

(i) $$\frac{d\xi^3}{d\xi^1} = \frac{\dot\gamma^{32}(\xi, t)}{\dot\gamma^{12}(\xi, t)}, \quad \text{or}$$

(ii) $$\frac{d\xi^3}{d\xi^1} = \frac{\gamma_{11}\dot\gamma_{32}}{\gamma_{33}\dot\gamma_{12}}.$$

The angle $\zeta(\mathrm{P}, t)$ which the line of shear that passes through P makes with the ξ^1 curve at P at time t is given by the equation

(iii) $\tan\zeta = (\gamma_{11}/\gamma_{33})^{1/2}(\dot\gamma_{32}/\dot\gamma_{12})$.

Proof As shown in Fig. 3.2(14), we consider the following three particles:

P_1, with coordinates ξ_1^i, on $\xi^2 = c + \delta c$;
P, with coordinates $\xi^i = \xi_1^i + \delta\xi_1^i$, on $\xi^2 = c$;
P′, with coordinates $\xi^i + d\xi^i = \xi_1^i + \delta\xi_1^i + d\xi^i$, on $\xi^2 = c$.

PP_1 is perpendicular to $\xi^2 = c$ at time t; $\mathrm{P'P}_1$ is perpendicular to $\xi^2 = c$ at time $t' = t + dt$. As δc and dt both tend to zero, P′P becomes part of a line of shear through P at time t. From 2.7(11), it follows that

$$\delta\xi_1^i = \lambda\gamma^{ij}(\xi_1, t)\,\partial\xi^2/\partial\xi^j = \lambda\gamma^{i2}(\xi_1, t),$$

for some λ; but $\delta\xi_1^2 = -\delta c$, and hence

(16) $$\delta\xi_1^i = -\gamma^{i2}(\xi_1, t)\,\delta c/\gamma^{22}(\xi_1).$$

γ^{22} is independent of t, from (9). Similarly, it follows that

$$\delta\xi_1^i + d\xi^i = -\gamma^{i2}(\xi_1, t + dt)\, \delta c/\gamma^{22}(\xi_1),$$

and hence, from (15), we have

$$d\xi^i = -\dot{\gamma}^{i2}(\xi_1, t)\, dt\, \delta c/\gamma^{22}(\xi_1) + \cdots \qquad (i = 1, 2, 3). \tag{17}$$

Moreover, in this equation, we can replace ξ_1 in the arguments by ξ, since the difference is of higher order than the leading term. The resulting equation with $i = 2$ gives nothing new, but with $i = 1$ and 3 gives the required result (15)(i) on taking the ratio of $d\xi^1$ and $d\xi^3$. Equation (15)(ii) then follows immediately from (4) and (8).

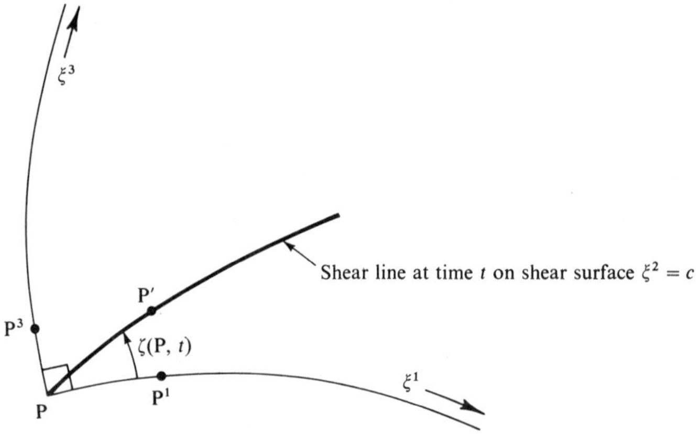

Figure 3.2(18) Sign convention for the angle $\zeta(P, t)$ between a shear line through P at time t and the ξ^1 curve of a shear flow coordinate system having ξ^2 surfaces as shear surfaces.

(18) To determine the angle ζ between the contravariant body vectors $\overrightarrow{PP'}$ and $\overrightarrow{PP^1}$, where P^1 is on the ξ^1 curve through P [Fig. 3.2(18)], we use the definition 2.7(7). It is convenient also to use P^3, on the ξ^3 curve through P. We then have the following components, with $d\xi^3$ and $d\xi^1$ related by (15)(ii):

$$\overrightarrow{PP^1}B = \begin{bmatrix} d\xi^1 \\ 0 \\ 0 \end{bmatrix}; \quad \overrightarrow{PP^3}B = \begin{bmatrix} 0 \\ 0 \\ d\xi^3 \end{bmatrix}; \quad \overrightarrow{PP'}B = \begin{bmatrix} d\xi^1 \\ 0 \\ d\xi^3 \end{bmatrix}.$$

Since $\gamma_{13} = 0$, from (5), it follows from 2.7(7) that

$$PP^1 \cdot PP' \cos\zeta = \overrightarrow{PP^1} \cdot \boldsymbol{\gamma} \cdot \overrightarrow{PP'} = \gamma_{11}(d\xi^1)^2,$$

and

$$PP^3 \cdot PP' \sin\zeta = \overrightarrow{PP^3} \cdot \boldsymbol{\gamma} \cdot \overrightarrow{PP'} = \gamma_{33}(d\xi^3)^2.$$

Since $(\mathrm{PP}^1)^2 = \gamma_{11}(d\xi^1)^2$ and $(\mathrm{PP}^3)^2 = \gamma_{33}(d\xi^3)^2$, the required result (15)(iii) follows at once on using (15)(ii).

(19) **Definition** A shear flow is *unidirectional* if, for all particles P, $\zeta(\mathrm{P}, t)$ is independent of t. From (4) and (15)(iii), it follows that

(20) $$(\partial/\partial t)(\dot{\gamma}_{32}/\dot{\gamma}_{12}) = 0 \qquad \text{(for unidirectional shear flow).}$$

Throughout any time interval in unidirectional shear flow, it follows from (15) that through each particle P there passes one and only one line of shear. Hence, *in unidirectional shear flow, the lines of shear on any given shearing surface form a one-parameter family of nonintersecting material lines.* They could therefore be used as the ξ^1 lines of a shear-flow coordinate system. Such a coordinate system can always be chosen for any given unidirectional shear flow, as the following problem indicates.

(21) **Problem** Suppose a body coordinate system $B:\mathrm{P} \to \xi$ is given to the extent that the ξ^2 surfaces are specified (e.g., as shear surfaces) and the ξ^1 lines are specified (e.g., as shear lines in a unidirectional shear flow). Show that it is always possible to choose the ξ^1 surfaces so that the ξ^3 lines are everywhere orthogonal to the ξ^1 lines at any one instant.

It follows from (15)(ii) that the ξ^1 lines are lines of shear at time t if

(22) $$\dot{\gamma}_{32}(\mathrm{P}, t) = 0 \qquad (\xi^1 \text{ lines are lines of shear at time } t).$$

If this holds throughout a time interval, then the shear flow is unidirectional with the ξ^1 lines as lines of shear throughout the interval. Similarly,

(23) $$\dot{\gamma}_{12}(\mathrm{P}, t) = 0 \qquad (\text{if the } \xi^3 \text{ lines are lines of shear at } t).$$

Equations (13) and (15)(iii) substantiate our earlier statement that when the shearing surfaces are known, a shear flow is completely characterized by two functions of time: the shear rate $\dot{s}$ and the angle ζ.

(24) **Definition** A shear flow is *steady* if $\dot{s}$ and ζ are independent of time.

A *steady shear flow is necessarily unidirectional* according to this definition; it is not sufficient to have the shear rate constant. We shall see that the orthogonal rheometer flow has constant shear rate but varying direction ζ, and is therefore not steady. Since this flow is usually steady in the hydrodynamic sense [i.e., $\partial v^i(x, t)/\partial t = 0$], it follows that our use of the word *steady* differs from that common in hydrodynamics. The hydrodynamic usage involves reference to directions fixed in space and is therefore not of immediate interest in rheology when one is concerned with the description of rheological properties of a material.

A previous treatment of shear flow (Lodge, 1964, pp. 337–344) was similar to that given above but was unnecessarily restrictive in using an assumption that a shear-flow body coordinate system could be chosen such that the ξ^3 surfaces were orthogonal to the shearing surfaces $\xi^2 = \text{const}$. This is not valid, in particular, for helical flow. The present treatment makes no such assumption but achieves sufficient simplification by taking the ξ^3 lines to be orthogonal to the ξ^1 lines; Problem (21) shows that this involves no restriction, and we shall see later that helical flow can be treated with the present formalism.

It is shown in Section 11.3 that the term unidirectional shear flow is synonymous with the term viscometric flow.

3.3 Base vectors and strain tensors for shear flow

Let $B: \mathrm{P} \to \xi$ be any shear-flow body coordinate system for an arbitrary shear flow. The associated base vectors $\boldsymbol{\beta}_i(B, \mathrm{P})$ and $\boldsymbol{\beta}^i(B, \mathrm{P})$ are independent of time but are neither orthogonal nor orthonormal throughout any time interval, with the exception of $\boldsymbol{\beta}_1$ and $\boldsymbol{\beta}_3$, which are always orthogonal, by 3.2(5). It is convenient to construct from these base vectors two reciprocal sets, $\boldsymbol{\alpha}_i(\mathrm{P}, t)$ and $\boldsymbol{\alpha}^i(\mathrm{P}, t)$, of time-dependent vectors, *each set being instantaneously orthonormal at time t*, as follows:

(1) $\boldsymbol{\alpha}_i(\mathrm{P}, t)$ $(i = 1, 2, 3)$ are orthonormal at time t.

(2) $\boldsymbol{\alpha}_i(\mathrm{P}, t) \cdot \boldsymbol{\alpha}^j(\mathrm{P}, t) = \delta_i^j$.

(3) $\boldsymbol{\alpha}^2(\mathrm{P}, t)$ is always normal to the shearing surface $\xi^2 = c$ at P.

(4) $\boldsymbol{\alpha}_1(\mathrm{P}, t)$ is always tangential to the ξ^1 curve at P.

(5) $\boldsymbol{\alpha}_3(\mathrm{P}, t)$ is always tangential to the ξ^3 curve at P.

(6) These vectors are illustrated in Fig. 3.3(6). Equation (2) states that the two sets are reciprocal. As the notation indicates, the $\boldsymbol{\alpha}_i$ are contravariant and the $\boldsymbol{\alpha}^i$ are covariant body vectors. Property (5) follows from (1), (4), and the fact that the ξ^1 curves are orthogonal to the ξ^3 curves. We will call $\boldsymbol{\alpha}_i$ a *contravariant shear flow basis* and $\boldsymbol{\alpha}^i$ a *covariant shear flow basis*.

The relations between the orthonormal bases and the coordinate system base vectors are given by the following equations:

(7) (i) $\boldsymbol{\alpha}_1 = (\gamma_{11})^{-1/2}\boldsymbol{\beta}_1$;

(ii) $\boldsymbol{\alpha}_2 = (\gamma^{22})^{-1/2}(\gamma^{21}\boldsymbol{\beta}_1 + \gamma^{22}\boldsymbol{\beta}_2 + \gamma^{23}\boldsymbol{\beta}_3)$;

(iii) $\boldsymbol{\alpha}_3 = (\gamma_{33})^{-1/2}\boldsymbol{\beta}_3$.

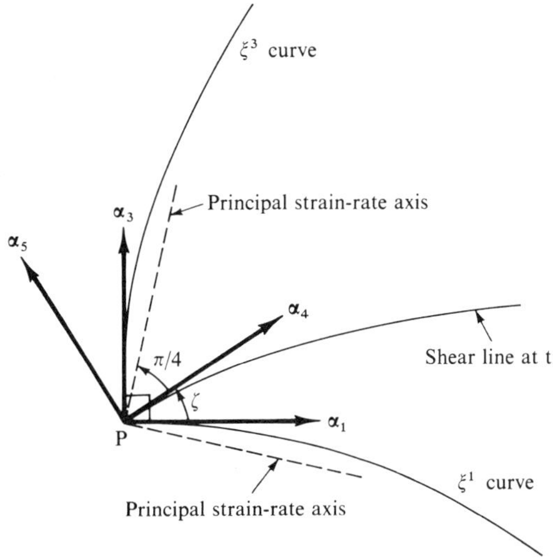

Figure 3.3(6) Contravariant body vectors $\boldsymbol{\alpha}_i = \boldsymbol{\alpha}_i(\mathrm{P}, t)$ tangential at P to a shear surface $\xi^2 = c$. ($\boldsymbol{\alpha}_1$, $\boldsymbol{\alpha}_2$, $\boldsymbol{\alpha}_3$) are orthonormal at time t and are tangential to the coordinate curves passing through P of a shear-flow body coordinate system $B: \mathrm{P} \to \xi$ ($\boldsymbol{\alpha}_2$ is not shown). ($\boldsymbol{\alpha}_4$, $\boldsymbol{\alpha}_2$, $\boldsymbol{\alpha}_5$) are orthonormal at time t, with $\boldsymbol{\alpha}_4$ tangential to the shear line through P at t.

(8) (i) $\boldsymbol{\alpha}^1 = (\gamma_{11})^{-1/2}(\gamma_{11}\boldsymbol{\beta}^1 + \gamma_{12}\boldsymbol{\beta}^2)$;

(ii) $\boldsymbol{\alpha}^2 = (\gamma^{22})^{-1/2}\boldsymbol{\beta}^2$;

(iii) $\boldsymbol{\alpha}^3 = (\gamma_{33})^{-1/2}(\gamma_{23}\boldsymbol{\beta}^2 + \gamma_{33}\boldsymbol{\beta}^3)$.

(9) $\boldsymbol{\alpha}_i \equiv \boldsymbol{\alpha}_i(\mathrm{P}, t)$; $\quad \boldsymbol{\alpha}^i \equiv \boldsymbol{\alpha}^i(\mathrm{P}, t)$; $\quad \boldsymbol{\beta}_i \equiv \boldsymbol{\beta}_i(B, \mathrm{P})$; $\quad \boldsymbol{\beta}^i \equiv \boldsymbol{\beta}^i(B, \mathrm{P})$.

To derive (7) and (8), we first note that, from (3) and 2.4(17), $\boldsymbol{\alpha}^2$ must be a scalar multiple of $\boldsymbol{\beta}^2$; this scalar is determined from 2.7(4)(i) and the fact that $\boldsymbol{\alpha}^2$ is of unit magnitude; thus (8)(ii) is obtained. In a similar fashion, we obtain (7)(i) and (7)(iii) from (4) and (5), using 2.7(4)(i) and the fact that the $\boldsymbol{\beta}_i$ are tangents to coordinate curves. Next, we obtain (8)(i) from (7)(i), by using (2), 3.2(5), 2.7(16)(ii), and the equations

(10) $$\boldsymbol{\gamma} \cdot \boldsymbol{\beta}_k = \gamma_{ik}\boldsymbol{\beta}^i,$$

which come from 2.6(16) and 2.4(26)(i). Similarly, we obtain (8)(iii) from (7)(iii). Finally, in a similar way, we obtain (7)(ii) from (8)(ii) by using 2.7(16)(ii) and the equations

(11) $$\boldsymbol{\gamma}^{-1} \cdot \boldsymbol{\beta}^k = \gamma^{ik}\boldsymbol{\beta}_i,$$

which come from 2.6(17) and 2.4(26)(i). This completes the proof of (7) and (8).

It is convenient to define scalars $\dot{s}_1$ and $\dot{s}_3$ as follows:

(12) $$\dot{s}_1 = \dot{s}\cos\zeta, \qquad \dot{s}_3 = \dot{s}\sin\zeta;$$

the shear rate $\dot{s}$ and the shear orientation angle ζ are given by 3.2(13) and 3.2(15).

(13) **Problem** Prove the following relations:

(i) $\dot{s}_1 = (\gamma_{33}/|\gamma|)^{1/2}\dot{\gamma}_{12}, \qquad \dot{s}_3 = (\gamma_{11}/|\gamma|)^{1/2}\dot{\gamma}_{32};$

(ii) $\dot{s}\boldsymbol{\alpha}_4 = \dot{s}_1\boldsymbol{\alpha}_1 + \dot{s}_3\boldsymbol{\alpha}_3;$

(iii) $\dot{s}\boldsymbol{\alpha}_5 = -\dot{s}_3\boldsymbol{\alpha}_1 + \dot{s}_1\boldsymbol{\alpha}_3.$

$\boldsymbol{\alpha}_4$ and $\boldsymbol{\alpha}_5$ are contravariant unit vectors, respectively tangential and normal to the line of shear at P, and normal to $\boldsymbol{\alpha}^2$ [Fig. 3.3(6)]. A choice of sign has been made arbitrarily in the first of equations (i).

By differentiating (7) and (8), using 3.2(4), 3.2(8), 3.2(9), and the fact that $\boldsymbol{\beta}_i$ and $\boldsymbol{\beta}^i$ are independent of time, it is a straightforward matter to obtain the following expressions for the *time derivatives of the shear-flow base vectors*:

(14) $$\dot{\boldsymbol{\alpha}}_1 = \mathbf{0}; \qquad \dot{\boldsymbol{\alpha}}_2 = -\dot{s}_1\boldsymbol{\alpha}_1 - \dot{s}_3\boldsymbol{\alpha}_3; \qquad \dot{\boldsymbol{\alpha}}_3 = \mathbf{0};$$

(15) $$\dot{\boldsymbol{\alpha}}^1 = \dot{s}_1\boldsymbol{\alpha}^2; \qquad \dot{\boldsymbol{\alpha}}^2 = \mathbf{0}; \qquad \dot{\boldsymbol{\alpha}}^3 = \dot{s}_3\boldsymbol{\alpha}^2.$$

Integrating these equations with respect to time, we obtain the equations

(16) $$\begin{aligned} &\boldsymbol{\alpha}'_1 = \boldsymbol{\alpha}_1; && \boldsymbol{\alpha}'_2 = \boldsymbol{\alpha}_2 - s_1\boldsymbol{\alpha}_1 - s_3\boldsymbol{\alpha}_3; && \boldsymbol{\alpha}'_3 = \boldsymbol{\alpha}_3; \\ &\boldsymbol{\alpha}'^1 = \boldsymbol{\alpha}^1 + s_1\boldsymbol{\alpha}^2; && \boldsymbol{\alpha}'^2 = \boldsymbol{\alpha}^2; && \boldsymbol{\alpha}'^3 = \boldsymbol{\alpha}^3 + s_3\boldsymbol{\alpha}^2. \end{aligned}$$

We here use the notation

(17) $$s_k = s_k(\mathrm{P}, t', t) = \int_t^{t'} \dot{s}_k(\mathrm{P}, t'')\,dt'' \qquad (k = 1, 3);$$

(18) $$\boldsymbol{\alpha}'_i = \boldsymbol{\alpha}_i(\mathrm{P}, t'), \qquad \boldsymbol{\alpha}'^i = \boldsymbol{\alpha}^i(\mathrm{P}, t'); \qquad \boldsymbol{\alpha}_i = \boldsymbol{\alpha}_i(\mathrm{P}, t), \qquad \boldsymbol{\alpha}^i = \boldsymbol{\alpha}^i(\mathrm{P}, t).$$

Since $\boldsymbol{\alpha}^i$ and $\boldsymbol{\alpha}_i$ are orthonormal bases at time t, and $\boldsymbol{\alpha}'^i$ and $\boldsymbol{\alpha}'_i$ are orthonormal bases at time t', it follows from 2.7(16) that

(19) $$\boldsymbol{\gamma} = \boldsymbol{\alpha}^i\boldsymbol{\alpha}^i, \qquad \boldsymbol{\gamma}^{-1} = \boldsymbol{\alpha}_i\boldsymbol{\alpha}_i, \qquad \boldsymbol{\gamma}' = \boldsymbol{\alpha}'^i\boldsymbol{\alpha}'^i, \qquad \boldsymbol{\gamma}'^{-1} = \boldsymbol{\alpha}'_i\boldsymbol{\alpha}'_i,$$

where $\boldsymbol{\gamma} = \boldsymbol{\gamma}(\mathrm{P}, t)$, $\boldsymbol{\gamma}^{-1} = \boldsymbol{\gamma}^{-1}(\mathrm{P}, t)$, $\boldsymbol{\gamma}' = \boldsymbol{\gamma}(\mathrm{P}, t')$, and $\boldsymbol{\gamma}'^{-1} = \boldsymbol{\gamma}^{-1}(\mathrm{P}, t')$.

From (16) and (19), we immediately obtain the following equations for the strain tensors relating states t and t':

$$\gamma' - \gamma = s_k(\boldsymbol{\alpha}^2\boldsymbol{\alpha}^k + \boldsymbol{\alpha}^k\boldsymbol{\alpha}^2) + s_k s_k \boldsymbol{\alpha}^2\boldsymbol{\alpha}^2; \tag{20}$$

$$\gamma'^{-1} - \gamma^{-1} = -s_k(\boldsymbol{\alpha}_2\boldsymbol{\alpha}_k + \boldsymbol{\alpha}_k\boldsymbol{\alpha}_2) + s_k{}^2\boldsymbol{\alpha}_k\boldsymbol{\alpha}_k + s_1 s_3(\boldsymbol{\alpha}_1\boldsymbol{\alpha}_3 + \boldsymbol{\alpha}_3\boldsymbol{\alpha}_1). \tag{21}$$

In each equation, there is summation for $k = 1$ and 3; in (20), we have written $s_k s_k$ (instead of $s_k{}^2$) so that summation will be implied by our convention. These equations are valid for an arbitrary shear flow history; $\boldsymbol{\alpha}^2$ is normal to a shearing surface, the bases $\boldsymbol{\alpha}^i$ and $\boldsymbol{\alpha}_i$ are orthonormal at time t, and the shear rate $\dot{s}$ is given by the equation

$$\dot{s}^2 = \dot{s}_k \dot{s}_k = \dot{s}_1{}^2 + \dot{s}_3{}^2, \tag{22}$$

which comes from (12).

For unidirectional shear flow, in which, say, the ξ^1 curves are the lines of shear, it follows from (12) and 3.2(22) that $\zeta = 0$ and

$$\dot{s} = \dot{s}_1, \qquad \dot{s}_3 = 0 \qquad \begin{pmatrix}\text{unidirectional shear flow with } \xi^1 \text{ curves}\\ \text{as lines of shear;} \quad \dot{\gamma}_{32} = 0\end{pmatrix}. \tag{23}$$

Similarly, if the ξ^3 curves are lines of shear, then $\zeta = \pi/2$ and

$$\dot{s} = \dot{s}_3, \qquad \dot{s}_1 = 0 \qquad \begin{pmatrix}\text{unidirectional shear flow with } \xi^3 \text{ curves}\\ \text{as lines of shear;} \quad \dot{\gamma}_{12} = 0)\end{pmatrix}. \tag{24}$$

It now follows from (20) that $\gamma' - \gamma$ for an arbitrary shear flow is the sum of similar strain tensors for two unidirectional shear flows: one with shear rate $\dot{s}_1$ and ξ^1 curves as lines of shear, and the other with shear rate $\dot{s}_3$ and ξ^3 curves as lines of shear. Since these two families of coordinate curves are orthogonal (but otherwise arbitrarily chosen on the shearing surfaces), we may refer to this situation as one of orthogonal superposition of unidirectional shear flows. Furthermore, the covariant strain tensor $\gamma' - \gamma$, when given for all pairs of values of t and t', completely and uniquely describes the strain history. We therefore have established the following important result.

(25) **Theorem** Insofar as the strain history at any given particle P is concerned, *an arbitrary shear flow can be regarded as the orthogonal superposition of two unidirectional shear flows whose lines of shear are orthogonal but otherwise arbitrary.* The shear rates are compounded according to (22), and the covariant strain tensors according to (20).

Insofar as the variation of strain history from one particle to another is concerned, we shall see in Section 8.8 that in general, the sum of two strain tensor fields is not a possible strain tensor field, because metric tensor fields necessarily satisfy the condition that the Riemann–Christoffel curvature tensor constructed from them must vanish everywhere; this is a condition involving covariant derivatives of the metric tensor and is nonlinear in the metric tensor. In order that Theorem (25) shall be applicable globally, therefore, certain further conditions must be satisfied; but we shall not need to consider these here.

We may summarize Theorem (25) by saying that the *covariant strain tensors are additive when one expresses an arbitrary shear flow as the orthogonal superposition of two unidirectional shear flows.* It follows also that $\partial^n\gamma(\mathrm{P}, t)/\partial t^n$ is additive; for the first two derivatives, we have

(26) $$\dot{\gamma} = \dot{s}_k(\boldsymbol{\alpha}^2\boldsymbol{\alpha}^k + \boldsymbol{\alpha}^k\boldsymbol{\alpha}^2);$$

(27) $$\ddot{\gamma} = \ddot{s}_k(\boldsymbol{\alpha}^2\boldsymbol{\alpha}^k + \boldsymbol{\alpha}^k\boldsymbol{\alpha}^2) + 2\dot{s}_k\,\dot{s}_k\,\boldsymbol{\alpha}^2\boldsymbol{\alpha}^2.$$

These equations can be obtained simply by differentiating the first of equations (19) and using (15). Equation (27) shows that $\ddot{\gamma} \neq 0$ in steady shear flow.

It is interesting to note that due to the presence of the cross term involving $s_1 s_3$ in (21), *the contravariant strain tensors are not additive.* There is no reason to expect them to be additive; covariant and contravariant metric tensors and strain tensors are in a sense dual descriptions of the same property (i.e., time-dependent separations of pairs of neighboring particles and of neighboring surfaces of a deforming geometric manifold), and so one might perhaps have expected additivity in the one to be reflected as additivity in the other. However, in defining shear flow, we necessarily treat surfaces and lines quite differently (e.g., we postulate the existence of *surfaces* which move isometrically) and thereby lose any symmetry between a description and the dual description which might have otherwise existed. It should also be noted that it is the covariant *strain* tensors which are additive; *the covariant metric tensors are not additive.* For the contravariant metric tensor, the first time derivative is additive but the second is not, as shown by the following equations, which can be obtained from (14) and (19):

(28) $$\partial\gamma^{-1}/\partial t = -\dot{s}_k(\boldsymbol{\alpha}_2\,\boldsymbol{\alpha}_k + \boldsymbol{\alpha}_k\,\boldsymbol{\alpha}_2);$$

(29) $$\partial^2\gamma^{-1}/\partial t^2 = -\ddot{s}_k(\boldsymbol{\alpha}_2\,\boldsymbol{\alpha}_k + \boldsymbol{\alpha}_k\,\boldsymbol{\alpha}_2) + \dot{s}_k\,\dot{s}_m(\boldsymbol{\alpha}_k\,\boldsymbol{\alpha}_m + \boldsymbol{\alpha}_m\,\boldsymbol{\alpha}_k).$$

(k, $m = 1$, 3 in these summations). The term in $\dot{s}_k\,\dot{s}_m$ in (29) shows that the second time derivative is not additive.

(30) **Problem** For an arbitrary shear flow of shear rate $\dot{s}$, prove that at any particle P, the principal strain-rate axes corresponding to the principal strain rates $\pm\dot{s}$ are orthogonal to $\boldsymbol{\alpha}^3$ and make angles $\pm\pi/4$ with the line of shear at P.

(31) **Problem** For an arbitrary shear flow, prove that

$$\dot{\boldsymbol{\gamma}} \cdot \boldsymbol{\gamma}^{-1} \cdot \dot{\boldsymbol{\gamma}} \cdot \boldsymbol{\gamma}^{-1} \cdot \dot{\boldsymbol{\gamma}} = \dot{s}^2 \dot{\boldsymbol{\gamma}}.$$

(32) **Problem** For an arbitrary *unidirectional* shear flow (taking the ξ^1 curves as lines of shear), prove the following results when the shear rate $\dot{s}$ ($= \dot{s}_1$) is not zero:

(i) $\boldsymbol{\alpha}^1\boldsymbol{\alpha}^1 = \dot{s}^{-2}(\dot{\boldsymbol{\gamma}} \cdot \boldsymbol{\gamma}^{-1} \cdot \dot{\boldsymbol{\gamma}} - \tfrac{1}{2}\ddot{\boldsymbol{\gamma}}) + \tfrac{1}{2}\dot{s}^{-3}\ddot{s}\dot{\boldsymbol{\gamma}}$;

(ii) $\boldsymbol{\alpha}_1\boldsymbol{\alpha}_1 = \tfrac{1}{2}\dot{s}^{-2}\ddot{\boldsymbol{\gamma}}^{-1} - \tfrac{1}{2}\dot{s}^{-3}\ddot{s}\dot{\boldsymbol{\gamma}}^{-1}$; (see 2.6(32)(ii);

(iii) $\boldsymbol{\alpha}^3\boldsymbol{\alpha}^3 = \boldsymbol{\gamma} - \dot{s}^{-2}\dot{\boldsymbol{\gamma}} \cdot \boldsymbol{\gamma}^{-1} \cdot \dot{\boldsymbol{\gamma}}$;

(iv) $\boldsymbol{\alpha}_3\boldsymbol{\alpha}_3 = \boldsymbol{\gamma}^{-1} - \dot{s}^{-2}\dot{\boldsymbol{\gamma}}^{-1} \cdot \boldsymbol{\gamma} \cdot \dot{\boldsymbol{\gamma}}^{-1}$.

Important particular cases of the preceding formalism, useful for analysis of various laboratory experiments, are summarized as follows.

(33) *Steady shear flow*: unidirectional, with constant shear rate $\dot{s} = \dot{s}_0$, say; hence $s(\mathrm{P}, t', t) = (t' - t)\dot{s}_0(\mathrm{P})$.

(34) *Oscillatory shear*: unidirectional, with $\dot{s} = \alpha\omega \cos \omega t$ (α, ω consts.); hence $s(\mathrm{P}, t', t) = \alpha(\sin \omega t' - \sin \omega t)$.

(35) *Oscillatory shear on steady shear flow, parallel superposition*: unidirectional, with $\dot{s} = \dot{s}_0 + \alpha\omega \cos \omega t$. Hence

$$s(\mathrm{P}, t', t) = (t' - t)\dot{s}_0 + \alpha(\sin \omega t' - \sin \omega t).$$

(36) *Oscillatory shear on steady shear flow, orthogonal superposition*: not unidirectional; given by preceding equations with

$\dot{s}_1 = \dot{s}_0$ (ξ^1 curves are shear lines for the steady shear flow);

$\dot{s}_3 = \alpha\omega \cos \omega t$ (ξ^3 curves are shear lines for the oscillatory shear).

(37) **Problem** Show that in an arbitrary unidirectional shear flow, $\dddot{\boldsymbol{\gamma}}$ depends linearly on $\dot{\boldsymbol{\gamma}}$ and $\ddot{\boldsymbol{\gamma}}$ according to the equation

$$\dot{s}^2\dddot{\boldsymbol{\gamma}} = (\dddot{s}\dot{s} - 3\ddot{s}^2)\dot{\boldsymbol{\gamma}} + 3\dot{s}\ddot{s}\ddot{\boldsymbol{\gamma}},$$

where $\dot{s}$, the shear rate, is a root of $|\dot{\boldsymbol{\gamma}} - \dot{s}\boldsymbol{\gamma}| = 0$.

(38) **Problem** Show that in an arbitrary unidirectional shear flow, the covariant strain tensor $\boldsymbol{\gamma}' - \boldsymbol{\gamma}$, for any two instants t' and t, is related to $\dot{\boldsymbol{\gamma}}$ and $\ddot{\boldsymbol{\gamma}}$ by the equation

$$2\dot{s}^3(\boldsymbol{\gamma}' - \boldsymbol{\gamma}) = s(2\dot{s}^2 - s\ddot{s})\dot{\boldsymbol{\gamma}} + \dot{s}s^2\ddot{\boldsymbol{\gamma}}, \qquad \text{where} \quad s = \int_t^{t'} \dot{s}''\, dt'',$$

and $\dot{s}$ and $\dot{s}''$ denote values of shear rate at times t and t'', respectively. Show also that the contravariant strain tensor satisfies an equation of the same form, namely

$$2\dot{s}^3(\gamma'^{-1} - \gamma^{-1}) = s(2\dot{s}^2 - s\ddot{s})\frac{\partial \gamma^{-1}}{\partial t} + \dot{s}s^2\frac{\partial^2 \gamma^{-1}}{\partial t^2}.$$

3.4 Torsional flow between circular parallel plates in relative rotation about a common axis

Let $C: \mathrm{Q} \rightarrow y$ be a rectangular space Cartesian coordinate system, defined in 2.7(15); then for neighboring places of coordinates y^i and $y^i + dy^i$, we have

(1) $$(ds)^2 = dy^i\, dy^i.$$

Let us define a *cylindrical polar coordinate system* $S: \mathrm{Q} \rightarrow x$ for space by means of the coordinate transformation

(2) $$y^1 = r\cos\phi, \qquad y^2 = r\sin\phi, \qquad y^3 = z,$$

where we have written

(3) $$x^1 = \phi, \qquad x^2 = z, \qquad x^3 = r \geq 0.$$

From (1) and 2.2(11), it is easy to obtain the following values for the metric tensor components in S:

(4) $$[g_{ij}] = \mathrm{diag}[r^2, 1, 1], \qquad |g| = r^2.$$

From (2), the coordinate transformation matrix defined in 2.3(31) is found to have a determinant $|L_{\mathrm{Q}}(S, C)| = r$; provided that ϕ is restricted to a suitable interval (say $0 \leq \phi \leq 2\pi$) to make the transformation $x \leftrightarrow y$ one-to-one, the transformation (2) satisfies the conditions laid down for a coordinate transformation, except at the point $r = 0$.

Let us now choose a body coordinate system $B: \mathrm{P} \rightarrow \xi$ which is congruent to the cylindrical polar coordinate system S at an arbitrary but definite time t_0:

(5) $$B \overset{t_0}{\equiv} S; \qquad \xi^i = x^i \quad \text{at} \quad t_0.$$

Using a superior dot to denote $\partial/\partial t$ at constant ξ, we consider a flow defined by the equations

(6) $$\dot{z} = \dot{r} = 0, \qquad \dot{\phi} = \Omega(z, t),$$

where $\Omega(z, t)$ is a given function. The equations 2.2(4) describing the flow are readily found to be as follows:

(7) $$\phi = \xi^1 + \int_{t_0}^{t} \Omega(\xi^2, t')\, dt', \qquad z = \xi^2, \quad r = \xi^3.$$

This flow will be called a *torsional flow between parallel plates.*

Using 2.2(13) with (4) and (7), it is easy to show that

(8) $$[\gamma_{rc}] = \begin{bmatrix} r^2 & r^2\phi_2 & 0 \\ r^2\phi_2 & 1 + (r\phi_2)^2 & 0 \\ 0 & 0 & 1 \end{bmatrix}, \qquad [\dot{\gamma}_{rc}] = \begin{bmatrix} 0 & r^2\Omega_2 & 0 \\ r^2\Omega_2 & 2r^2\phi_2\Omega_2 & 0 \\ 0 & 0 & 0 \end{bmatrix},$$

where

(9) $$\phi_2 \equiv \partial\phi/\partial\xi^2, \qquad \Omega_2 \equiv \partial^2\phi/\partial t\, \partial\xi^2 = \partial\Omega(z, t)/\partial z; \qquad \gamma_{rc} \equiv \gamma_{rc}(\xi, t).$$

Hence $|\gamma| = r^2$ is independent of time, so the flow is at constant volume, and γ_{11}, γ_{13}, and γ_{33} are independent of time, so the flow is a shear flow with $\xi^2 = c$ (i.e., planes $z = c$) as shearing surfaces. Furthermore, $\dot{\gamma}_{32} = 0$, so the shear flow is unidirectional with the ξ^1 curves (i.e., the circles given by the intersection of the cylinders $r = c$ and the planes $z = c'$) as lines of shear [from 3.2(15) or 3.3(23)]. From 3.2(13) and (8), we find that

(10) $$\dot{s} = \dot{s}_1 = \pm r\, \partial\Omega(z, t)/\partial z \qquad (\dot{s}_3 = 0, \quad \zeta = 0).$$

3.5 Torsional flow between a cone and a touching plate in relative rotation about a common axis

We define a *spherical polar coordinate system* $S: Q \to x$ for space by the coordinate transformation

(1) $$y^1 = r \sin\theta \cos\phi, \qquad y^2 = r \sin\theta \sin\phi, \qquad y^3 = r \cos\theta,$$

where

(2) $$x^1 = \phi, \qquad x^2 = \theta, \qquad x^3 = r,$$

and y^i are rectangular Cartesian coordinates. With suitable restrictions (e.g., $0 \le \phi \le 2\pi$, $0 \le \theta \le \pi$; $r = 0$ excluded), these equations satisfy the conditions for a coordinate transformation. From 2.2(11) and 3.4(1), we find that

(3) $$[g_{rc}] = \mathrm{diag}[r^2 \sin^2\theta, r^2, 1], \qquad |g| = r^4 \sin^2\theta.$$

Let us choose a body coordinate system $B: P \to \xi$ congruent to S at some arbitrary but definite instant t_0, and consider a *torsional flow between cone and plate* defined by the equations

(4) $$\dot{\phi} = \Omega(\theta, t), \qquad \dot{\theta} = \dot{r} = 0,$$

where $\Omega(\theta, t)$ is a given function. On integration with respect to time, we find that

(5) $$\phi = \xi^1 + \int_{t_0}^{t} \Omega(\xi^2, t')\, dt', \qquad \theta = \xi^2, \quad r = \xi^3.$$

From 2.2(13), (3), and (7), we obtain the equations

(6) $$[\gamma_{rc}] = \begin{bmatrix} r^2 \sin^2\theta & r^2\phi_2 \sin^2\theta & 0 \\ r^2\phi_2 \sin^2\theta & r^2(1 + \phi_2{}^2 \sin^2\theta) & 0 \\ 0 & 0 & 1 \end{bmatrix},$$

(7) $$[\dot{\gamma}_{rc}] = \begin{bmatrix} 0 & r^2\Omega_2 \sin^2\theta & 0 \\ r^2\Omega_2 \sin^2\theta & 2r^2\phi_2\Omega_2 \sin^2\theta & 0 \\ 0 & 0 & 0 \end{bmatrix},$$

where

(8) $$\phi_2 \equiv \partial\phi/\partial\xi^2, \qquad \Omega_2 \equiv \partial\Omega(\theta, t)/\partial\theta = \partial^2\phi/\partial t\, \partial\xi^2.$$

Hence $|\gamma| = r^4 \sin^2\theta$ is independent of time, and so the flow is at constant volume. Also, $\dot{\gamma}_{11} = \dot{\gamma}_{13} = \dot{\gamma}_{33} = 0$, and so the flow is a shear flow with surfaces $\xi^2 = c$ (i.e., the cones $\theta = c$) as shearing surfaces. Further, $\dot{\gamma}_{32} = 0$, so the shear flow is unidirectional with the ξ^1 curves as lines of shear; these are circles given by the intersection of the cones $\theta = c$ and the spheres $r = c'$. From (7) and 3.2(13), the shear rate is given by the equation

(9) $$\dot{s} = \dot{s}_1 = \pm(\sin\theta)\, \partial\Omega(\theta, t)/\partial\theta \qquad (\dot{s}_3 = 0, \quad \zeta = 0).$$

3.6 Helical flow between coaxial right circular cylinders

Let us use a cylindrical polar coordinate system $S: Q \to x$ for space, as defined by 3.4(2), but with

(1) $$x^1 = \phi, \qquad x^2 = r, \qquad x^3 = z,$$

so that [compared with 3.4(3)] x^2 and x^3 have been interchanged. It follows that 3.4(4), namely,

(2) $$[g_{rc}] = \mathrm{diag}[r^2, 1, 1], \qquad |g| = r^2,$$

still applies.

Let us use a body coordinate system $B: P \to \xi$ which is congruent to S at some arbitrary but definite instant t_0, and consider the *helical flow* defined by the equations

$$(3) \qquad \dot{\phi} = \Omega(r, t), \qquad \dot{z} = v(r, t), \qquad \dot{r} = 0,$$

where $\Omega(r, t)$ and $v(r, t)$ are given functions. This evidently differs from torsional flow between parallel plates in that an axial velocity $v(r, t)$ has been added and the angular velocity now depends on r instead of on z. Integrating (3) with respect to time at constant ξ, we obtain the results

$$(4) \qquad \phi = \xi^1 + \int_{t_0}^{t} \Omega(\xi^2, t')\, dt', \qquad r = \xi^2, \qquad z = \xi^3 + \int_{t_0}^{t} v(\xi^2, t')\, dt'.$$

Using these results with 2.2(13), we obtain the following equations for the body metric tensor components and the strain-rate tensor components:

$$(5) \qquad [\gamma_{rc}] = \begin{bmatrix} r^2 & r^2\phi_2 & 0 \\ r^2\phi_2 & 1 + (r\phi_2)^2 + z_2{}^2 & z_2 \\ 0 & z_2 & 1 \end{bmatrix}, \qquad |\gamma| = r^2,$$

$$(6) \qquad [\dot{\gamma}_{rc}] = \begin{bmatrix} 0 & r^2\Omega_2 & 0 \\ r^2\Omega_2 & 2r^2\phi_2\Omega_2 + 2z_2 v_2 & v_2 \\ 0 & v_2 & 0 \end{bmatrix},$$

where

$$(7) \qquad \phi_2 = \partial\phi/\partial\xi^2, \qquad z_2 = \partial z/\partial\xi^2, \qquad \Omega_2 = \partial\Omega(r, t)/\partial r, \qquad v_2 = \partial v(r, t)/\partial r.$$

Thus $|\gamma|$ is independent of time, so the flow is at constant volume. Also, $\dot{\gamma}_{11} = \dot{\gamma}_{13} = \dot{\gamma}_{33} = 0$, so the flow is a shear flow with the surfaces $\xi^2 = c$ (i.e., the cylinders $r = c$) as shearing surfaces. From 3.2(15), the angle ζ which the lines of shear make with the ξ^1 curves is given by

$$(8) \qquad \tan\zeta = v_2/r\Omega_2,$$

which, in general, varies with time, and so *helical flow is not necessarily unidirectional.* From 3.2(15), the differential equations of the lines of shear (on the shearing surfaces $\xi^2 = c$) become

$$(9) \qquad \frac{d\xi^3}{d\xi^1} = \frac{v_2(\xi^2, t)}{\Omega_2(\xi^2, t)},$$

which evidently define circular helices inclined at a time-dependent angle $\frac{1}{2}\pi - \zeta$ to the cylinder generators (ξ^3 lines).

From 3.3(13)(i) with (5) and (6), we find that

$$(10) \qquad \dot{s}_1 = r\Omega_2, \qquad \dot{s}_3 = v_2, \qquad \dot{s}^2 = (r\Omega_2)^2 + v_2{}^2.$$

From (8), it follows that *helical flow is unidirectional if and only if v_2/Ω_2 is independent of time.* Important particular cases of this are the following:

(11) **Telescopic flow** through an annulus or a circular tube is a helical flow with $\Omega = 0$; the lines of shear are the straight lines along which z alone varies; the shear rate is $\dot{s} = \dot{s}_3 = \pm\partial v(r, t)/\partial r$.

(12) **Couette flow** between coaxial circular cylinders in relative rotation is a helical flow with $v = 0$; the lines of shear are the circles round which ϕ alone varies; the shear rate is $\dot{s} = \dot{s}_1 = \pm r\,\partial\Omega(r, t)/\partial r$.

The application of the superposition theorem 3.3(25) with (11) and (12) shows that *helical flow can be regarded as an orthogonal superposition of telescopic flow and Couette flow*; moreover, the superposition is global because telescopic flow and Couette flow are both kinematically possible, i.e., the separate covariant strain tensors are derivable from velocity fields, and we have shown that the superposed flow is also derivable from a velocity field.

3.7 Orthogonal rheometer flow

(1) We now consider a flow in which each circular disk of liquid rotates rigidly about an axis normal to its plane through its center [Fig. 3.7(1)] at an angular velocity Ω; if the axes of rotation were coaxial, the motion would be a rigid rotation; but the axes are displaced by an amount proportional to the separation of the circular disks, and are parallel. It is convenient to use two space coordinate systems C and S, and one body coordinate system B.

$C: \mathrm{Q} \to y$ is a rectangular Cartesian space coordinate system with the y^2 axis parallel to the rotation axes, and the y^1 plane containing the line of centers O ··· O′ [Fig. 3.7(1)]. The line of centers makes an angle $\tan^{-1} a$, say, with the y^2 axis.

$S: \mathrm{Q} \to x$ is a space coordinate system defined by the transformation

(2) $$y^1 = r \sin\phi, \qquad y^2 = x^2, \qquad y^3 = r\cos\phi + ax^2,$$

where

(3) $$x^1 = r, \qquad x^2 = y^2, \qquad x^3 = \phi.$$

The x^2 lines are thus parallel to the axes of rotation, and in each x^2 plane, the coordinates (x^1, x^3) are plane polar coordinates with the center of rotation of that plane as pole ($x^1 = 0$). It is easy to obtain the following results for the space metric tensor components in S:

(4) $$[g_{rc}] = \begin{bmatrix} 1 & a\cos\phi & 0 \\ a\cos\phi & 1 + a^2 & -ar\sin\phi \\ 0 & -ar\sin\phi & r^2 \end{bmatrix}, \qquad |g| = r^2.$$

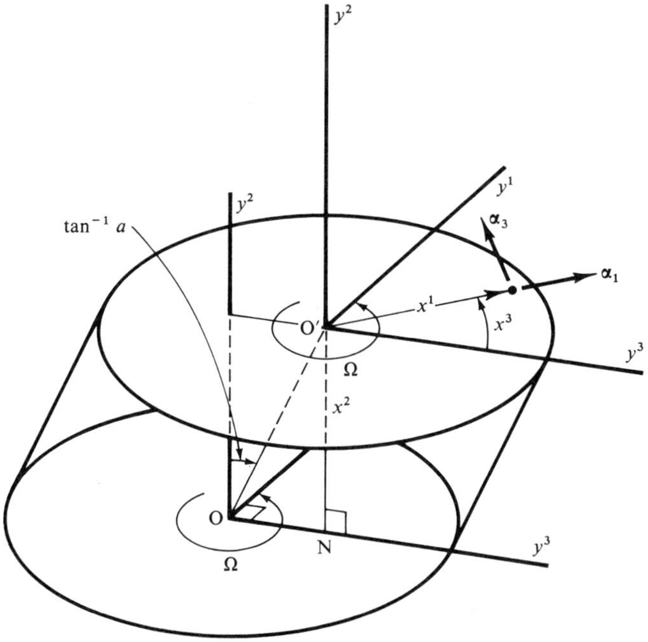

Figure 3.7(1) Orthogonal rheometer flow: Parallel circular disks rotate rigidly at the same angular velocity Ω about parallel axes Oy^2 and $O'y^2$. Here, $Oy^1y^2y^3$ and $O'y^1y^2y^3$ are rectangular space coordinate systems differing only in origin. $S: Q \to x$ is a nonorthogonal space coordinate system having $x^2 = O'N$ and (x^1, x^3) as plane polar coordinates (r, ϕ) in a plane $O'y^3y^1$.

When $a \neq 0$, it is evident that S is not orthogonal; when $a = 0$, S is a cylindrical polar coordinate system.

Let us choose $B: P \to \xi$ congruent to S at $t = 0$ (if instead we took $B = S$ at t_0, then we would simply have to replace t in the following equations by $t - t_0$). We define *orthogonal rheometer flow* by the equations

$$\text{(5)} \qquad \dot{r} = 0, \qquad \dot{x}^2 = 0, \qquad \dot{\phi} = \Omega,$$

where Ω is a constant, independent of position and time. Integrating these equations with respect to time, we see that

$$\text{(6)} \qquad r = \xi^1, \qquad x^2 = \xi^2, \qquad \phi = \xi^3 + \Omega t.$$

Using 2.2(13), with (4) and (6), we see that (since $\partial x^i/\partial \xi^j$ is independent of t)

$$\text{(7)} \qquad \gamma_{rc} = g_{rc}, \qquad [\dot{\gamma}_{rc}] = \begin{bmatrix} 0 & -a\Omega \sin\phi & 0 \\ -a\Omega \sin\phi & 0 & -ar\Omega\cos\phi \\ 0 & -ar\Omega\cos\phi & 0 \end{bmatrix}.$$

Hence, since $|\gamma|$, γ_{11}, γ_{13}, and γ_{33} are independent of time, the flow is a shear flow with the planes $\xi^2 = c$ (i.e., $y^2 = c$) as shearing surfaces. Since $\gamma_{13} = 0$, the ξ^1 curves and the ξ^3 curves are orthogonal.

From 3.2(15), it follows that

$$\text{(8)} \qquad \tan\zeta = \cot\phi, \qquad \zeta = \tfrac{1}{2}\pi - \phi = \tfrac{1}{2}\pi - \xi^3 - \Omega t.$$

Thus $\zeta(\mathrm{P}, t)$, *the angle between a shear line and a* ξ^1 *curve, varies with time, and therefore the flow is not undirectional.* From 3.2(15), the differential equation for the lines of shear at time t takes the form

$$\text{(9)} \qquad \xi^1\, d\xi^3/d\xi^1 = \cot(\xi^3 + \Omega t) \qquad (\xi^2 = c),$$

whose solution is

$$\text{(10)} \qquad \xi^1 \cos(\xi^3 + \Omega t) = A \qquad (\xi^2 = c),$$

where A is a constant of integration. Since $B \equiv S$ at $t = 0$ and the surfaces $\xi^2 = c$ move rigidly, it follows that (ξ^1, ξ^3) *are plane polar coordinates in each shearing surface at all times.* Equation (10) therefore describes a family of parallel straight lines (with A varying from one line to another) inclined at an angle $\frac{1}{2}\pi - \Omega t$ to the material line $\xi^3 = 0$. *These lines are in fact always parallel to the* y^1 *axis* (which is fixed in space).

From 3.3(13), (4), and (7), we see that

$$\text{(11)} \qquad \dot{s}_1 = -a\Omega \sin(\xi^3 + \Omega t), \qquad \dot{s}_3 = -a\Omega \cos(\xi^3 + \Omega t),$$

and hence

$$\text{(12)} \qquad \dot{s} = \pm a\Omega.$$

It follows that *orthogonal rheometer flow is a shear flow at constant and uniform shear rate* (*equal to* $\pm a\Omega$) *with straight lines of shear which rotate* (*relative to the material*) *at constant angular velocity* Ω. At any particle P, the covariant strain tensor $\boldsymbol{\gamma}' - \boldsymbol{\gamma}$ and the strain-rate tensor $\dot{\boldsymbol{\gamma}}$ are equal to the sum of similar tensors which, from (11), would arise from oscillatory unidirectional shears of angular frequency Ω and phase (ξ^3) which varies from one particle to another. It is not known, however, whether these individual tensors can be obtained from velocity fields, i.e., whether the oscillatory shears are *globally* superposable.

Commercial instruments are available which involve orthogonal rheometer flow, at least as an approximation. An orthogonal rheometer was developed by Maxwell and Chartoff (1965); a mechanical spectrometer was developed by Macosko and Starita (1971).

3.8 Balance rheometer flow

(1) We now consider a state of flow in which infinitesimally thin, concentric, material spherical shells rotate rigidly at the same constant angular velocity about axes, fixed in space, which pass through a common center [Fig. 3.8(1)]. The axes lie in a fixed vertical plane; the axis for a shell of radius r makes an angle $\varepsilon(r)$ with the vertical axis, which is the axis of rotation for the largest shell. Flows of this type have been considered by Kaelble (1969), and by Walters (1970b) in connection with a commercial apparatus, the balance rheometer, designed by Képès and marketed by Contraves and Company,

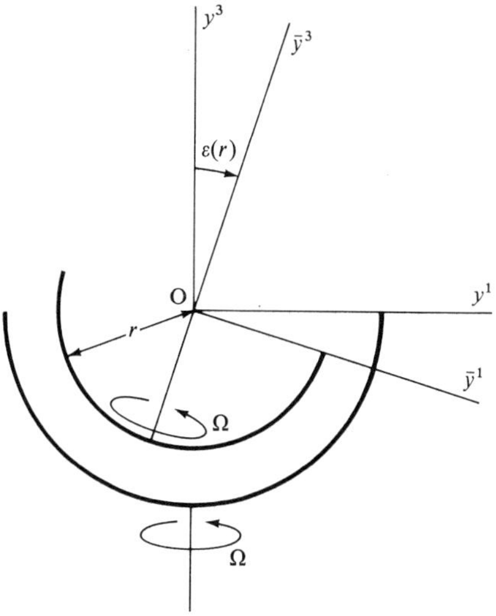

Figure 3.8(1) Balance rheometer flow: Concentric spheres (or hemispheres) rotate rigidly at the same angular velocity Ω about axes Oy^3 and $O\bar{y}^3$ inclined at an angle $\varepsilon(r)$. $Oy^1y^2y^3$ and $O\bar{y}^1\bar{y}^2\bar{y}^3$ are axes of rectangular space coordinate systems having Oy^2 and $O\bar{y}^2$ coincident. A nonorthogonal space coordinate system $S: Q \to x$ is used in which (x^1, x^2, x^3) [$\equiv (\theta, r, \phi)$] are related to $\bar{y}^i$ by a spherical polar coordinate transformation.

Switzerland; this apparatus has the material contained between two concentric hemispheres which rotate; components of the couple on one hemisphere are measured.

It is convenient to use three coordinate systems: a rectangular Cartesian space coordinate system $C: Q \to y$, having its origin O at the common center of the spherical shells, its axis Oy^3 vertically upward, and its Oy^1y^3 plane containing the other axes of rotation; a nonorthogonal space coordinate system $S: Q \to x$, to be defined; and a body coordinate system $B: P \to \xi$ congruent to S at an arbitrary time $t = 0$, say.

To define S, we consider a shell of radius r and axis of rotation $O\bar{y}^3$, say; we use (temporarily) a rectangular Cartesian coordinate system $\bar{C}:Q \rightarrow \bar{y}$ for this shell, such that $O\bar{y}^3$ is its axis of rotation and axes $O\bar{y}^2$ and Oy^2 coincide. Thus the axes $O\bar{y}^1\bar{y}^2\bar{y}^3$ are obtained from axes $Oy^1y^2y^3$ by rotation through an angle $\varepsilon(r)$ about Oy^2 [Fig. 3.8(1)], and the coordinate transformation $y \rightarrow \bar{y}$ can be expressed in complex variable notation in the form

$$(2) \qquad y^1 + iy^3 = (\bar{y}^1 + i\bar{y}^3)e^{-i\varepsilon(r)}, \qquad y^2 = \bar{y}^2 \qquad (i \equiv \sqrt{-1}).$$

The coordinates $(x^1, x^2, x^3) \equiv (\theta, r, \phi)$ are now defined by the spherical polar coordinate transformation from $\bar{y}^i$, namely

$$(3) \qquad \bar{y}^1 + i\bar{y}^3 = r(\sin\theta\cos\phi + i\cos\theta), \qquad \bar{y}^2 = r\sin\theta\sin\phi.$$

Combining (2) and (3), we obtain the required transformation $x \rightarrow y$:

$$(4) \qquad y^1 + iy^3 = (\sin\theta\cos\phi + i\sin\theta)re^{-i\varepsilon(r)}, \qquad y^2 = r\sin\theta\sin\phi.$$

From this equation and the equations

$$(dy^2)^2 + |dy^1 + i\,dy^3|^2 = g_{ij}(x)\,dx^i\,dx^j,$$

after some straightforward reduction, we obtain the result

$$(5) \quad [g_{rc}] = \begin{bmatrix} r^2 & r^2\varepsilon'\cos\phi & 0 \\ r^2\varepsilon'\cos\phi & 1 + (r\varepsilon')^2(\cos^2\theta + \cos^2\phi\sin^2\theta) & -r^2\varepsilon'\cos\theta\sin\theta\sin\phi \\ 0 & -r^2\varepsilon'\cos\theta\sin\theta\sin\phi & r^2\sin^2\theta \end{bmatrix},$$

where $\varepsilon' \equiv d\varepsilon(r)/dr$, and hence

$$(6) \qquad |g| = r^4\sin^2\theta.$$

These results are exact, without restriction on ε.

The advantage of using the coordinate system S is that it enables us to write the equations for the assumed state of flow in the simple form $\dot{r} = \dot{\theta} = 0$, $\dot{\phi} = \Omega$. Since $B \overset{t=0}{\equiv} S$, these integrate to give

$$(7) \qquad x^1 \equiv \theta = \xi^1; \qquad x^2 \equiv r = \xi^2; \qquad x^3 \equiv \phi = \xi^3 + \Omega t \qquad (\text{all } t).$$

Because Ω is constant, Eqs. (7) give $\partial x^i/\partial\xi^j = \delta^i_j$, and hence, from 2.2(13), we have

$$(8) \qquad \gamma_{ij}(\xi, t) = g_{ij}(x), \qquad |\gamma| = r^4\sin^2\theta \qquad [x^i \text{ given by (7)}].$$

Differentiating with respect to time, using (5) and (7), we have

$$(9) \quad [\dot{\gamma}_{rc}(\xi, t)] = \begin{bmatrix} 0 & \sin\phi & 0 \\ \sin\phi & O(\varepsilon') & \sin\theta\cos\theta\cos\phi \\ 0 & \sin\theta\cos\theta\cos\phi & 0 \end{bmatrix}(-\Omega r^2\varepsilon'),$$

where (7) holds.

Using (7)–(9), together with the definitions 3.2(13), 3.3(13), and 3.2(15), for shear rate and shear lines, we obtain the following results to $O(\varepsilon')$:

(10) $$\dot{s} = \pm\Omega r\varepsilon' (\sin \xi^1)[1 - (\sin^2 \xi^1) \sin^2(\xi^3 + \Omega t)]^{1/2};$$

(11) $$\dot{s}_1 = \Omega r\varepsilon'(\sin \xi^1) \sin(\xi^3 + \Omega t); \qquad \dot{s}_3 = \Omega r\varepsilon'(\cos \xi^1) \cos(\xi^3 + \Omega t);$$

(12) $$\tan \zeta = (\cos \xi^1) \cot(\xi^3 + \Omega t) \qquad \text{(shear angle } \zeta);$$

(13) $$d\xi^3/d\xi^1 = (\cot \xi^1) \cot(\xi^3 + \Omega t) \qquad \text{(shear lines at } t).$$

The differential equations (13) for the shear lines at time t integrate to give

(14) $$\cos(\xi^3 + \Omega t) \sin \xi^1 = c \quad \text{(const)}; \qquad \xi^2 = \text{const} \qquad \text{(shear lines)}.$$

These are circles (on the shell of radius $r = \xi^2$) in a plane

(15) $$\bar{y}^1 = cr, \qquad r = \text{const}.$$

We can summarize the kinematic properties of balance rheometer flow, given by the preceding equations, as follows. The shear rate and shear angle are time dependent, so the flow is "rheologically unsteady" and differs from orthogonal rheometer flow, for which the shear rate is constant. The shear lines (at any instant) are circles. While the shear rate $\dot{s}$ has a complicated dependence on time, the components $\dot{s}_1$ and $\dot{s}_3$ vary sinusoidally with time. These results are valid for small values of $d\varepsilon/dr$.

3.9 Shear-free flow

(1) **Definition** A flow is *shear free* if there exists a body coordinate system $B: \mathrm{P} \to \xi$ that is always orthogonal.

Uniform elongation of a cylinder is a shear-free flow; B can be taken congruent at any instant to a cylindrical polar coordinate system with axis $r = 0$ along the cylinder axis. Uniform inflation of a spherical shell is a shear-free flow; B can be taken congruent at any instant to a spherical polar coordinate system with pole $r = 0$ at the center of the spherical shell. A body coordinate system can always be chosen to be orthogonal at any one instant, whatever the flow, but in general it will not be orthogonal at other instants; it is essential to the definition of shear-free flow that B shall be orthogonal throughout a nonzero time interval. It should be noted that there is no point in attempting to use a similar method in order to define a shear-free *strain* (relating two, and only two, states t, t'), because, through any particle P, there are three material surfaces that are locally orthogonal in both states, by 2.8(22), for every strain $t \to t'$.

From 2.7(9) and (1), it follows that

(2) $$\gamma^{ij}(\xi, t) = 0 \qquad (i \neq j, \quad \text{all } t),$$

and hence that

(3) $$\gamma_{ij}(\xi, t) = \dot{\gamma}_{ij}(\xi, t) = 0 \qquad (i \neq j, \quad \text{all } t).$$

It follows from (3) and 2.8(28) that

(4) $$\dot{\gamma}_{\hat{\imath}\hat{\imath}} = \dot{\kappa}_{\hat{\imath}} \gamma_{\hat{\imath}\hat{\imath}},$$

where the $\dot{\kappa}_i$ denote the principal strain rates in the shear-free flow. By integrating (4) with respect to time at constant P, we find that

(5) $$\gamma'_{\hat{\imath}\hat{\imath}} = e_{\hat{\imath}}^2 \gamma_{\hat{\imath}\hat{\imath}}, \qquad \text{where} \quad e_i^2 = \exp \int_t^{t'} \dot{\kappa}_i'' \, dt''.$$

It follows from (5) and 2.8(21) that $e_i \equiv e_i(\text{P}, t, t')$ are the principal elongation ratios for the strain $t' \to t$.

Since B is orthogonal, it follows that the covariant base vectors $\boldsymbol{\beta}^i(B, \text{P})$ are orthogonal and that the contravariant base vectors $\boldsymbol{\beta}_i(B, \text{P})$ are orthogonal. It is convenient to introduce the corresponding unit vectors (which have unit length at time t):

(6) $$\boldsymbol{\alpha}^i = (\gamma^{\hat{\imath}\hat{\imath}})^{-1/2} \boldsymbol{\beta}^{\hat{\imath}} = (\gamma_{\hat{\imath}\hat{\imath}})^{1/2} \boldsymbol{\beta}^{\hat{\imath}},$$

and

(7) $$\boldsymbol{\alpha}_i = (\gamma_{\hat{\imath}\hat{\imath}})^{-1/2} \boldsymbol{\beta}_{\hat{\imath}}:$$

we have used 2.7(4)(i) and the equation $\gamma^{\hat{\imath}\hat{\imath}} = (\gamma_{\hat{\imath}\hat{\imath}})^{-1}$, which follows from the fact that B is orthogonal. Differentiating (6) and using the fact that the $\boldsymbol{\beta}^i$ are independent of time, it follows from (4) that

(8) $$\dot{\boldsymbol{\alpha}}^i = \tfrac{1}{2} \dot{\kappa}_{\hat{\imath}} \boldsymbol{\alpha}^{\hat{\imath}}.$$

Since $\boldsymbol{\alpha}^i$ are orthonormal at time t, it follows from 2.7(16)(iii) that

(9) $$\boldsymbol{\gamma} = \boldsymbol{\alpha}^i \boldsymbol{\alpha}^i,$$

and on differentiating this equation with respect to t at constant P, using (8) and 2.7(16), we obtain the equations

(10) $$\dot{\boldsymbol{\gamma}} = \dot{\kappa}_i \boldsymbol{\alpha}^i \boldsymbol{\alpha}^i,$$

(11) $$\dot{\boldsymbol{\gamma}} \cdot \boldsymbol{\gamma}^{-1} \cdot \dot{\boldsymbol{\gamma}} = \dot{\kappa}_i{}^2 \boldsymbol{\alpha}^i \boldsymbol{\alpha}^i.$$

From (5), (6), and 2.6(16), we obtain the equation

(12) $$\boldsymbol{\gamma}' = e_i^2 \boldsymbol{\alpha}^i \boldsymbol{\alpha}^i \qquad [e_i = e_i(\text{P}, t, t'); \quad \boldsymbol{\alpha}^i \text{ orthonormal at } t].$$

For later use in obtaining reduced constitutive equations (Section 7.6), we solve (9)–(11) to express $\boldsymbol{\alpha}^1 \boldsymbol{\alpha}^1$, etc., in terms of $\boldsymbol{\gamma}$ and $\dot{\boldsymbol{\gamma}}$. The results are as follows.

Case (i) $\dot{\kappa}_1, \dot{\kappa}_2, \dot{\kappa}_3$ *all different*

(13) $\boldsymbol{\alpha}^1\boldsymbol{\alpha}^1(\dot{\kappa}_1 - \dot{\kappa}_2)(\dot{\kappa}_1 - \dot{\kappa}_3) = \dot{\kappa}_2\dot{\kappa}_3\boldsymbol{\gamma} - (\dot{\kappa}_2 + \dot{\kappa}_3)\dot{\boldsymbol{\gamma}} + \dot{\boldsymbol{\gamma}}\cdot\boldsymbol{\gamma}^{-1}\cdot\dot{\boldsymbol{\gamma}},$ etc.

Case (ii) $\dot{\kappa}_2 = \dot{\kappa}_3 \neq \dot{\kappa}_1$

(14) $\boldsymbol{\alpha}^1\boldsymbol{\alpha}^1(\dot{\kappa}_2 - \dot{\kappa}_1) = \dot{\kappa}_2\boldsymbol{\gamma} - \dot{\boldsymbol{\gamma}};\qquad (\boldsymbol{\alpha}^2\boldsymbol{\alpha}^2 + \boldsymbol{\alpha}^3\boldsymbol{\alpha}^3)(\dot{\kappa}_2 - \dot{\kappa}_1) = \dot{\boldsymbol{\gamma}} - \dot{\kappa}_1\boldsymbol{\gamma};$

(15) $$\dot{\boldsymbol{\gamma}}\cdot\boldsymbol{\gamma}^{-1}\cdot\dot{\boldsymbol{\gamma}} = (\dot{\kappa}_1 + \dot{\kappa}_2)\dot{\boldsymbol{\gamma}} - \dot{\kappa}_1\dot{\kappa}_2\boldsymbol{\gamma}.$$

Case (iii) $\dot{\kappa}_1 = \dot{\kappa}_2 = \dot{\kappa}_3$

(16) $$\dot{\kappa}_1(\boldsymbol{\alpha}^1\boldsymbol{\alpha}^1 + \boldsymbol{\alpha}^2\boldsymbol{\alpha}^2 + \boldsymbol{\alpha}^3\boldsymbol{\alpha}^3) = \dot{\kappa}_1\boldsymbol{\gamma} = \dot{\boldsymbol{\gamma}}.$$

Equations (8) and (10)–(12) (for shear-free flow) may be contrasted with the corresponding equations 3.3(14), 3.3(26), 3.3(32)(iii), and 3.3(20) (for shear flow). For flow at constant volume, it follows from 2.8(28) and 5.5(4) that

(17) $$\dot{\kappa}_1 + \dot{\kappa}_2 + \dot{\kappa}_3 = 0 \qquad \text{(constant volume)}.$$

We therefore have the following important subclass of Case (ii):

Case (iib) $\dot{\kappa}_2 = \dot{\kappa}_3 = -\dot{\kappa}_1/2$ *(constant-volume elongation)*

(18) $$\boldsymbol{\alpha}^1\boldsymbol{\alpha}^1 = \tfrac{1}{3}\boldsymbol{\gamma} - (2/3\dot{\kappa}_1)\dot{\boldsymbol{\gamma}};\qquad \boldsymbol{\alpha}^2\boldsymbol{\alpha}^2 + \boldsymbol{\alpha}^3\boldsymbol{\alpha}^3 = \tfrac{2}{3}\boldsymbol{\gamma} + (2/3\dot{\kappa}_1)\dot{\boldsymbol{\gamma}}.$$

(19) **Problem** Show that the only shear-free shear flow is a rigid motion.

4 CARTESIAN VECTOR AND TENSOR FIELDS

4.1 Rectangular Cartesian coordinate systems

From the definition 2.7(15), a space coordinate system $C: \mathrm{Q} \rightarrow y$ is *rectangular Cartesian* if the separation ds for an *arbitrary* pair Q, Q′ of neighboring places is given by the equation

$$(ds)^2 = \delta_{ij}\, dy^i\, dy^j, \tag{1}$$

where the dy^i are the coordinate differences for Q and Q′ in C. It is assumed that such coordinate systems exist for the space manifold, which is then said to be *Euclidean*. The essential feature of the definition is that an equation of the form (1) shall hold for *all* pairs of neighboring places, so that the metric tensor components δ_{ij} in C are independent of position.

In defining contravariant and covariant vectors and tensors in Chapter 2, we used the set $\mathscr{S} = \{S\}$ of all space coordinate systems. In this chapter, *we shall define Cartesian vectors and tensors by using the subset $\mathscr{C} = \{C\}$ of all rectangular Cartesian space coordinate systems.* The methods of definition are otherwise similar. Although $\mathscr{C}$ is a subset of $\mathscr{S}$, we shall see that *Cartesian* *vectors and tensors are not subsets of general vectors and tensors.* We consider space fields only, because the space metric tensor field $\boldsymbol{g}(\mathrm{Q})$ is independent of time, and so a space coordinate system which is rectangular Cartesian at time t will always be rectangular Cartesian. The body metric tensor field $\gamma(\mathrm{P}, t)$, on the other hand, varies with t (except in the case of a rigid motion, which does not concern us in rheological applications), and so a body coordinate system which is rectangular Cartesian at time t will not, in general, be rectangular Cartesian at other times. *Cartesian body tensor fields cannot, therefore, be defined throughout any nonzero time interval during which the body deforms.*

(2) **Theorem** For every pair C, $\bar{C}$ of rectangular Cartesian coordinate systems, the transformation matrix $L \equiv L_{\mathrm{Q}}(C, \bar{C}) \equiv [\partial \bar{y}^r/\partial y^c]$ is constant and orthogonal, and the coordinate transformation is of the form

$$\bar{y} = Ly + a \qquad (\text{i.e., } \bar{y}^i = L^i{}_j y^j + a^i), \tag{3}$$

where the matrices L and $a = [a^r]$ are independent of y and $\bar{y}$.
Proof Since C and $\bar{C}$ are rectangular Cartesian, it follows from (1) that

$$\widetilde{dy}\, dy = \widetilde{d\bar{y}}\, d\bar{y} = \widetilde{dy}\, \tilde{L}L\, dy \qquad (\text{since} \quad d\bar{y} = L\, dy). \tag{4}$$

Since this equation holds for all infinitesimal matrices dy, and $\tilde{L}L$ is symmetric, it follows that

$$\tilde{L}L = I \qquad (\text{i.e., } L^i{}_k L^i{}_m = \delta_{km}), \tag{5}$$

proving that L is orthogonal, as stated.

To prove that L is constant, we differentiate (5), and obtain the equations

$$L^i{}_k \bar{y}^i{}_{,mj} + L^i{}_m \bar{y}^i{}_{,kj} = 0 \qquad (j, k, m = 1, 2, 3), \tag{6}$$

where $\bar{y}^i{}_{,mj} \equiv \partial^2 \bar{y}^i/\partial y^m\, \partial y^j$. This is a set of simultaneous equations, linear and homogeneous in the second derivatives $\bar{y}^i{}_{,mj}$; we make use of the fact that the coefficients $L^i{}_k$ are elements of an othogonal (and therefore nonsingular) matrix to show that the only solution of (6) is that in which all second derivatives $\bar{y}^i{}_{,mj}$ are zero; it will then follow that L is independent of y, and that (3) follows on integrating the equation $d\bar{y} = L\, dy$, which will complete the proof of the theorem.

Let us first select from (6) the nine equations obtained by putting (mkj) equal to (112), (113), (221), (223), (331), (332), (123), (132), and (231) in turn. These nine equations are linear and homogeneous in the second derivatives $\bar{y}^i{}_{,mj}$, of which at most nine, namely

$$\bar{y}^i{}_{,23}, \quad \bar{y}^i{}_{,31}, \qquad \text{and} \qquad \bar{y}^i{}_{,12} \qquad (i = 1, 2, 3), \tag{7}$$

are independent, because $\bar{y}^i{}_{,mj} = \bar{y}^i{}_{,jm}$. A necessary and sufficient condition that a solution exist in which not all the derivatives (7) are zero is that the determinant D (a 9 × 9 determinant whose elements are either zero or $L^i{}_j$) of the coefficients shall vanish. It is a straightforward matter to show that $D = \pm |L|^3$, by making certain obvious interchanges of rows in D followed by a Laplace expansion by third minors. Since L is orthogonal, $|L| = \pm 1$; this follows on taking the determinant of (5). Hence $D \neq 0$, and therefore all second derivatives in the set (7) vanish. Finally, on taking (mkj) equal to (121), (131), (111), (232), (212), (222), (313), (323), and (333) in turn and following a similar argument, it is easy to show that the remaining second derivatives, namely

$$(8) \qquad \bar{y}^i{}_{,11}, \quad \bar{y}^i{}_{,22}, \quad \text{and} \quad \bar{y}^i{}_{,33} \qquad (i = 1, 2, 3),$$

all vanish. This completes the proof of the theorem.

An alternative proof (valid for an n-dimensional space) can be given by making use of a theorem which states that the general transformation of a quadratic differential form in n variables into another such form contains at most $n(n + 1)/2$ arbitrary constants [cf., e.g., Eisenhart (1926, p. 24)]. Because of the fundamental importance of Theorem (2) in our formalism, we have considered it worthwhile to offer the preceding elementary proof for the case $n = 3$.

4.2 The definition of Cartesian vector fields

It follows from Theorem 4.1(2) that the set $\mathscr{C}$ of all rectangular Cartesian coordinate systems is an infinite set, in which all members are obtainable from any given member C by allowing the square matrix L and the column matrix a in 4.1(3) to assume all possible real values [consistent with the orthogonality condition 4.1(5)]. Since $\mathscr{C}$ is a subset of $\mathscr{S}$, it follows from 2.3(32) that the rectangular Cartesian transformation matrices also possess the transitive propeıty

$$(1) \qquad L^{-1}(C, \bar{C})L^{-1}(\bar{C}, \bar{\bar{C}}) = L^{-1}(C, \bar{\bar{C}}) \qquad \text{for all} \quad C, \bar{C}, \bar{\bar{C}} \quad \text{in} \quad \mathscr{C}.$$

We have omitted the subscript Q here because Theorem 4.1(2) shows that these L matrices are independent of place. The analogous equation 2.3(32) played a fundamental role in our method of defining contravariant vectors in 2.3(29) and in proving their independence of the choice of coordinate system. We can therefore use a similar method of definition, with $\mathscr{S}$ replaced by the subset $\mathscr{C}$, to define a new kind of vector, which we call a Cartesian vector.

(2) **Definition** Let $\bar{v}$ be any 3×1 column matrix associated with any given rectangular Cartesian space coordinate system $\bar{C}$. A *Cartesian space vector* $\mathbf{v}$ is a function $\mathbf{v}: \mathscr{C} \to A_{3\times 1}$ which associates with any rectangular Cartesian coordinate system C a 3×1 column matrix v, where

(3) $$\mathbf{v}C = v = L^{-1}(C, \bar{C})\bar{v}.$$

It is easy to verify that the vector $\mathbf{v}$ thus defined is independent of the choice of $\bar{C}$ in the usual sense, and that, by allowing the representative matrix $\bar{v}$ to assume all possible real values, one obtains an infinite set $\{\mathbf{v}\}$ of Cartesian vectors which form a linear space in which operations of addition and multiplication by numbers are defined by the equation

(4) $$(a_1\mathbf{v}_1 + a_2\mathbf{v}_2)C = a_1 v_1 + a_2 v_2 = L^{-1}(C, \bar{C})(a_1\bar{v}_1 + a_2\bar{v}_2),$$

which is similar in form to the corresponding equation 2.3(37) for contravariant vectors.

Cartesian vectors differ from general (i.e., contravariant and covariant) *vectors in three important ways.* First, because $L(C, \bar{C})$ is independent of Q, *Cartesian vectors associated with different places may be added and are elements of the same linear space*; general vectors, on the other hand, which are associated with different places, are elements in disjoint linear spaces and addition is not defined for them. Associated with a given manifold, there is one and only one set of Cartesian vectors, but there is an infinite number of disjoint general vector spaces—one covariant and one contravariant space for each point of the manifold.

Second, because $L(C, \bar{C})$ is orthogonal, *there is no distinction for Cartesian vectors which is comparable to the distinction between contravariant and covariant* (*general*) *vectors*; when the coordinate systems used in the definitions of vectors are restricted to be rectangular Cartesian, the equations corresponding to 2.3(30) (for contravariant vectors) and 2.4(19) (for covariant vectors) become identical.

Third, *any given Cartesian vector necessarily has a unique magnitude*; a given general vector, on the other hand, can have a magnitude which varies with time if the metric tensor varies with time. This property of a Cartesian vector accords with the property inherent in the familiar definition of a vector (in elementary vector analysis) as an entity which has a unique magnitude and a unique direction. The magnitude $\|\mathbf{v}\|$ of a Cartesian vector $\mathbf{v}$ is defined by the equation

(5) $$\|\mathbf{v}\|^2 = \tilde{v}v = \tilde{\bar{v}}\bar{v} = \cdots,$$

and is evidently a unique (positive) number independent of the choice of coordinate system, because $\tilde{\bar{v}}\bar{v} = \tilde{v}\tilde{L}Lv = \tilde{v}Iv = \tilde{v}v$, since L is orthogonal.

Similarly, the scalar product $\mathbf{u} \cdot \mathbf{v}$ of any two Cartesian vectors may be defined by the equation

$$(6) \qquad \mathbf{u} \cdot \mathbf{v} = \tilde{u}v = \tilde{\bar{u}}\bar{v} = \cdots,$$

and is a unique number, independent of the choice of coordinate system.

Further, if Q and Q′ are any two places (neighboring or nonneighboring), we can define a *Cartesian displacement vector* QQ′ by the equation

$$(7) \qquad \overrightarrow{\mathrm{QQ'}}C = y' - y = L^{-1}(C, \bar{C})(\bar{y}' - \bar{y}) \qquad \text{[by 4.1(3)]},$$

where y' and y are respectively the rectangular Cartesian *coordinate* matrices for Q and Q′ in C. It is easy to verify, from this definition, that displacement vectors satisfy the triangle law of addition of elementary vector analysis. *We may therefore identify our Cartesian vectors with the vectors of elementary vector analysis.*

Because Cartesian vectors have extra properties (and operations) not possessed (or defined) for any general vector, it follows that *Cartesian vectors are not a subset of any set of general vectors*, and that, associated with any given Euclidean manifold, *three different kinds of vector field* (*Cartesian, covariant and contravariant*) *can be defined.*

4.3 The definition of Cartesian second-rank tensor fields

(1) **Definition** A *second-rank Cartesian tensor* $\mathbf{p}$ is a function $\mathbf{p}: \mathscr{C} \to A_{3 \times 3}$ defined by the equation

$$(2) \qquad \mathbf{p}C = \mathsf{p} = L^{-1}(C, \bar{C})\bar{\mathsf{p}}L(C, \bar{C}) \qquad (\text{for all } C \text{ in } \mathscr{C}),$$

where $\mathsf{p} = [\mathsf{p}^{rc}]$ is an arbitrary square matrix associated with any given rectangular Cartesian coordinate system C, and $\bar{\mathsf{p}} = [\bar{\mathsf{p}}^{rc}]$ is the corresponding matrix associated with $\bar{C}$.

It is easy to verify that second-rank Cartesian tensors have the following properties (in common with general second-rank tensors): $\mathbf{p}$ is independent of the choice of $(\bar{\mathsf{p}}, \bar{C})$ in the usual sense; the set $\{\mathbf{p}\}$ obtained by letting the matrix $\bar{\mathsf{p}}$ assume all possible real values is a linear space in which addition and multiplication by numbers are defined by the equation

$$(3) \qquad (a_1\mathbf{p}_1 + a_2\mathbf{p}_2)C = a_1\mathsf{p}_1 + a_2\mathsf{p}_2 = L^{-1}(C, \bar{C})(a_1\bar{\mathsf{p}}_1 + a_2\bar{\mathsf{p}}_2)L(C, \bar{C});$$

the Cartesian unit tensor $\mathbf{U}$ has the same matrix representative $I = [\delta_{rc}]$ in every rectangular Cartesian coordinate system; the outer product $\mathbf{uv}$ of two Cartesian vectors $\mathbf{u}$ and $\mathbf{v}$ is a second-rank Cartesian tensor defined by the equation

$$(4) \qquad \mathbf{uv}C = \mathsf{u}\tilde{\mathsf{v}} = [\mathsf{u}^r\mathsf{v}^c] = L^{-1}(C, \bar{C})\bar{\mathsf{u}}\tilde{\bar{\mathsf{v}}}L(C, \bar{C}) \qquad (\text{for all } C \text{ in } \mathscr{C});$$

the inner product (denoted by a dot) is formed by contraction, as in the examples

(5) $\operatorname{tr} \mathbf{p} = \operatorname{tr} \mathsf{p} = \mathsf{p}^{ii}$ (scalar)

(6) $\mathbf{p} \cdot \mathbf{u}C = \mathsf{pu} = [\mathsf{p}^{ri}\mathsf{u}^{i}]$; (vector)

(7) $\mathbf{p} \cdot \mathbf{q}C = \mathsf{pq} = [\mathsf{p}^{ri}\mathsf{q}^{ic}]$ (second rank tensor);

(8) $\mathbf{p}:\mathbf{q} = \operatorname{tr} \mathbf{p} \cdot \mathbf{q} = \operatorname{tr} \mathsf{pq} = [\mathsf{p}^{ij}\mathsf{q}^{ji}]$ (scalar);

(9) $\mathbf{u} \cdot \mathbf{p}C = \tilde{\mathsf{u}}\mathsf{p} = [\mathsf{u}^{i}\mathsf{p}^{ir}]$ (vector).

The transpose $\tilde{\mathbf{p}}$ and the reciprocal $\mathbf{p}^{-1}$ are Cartesian second-rank tensors defined by the equations

(10) $$\tilde{\mathbf{p}}C = \tilde{\mathsf{p}}, \qquad \mathbf{p}^{-1}C = \mathsf{p}^{-1}.$$

Because $L(C, \bar{C})$ is orthogonal, there are not different kinds of second-rank Cartesian tensor; *the distinction among covariant, contravariant, and mixed general tensors of second rank has no counterpart for Cartesian tensors.* Because $L(C, \bar{C})$ is independent of place, Cartesian tensors of second rank associated with different places belong to the same linear space.

In general tensor analysis, the unit tensor $\boldsymbol{I}$ and the metric tensor $\boldsymbol{g}$ are different. *In Cartesian tensor analysis, the same Cartesian tensor* $\mathbf{U}$ *plays the role of a unit tensor and a metric tensor*: we have

(11) $\mathbf{U} \cdot \mathbf{v} = \mathbf{v}, \qquad \mathbf{U} \cdot \mathbf{p} = \mathbf{p}$ ($\mathbf{U}$ as a unit tensor),

(12) $\|\mathbf{v}\|^2 = \mathbf{v} \cdot \mathbf{U} \cdot \mathbf{v}$ ($\mathbf{U}$ as a metric tensor),

for any Cartesian vector $\mathbf{v}$ and Cartesian tensor $\mathbf{p}$. Equation (12) includes the case

(13) $(ds)^2 = d\mathbf{y} \cdot \mathbf{U} \cdot d\mathbf{y}$ ($\mathbf{U}$ as a metric tensor),

where $d\mathbf{y} = \overrightarrow{QQ'}$ is the Cartesian displacement vector for two neighboring places Q and Q′; this equation follows from (1) and the definition of $\mathbf{U}$:

(14) $$\mathbf{U}C = I.$$

4.4 Relations between Cartesian and general tensor fields

For a given Euclidean manifold, a tensor field is uniquely determined when we specify its kind (i.e., Cartesian, covariant, contravariant, mixed), its rank (first- and second-rank tensors have been defined here), and its representative component functions in any one specified coordinate system. In particular, we can choose a rectangular Cartesian coordinate system for specifying the component functions of any *general* tensor field of given kind and rank. But the same component functions also determine uniquely a *Cartesian* tensor field of the same rank. There is, therefore, *a one-to-one correspondence between general tensor fields of specified kind and rank and Cartesian tensor fields of the same rank.*

Suppose that a Cartesian vector field $\mathbf{v}(\mathrm{Q})$ is given, and that $v(y)$ is its representative matrix function in an arbitrary rectangular Cartesian coordinate system $C: \mathrm{Q} \rightarrow y$. We can now define a covariant vector field $\boldsymbol{u}(\mathrm{Q})$ and a contravariant vector field $\boldsymbol{v}(\mathrm{Q})$ which have the same representative matrix function $v(y)$ in C:

(1) $$\mathbf{v}(\mathrm{Q})C = v(y); \qquad \boldsymbol{u}(\mathrm{Q})C = v(y); \qquad \boldsymbol{v}(\mathrm{Q})C = v(y).$$

Similar equations will hold for all rectangular Cartesian coordinate systems, so that

(2) $$\mathbf{v}(\mathrm{Q})C = \boldsymbol{u}(\mathrm{Q})C = \boldsymbol{v}(\mathrm{Q})C \qquad (\text{for all } C \text{ in } \mathscr{C}).$$

However, if $S: \mathrm{Q} \rightarrow x$ is any coordinate system which is not rectangular Cartesian, then we have

(3) $$\boldsymbol{u}(\mathrm{Q})S \neq \boldsymbol{v}(\mathrm{Q})S \quad (\text{in general}); \qquad \mathbf{v}(\mathrm{Q})S \text{ is not defined.}$$

(4) A schematic illustration of the relation between the three corresponding vectors $\mathbf{v}(\mathrm{Q})$, $\boldsymbol{v}(\mathrm{Q})$, and $\boldsymbol{u}(\mathrm{Q})$ (for one place Q) is given in Fig 4.4(4). This figure also emphasizes the nature of the distinction between Cartesian vectors and general vectors; according to the definitions used here, *Cartesian vectors and general vectors are functions with the same range* $A_{3\times 1}$ *but different domains:* $\mathscr{C}$ for Cartesian vectors, and $\mathscr{S}$ for general vectors.

For a Euclidean manifold whose metric tensor field is independent of time, Cartesian tensors and general tensors represent two different sets of tools, either of which can be used in applications. For most purposes, Cartesian tensors are simpler to use, and have been widely used in the literature on elasticity, hydrodynamics, and continuum mechanics. The distinction which we have made in the definitions of Cartesian tensors and general tensors is not usually made explicitly in the continuum mechanics literature and the question arises in some cases as to what kind of tensor field the boldface symbols used in the literature represent. Examination of various texts usually shows, however, that (i) vectors or tensors at different places can be added, and (ii) no distinction is made between covariant vectors and contravariant vectors; in such texts, therefore, *the vectors and tensors can be regarded as Cartesian* in the sense of our definition (Lodge, 1972).

The body manifold, however, is a Euclidean manifold whose metric tensor field varies with time; therefore Cartesian body tensor fields cannot be defined usefully, but general body tensor fields can.

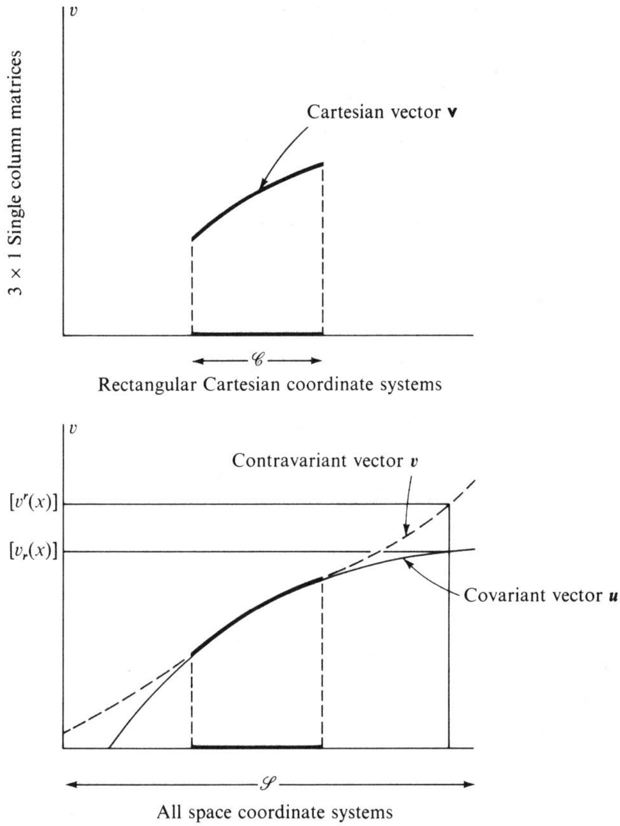

Figure 4.4(4) Schematic diagram to illustrate the relation between a Cartesian vector **v**(Q) and the *corresponding* contravariant vector $\boldsymbol{v}$(Q) and covariant vector $\boldsymbol{u}$(Q). The three vectors have equal representative matrices in an arbitrary rectangular space coordinate system. In an arbitrary coordinate system $S: \mathrm{Q} \rightarrow x$, the components $v^i(x)$ of the contravariant vector $\boldsymbol{v}$(Q) are equal to the contravariant components of the Cartesian vector **v**(Q) in S, and the components $v_i(x)$ of the covariant vector $\boldsymbol{u}$(Q) are equal to the covariant components of the Cartesian vector **v**(Q) in S.

4.5 Cartesian base vectors for a curvilinear coordinate system

Cartesian vector and tensor *fields* can be defined over the space manifold; any such field is simply a correspondence between Cartesian vectors or tensors and places in space. In particular, if we are given any space coordinate system $S: \mathrm{Q} \rightarrow x$ (which we shall here refer to as a curvilinear coordinate system), the coordinate curves form sets of curves in space, and we can define *Cartesian base vector fields* $\mathbf{b}_i(S, \mathrm{Q})$ and $\mathbf{b}^i(S, \mathrm{Q})$ which, at each place Q, are respectively tangential to coordinate curves and normal to coordinate surfaces of S. The normalization of these vectors is arbitrary; to follow as closely as possible the definition of contravariant and covariant base vectors $\boldsymbol{b}_i(S, \mathrm{Q})$ and $\boldsymbol{b}^i(S, \mathrm{Q})$ given in 2.3(35) and 2.4(21), we proceed as follows.

Let Q_0 be an arbitrary but definite place, which we shall call an *origin* for Cartesian displacement vectors

(1) $$\mathbf{r} = \overrightarrow{\mathrm{Q}_0\mathrm{Q}}, \qquad \mathbf{r}' = \overrightarrow{\mathrm{Q}_0\mathrm{Q}'}, \ldots,$$

where Q, Q′, ... are other places. If Q and Q′ are any two neighboring places on a given coordinate curve, say an x^1 curve of S, and Q has coordinates (x^1, x^2, x^3) while Q′ has coordinates $(x^1 + dx^1, x^2, x^3)$ in S, then as dx^1 tends to zero, the Cartesian vector

(2) $$\mathsf{d}\mathbf{r} = \mathbf{r}' - \mathbf{r} = \overrightarrow{\mathrm{QQ}'} = (\partial\mathbf{r}/\partial x^1)\, dx^1$$

becomes a tangent at Q to the x^1 curve through Q. We define, then, the *Cartesian tangent base vectors* $\mathbf{b}_i(S, \mathrm{Q})$ for S at Q by the equations

(3) $$\mathbf{b}_i(S, \mathrm{Q}) = \partial\mathbf{r}(x^1, x^2, x^3)/\partial x^1 \qquad (i = 1, 2, 3),$$

where $\mathbf{r}(x^1, x^2, x^3) = \overrightarrow{\mathrm{Q}_0\mathrm{Q}}$ is a Cartesian vector function of the S coordinates of Q. *There is no general-vector analog for this function or for Eq.* (3). In 2.3(35), a different method is used to define the contravariant base vectors $\boldsymbol{b}_i(S, \mathrm{Q})$. The *Cartesian normal base vectors* $\mathbf{b}^i(S, \mathrm{Q})$ can most simply be defined as the set of Cartesian vectors reciprocal to the set $\mathbf{b}_i(S, \mathrm{Q})$, i.e., the vectors that satisfy the equations

(4) $$\mathbf{b}_i(S, \mathrm{Q}) \cdot \mathbf{b}^j(S, \mathrm{Q}) = \delta_i^j \qquad (i, j = 1, 2, 3).$$

The Cartesian tangent base vectors $\mathbf{b}_i(S, \mathrm{Q})$ are often called *contravariant* base vectors in the literature, while the Cartesian normal base vectors $\mathbf{b}^i(S, \mathrm{Q})$ are called *covariant* base vectors. In the present context, it is clear that such a terminology is confusing, because these vectors are all Cartesian vectors, and we have shown that the distinction between contravariant and covariant Cartesian vectors is nonexistent.

Let Q, with coordinates x^i, and Q′, with coordinates $x^i + dx^i$, be any two neighboring places (not necessarily on the same coordinate curve). These two places give rise to two distinct displacement vectors: a contravariant vector $\boldsymbol{dx}$ and a Cartesian vector $\mathsf{d}\mathbf{r}$ given by the equations

(5) $\boldsymbol{dx} = dx^i\, \boldsymbol{b}_i(S, \mathrm{Q})$ [contravariant; from 2.3(38)];

(6) $\mathsf{d}\mathbf{r} = dx^i\, \mathbf{b}_i(S, \mathrm{Q})$ [Cartesian; from (3) and $\mathsf{d}\mathbf{r} = dx^i\, \partial\mathbf{r}(x)/\partial x^i$]

The d notation in (6) is chosen to emphasize the fact that $\mathsf{d}\mathbf{r}$ is obtained from the difference of two $\mathbf{r}$ vectors, while $\boldsymbol{dx}$ is not obtained from the difference of two $\boldsymbol{x}$ vectors, since these do not exist in general vector analysis. From (5), (6), and the fact that $\boldsymbol{dx}\ C = \mathbf{dr}\ C$, it follows that

(7) $$\boldsymbol{b}_i(S, \mathrm{Q})C = \mathbf{b}_i(S, \mathrm{Q})C.$$

At any place Q, the Cartesian tangent base vectors $\mathbf{b}_i(S, \mathrm{Q})$ form a basis for all Cartesian vectors, and so also do the Cartesian normal base vectors $\mathbf{b}^i(S, \mathrm{Q})$. For any Cartesian vector field $\mathbf{v}(\mathrm{Q})$, we can therefore write equations of the form

(8) $$\mathbf{v}(\mathrm{Q}) = v^i(x)\mathbf{b}_i(S, \mathrm{Q}) = v_i(x)\mathbf{b}^i(S, \mathrm{Q}).$$

The coefficients $v^i(x)$ are usually called *contravariant components* of $\mathbf{v}$, and the coefficients $v_i(x)$ are called *covariant components* of $\mathbf{v}$ in S. *The second of Eqs.* (8) *has no counterpart in general vector analysis*; the contravariant base vectors $\boldsymbol{b}_i(S, \mathrm{Q})$ which occur in a related context in 2.3(36) belong to a different linear space from that which contains the covariant base vectors $\boldsymbol{b}^i(S, \mathrm{Q})$, which occur in 2.4(22).

(9) **Problem** With the notation of (8), let $\boldsymbol{v}(\mathrm{Q})$ and $\boldsymbol{u}(\mathrm{Q})$ denote the contravariant vector and the covariant vector that correspond to the given Cartesian vector $\mathbf{v}(\mathrm{Q})$ as illustrated in Fig. 4.4(4). Prove that, in S, $\boldsymbol{v}(\mathrm{Q})$ and $\boldsymbol{u}(\mathrm{Q})$ have components respectively equal to $v^i(x)$ and $v_i(x)$.

(10) **Problem** Prove that $\mathbf{b}_i(S, \mathrm{Q})\mathbf{b}^i(S, \mathrm{Q}) = \mathbf{b}^i(S, \mathrm{Q})\mathbf{b}_i(S, \mathrm{Q}) = \mathbf{U}$.

(11) **Problem** Prove that $\mathbf{b}_i(S, \mathrm{Q}) \cdot \mathbf{b}_j(S, \mathrm{Q}) = g_{ij}(x)$ and that $\mathbf{b}^i(S, \mathrm{Q}) \cdot \mathbf{b}^j(S, \mathrm{Q}) = g^{ij}(x)$, where $g_{ij}(x)$ and $g^{ij}(x)$ are respectively the components in S of the covariant metric-tensor $\boldsymbol{g}(\mathrm{Q})$ and its reciprocal $\boldsymbol{g}^{-1}(\mathrm{Q})$.

A second-rank Cartesian tensor field $\mathbf{p}(\mathrm{Q})$ can be expressed in terms of outer products of the Cartesian base vectors as follows:

(12) $$\begin{aligned}\mathbf{p}(\mathrm{Q}) &= p^{ij}(x)\mathbf{b}_i(S, \mathrm{Q})\mathbf{b}_j(S, \mathrm{Q}) \\ &= p_{ij}(x)\mathbf{b}^i(S, \mathrm{Q})\mathbf{b}^j(S, \mathrm{Q}) = p^i{}_j(x)\mathbf{b}_i(S, \mathrm{Q})\mathbf{b}^j(S, \mathrm{Q}).\end{aligned}$$

p^{ij}, p_{ij}, and $p^i{}_j$ are called, respectively, the *contravariant, covariant, and mixed components of the Cartesian tensor* $\mathbf{p}$ in S. The second and third equations in (12) have no counterpart in general tensor analysis.

(13) **Problem** Let $\boldsymbol{p}(\mathrm{Q})$, $\boldsymbol{q}(\mathrm{Q})$, and $\boldsymbol{m}(\mathrm{Q})$ denote respectively the contravariant, covariant, and right-covariant mixed tensors that "correspond" to the given Cartesian tensor $\mathbf{p}(\mathrm{Q})$ under the same type of correspondence as that

illustrated for vectors in Fig. 4.4(4); i.e., $\boldsymbol{p}$(Q) is a contravariant tensor which has the same representative matrix as $\mathbf{p}$(Q) has in any given rectangular Cartesian coordinate system C, etc. Prove that, in S, the components of $\boldsymbol{p}$, $\boldsymbol{q}$, and $\boldsymbol{m}$ are respectively equal to p^{ij}, p_{ij}, and $p^{i}{}_{j}$ [as defined in (12)].

(14) **Problem** Prove the following relations:

(i) $\mathbf{U} = g_{ij}(x)\mathbf{b}^i(S, \mathrm{Q})\mathbf{b}^j(S, \mathrm{Q}) = g^{ij}(x)\mathbf{b}_i(S, \mathrm{Q})\mathbf{b}_j(S, \mathrm{Q})$;

(ii) $\mathbf{b}_i(S, \mathrm{Q}) = g_{ij}(x)\mathbf{b}^j(S, \mathrm{Q}), \qquad \mathbf{b}^i(S, \mathrm{Q}) = g^{ij}(x)\mathbf{b}_j(S, \mathrm{Q})$.

(15) **Problem** Prove the following relations:

$$\mathbf{b}_i(S, \mathrm{Q}) = \left(\frac{\partial y^k}{\partial x^i}\right)_{\mathrm{Q}} \mathbf{b}_k(C), \qquad \mathbf{b}^i(S, \mathrm{Q}) = \left(\frac{\partial x^i}{\partial y^k}\right)_{\mathrm{Q}} \mathbf{b}^k(C)$$

$$(C : \mathrm{Q} \to y; \quad S : \mathrm{Q} \to x).$$

(16) **Problem** Show that $\mathbf{r}(\mathrm{Q}) = (y^i - y_0^i)\mathbf{b}_i(C)$, where the $\mathbf{b}_i(C)$ are the Cartesian base vectors for any rectangular Cartesian coordinate system $C : \mathrm{Q} \to y$, $\mathbf{r}(\mathrm{Q}) = \overrightarrow{\mathrm{Q}_0\,\mathrm{Q}}$ is the position vector of a place Q with respect to any given place Q_0, and y^i and y_0^i denote the C coordinates of Q and Q_0, respectively.

5 RELATIVE TENSORS, FIELD TRANSFER, AND THE BODY STRESS TENSOR FIELD

5.1 Relative tensors

The properties of vectors and tensors derive essentially from two facts: the linearity of the transformation relating representative matrices in two coordinate systems [e.g., 2.3(13) for a contravariant body vector], and the transitivity property for successive coordinate transformations [expressed by 2.3(5) and 2.3(6) for contravariant and covariant body vectors, respectively]. On taking the determinants of both sides of the matrix equation 2.3(5), we see that the determinants also possess a similar transitive property, namely

$$(1) \qquad |\Lambda_P^{-1}(B, \bar{B})| \; |\Lambda_P^{-1}(\bar{B}, \bar{\bar{B}})| = |\Lambda_P^{-1}(B, \bar{\bar{B}})| .$$

It follows from this equation that

$$(2) \qquad |\Lambda_P(B, \bar{B})|^n |\Lambda_P(\bar{B}, \bar{\bar{B}})|^n = |\Lambda_P(B, \bar{\bar{B}})|^n,$$

where n is any integer. The restriction to integral values for n is not necessary for transitivity, but it gives us sufficient generality for our applications.

Thus *the nth power of the transformation matrix determinant possesses the transitivity property*. We may therefore define new geometric objects by using our previous definitions of tensors with the transformation matrix products $\Lambda \cdots \tilde{\Lambda}$, etc., multiplied by $|\Lambda|^{-n}$; the objects so defined are called *relative tensors of weight n*. Tensors previously defined have $n = 0$ and are called *absolute tensors* when it is necessary to emphasize the distinction between them and relative tensors. When $n \neq 0$, relative tensors of ranks zero and one are called *relative scalars* and *relative vectors*, respectively.

The preceding discussion has been given for body tensors, but it is clear that it can be repeated for general space tensors. For *Cartesian* space tensors, however, the transformation matrix $L(C, \bar{C})$ used in the definition of tensors is orthogonal, and hence

(3) $$|L(C, \bar{C})| = \pm 1 \qquad \text{(Cartesian tensors).}$$

If, therefore, n is even, the relative Cartesian tensors are the same as absolute Cartesian tensors; there is, in fact, only one kind of relative Cartesian tensor, namely that given by n odd, or by $n = 1$. All odd powers of n give the same transformation law. Cartesian vectors are often called *polar vectors* if $n = 0$ and *axial vectors* if $n = 1$. Relative tensors are sometimes also called *pseudotensors*, but the prefix "pseudo" suggests a lowering of status that is hardly justified.

Relative tensors are convenient to use in discussing surface and volume elements, the alternating tensor and vector products, and the invariant differential operators curl and div. Moreover, recognition of the fact that all terms in a vector equation must be of equal weight can sometimes help one to write down possible forms for the solution of a set of partial differential equations [see, e.g., Landau and Lifschitz (1959, p. 64)].

(4) The definitions of relative tensors of first and second rank are given in Table 5.1(4). It is easy to see from the definitions that these tensors possess most of the properties that absolute tensors possess: e.g., relative tensors of given weight, kind, and rank at a given point form a linear space, but relative tensors of different weights belong to disjoint spaces. In outer products or inner products, the weights of the factors are additive. Taking the transpose does not alter the weight, but taking the reciprocal does: A nonsingular second-rank tensor of weight n has a reciprocal of weight $-n$.

Following the terminology introduced in previous chapters, we shall continue to use the word tensor to mean absolute tensor, unless the contrary is explicitly stated. When a tensor is not an absolute tensor, we shall refer to it as a relative tensor, or give its weight, or both. The unit tensors $\boldsymbol{\delta}$, $\boldsymbol{I}$, and $\mathbf{U}$, are all absolute tensors, but there is a null tensor $\mathbf{0}$ of any given weight.

Table 5.1(4) Definitions of Tensors of Weight n

	Body tensors at P	Space tensors at Q	Cartesian tensors
Contravariant vector	$\boldsymbol{\theta} B = \theta = \lvert\Lambda\rvert^n \Lambda^{-1} \bar{\theta}$	$\boldsymbol{v} S = v = \lvert L\rvert^n L^{-1} \bar{v}$	Vector: $\mathbf{v} C = v = \lvert L\rvert^n L^{-1} \bar{v}$
Covariant vector	$\boldsymbol{\psi} B = \psi = \lvert\Lambda\rvert^n \tilde{\Lambda} \bar{\psi}$	$\boldsymbol{u} S = u = \lvert L\rvert^n \tilde{L} \bar{u}$	(Polar vector, $n = 0$; axial vector, $n = 1$)
Contravariant second-rank tensor	$\boldsymbol{\Theta} B = \Theta = \lvert\Lambda\rvert^n \Lambda^{-1} \bar{\Theta} \tilde{\Lambda}^{-1}$	$\boldsymbol{p} S = p = \lvert L\rvert^n L^{-1} \bar{p} \tilde{L}^{-1}$	Second-rank tensor:
Covariant second-rank tensor	$\boldsymbol{\Psi} B = \Psi = \lvert\Lambda\rvert^n \tilde{\Lambda} \bar{\Psi} \Lambda$	$\boldsymbol{q} S = q = \lvert L\rvert^n \tilde{L} \bar{q} L$	$\mathbf{p} C = p = \lvert L\rvert^n L^{-1} \bar{p} L$
Right-covariant mixed tensor	$\boldsymbol{\mu} B = \mu = \lvert\Lambda\rvert^n \Lambda^{-1} \bar{\mu} \Lambda$	$\boldsymbol{m} S = m = \mathbf{p}\, \lvert L\rvert^n L^{-1} \bar{m} L$	
Left-covariant mixed tensor	$\tilde{\boldsymbol{\mu}} B = \tilde{\mu} = \lvert\Lambda\rvert^n \tilde{\Lambda} \tilde{\bar{\mu}} \tilde{\Lambda}^{-1}$	$\tilde{\boldsymbol{m}} S = \tilde{m} = \lvert L\rvert^n \tilde{L} \tilde{\bar{\boldsymbol{m}}} \tilde{L}^{-1}$	
	Body coordinate systems:	Space coordinate systems:	Rectangular Cartesian space coordinate systems:
	$B : \mathrm{P} \to \xi,\ \bar{B} : \mathrm{P} \to \bar{\xi};$	$S : \mathrm{Q} \to x,\ \bar{S} : \mathrm{Q} \to \bar{x};$	$C : \mathrm{Q} \to y,\ \bar{C} : \mathrm{Q} \to \bar{y};$
	$\Lambda \equiv \Lambda_{\mathrm{P}}(B, \bar{B}) = [\partial \bar{\xi}^r / \partial \xi^c]_{\mathrm{P}}$	$L \equiv L_{\mathrm{Q}}(S, \bar{S}) = [\partial \bar{x}^r / \partial x^c]_{\mathrm{Q}}$	$L \equiv L(C, \bar{C}) = [\partial \bar{y}^r / \partial y^c] = \tilde{L}^{-1}$

On taking the determinant of the transformation law 2.5(3) for an absolute contravariant second-rank tensor, we obtain the equation

(5) $$|\Theta| = |\Lambda|^{-2}|\bar{\Theta}|,$$

which is the transformation law for a relative scalar of weight $n = -2$; the tensor field $\mathbf{\Theta}(\mathrm{P})$ therefore gives rise to a relative scalar field, which we may denote by $|\mathbf{\Theta}(\mathrm{P})|$. Similarly, an absolute covariant field $\mathbf{\Psi}(\mathrm{P})$ of second rank gives rise to a relative scalar field $|\mathbf{\Psi}(\mathrm{P})|$ of weight $n = +2$; this follows on taking the determinant of the covariant transformation law 2.5(8)(ii).

More generally, we have the following results, which are readily verified:

(6)
$$|\mathbf{\Phi}(\mathrm{P})| \quad \text{is of weight} \quad \begin{Bmatrix} n-2 \\ n \\ n+2 \end{Bmatrix} \quad \text{according as} \quad \mathbf{\Phi}(\mathrm{P}) \quad \text{is} \quad \begin{Bmatrix} \text{contravariant} \\ \text{mixed} \\ \text{covariant} \end{Bmatrix},$$

where $\mathbf{\Phi}(\mathrm{P})$ is a second-rank tensor of weight n.

5.2 Tensors of third and higher rank

The matrix notation has been convenient for defining tensors of first and second rank; for tensors of third and higher rank, however, it is easier to revert to the more conventional component notation. We shall make only a limited use of third- and fourth-rank tensors, and it will suffice to give only a brief treatment of them, because their definition and properties follow closely those already presented for tensors of first and second rank.

As a guide to the definition of higher-rank tensors, we rewrite the transformation laws involved in 2.3(11), 2.4(11), and 2.5(8) as follows:

(1) $$\sigma(\xi) = \bar{\sigma}(\bar{\xi}) \qquad \text{(zeroth-rank tensor, or scalar);}$$

(2) $$\theta^i(\xi) = \frac{\partial \xi^i}{\partial \bar{\xi}^r}\bar{\theta}^r(\bar{\xi}) \qquad \text{(first-rank contravariant);}$$

(3) $$\Theta^{ij}(\xi) = \frac{\partial \xi^i}{\partial \bar{\xi}^r}\frac{\partial \xi^j}{\partial \bar{\xi}^s}\bar{\Theta}^{rs}(\bar{\xi}) \qquad \text{(second-rank contravariant).}$$

Accordingly, we define the transformation law for a third-rank contravariant tensor as follows:

(4) $$\Phi^{ijk}(\xi) = \frac{\partial \xi^i}{\partial \bar{\xi}^r}\frac{\partial \xi^j}{\partial \bar{\xi}^s}\frac{\partial \xi^k}{\partial \bar{\xi}^t}\bar{\Phi}^{rst}(\bar{\xi}) \qquad \text{(third-rank contravariant).}$$

The third-rank contravariant tensor itself, $\mathbf{\Phi}(\mathrm{P})$, say, at P, is defined as a mapping $\mathbf{\Phi}: \mathscr{B} \to A_{3\times3\times3}$ from the set $\mathscr{B}$ of all body coordinate systems in to $A_{3\times3\times3}$, the space of all *ordered* $3 \times 3 \times 3$ arrays of real numbers, with the condition that the arrays $\Phi^{ijk}(\xi)$ and $\overline{\Phi}^{rst}(\bar{\xi})$ associated with any two body coordinate systems B and $\bar{B}$, respectively, shall be related by a transformation of the form (4).

In a similar way, we can define a body tensor $\mathbf{\Psi}(\mathrm{P})$ of arbitrary rank, kind, and weight; the transformation law is of the form

$$\Psi^{ijk\ldots}_{abc\ldots}(\xi) = |\Lambda|^n \frac{\partial\xi^i}{\partial\bar{\xi}^r} \cdots \frac{\partial\bar{\xi}^u}{\partial\xi^a} \cdots \overline{\Psi}^{rst\ldots}_{uvw\ldots}(\bar{\xi}), \tag{5}$$

where n is the weight of the tensor and $|\Lambda| = |\partial\bar{\xi}^u/\partial\xi^v|$; the transformation law contains one factor of type $\partial\xi^i/\partial\bar{\xi}^r$ for each "contravariant index" i, and one factor of type $\partial\bar{\xi}^u/\partial\xi^a$ for each "covariant index" a.

The transformation laws are all linear and homogeneous in the representative arrays, and the individual factors Λ^{-1}, $\tilde{\Lambda}$ each possess the transitive property 2.3(4). It is easy to see that tensors of any given rank, weight, and kind possess properties and admit operations similar to those already presented for tensors of first and second rank. In particular, tensors of given rank, weight, and kind at P form a linear space; tensors that differ in at least one of the items in the list (rank, weight, kind, P) belong to disjoint linear spaces; outer products of tensors at P can be defined, and give tensors of higher rank; inner products of tensors at P can be defined by contracting any contravariant index with any covariant index in the representative arrays; the transpose is defined; the reciprocal is defined only for second-rank tensors.

These definitions sketched for body tensors can evidently be repeated for general space tensors, and also for Cartesian space tensors with the proviso that the distinction between covariance and contravariance lapses, and that we need consider only the cases $n = 0$ or 1.

(6) **Theorem** The quantities e^{ijk} and e_{ijk}, defined in 1.3(25), are components of *numerical tensors* (i.e., of tensors represented in every coordinate system by the same ordered sets of numbers); these tensors are of third rank and of the following kinds:

$$\begin{aligned} e^{ijk}&: \text{ contravariant, weight } 1;\\ e_{ijk}&: \text{ covariant, weight } -1. \end{aligned}$$

Such tensors can be defined at any point of a three-dimensional manifold (e.g., body or space). They are called the *alternating tensors.*

Proof Let us associate e^{ijk} with any space coordinate system S, and $\bar{e}^{ijk} = e^{ijk}$ with any other space coordinate system $\bar{S}$. Applying 1.3(27) with e_{rst}

replaced by e^{rst}, e^{ijk} replaced by $\bar{e}^{ijk}$, and $A_{ri} = \partial x^i/\partial \bar{x}^r$, so that $A = L^{-1}$ and $|A| = |L|^{-1}$, we obtain the result

$$e^{rst} = |L| \frac{\partial x^r}{\partial \bar{x}^i} \frac{\partial x^s}{\partial \bar{x}^j} \frac{\partial x^t}{\partial \bar{x}^k} \bar{e}^{ijk}. \tag{7}$$

Apart from changes of letters used for indices, this equation is evidently of the form (4) with an extra factor $|L|$; it follows that the e^{ijk} are components of a third-rank contravariant numerical tensor of weight 1. The other result is proved in a similar way; by taking $A = L$ in 1.3(27), we obtain the equation

$$e_{rst} = |L|^{-1} \frac{\partial \bar{x}^i}{\partial x^r} \frac{\partial \bar{x}^j}{\partial x^s} \frac{\partial \bar{x}^k}{\partial x^t} \bar{e}_{ijk}, \tag{8}$$

which [on comparison with (5)] proves the result stated.

In an n-dimensional manifold, alternating tensors of nth rank can be defined. They have similar transformation properties, because (7) and (8) are merely different ways of writing the determinant $|L|$ and can evidently be generalized for $n \times n$ determinants.

(9) **Definition** Let $\boldsymbol{v}_1(Q)$ and $\boldsymbol{v}_2(Q)$ be contravariant vectors, with components $v^i_{(1)}$ and $v^i_{(2)}$, respectively, in an arbitrary coordinate system S. The *vector product* $\boldsymbol{v}_1 \times \boldsymbol{v}_2 \equiv \boldsymbol{u}$, say, is a covariant vector of weight -1, whose components in S are $u_i = e_{ijk} v^j_{(1)} v^k_{(2)}$. If $\boldsymbol{v}_1$ and $\boldsymbol{v}_2$ are relative vectors of weights n_1 and n_2, then $\boldsymbol{u}$ is of weight $n_1 + n_2 - 1$. The vector product $\boldsymbol{u}^1 \times \boldsymbol{u}^2 \equiv \boldsymbol{v}$, say, of covariant vectors $\boldsymbol{u}^1$ and $\boldsymbol{u}^2$ is a contravariant vector of weight $+1$ whose components are $v^i = e^{ijk} u^{(1)}_j u^{(2)}_k$. If $\boldsymbol{u}^1$ and $\boldsymbol{u}^2$ have weights n_1 and n_2, then $\boldsymbol{v}$ has weight $n_1 + n_2 + 1$.

The stated properties follow at once from (6) on contraction. From 1.3(25), it follows that the vector product components are as follows:

$$\begin{aligned} u_1 &= v^2_{(1)} v^3_{(2)} - v^3_{(1)} v^2_{(2)}, \\ u_2 &= v^3_{(1)} v^1_{(2)} - v^1_{(1)} v^3_{(2)}, \\ u_3 &= v^1_{(1)} v^2_{(2)} - v^2_{(1)} v^1_{(2)}. \end{aligned} \tag{10}$$

These equations are of the same form as the corresponding equations for the vector product in elementary vector analysis. The familiar properties $\boldsymbol{v} \times \boldsymbol{v} = \boldsymbol{0}$ and $\boldsymbol{v} \times \boldsymbol{v}' = -\boldsymbol{v}' \times \boldsymbol{v}$ follow at once from (9).

(11) **Definition** The *scalar triple product* of contravariant vectors $\boldsymbol{v}_i$ of weights n_i is the scalar $\boldsymbol{v}_1 \times \boldsymbol{v}_2 \cdot \boldsymbol{v}_3$ of weight $n_1 + n_2 + n_3 - 1$. The scalar triple product of covariant vectors $\boldsymbol{u}^i$ of weights n_i is the scalar $\boldsymbol{u}^1 \times \boldsymbol{u}^2 \cdot \boldsymbol{u}^3$ of weight $n_1 + n_2 + n_3 + 1$.

(12) **Problem** From the definition (11), derive the following properties of the scalar triple product: (i) $\boldsymbol{v}_1 \times \boldsymbol{v}_2 \cdot \boldsymbol{v}_3 = \boldsymbol{v}_1 \cdot \boldsymbol{v}_2 \times \boldsymbol{v}_3$; (ii) cyclic permutation of factors leaves a scalar triple product unchanged in value; (iii) a necessary and sufficient condition that $\boldsymbol{v}_1$, $\boldsymbol{v}_2$, and $\boldsymbol{v}_3$ (having equal weights) be linearly independent is $\boldsymbol{v}_1 \times \boldsymbol{v}_2 \cdot \boldsymbol{v}_3 \neq 0$; (iv) $\boldsymbol{v}_1 \times \boldsymbol{v}_2 \cdot \boldsymbol{v}_3 = |v^j_{(i)}|$, where the $v^j_{(i)}$ are the components of $\boldsymbol{v}_i$ in any S; (v) if $\boldsymbol{v}_1$, $\boldsymbol{v}_2$, and $\boldsymbol{v}_3$ are linearly independent contravariant vectors of weight n, the reciprocal set $\boldsymbol{u}^i$ are covariant vectors of weight $-n$ given by the equations

(13) $$\boldsymbol{u}^1 = s\boldsymbol{v}_2 \times \boldsymbol{v}_3, \qquad \boldsymbol{u}^2 = s\boldsymbol{v}_3 \times \boldsymbol{v}_1, \qquad \boldsymbol{u}^3 = s\boldsymbol{v}_2 \times \boldsymbol{v}_1,$$

where

$$s^{-1} = \boldsymbol{v}_1 \times \boldsymbol{v}_2 \cdot \boldsymbol{v}_3 .$$

5.3 Quotient theorems

We have seen that a single contraction in a tensor product yields a tensor of a rank that is lower by two than the rank of the original product. There are various theorems, known as *quotient theorems*, which in a sense are converse theorems to the contraction theorems, and are sometimes useful in enabling one to prove that a given array of components associated with an arbitrary but definite coordinate system represents a tensor. If the given components can be contracted one or more times with the components of an arbitrary tensor in such a way as to yield components of a tensor, then the given components do represent a tensor of known kind. It will be sufficient to illustrate the theorems by stating and proving them for one or two examples.

(1) **Theorem** Let $\boldsymbol{p}$: $\mathscr{S} \to A_{3\times 3}$ be a function such that pq represents in an arbitrary S a right-covariant mixed tensor at Q, where $\boldsymbol{p}S = p$, $\boldsymbol{q}S = q$, and $\boldsymbol{q}$(Q) is an *arbitrary* covariant second-rank tensor. Then $\boldsymbol{p}$ is a contravariant second-rank tensor at Q.

Proof For any S and $\bar{S}$, it follows from 2.5(18)(iii) that $pq = L^{-1}\bar{p}\bar{q}L$, where $\bar{p} = \boldsymbol{p}\bar{S}$, $\bar{q} = \boldsymbol{q}\bar{S}$, and $L \equiv L_{\mathrm{Q}}(S, \bar{S})$. From 2.5(18)(ii), we have $q = \tilde{L}\bar{q}L$, and hence, on substituting for q and multiplying by L^{-1} from the right, we have $p\tilde{L}\bar{q} = L^{-1}\bar{p}\bar{q}$. But $\boldsymbol{q}$ is an arbitrary tensor, and therefore the matrix $\bar{q}$ can be assigned arbitrary values. It follows that $p\tilde{L} = L^{-1}\bar{p}$, and hence $p = L^{-1}\bar{p}\tilde{L}^{-1}$, which is just the transformation law 2.5(18)(i) for a contravariant second-rank tensor at Q. This proves that $\boldsymbol{p}$ is a second-rank contravariant tensor at Q, as stated.

(2) **Problem** Use a quotient theorem to give an alternative proof of the statement that the 3×3 unit matrix I is the representative in every S of a second-rank, absolute, mixed tensor.

Inspection of the proof of (1) shows that the contracted tensor $\boldsymbol{q}$ has to be arbitrary in order that its representative matrix $\bar{q}$ can be "removed" from the equation $p\tilde{L}\bar{q} = L^{-1}\bar{p}\bar{q}$ in order to derive the transformation law for $\boldsymbol{p}$. If the contracted tensor $\boldsymbol{q}$ is symmetric (or antisymmetric), then $\boldsymbol{q}$ is not completely arbitrary, and a modified form of quotient theorem must be used. To illustrate this, we give an alternative proof that $\boldsymbol{g}$ is a covariant tensor.

(3) **Theorem** Let $\boldsymbol{g}: \mathscr{S} \to A_{3\times 3}$ be a function such that $g = \tilde{g}$ and $\widetilde{dx}\, g\, dx$ is an absolute scalar, where $g = \boldsymbol{g}S$, $dx = \boldsymbol{dx}\, S$, and $\boldsymbol{dx}$ is an arbitrary, infinitesimal, absolute contravariant vector at Q. Then $\boldsymbol{g}$ is an absolute, covariant, second-rank tensor at Q.

Proof Since $\widetilde{dx}\, g\, dx$ is an absolute scalar, we have $\widetilde{dx}\, g\, dx = \widetilde{d\bar{x}}\, \bar{g}\, d\bar{x}$, where $d\bar{x} = \boldsymbol{dx}\, \bar{S}$ and $\bar{g} = \boldsymbol{g}\bar{S}$, for any two coordinate systems S and $\bar{S}$. Since $\boldsymbol{dx}$ is an absolute, contravariant vector, we have $dx = L^{-1}\, d\bar{x}$, from 2.3(30). Substituting for $d\bar{x}$, we have $\widetilde{dx}\, g\, dx = \widetilde{dx}\, \tilde{L}\, \bar{g} L\, dx$. Using the fact that $\boldsymbol{dx}$ is arbitrary (so that $\boldsymbol{dx}\, \boldsymbol{dx}$ is an arbitrary, *symmetric* tensor), we can deduce only that the symmetric parts of the coefficient matrices must be equal, i.e., that $g + \tilde{g} = \tilde{L}\bar{g}L + \widetilde{\tilde{L}\bar{g}L} = \tilde{L}(\bar{g} + \tilde{\bar{g}})L$. But it is also given that $g = \tilde{g}$. Hence $g = \tilde{L}\bar{g}L$, which proves that $\boldsymbol{g}$ is an absolute, covariant, second-order tensor at Q. In this result, we are contracting the given function $\boldsymbol{g}$ (whose nature is to be determined) twice with the tensor $\boldsymbol{dx}\, \boldsymbol{dx}$, which is symmetric (and infinitesimal). If it were not given that $g = \tilde{g}$, we could deduce only that $\boldsymbol{g} + \tilde{\boldsymbol{g}}$ is a covariant tensor.

5.4 Correspondence between body and space fields at time t

(1) Any given pair P and P_1 of neighboring particles determines a unique contravariant body vector $\boldsymbol{d\xi} = \overrightarrow{PP_1}$, as defined in 2.3(9), which does not change with time even though the body is moving and deforming. Let P and P_1 occupy places Q and Q_1 at time t, Q' and Q_1' at time t', etc. Then there is a unique contravariant space vector $\boldsymbol{dx} = \overrightarrow{QQ_1}$ which corresponds to $\boldsymbol{d\xi}$ at time t, and there is another contravariant space vector $\boldsymbol{dx}' = \overrightarrow{Q'Q_1'}$ which corresponds to $\boldsymbol{d\xi}$ at time t', and so on [see Fig. 5.4(1)]. Thus *a given body vector* $\boldsymbol{d\xi}$ *generates a one-parameter family of space vectors* $\boldsymbol{dx}$, $\boldsymbol{dx}'$, ... *whenever a motion of the body through space is given.* We shall introduce the equivalent symbols $\mathbb{T}(t)$ and $\overset{t}{\to}$ to represent a correspondence of this kind between body vectors and space vectors which is generated by a state t, so that we may write the preceding example in the form

(2) $$\boldsymbol{d\xi} \overset{t}{\to} \boldsymbol{dx} = \mathbb{T}(t)\, \boldsymbol{d\xi}, \qquad \boldsymbol{d\xi} \overset{t'}{\to} \boldsymbol{dx}' = \mathbb{T}(t')\, \boldsymbol{d\xi}.$$

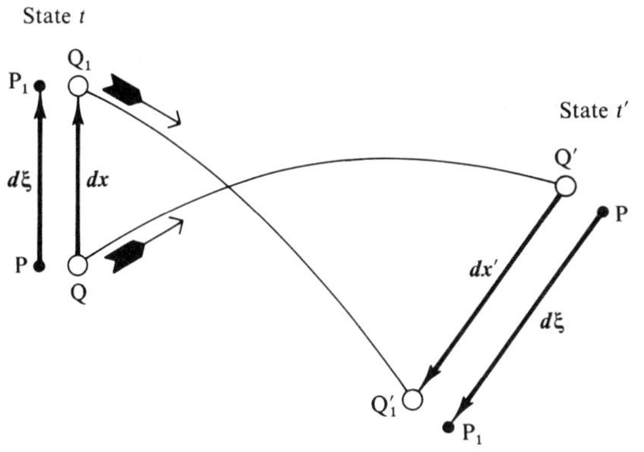

Figure 5.4(1) Neighboring particles P and P_1 move from places Q and Q_1 to places Q' and Q_1' (the particles have been drawn slightly away from the places they occupy for clarity). The constant body vector $\boldsymbol{d\xi}$ gives rise to two different space vectors $\boldsymbol{dx} = \mathbb{T}(t)\ \boldsymbol{d\xi}$ and $\boldsymbol{dx}' = \mathbb{T}(t')\ \boldsymbol{d\xi}$.

We define similar correspondences for body fields and space fields of any type as follows.

(3) **Definition** $\mathbb{T}(t)$ is a correspondence between space fields and body fields of the same rank, kind, and weight such that, for any given body tensor field $\boldsymbol{\Phi}(\mathrm{P})$, the corresponding space tensor field $\boldsymbol{F}(\mathrm{Q}, t)$ satisfies the equation

(4) $$FS = \boldsymbol{\Phi}B \qquad (S \stackrel{t}{\equiv} B;\quad \mathrm{Q} \stackrel{t}{\leftrightarrow} \mathrm{P}).$$

The notation $\mathrm{Q} \stackrel{t}{\leftrightarrow} \mathrm{P}$ means "particle P occupies place Q in state t." The correspondence will be written in the form

(5) $$\boldsymbol{\Phi}(\mathrm{P}) \stackrel{t}{\rightarrow} \boldsymbol{F}(\mathrm{Q}, t) = \mathbb{T}(t)\boldsymbol{\Phi}(\mathrm{P}).$$

Equation (4) means that $\boldsymbol{F}$ and $\boldsymbol{\Phi}$ *have the same representative matrix (or array) in coordinate systems S and B which are congruent in state t* [see 2.2(23)].

It is evident from the definition, coupled with the fact that a tensor field of given rank, kind, and weight defined over a given manifold is uniquely determined when its representative matrix (or array) is given in any one specified coordinate system, that *the correspondence* (3) *is one-to-one*; if the space field $\boldsymbol{F}(\mathrm{Q})$ is given, then we may write

(6) $$\boldsymbol{F}(\mathrm{Q}) \stackrel{t}{\rightarrow} \boldsymbol{\Phi}(\mathrm{P}, t) = \mathbb{T}^{-1}(t)\boldsymbol{F}(\mathrm{Q}),$$

where Eq. (4) serves to define the corresponding body field $\boldsymbol{\Phi}(\mathrm{P}, t)$.

(7) **Theorem** All "one-state" relations between, and operations with, body tensor fields are reproduced for the space fields generated by $\mathbb{T}(t)$.

Proof. The one-state operations defined so far are the following:

(8) addition, multiplication by scalars, formation of inner and outer products, transpose, reciprocal, and determinant.

The tensor nature of these operations derives entirely from the form of the transformation laws which relate representative matrices (or arrays) in any two body coordinate systems B and $\bar{B}$. It follows from (4) that

(9) $$\boldsymbol{F}S = \boldsymbol{\Phi}B, \qquad \boldsymbol{F}\bar{S} = \boldsymbol{\Phi}\bar{B}, \qquad \text{where} \quad S \stackrel{t}{\equiv} B \quad \text{and} \quad \bar{S} \stackrel{t}{\equiv} \bar{B}.$$

Hence any relation between the representative body arrays $\boldsymbol{\Phi}B$ and $\boldsymbol{\Phi}\bar{B}$ is reproduced between the corresponding space arrays $\boldsymbol{F}S$ and $\boldsymbol{F}\bar{S}$. Moreover, this statement holds for every pair S and $\bar{S}$ of space coordinate systems, because the appropriate body coordinate systems B and $\bar{B}$ can always be chosen such that $B \stackrel{t}{\equiv} S$ and $\bar{B} \stackrel{t}{\equiv} \bar{S}$. Thus the operations (8) on body tensor fields give rise to corresponding relations between space tensor field arrays for arbitrary space coordinate systems, and hence give rise to corresponding operations with the space tensor fields themselves. This proves the theorem.

It follows, in particular, that $\mathbb{T}(t)$ is an additive operator, in the sense of the definition 1.2(8)(i) (extended so as to apply to two different spaces), for, if $\boldsymbol{\Phi}_1$ and $\boldsymbol{\Phi}_2$ are any two body tensor fields of the same rank, kind, and weight, it follows from (7) that

(10) $$\boldsymbol{F}_1 + \boldsymbol{F}_2 = \mathbb{T}(t)(\boldsymbol{\Phi}_1 + \boldsymbol{\Phi}_2), \qquad \text{where} \quad \boldsymbol{F}_i = \mathbb{T}(t)\boldsymbol{\Phi}_i \qquad (i = 1, 2).$$

Another way of stating Theorem (7) is to say that *the operator* $\mathbb{T}(t)$ *commutes with the operations* (8). We shall see presently that $\mathbb{T}(t)$ does not commute with the operator $\partial/\partial t$, which is a "two-state" operator since it involves values $\boldsymbol{\Phi}(\mathrm{P}, t)$ and $\boldsymbol{\Phi}(\mathrm{P}, t')$ of a body tensor field, taken at two times t and t', and a limiting process in which $t' \to t$.

(11) **Theorem** The body and space metric tensor fields are related by the correspondence $\mathbb{T}(t)$ as follows:

(i) $\boldsymbol{\gamma}(\mathrm{P}, t) \stackrel{t}{\to} \boldsymbol{g}(\mathrm{Q}) = \mathbb{T}(t)\boldsymbol{\gamma}(\mathrm{P}, t)$;

(ii) $\boldsymbol{\gamma}^{-1}(\mathrm{P}, t) \stackrel{t}{\to} \boldsymbol{g}^{-1}(\mathrm{Q}) = \mathbb{T}(t)\boldsymbol{\gamma}^{-1}(\mathrm{P}, t)$.

Proof (a) On taking $B \stackrel{t}{\equiv} S$, we have $x^i = f^i(\xi, t) = \xi^i$, and hence 2.2(13) reduces to

(12) $$\gamma_{ij}(\xi, t) = g_{ij}(x) \qquad (B \stackrel{t}{\equiv} S),$$

which can be rewritten in the form

(13) $$\boldsymbol{g}(\mathrm{Q})S = \gamma(\mathrm{P}, t)B \qquad (S \overset{t}{\equiv} B).$$

This is of the form (4), with $\boldsymbol{F} \equiv \boldsymbol{g}$ and $\boldsymbol{\Phi} \equiv \gamma$; but $\boldsymbol{g}$ and γ are tensors of the same rank, kind, and weight, and therefore the relation between them satisfies all the conditions of the definition (3). This proves (11)(i). Then (11)(ii) follows from (7), which includes the fact that the operator $\mathbb{T}(t)$ commutes with the operation of taking the reciprocal. This proves (11). An alternative proof, which is instructive in illustrating various uses of Theorem (7), is as follows.

Proof (b) The validity of (2) follows from 2.2(7) on taking $B \overset{t}{\equiv} S$ and using the fact that $\boldsymbol{dx}$ and $\boldsymbol{d\xi}$ are both absolute, contravariant vectors. From 2.6(6) and 2.6(14), we have

$$\begin{aligned} \boldsymbol{g}(\mathrm{Q}) : (\boldsymbol{dx}\,\boldsymbol{dx}) = (ds_t)^2 &= \gamma(\mathrm{P}, t) : (\boldsymbol{d\xi}\,\boldsymbol{d\xi}) \\ &= \mathbb{T}(t)[\gamma(\mathrm{P}, t) : (\boldsymbol{d\xi}\,\boldsymbol{d\xi})] && \text{[by (7) applied to scalars]} \\ &= [\mathbb{T}(t)\gamma(\mathrm{P}, t)] : [\mathbb{T}(t)(\boldsymbol{d\xi}\,\boldsymbol{d\xi})] && \text{[by (7)]} \\ &= [\mathbb{T}(t)\gamma(\mathrm{P}, t)] : (\boldsymbol{dx}\,\boldsymbol{dx}) && \text{[by (2) and (7)].} \end{aligned}$$

This is valid for arbitrary infinitesimal $\boldsymbol{dx}$; $\boldsymbol{g}$ is symmetric; and $\mathbb{T}(t)\gamma(\mathrm{P}, t)$ is symmetric [from (7) and the fact that γ is symmetric]. Hence we can equate the coefficients of $\boldsymbol{dx}\,\boldsymbol{dx}$, and so obtain the required result (11)(i). Then (11)(ii) follows as in Proof (a).

(14) **Definition** The *Cauchy strain tensor field* $\boldsymbol{C}_t(\mathrm{Q}, t')$ and the *Finger strain tensor field* $\boldsymbol{B}_t(\mathrm{Q}, t')$ are obtained from the body metric tensor field as follows:

(i) $\gamma(\mathrm{P}, t') \overset{t}{\rightarrow} \boldsymbol{C}_t(\mathrm{Q}, t') = \mathbb{T}(t)\gamma(\mathrm{P}, t') \qquad (\mathrm{P} \overset{t}{\leftrightarrow} \mathrm{Q})$;

(ii) $\gamma^{-1}(\mathrm{P}, t') \overset{t}{\rightarrow} \boldsymbol{B}_t(\mathrm{Q}, t') = \mathbb{T}(t)\gamma^{-1}(\mathrm{P}, t') \qquad (\mathrm{P} \overset{t}{\leftrightarrow} \mathrm{Q})$.

These strain fields are therefore absolute, second-rank, space tensor fields; $\boldsymbol{C}$ is covariant, and $\boldsymbol{B}$ is contravariant. They have the following properties:

(15) $$\boldsymbol{C}_t(\mathrm{Q}, t) = \boldsymbol{g}(\mathrm{Q}); \qquad \boldsymbol{B}_t(\mathrm{Q}, t) = \boldsymbol{g}^{-1}(\mathrm{Q}) \qquad \text{[from (11)];}$$

(16) $$\boldsymbol{C}_t(\mathrm{Q}, t') = \boldsymbol{B}_t^{-1}(\mathrm{Q}, t') \qquad \text{[from (7)];}$$

(17) $$(ds')^2 = \boldsymbol{C}_t(\mathrm{Q}, t') : (\boldsymbol{dx}\,\boldsymbol{dx}).$$

Here, ds' denotes the separation at time t' of neighboring particles P and P_1 which occupy places Q and Q_1 at time t, where $\overrightarrow{\mathrm{QQ}_1} = \boldsymbol{dx}$; (17) follows from the equation

(18) $$(ds')^2 = \gamma(\mathrm{P}, t') : (\boldsymbol{d\xi}\,\boldsymbol{d\xi})$$

on applying the transfer operator $\mathbb{T}(t)$ and using (7), (2), and (14)(i).

We use the notation $\boldsymbol{C}_t(\mathrm{Q}, t')$ [rather than $\boldsymbol{C}(\mathrm{Q}, t', t)$, say] to emphasize the fact that t and t' *play different roles*: t is the time at which the body metric tensor field $\boldsymbol{\gamma}(\mathrm{P}, t')$ is transferred to space; $\boldsymbol{C}_t(\mathrm{Q}, t')$ is not symmetric (or antisymmetric) in t and t'. The same is true for the space strain tensor field $\boldsymbol{C}_t(\mathrm{Q}, t') - \boldsymbol{g}(\mathrm{Q})$, which satisfies the equation

$$(ds')^2 - (ds)^2 = [\boldsymbol{C}_t(\mathrm{Q}, t') - \boldsymbol{g}(\mathrm{Q})]:(d\boldsymbol{x}\, d\boldsymbol{x}). \tag{19}$$

The corresponding body tensor equation

$$(ds')^2 - (ds)^2 = [\boldsymbol{\gamma}(\mathrm{P}, t') - \boldsymbol{\gamma}(\mathrm{P}, t)]:(d\boldsymbol{\xi}\, d\boldsymbol{\xi}), \tag{20}$$

on the other hand, involves a body strain tensor $\boldsymbol{\gamma}(\mathrm{P}, t') - \boldsymbol{\gamma}(\mathrm{P}, t)$ which does treat the times t and t' on an equal footing, as noted in 2.6(25).

Let $\boldsymbol{\Phi}(\mathrm{P}, t')$ be any time-dependent body tensor field and let

$$\boldsymbol{F}_t(\mathrm{Q}, t') = \mathbb{T}(t)\boldsymbol{\Phi}(\mathrm{P}, t') \qquad (\mathrm{P} \overset{t}{\leftrightarrow} \mathrm{Q}) \tag{21}$$

denote the corresponding space tensor field obtained by transfer at time t. On subtracting from (21) the corresponding equation obtained by putting t'' for t', dividing by $t' - t''$, and taking the limit $t'' \to t'$, we see that

$$\frac{\partial \boldsymbol{F}_t(\mathrm{Q}, t')}{\partial t'} = \mathbb{T}(t)\frac{\partial \boldsymbol{\Phi}(\mathrm{P}, t')}{\partial t'} \qquad (\mathrm{P} \overset{t}{\leftrightarrow} \mathrm{Q}), \tag{22}$$

which shows that $\mathbb{T}(t)$ *commutes with* $\partial/\partial t'$ *when* $t' \neq t$. Taking the limit $t' \to t$, we obtain the result

$$\left[\frac{\partial \boldsymbol{F}_t(\mathrm{Q}, t')}{\partial t'}\right]_{t'=t} = \mathbb{T}(t)\frac{\partial \boldsymbol{\Phi}(\mathrm{P}, t)}{\partial t}. \tag{23}$$

Since

$$[\partial \boldsymbol{F}_t(\mathrm{Q}, t')/\partial t']_{t'=t} \neq \partial \boldsymbol{F}_t(\mathrm{Q}, t)/\partial t, \tag{24}$$

it follows that *the operator* $\mathbb{T}(t)$ *does not commute with* $\partial/\partial t$. In applications of the formalism to strain-rate tensors, it is customary to write $\boldsymbol{F}_t(\mathrm{Q}, t) = \boldsymbol{A}^m(\mathrm{Q}, t)$, thus losing sight of the two different sources of time dependence, namely from t in $\boldsymbol{F}_t(\mathrm{Q}, t')$ and from t' in $\boldsymbol{F}_t(\mathrm{Q}, t')$, as $t' \to t$. An extra calculation is required in order to obtain the relation between $\mathbb{T}(t)\, \partial\boldsymbol{\Phi}(\mathrm{P}, t)/\partial t$ and $\partial \boldsymbol{F}_t(\mathrm{Q}, t)/\partial t$, as we now show.

(25) **Theorem** Let $\boldsymbol{\Phi}(\mathrm{P}, t)$ be any time-dependent body tensor field, and let $\boldsymbol{F}(\mathrm{Q}, t) = \mathbb{T}(t)\boldsymbol{\Phi}(\mathrm{P}, t)$ denote the corresponding space tensor field obtained by transfer at time t. We define the *convected derivative* $\mathscr{D}\boldsymbol{F}/\mathscr{D}t$ of $\boldsymbol{F}$ by the equation

$$\mathring{\boldsymbol{F}} \equiv \mathscr{D}\boldsymbol{F}(\mathrm{Q}, t)/\mathscr{D}t = \mathbb{T}(t)\, \partial\boldsymbol{\Phi}(\mathrm{P}, t)/\partial t \qquad (\mathrm{P} \overset{t}{\leftrightarrow} \mathrm{Q}). \tag{26}$$

Then, in an arbitrary space coordinate system $S: \mathrm{Q} \to x$, the components $\mathring{F}^{ij\cdots}_{rs\cdots}(x, t)$ of the convected derivative $\mathring{\boldsymbol{F}}$ are given by the equation

$$\mathring{F}^{ij\cdots}_{rs\cdots} = \left(\frac{\partial}{\partial t} + v^k \frac{\partial}{\partial x^k} + n \frac{\partial v^k}{\partial x^k}\right) F^{ij\cdots}_{rs\cdots}(x, t) + \sum \frac{\partial v^k}{\partial x^r} F^{ij\cdots}_{ks\cdots} - \sum' \frac{\partial v^i}{\partial x^k} F^{kj\cdots}_{rs\cdots}, \tag{27}$$

where the $v^k(x, t) = \boldsymbol{v}(\mathrm{Q}, t)S$ are the components in S of the contravariant velocity field [defined in 2.3(8) and 2.3(39)], n is the weight of $\boldsymbol{\Phi}$ [and therefore of $\boldsymbol{F}$ and $\mathring{\boldsymbol{F}}$, by (3)], the $F^{ij\cdots}_{rs\cdots}(x, t)$ are the components in S of $\boldsymbol{F}(\mathrm{Q}, t)$, $\sum$ denotes a sum of terms containing one term of the given form for each covariant index, and $\sum'$ denotes a sum containing one term for each contravariant index.

Proof We note first that from the definitions (26) and (3), it follows that the convected derivative $\mathring{\boldsymbol{F}}$ is a space tensor field of the same rank, kind, and weight as $\boldsymbol{F}$ (and $\boldsymbol{\Phi}$). The combination of the operations $\mathbb{T}(t)$ and $\partial/\partial t$ involved in the definition of the convected derivative thus represents a new method of obtaining a space tensor field $\mathring{\boldsymbol{F}}(\mathrm{Q}, t)$ from any given space tensor field $\boldsymbol{F}(\mathrm{Q}, t)$, by means of the following sequence of operations:

$$\begin{aligned} \boldsymbol{F}(\mathrm{Q}, t) &\xrightarrow{t} \boldsymbol{\Phi}(\mathrm{P}, t) = \mathbb{T}^{-1}(t)\boldsymbol{F}(\mathrm{Q}, t); \\ \dot{\boldsymbol{\Phi}}(\mathrm{P}, t) &\xrightarrow{t} \mathring{\boldsymbol{F}}(\mathrm{Q}, t) = \mathbb{T}(t)\dot{\boldsymbol{\Phi}}(\mathrm{P}, t). \end{aligned} \tag{28}$$

The relation between $\mathring{\boldsymbol{F}}$ and $\boldsymbol{F}$ depends on the motion of the body through the velocity field $\boldsymbol{v}(\mathrm{Q}, t)$. In agreement with the notation 2.6(23), we write $\dot{\boldsymbol{\Phi}}(\mathrm{P}, t)$ for $\partial\boldsymbol{\Phi}(\mathrm{P}, t)/\partial t$.

To prove (27), we differentiate with respect to t the relations between $\boldsymbol{F}(\mathrm{Q}, t)S$ and $\boldsymbol{\Phi}(\mathrm{P}, t)B$, where S and B are arbitrary space and body coordinate systems (not restricted to be congruent in state t). The various terms in (27) come from the time derivatives of the various factors in these relations.

Let $\bar{S} \overset{t}{\equiv} B$; then $\bar{x}^i = \xi^i$ at t, and, from (4) with $\bar{S}$ for S, we have $\boldsymbol{\Phi}(\mathrm{P},t)\, B = \boldsymbol{F}(\mathrm{Q}, t)\bar{S}$, i.e.,

$$\Phi^{ij\cdots}_{rs\cdots}(\xi, t) = \bar{F}^{ij\cdots}_{rs\cdots}(\bar{x}, t) = \left|\frac{\partial x}{\partial \bar{x}}\right|^n \frac{\partial \bar{x}^i}{\partial x^a} \cdots \frac{\partial x^u}{\partial \bar{x}^r} \cdots F^{a\cdots}_{u\cdots}(x, t);$$

this follows from the space analog of the transformation law 5.2(5) for a general body tensor. Since $\bar{x}^i = \xi^i$, we can rewrite this in the form

$$\Phi^{ij\cdots}_{rs\cdots}(\xi, t) = \left|\frac{\partial x}{\partial \xi}\right|^n \frac{\partial \xi^i}{\partial x^a} \cdots \frac{\partial x^u}{\partial \xi^r} \cdots F^{ab\cdots}_{uv\cdots}(x, t) \qquad [x^k = f^k(\xi, t)], \tag{29}$$

where

(30)
$$\partial x^u/\partial \xi^r \equiv \partial f^u(\xi, t)/\partial \xi^r,$$
$$\partial \xi^i/\partial x^a \equiv \partial \varphi^i(x, t)/\partial x^a, \qquad |\partial x/\partial \xi| \equiv |\partial x^k/\partial \xi^c|.$$

$f^u(\xi, t)$ and their inverses $\varphi^i(x, t)$ describe the motion, as defined in 2.2(4) and 2.2(5), when referred to arbitrary B and S.

We have thus obtained the important result that the condition (29) is necessary (for arbitrary $B\colon \mathrm{P} \to \xi$ and $S\colon \mathrm{Q} \to x$) if $\boldsymbol{F}(\mathrm{Q}, t) = \mathbb{T}(t)\boldsymbol{\Phi}(\mathrm{P}, t)$. It is easy to see that the argument leading to (29) is reversible. Furthermore, the argument is evidently unaffected if either $\boldsymbol{\Phi}$ or $\boldsymbol{F}$ or both depend on any parameter t', say, other than t. We have therefore proved the following result.

(31) **Theorem** Let $\boldsymbol{F}(\mathrm{Q}, t, t')$ and $\boldsymbol{\Phi}(\mathrm{P}, t, t')$ denote space and body tensor fields of the same rank, kind, and weight $(=n)$. In order that they shall be related by transfer at time t, i.e., that

(i) $\boldsymbol{\Phi}(\mathrm{P}, t, t') \xrightarrow{t} \boldsymbol{F}(\mathrm{Q}, t, t') = \mathbb{T}(t)\boldsymbol{\Phi}(\mathrm{P}, t, t') \qquad (\mathrm{P} \overset{t}{\leftrightarrow} \mathrm{Q}),$

it is necessary and sufficient that their components in *arbitrary* space and body coordinate systems $S\colon \mathrm{Q} \to x$ and $B\colon \mathrm{P} \to \xi$ shall satisfy the equations

(ii) $$\Phi^{ij\cdots}_{rs\ldots}(\xi, t, t') = \left|\frac{\partial x}{\partial \xi}\right|^n \frac{\partial \xi^i}{\partial x^a} \cdots \frac{\partial x^u}{\partial \xi^r} \cdots F^{ab\cdots}_{uv\ldots}(x, t, t'),$$

or the equations

(iii) $$F^{ij\cdots}_{rs\ldots}(x, t, t') = \left|\frac{\partial \xi}{\partial x}\right|^n \frac{\partial x^i}{\partial \xi^a} \cdots \frac{\partial \xi^u}{\partial x^r} \cdots \Phi^{ab\cdots}_{uv\ldots}(\xi, t, t'),$$

where

(iv) $x^k = f^k(\xi, t), \qquad \xi^k = \varphi^k(x, t)$

are the equations describing the motion, when referred to S and B.

As a mnemonic, it is evident that equations (ii) and (iii) are just those transformation equations that would be obtained if $\Phi^{ij\cdots}_{rs\ldots}$ and $F^{ab\cdots}_{uv\ldots}$ were regarded as components of a space field referred to two space coordinate systems, $\mathrm{Q} \to \xi$ and $\mathrm{Q} \to x$, respectively.

We now differentiate (29) with respect to t at constant ξ; we use the notation $\partial/\partial t$, or $(\partial/\partial t)_\xi$ when necessary, for this differentiation. On the right-hand side of (29), there are factors of four different kinds. We differentiate them in turn. For the determinant factor, we have

$$(32)\qquad \left(\frac{\partial}{\partial t}\right)_\xi \left|\frac{\partial x}{\partial \xi}\right|^n = n\left|\frac{\partial x}{\partial \xi}\right|^n \frac{\partial}{\partial t}\log_e\left|\frac{\partial x}{\partial \xi}\right|$$

$$= n\left|\frac{\partial x}{\partial \xi}\right|^n \frac{\partial \xi^i}{\partial x^k}\frac{\partial}{\partial t}\frac{\partial x^k}{\partial \xi^i} \qquad \left[\text{from 1.3(33) with } A_{rc} = \frac{\partial x^r}{\partial \xi^c}\right]$$

$$= n\left|\frac{\partial x}{\partial \xi}\right|^n \frac{\partial v^k(x, t)}{\partial x^k},$$

since

$$(33)\qquad \frac{\partial}{\partial t}\frac{\partial x^k}{\partial \xi^i} = \frac{\partial^2 f^k(\xi, t)}{\partial t\, \partial \xi^i} = \frac{\partial^2 f^k(\xi, t)}{\partial \xi^i\, \partial t} = \frac{\partial v^k}{\partial \xi^i} \qquad \text{[from 2.2(8)].}$$

For the tensor component factor, we have

$$(34)\qquad \left(\frac{\partial}{\partial t}\right)_\xi F^{ab\cdots}_{uv\ldots}(x, t) = \frac{\partial F^{ab\cdots}_{uv\ldots}(x, t)}{\partial t} + v^k \frac{\partial F^{ab\cdots}_{uv\ldots}(x, t)}{\partial x^k},$$

since 2.2(8) may be rewritten in the form

$$(35)\qquad v^k = \left(\frac{\partial}{\partial t}\right)_\xi x^k.$$

For the covariant transformation factor, we have, from (33),

$$(36)\qquad \left(\frac{\partial}{\partial t}\right)_\xi \frac{\partial x^u}{\partial \xi^r} = \frac{\partial v^u}{\partial \xi^r} = \frac{\partial x^k}{\partial \xi^r}\frac{\partial v^u}{\partial x^k}.$$

For the contravariant transformation factor, we have

$$(37)\qquad \frac{\partial \xi^i}{\partial x^a}\frac{\partial x^k}{\partial \xi^i} = \delta^k_a;$$

on differentiating this equation, using (36), and then reusing (37), we obtain the result

$$(38)\qquad \left(\frac{\partial}{\partial t}\right)_\xi \frac{\partial \xi^i}{\partial x^a} = -\frac{\partial \xi^i}{\partial x^k}\frac{\partial v^k}{\partial x^a}.$$

Finally, it is evident that for each of the four types of factor on the right-hand side of (29), the time derivative [given by (32), (34), (36), and (38)] contains a factor of the type differentiated. When the results are substituted in the time derivative of (29), therefore, we obtain an equation of the same form as (29) in which $\Phi^{ij\cdots}_{rs\ldots}(\xi, t)$ is replaced by $\partial\Phi^{ij\cdots}_{rs\ldots}(\xi, t)/\partial t$, and $F^{ab\cdots}_{uv\ldots}(x, t)$ is replaced by a sum of terms which are readily found to be just those given by $\mathring{F}^{ab\cdots}_{uv\ldots}(x, t)$ as defined in (27). By Theorem (31), it then follows that (26) is true, which completes the proof of Theorem (25).

It is noteworthy that, with the exception of the term $\partial F^{ij\cdots}_{rs\ldots}/\partial t$, the individual terms on the right-hand side of (27) are not components of tensors; this is readily verified by differentiating the appropriate transformation laws with respect to x^m. Theorem (25) shows, however, that the particular combination of terms on the right-hand side of (27) are components of a tensor. The "nontensor" terms in the differentiated transformation law must therefore cancel one another. The reader may readily verify this in the following simple example.

(39) **Problem** Let $\boldsymbol{X}(\mathrm{Q})$ and $\boldsymbol{Y}(\mathrm{Q})$ be absolute, contravariant, space vector fields. For any two space coordinate systems $S:\mathrm{Q} \to x$ and $\bar{S}:\mathrm{Q} \to \bar{x}$, prove that $Y^k\,\partial X^i/\partial x^k - X^k\,\partial Y^i/\partial x^k = (\partial x^i/\partial \bar{x}^j)(\bar{Y}^k\,\partial \bar{X}^j/\partial \bar{x}^k - \bar{X}^k\,\partial \bar{Y}^j/\partial \bar{x}^k)$, where $\boldsymbol{X}S = [X^r]$, $\boldsymbol{Y}S = [Y^r]$, $\boldsymbol{X}\bar{S} = [\bar{X}^r]$, and $\boldsymbol{Y}\bar{S} = [\bar{Y}^r]$. (It is conventional to express a statement of this kind as follows: $Y^k\,\partial X^i/\partial x^k - X^k\,\partial Y^i/\partial x^k$ are components of an absolute, contravariant vector.)

An important application of Theorem (25) enables us to write down the relations between components of certain covariant, absolute, second-rank space tensor fields $\boldsymbol{A}^m(\mathrm{Q}, t)$ $(m = 1, 2, \ldots)$ called *strain-rate fields*, which may be defined as follows:

(40) $$\boldsymbol{A}^m(\mathrm{Q}, t) = \mathbb{T}(t)\,\partial^m\gamma(\mathrm{P}, t)/\partial t^m \qquad (\mathrm{P} \overset{t}{\leftrightarrow} \mathrm{Q}).$$

From 2.6(20) and Theorem (7), it follows that

(41) $$\frac{d^m}{dt^m}(ds)^2 = (\boldsymbol{dx}\,\boldsymbol{dx}) : \boldsymbol{A}^m(\mathrm{Q}, t).$$

From Theorem (25), with $\boldsymbol{F} = \boldsymbol{A}^m$ and $\mathring{\boldsymbol{F}} = \boldsymbol{A}^{m+1}$, it follows that

(42) $$A^{(m+1)}_{rs}(x, t) = \left(\frac{\partial}{\partial t} + v^k\frac{\partial}{\partial x^k}\right)A^{(m)}_{rs}(x, t) + \frac{\partial v^k}{\partial x^r}A^{(m)}_{ks} + \frac{\partial v^k}{\partial x^s}A^{(m)}_{rk} \qquad (m = 1, 2, \ldots).$$

From Theorem (25), with $\boldsymbol{F} = \boldsymbol{g}$ and $\mathring{\boldsymbol{F}} = \boldsymbol{A}^1$, we have

(43) $$A^{(1)}_{rs}(x, t) = v^k\frac{\partial}{\partial x^k}g_{rs}(x) + \frac{\partial v^k}{\partial x^r}g_{ks} + \frac{\partial v^k}{\partial x^s}g_{rk}\,.$$

Consistent with our previous notation, we use $A^{(m)}_{rs}$ to denote the components of $\boldsymbol{A}^m$ in S.

Equations (42), which relate components of the mth and $(m + 1)$th space strain-rate tensors, may be compared with the following equations:

(44) $$\frac{\partial^{m+1}}{\partial t^{m+1}} \gamma_{rs}(\xi, t) = \frac{\partial}{\partial t} \frac{\partial^m}{\partial t^m} \gamma_{rs}(\xi, t).$$

Equations (42) and (44) are related by the transfer operator $\mathbb{T}(t)$. It is evident that the body field equations (44) are much simpler than the corresponding space field equations (42).

(45) **Problem** A space tensor field is $\dot{\boldsymbol{C}}$ defined by the equation

(i) $\dot{\boldsymbol{C}}_t(\mathrm{Q}, t') = \partial \boldsymbol{C}_t(\mathrm{Q}, t')/\partial t'$.

Prove the following results:

(ii) $\dot{\boldsymbol{C}}_t(\mathrm{Q}, t') = \mathbb{T}(t)\, \partial \boldsymbol{\gamma}(\mathrm{P}, t')/\partial t' \qquad (\mathrm{P} \overset{t}{\leftrightarrow} \mathrm{Q});$

(iii) $\dot{\boldsymbol{C}}_t(\mathrm{Q}, t) = \boldsymbol{A}^1(\mathrm{Q}, t);$

(iv) $$\dot{\boldsymbol{C}}_t(\mathrm{Q}, t')S = \left[\frac{\partial x'^i}{\partial x^r} \frac{\partial x'^j}{\partial x^c} A_{ij}^{(1)}(x', t')\right],$$

where x'^i and x^i denote the coordinates in S of the places Q′ and Q that particle P occupies at times t' and t, respectively.

These space fields have been chosen to be covariant. It is obvious that similar *contravariant strain-rate fields* can be defined by the equation

(46) $$\boldsymbol{A}_m(\mathrm{Q}, t) = \mathbb{T}(t)\, \partial^m \boldsymbol{\gamma}^{-1}(\mathrm{P}, t)/\partial t^m \qquad (m = 1, 2, \ldots),$$

which is similar to the covariant equation (40). The reader may readily verify that there are equations, similar to those of (45), which relate $\boldsymbol{A}_1$ and $\dot{\boldsymbol{B}}_t(\mathrm{Q}, t') = \partial \boldsymbol{B}_t(\mathrm{Q}, t')/\partial t'$, where $\boldsymbol{B}$ is the Finger strain tensor defined in (14)(ii).

Let $\boldsymbol{\Phi}(\mathrm{P}, t, t')$ be any body tensor field which depends on t and t'. Since integration with respect to t' can be defined as a limit of a process which involves addition of body tensors at P and multiplication by scalars, it is evident that *such a process of integration commutes with the operation of transfer to space at time* t, i.e., that

(47) $$\mathbb{T}(t)\int_{t_1}^{t_2} \boldsymbol{\Phi}(\mathrm{P}, t, t')\, dt' = \int_{t_1}^{t_2} \mathbb{T}(t)\boldsymbol{\Phi}(\mathrm{P}, t, t')\, dt'.$$

A similar result is evidently valid for any number of integrations, when the body field depends on a number of parameters $t, t', t'', \ldots$:

(48) $$\mathbb{T}(t)\int dt' \int dt'' \cdots \boldsymbol{\Phi}(\mathrm{P}, t, t', t'', \ldots)$$
$$= \int dt' \int dt'' \cdots \mathbb{T}(t)\boldsymbol{\Phi}(\mathrm{P}, t, t'\ t'', \ldots).$$

In particular, if $\mu(\mathrm{P}, t, t')$ is any absolute scalar body field and $m(\mathrm{Q}, t, t')$ is the corresponding space field obtained by transfer at time t, so that

(49) $$\mu(\mathrm{P}, t, t') \xrightarrow{t} m(\mathrm{Q}, t, t') = \mu(\mathrm{P}, t, t') \qquad (\mathrm{P} \overset{t}{\leftrightarrow} \mathrm{Q}),$$

we have the following results, which will be used later:

(50) $$\mathbb{T}(t) \int \mu(\mathrm{P}, t, t')\gamma(\mathrm{P}, t')\, dt' = \int m(\mathrm{Q}, t, t')\boldsymbol{C}_t(\mathrm{Q}, t')\, dt';$$

(51) $$\mathbb{T}(t) \int \mu(\mathrm{P}, t, t')\gamma^{-1}(\mathrm{P}, t')\, dt' = \int m(\mathrm{Q}, t, t')\boldsymbol{B}_t(\mathrm{Q}, t')\, dt';$$

(52) $$\mathbb{T}(t) \int \mu(\mathrm{P}, t, t')(\partial/\partial t')\gamma(\mathrm{P}, t')\, dt' = \int m(\mathrm{Q}, t, t')\dot{\boldsymbol{C}}_t(\mathrm{Q}, t')\, dt';$$

(53) $$\mathbb{T}(t) \int \mu(\mathrm{P}, t, t')(\partial/\partial t')\gamma^{-1}(\mathrm{P}, t')\, dt' = \int m(\mathrm{Q}, t, t')\dot{\boldsymbol{B}}_t(\mathrm{Q}, t')\, dt'.$$

In these equations, $\mathrm{P} \overset{t}{\leftrightarrow} \mathrm{Q}$; and

(54) $$\dot{\boldsymbol{B}}_t(\mathrm{Q}, t') = \mathbb{T}(t)(\partial/\partial t')\gamma^{-1}(\mathrm{P}, t') \qquad (\mathrm{P} \overset{t}{\leftrightarrow} \mathrm{Q}).$$

It is evident that the range of integration is immaterial, provided only that the integrals exist. In particular, the upper limit can be set equal to t without affecting the validity of the results.

(55) **Problem** Prove the following results:

(i) $$\dot{\boldsymbol{B}}_t(\mathrm{Q}, t')S = \left[\frac{\partial x^r}{\partial x'^i}\frac{\partial x^c}{\partial x'^j} A^{ij}_{(1)}(x', t')\right];$$

(ii) $$A^{ij}_{(1)}(x', t') = -g^{ir}(x')g^{js}(x')A^{(1)}_{rs}(x', t');$$

(iii) $$\boldsymbol{C}_t(\mathrm{Q}, t'')S = [g_{rc}(x)] - \left[\int_{t''}^{t} \frac{\partial x'^i}{\partial x^r}\frac{\partial x'^j}{\partial x^c} A^{(1)}_{ij}(x', t')\, dt'\right].$$

Here, x^i and x'^i denote coordinates in S of the places Q and Q′ that particle P occupies at times t and t', respectively.

(56) **Problem** Prove that

(i) $$\int_{t''}^{t} \mu(t - t')\boldsymbol{B}_t(\mathrm{Q}, t')\, dt' = G(0)\boldsymbol{g}^{-1}(\mathrm{Q}) - G(t - t'')\boldsymbol{B}_t(\mathrm{Q}, t'') - \int_{t''}^{t} G(t - t')\dot{\boldsymbol{B}}_t(\mathrm{Q}, t')\, dt',$$

where

(ii) $G(t) = \int_t^{t''} \mu(s)\, ds.$

(57) **Problem** Prove the results

(i) $C_t(\mathrm{Q}, t')S = \left[\dfrac{\partial x'^i}{\partial x^r} \dfrac{\partial x'^j}{\partial x^c}\, g_{ij}(x')\right];$

(ii) $B_t(\mathrm{Q}, t')S = \left[\dfrac{\partial x^r}{\partial x'^i} \dfrac{\partial x^c}{\partial x'^j}\, g^{ij}(x')\right].$

For any given body tensor field, the transfer operation $\mathbb{T}(t)$, as defined in (3), gives rise to a unique *general* space tensor field. But this general space tensor field in turn gives rise to a unique *Cartesian* space tensor field (of the same rank and parity of weight), according to the correspondence defined in Section 4.4. *A given state t therefore generates a correspondence between body tensor fields and Cartesian space tensor fields*; we shall use the same symbol $\mathbb{T}(t)$ for this correspondence, which may be defined as follows.

(58) **Definition** Let $\boldsymbol{\Phi}(\mathrm{P})$ be any body tensor field of rank r and weight n. Define a Cartesian space tensor field $\mathbf{F}_t(\mathrm{Q}) = \mathbb{T}(t)\boldsymbol{\Phi}(\mathrm{P})$ of rank r and weight 0 if n is even, and 1 if n is odd, whose representative matrix (or array) in an arbitrary rectangular Cartesian space coordinate system $C : \mathrm{Q} \rightarrow y$ is given by the equation

(i) $F_t(\mathrm{Q})C = \Phi(\mathrm{P})B$, where $B \overset{t}{\equiv} C$ (and $\mathrm{P} \overset{t}{\leftrightarrow} \mathrm{Q}$).

Under the transformation $\mathbb{T}(t)$, a given body field always gives rise to a unique Cartesian space tensor field, but the converse is not true. There are always body tensor fields of different kinds and given rank whose representative matrices (or arrays) in any given body coordinate system, which is instantaneously rectangular Cartesian at time t, are equal. Such body fields will therefore give rise to the same Cartesian tensor field under the transformation $\mathbb{T}(t)$.

An important example of this is furnished by the body metric tensors $\gamma(\mathrm{P}, t)$ and $\gamma^{-1}(\mathrm{P}, t)$ and the body unit tensor $\boldsymbol{\delta}$, which are each represented by the unit matrix I in any body coordinate system that is instantaneously rectangular Cartesian at time t. It follows that

(59) $$\mathbb{T}(t)\gamma(\mathrm{P}, t) = \mathbb{T}(t)\gamma^{-1}(\mathrm{P}, t) = \mathbb{T}(t)\,\boldsymbol{\delta} = \mathbf{U},$$

where $\mathbf{U}$ is the Cartesian unit tensor.

It follows from this that *any two body tensor fields that are related to one another by contraction with the body metric tensor* $\gamma(P, t)$ *will give rise to the same Cartesian tensor field on transfer to space at time* t. The time derivatives of such body tensor fields, however, will give rise to different Cartesian tensor fields on transfer to space at time t, and *therefore a given Cartesian space tensor field will, in general, possess many different convected derivatives.* Convected derivatives of the Cartesian unit tensor $\mathbf{U}$, for example, include $\mathbf{A}^1$, $-\mathbf{A}^1$, and $\mathbf{0}$, because

(60) $$\mathbb{T}(t)\frac{\partial}{\partial t}\gamma(P, t) = \mathbf{A}^1; \quad \mathbf{A}^1 C = \left[\frac{\partial v^c}{\partial y^r} + \frac{\partial v^r}{\partial y^c}\right] \quad \text{[from (43) with } g_{rs} = \delta_{rs}\text{]};$$

(61) $$\mathbb{T}(t)\frac{\partial}{\partial t}\gamma^{-1}(P, t) = \mathbf{A}_1; \quad \mathbf{A}_1 C = -\left[\frac{\partial v^c}{\partial y^r} + \frac{\partial v^r}{\partial y^c}\right] \quad \text{[from (55)(ii)]};$$

(62) $$\mathbb{T}(t)\frac{\partial}{\partial t}\boldsymbol{\delta} = \mathbf{0}.$$

Other convected derivatives of $\mathbf{U}$ can be obtained by using relative tensors, e.g., by considering $\mathbb{T}(t)\,\partial[|\gamma|^n\gamma]/\partial t$, for various values of n.

Another important example arises with the body stress tensor field $\boldsymbol{\pi} = \boldsymbol{\pi}(P, t)$, an absolute, second-rank, contravariant body tensor field defined in Section 5.6. Writing $\gamma(P, t) = \gamma$, we have

(63) $$\mathbb{T}(t)\boldsymbol{\pi} = \mathbb{T}(t)\gamma \cdot \boldsymbol{\pi} = \mathbb{T}(t)\gamma \cdot \boldsymbol{\pi} \cdot \gamma = \mathbf{p},$$

where $\mathbf{p} = \mathbf{p}(Q, t)$ is the Cartesian stress tensor (of second rank and zero weight) obtained by transfer at time t of the contravariant tensor $\boldsymbol{\pi}$, the mixed tensor $\gamma \cdot \boldsymbol{\pi}$, or the covariant tensor $\gamma \cdot \boldsymbol{\pi} \cdot \gamma$. By taking the time derivatives of these body stress tensors and then transferring to space at time t, we obtain the following results:

(64) $$\left\{\mathbb{T}(t)\frac{\partial \boldsymbol{\pi}}{\partial t}\right\}C = \left[\left(\frac{\partial}{\partial t} + v^k\frac{\partial}{\partial y^k}\right)p^{rc}(y, t) - \frac{\partial v^r}{\partial y^k}p^{kc} - \frac{\partial v^c}{\partial y^k}p^{rk}\right];$$

(65) $$\left\{\mathbb{T}(t)\frac{\partial}{\partial t}\gamma \cdot \boldsymbol{\pi}\right\}C = \left[\left(\frac{\partial}{\partial t} + v^k\frac{\partial}{\partial y^k}\right)p^{rc}(y, t) + \frac{\partial v^k}{\partial y^r}p^{kc} - \frac{\partial v^c}{\partial y^k}p^{rk}\right];$$

(66) $$\left\{\mathbb{T}(t)\frac{\partial}{\partial t}\gamma \cdot \boldsymbol{\pi} \cdot \gamma\right\}C = \left[\left(\frac{\partial}{\partial t} + v^k\frac{\partial}{\partial y^k}\right)p^{rc}(y, t) + \frac{\partial v^k}{\partial y^r}p^{kc} + \frac{\partial v^k}{\partial y^c}p^{rk}\right].$$

The right-hand sides are the components in a rectangular Cartesian coordinate system $C: Q \to y$ of three of the *convected derivatives of the Cartesian space tensor field* $\mathbf{p}(Q, t)$. They have been obtained by using (27), whose validity for Cartesian fields is easy to see.

The transfer of base vectors is treated in Section 8.3.

If the contravariant body stress tensor $\boldsymbol{\pi}(\mathrm{P}, t)$ is transferred to space at any other time $\bar{t}$ and multiplied by a scalar factor equal to the volume ratio in states t and $\bar{t}$, we obtain the Kirchhoff stress tensor $\bar{\boldsymbol{S}}(\bar{\mathrm{Q}}, t, \bar{t})$:

$$(67) \qquad \bar{\boldsymbol{S}}(\bar{\mathrm{Q}}, t, \bar{t}) = \frac{\rho(\mathrm{P}, \bar{t})}{\rho(\mathrm{P}, t)} \mathbb{T}(\bar{t})\boldsymbol{\pi}(\mathrm{P}, t).$$

The notation $\bar{\boldsymbol{S}}$ used here is related to that used by Prager (1961, Eq. (4.6)); one can easily verify that the components $\bar{S}^{ij}$ of $\bar{\boldsymbol{S}}$ defined by (67) are the same as the components $\bar{S}_{ij}$ given in Prager's equation (4.6): one can use our equation (31)(iii) with t, t', x^i, F, Φ replaced by $\bar{t}$, t, a_i, $\bar{S}$, and π, respectively, to carry out the transfer operation at time $\bar{t}$, and then one can take $B \equiv \overset{t}{S}$ to express the result in terms of the space stress tensor at time t. $\bar{\boldsymbol{S}}$, as defined by (67), can be taken to be either a Cartesian space tensor or an absolute contravariant space tensor. It should be emphasized that *the Kirchhoff stress tensor $\bar{\boldsymbol{S}}(\bar{\mathrm{Q}}, t, \bar{t})$ is a space tensor which depends on two states, t and $\bar{t}$, and thus furnishes a description of the state of stress in state t which is more complicated than that furnished by the one-state body tensor $\boldsymbol{\pi}(\mathrm{P}, t)$*. The relation between the Kirchhoff stress tensor and the space stress tensor $\boldsymbol{p}$ (both absolute contravariant) is given by the equation

$$(68) \qquad \bar{\boldsymbol{S}}(\bar{\mathrm{Q}}, t, \bar{t}) = \frac{\rho(\mathrm{P}, \bar{t})}{\rho(\mathrm{P}, \bar{t})} \mathbb{T}(\bar{t})\mathbb{T}^{-1}(t)\boldsymbol{p}(\bar{\mathrm{Q}}, t),$$

which comes from (67) and the equation $\boldsymbol{p}(\mathrm{Q}, t) = \mathbb{T}^{-1}(t)\boldsymbol{\pi}(\mathrm{P}, t)$.

5.5 Volume and surface elements

It is well known that a change of variables in a multiple integral involves a factor in the integrand which is equal to the positive value of the Jacobian of the transformation. For our present purposes, it is convenient (and permissible) to use the Jacobian itself, instead of its positive value, so that for a triple integral, we may write symbolically

$$(1) \qquad d\bar{x}^1\, d\bar{x}^2\, d\bar{x}^3 = |\partial\bar{x}/\partial x|\, dx^1\, dx^2\, dx^3,$$

where $|\partial\bar{x}/\partial x| = \det \partial\bar{x}^i/\partial x^j = |L_{\mathrm{Q}}(S, \bar{S})|$, by 2.3(31), when the change of variables is that appropriate to a space coordinate transformation $S \to \bar{S}$. The quantity $dx^1\, dx^2\, dx^3$ may be regarded as a particular kind of differential form [which Rudin (1964, p. 210) denotes by $dx^1 \wedge dx^2 \wedge dx^3$], which is the component in S of a relative scalar of weight -1.

If, therefore, we use an integrand that is the component in S of a relative scalar of weight $+1$, the resulting triple integral (taken over coordinate intervals which include the image of a given three-dimensional region of space) will have a value which is independent of the choice of coordinate system S from the set $\mathscr{S}$. We may call such a quantity an *integral invariant*. Although its value is a number, it is not a scalar, according to our definition [preceding 2.4(24)], because the number depends on a set of places in space instead of on a single place.

An important example is furnished by the quantity $\sqrt{|g|}$, which is a component of a relative scalar of weight $+1$, by virtue of 5.1(6) applied to the second-rank absolute covariant tensor $\boldsymbol{g}(\mathrm{Q})$, *provided that we choose the sign of the square root so as to satisfy the equation*

$$\sqrt{|\bar{g}|} = |L|^{-1}\sqrt{|g|} \qquad [L = L_{\mathrm{Q}}(S, \bar{S})]. \tag{2}$$

We shall use the convention that $a^{1/2}$ denotes the positive root of a, while $\sqrt{a}$ can denote either $a^{1/2}$ or $-a^{1/2}$. *In the case of* $|g|$, *we shall always choose the sign of* $\sqrt{|g|}$ *so as to satisfy the transformation law* (2) *for a relative scalar of weight* $+1$. In any one coordinate system, S say, we can choose either sign for $\sqrt{|g|}$; let us choose it so that the "volume element" dV, defined by the equation

$$dV = \sqrt{|g|}\; dx^1\, dx^2\, dx^3, \tag{3}$$

is positive. Since dV is an absolute scalar, it follows that dV is positive for every coordinate system.

Equation (3) assigns a nonnegative number dV to an infinitesimal element in the space manifold. The justification for calling it the *volume* of the element is that it agrees with our definition $dy^1\, dy^2\, dy^3$ when the coordinate system is rectangular Cartesian [or "locally rectangular Cartesian" at Q, i.e., such that $g_{ij}(y) = \delta_{ij}$ at y]; moreover, dV, as defined by (3), is the only absolute scalar that is linear in the "coordinate element" $dx^1\, dx^2\, dx^3$ and that reduces to $dy^1\, dy^2\, dy^3$ when the coordinate system is locally rectangular Cartesian.

Similarly, we can define the volume element

$$dV = \sqrt{|\gamma(\mathrm{P}, t)|}\; d\xi^1\, d\xi^2\, d\xi^3 \tag{4}$$

which gives the volume at time t of the "material element" $d\xi^1\, d\xi^2\, d\xi^3$ that contains P.

A similar definition of the volume element can be given for a Riemannian manifold of any finite dimension, whether the manifold is Euclidean or not. When the dimension is greater than three, the term "extension" is sometimes used instead of volume. For a two-dimensional manifold, area is the appropriate term; for a one-dimensional manifold (or line), length is the appropriate term, and it is easy to verify that this element of length agrees with the separation ds defined in 2.2(11).

In our three-dimensional Euclidean space $\mathscr{Q}$ (or $\mathscr{P}$), a surface is a two-dimensional Riemannian manifold which is non-Euclidean unless the surface is plane. For an x^3-coordinate surface in any given space-coordinate system $S: \mathrm{Q} \to x$, we have $x^3 = c$ for all places on the surface, and hence the separation ds of any two neighboring places is given by the equation

(5) $$(ds)^2 = g_{11}(dx^1)^2 + 2g_{12}\, dx^1\, dx^2 + g_{22}(dx^2)^2,$$

which follows from 2.2(11) on putting $dx^3 = 0$. This proves that the surface is a Riemannian manifold, according to the definition 2.2(9). We can therefore define a *surface metric tensor field* $\mathbf{G}^*(\mathrm{Q})$, say, for this manifold, by rewriting (5) in the form

(6) $$(ds)^2 = G_{ij}^*(x^1, x^2)\, dx^i\, dx^j \qquad (i, j = 1, 2),$$

where $G_{ij}^*(x^1, x^2) = g_{ij}(x^1, x^2, c)$. The element of extension, or area, dA is therefore given by the equations

(7) $$dA = \sqrt{|G^*|}\, dx^1\, dx^2,$$

(8) $$|G^*| = g_{11}g_{22} - (g_{12})^2 = |g|g^{33}.$$

The x^1-coordinate curve is a one-dimensional Riemannian manifold for which the element of extension, or length, is given by the equation

(9) $$ds = \sqrt{g_{11}}\, dx^1,$$

which is obtained from 2.2(11) (with $dx^2 = dx^3 = 0$) and is evidently of the form appropriate to (3) applied to a one-dimensional manifold.

There is no loss of generality in choosing *coordinate* surfaces and curves for the foregoing discussion; for any sufficiently smooth surface, we can always construct a coordinate system that has that surface as one (or part of one) of its coordinate surfaces, and a similar statement is true of coordinate curves.

(10) **Problem** P, P′, and P″ are three neighboring particles on a material surface; the vectors $\boldsymbol{d\xi}' = \overrightarrow{\mathrm{PP}'}$ and $\boldsymbol{d\xi}'' = \overrightarrow{\mathrm{PP}''}$ are linearly independent. An absolute, covariant, unit vector $\mathbf{v} = \mathbf{v}(\mathrm{P}, t)$ is defined by the equation $\mathbf{v}\, dA = |\mathbf{v}(\mathrm{P}, t)|^{1/2}\, \boldsymbol{d\xi}' \times \boldsymbol{d\xi}''$, where dA is a positive, absolute scalar. Prove the following results:

(i) $\mathbf{v}(\mathrm{P}, t)$ is independent of the choice of P′ and P″;
(ii) $\mathbf{v}(\mathrm{P}, t)$ is normal to the material surface at P and t;
(iii) $dA = \mathrm{PP}'\ \mathrm{PP}''\ \sin \angle \mathrm{P}'\mathrm{PP}''$;
(iv) $dA = \sqrt{(|\gamma|\gamma^{33})}\, d\xi^1\, d\xi^2$ in a coordinate system $B: \mathrm{P} \to \xi$ for which $\widetilde{\boldsymbol{d\xi}'B} = [d\xi^1, 0, 0]$ and $\widetilde{\boldsymbol{d\xi}''B} = [0, d\xi^2, 0]$.

5.6 The body stress tensor field

In Chapter 2, we defined the body metric tensor field $\gamma(\mathrm{P}, t)$. We now define the body stress tensor field $\pi(\mathrm{P}, t)$. According to most current rheological theories, the properties of viscoelastic materials are describable by relations involving π and γ. For our purposes, it is convenient to define the stress tensor in the context of the following assumptions.

(1) **Contact-force assumptions** Let dA denote the area at time t of an arbitrarily small plane material element including a particle P; let $\mathbf{v} = \mathbf{v}(\mathrm{P}, t)$ be a unit normal (an absolute covariant body vector) to this element. It is assumed that material on the $+\mathbf{v}$ side of dA exerts on material on the $-\mathbf{v}$ side a *contact force* $d\boldsymbol{\varphi}$, which is an absolute, contravariant, body vector expressible in the form

$$d\boldsymbol{\varphi} = \boldsymbol{\pi} \cdot \mathbf{v}\, dA, \tag{2}$$

where $\boldsymbol{\pi} = \boldsymbol{\pi}(\mathrm{P}, t)$ is independent of $\mathbf{v}$ and dA, and

$$\boldsymbol{\pi} = \tilde{\boldsymbol{\pi}}. \tag{3}$$

From the form of (2), it is evident that the *stress tensor* $\boldsymbol{\pi}(\mathrm{P}, t)$ is an absolute, contravariant, body tensor field of second rank. The quantity $\boldsymbol{\pi} \cdot \mathbf{v}$, which is a contravariant body vector having dimensions of force/area, is called the *traction vector.*

The definition of the stress tensor involves the use of the concept *force*, which we have listed in 1.1(1) as a primitive concept. This seems to be the simplest way to treat force, namely, as an undefined element to which certain properties are assigned in the course of the analysis. The molecular origin of contact forces in actual materials is twofold: transport of *molecular* momentum across the area dA, and forces acting between molecules (or atoms) separated by dA; the contact force itself can, in principle, be obtained from suitable spatial and temporal averages of the rapidly fluctuating molecular contributions. The appropriate element dA in this context is called a physically infinitesimal element (Rosenfeld, 1965); it is small enough (on a macroscopic scale) for the ratio $d\boldsymbol{\varphi}/dA$ in (2) to be independent of dA, and is large enough (on a molecular scale) for the averaging processes to involve large numbers of molecules. Similarly, the time t in (1) represents a physically infinitesimal time interval on a molecular time scale. The task of performing such averages for any given molecular model of a condensed phase is difficult, and is the subject of current activity.

The assumption (3), that the stress tensor is symmetric, is a restriction (Dahler and Scriven, 1961) which is not appropriate in certain circumstances [e.g., in the kinetic theory of dense gases: Curtiss (1956); Livingston and

Curtiss (1959)]. For polymeric materials, however, the assumption appears to be valid: We know of no evidence, experimental or theoretical, which casts doubt on its validity, although, of course, the possibility of requiring a non-symmetric stress tensor must always be borne in mind.

In the standard textbooks on continuum mechanics, the space-tensor equations obtained from (2) and (3) by transfer at time t are usually derived from another set of assumptions (no contact couples, no point forces, no external body couples, plus Newton's laws of motion).

The linearity of the relation between $\boldsymbol{d\varphi}$ and $\mathbf{v}$ [expressed by (2)] is deduced from the resultant of contact forces acting across the faces of an infinitesimal tetrahedron; the symmetry of $\boldsymbol{\pi}$ [expressed by (3)] is deduced from an equation of moments for contact forces acting across the faces of an infinitesimal cubical element of material. Transfer of the space equations to the body [using $\mathbb{T}^{-1}(t)$] then yields (2) and (3), at least when one specifies that $\boldsymbol{d\varphi}$ shall be contravariant and $\boldsymbol{\pi}$ contravariant, second rank. There is no loss of generality in so specifying these body tensors, for any other alternative tensors can be obtained by contraction with $\boldsymbol{\gamma}$ and multiplication by any power of $|\boldsymbol{\gamma}|$. In particular, we may also consider the following additional stress tensors:

(4)
$$\begin{aligned} &\boldsymbol{\pi}\cdot\boldsymbol{\gamma} \quad \text{(right-covariant mixed)};\\ &\boldsymbol{\gamma}\cdot\boldsymbol{\pi} \quad \text{(left-covariant mixed)};\\ &\boldsymbol{\gamma}\cdot\boldsymbol{\pi}\cdot\boldsymbol{\gamma} \quad \text{(covariant)}. \end{aligned}$$

The unit normal $\mathbf{v}$ is a covariant vector whose magnitude is unity at time t; $\boldsymbol{\gamma}^{-1}\cdot\mathbf{v}$ is a contravariant vector whose magnitude is unity at time t. The contact force $\boldsymbol{d\varphi}$ can be expressed as a sum

(5)
$$\begin{aligned} \boldsymbol{d\varphi} &= \boldsymbol{\gamma}^{-1}\cdot[\mathbf{vv} + (\boldsymbol{\gamma} - \mathbf{vv})]\cdot\boldsymbol{d\varphi}\\ &= \boldsymbol{\gamma}^{-1}\cdot\mathbf{vv}\cdot\boldsymbol{d\varphi} + (\boldsymbol{\delta} - \boldsymbol{\gamma}^{-1}\cdot\mathbf{vv})\cdot\boldsymbol{d\varphi} \end{aligned}$$

of two contravariant vectors: the first is in the direction of the associated normal $\boldsymbol{\gamma}^{-1}\cdot\mathbf{v}$ and is called the normal component; the second is orthogonal at time t to this direction and is called the tangential component. The magnitude of the first vector is $\mathbf{v}\cdot\boldsymbol{d\varphi} = (\boldsymbol{\pi}:\mathbf{vv})\,dA$; the absolute scalar $\boldsymbol{\pi}:\mathbf{vv}$ is called the *normal component of traction*, and is positive for tension. For any given stress, there are always three directions $\mathbf{v}^1$, $\mathbf{v}^2$, and $\mathbf{v}^3$, say, for each of which the contact force is in the direction of the associated normal; it is consistent with 2.8(1) to call these the *principal directions of stress*, and to call the corresponding normal components of traction the *principal values of stress*. The principal directions do not have to be described by unit vectors. We thus have the following theorem, by application of 2.8(2) and 2.8(20):

(6) **Theorem** For any given state of stress at time t, there are three linearly

independent vectors $\boldsymbol{\psi}^i(\mathrm{P})$ at any given particle P such that

(i) $(\boldsymbol{\pi} - \sigma_i \boldsymbol{\gamma}^{-1}) \cdot \boldsymbol{\psi}^i = \mathbf{0} \qquad (i = 1, 2, 3),$

where the σ_i are absolute scalars. The vectors $\boldsymbol{\psi}^i$ are mutually orthogonal at time t. The $\boldsymbol{\psi}^i$ are called covariant *principal axes of stress*, and the σ_i are *called principal values of stress.* The $\boldsymbol{\theta}_i = \boldsymbol{\gamma}^{-1} \cdot \boldsymbol{\psi}^i$ are called contravariant principal axes of stress. The stress tensor $\boldsymbol{\pi} = \boldsymbol{\pi}(\mathrm{P}, t)$ can be expressed in the form

(ii) $\boldsymbol{\pi} = \sigma_i \|\boldsymbol{\theta}_i\|^{-2} \boldsymbol{\theta}_i \boldsymbol{\theta}_i, \qquad \text{where} \quad \|\boldsymbol{\theta}_i\|^2 = \boldsymbol{\gamma} : \boldsymbol{\theta}_i \boldsymbol{\theta}_i .$

(7) **Problem** If, throughout some time interval, the principal values of stress at P are independent of time and the contravariant principal axes of stress at P are tangential to the same three material lines through P, show that $\partial(\boldsymbol{\pi} \cdot \boldsymbol{\gamma})/\partial t = \mathbf{0}$.

(8) **Problem** Prove that $\sigma_1 \sigma_2 \sigma_3 = |\boldsymbol{\pi}|\,|\boldsymbol{\gamma}|$ and that $\sigma_1 + \sigma_2 + \sigma_3 = \boldsymbol{\pi} : \boldsymbol{\gamma}$.

This result shows that *the stress tensor does not always possess an inverse*; if one (or more) of the principal values of stress is zero, then it follows from (8) that $|\boldsymbol{\pi}| = 0$, i.e., that $\boldsymbol{\pi}$ is singular, and so $\boldsymbol{\pi}^{-1}$ does not exist. *If no principal value of stress is zero, then* $\boldsymbol{\pi}^{-1}$ *exists.*

We may call the totality of contact forces $d\boldsymbol{\varphi}$ for all material plane elements dA *the state of stress in the body at time* t. From (2), it is evident that the state of stress at time t is uniquely determined when the stress tensor field $\boldsymbol{\pi}(\mathrm{P}, t)$ is given. Since $\boldsymbol{\pi}$ is a symmetric, second-rank tensor field, it is determined by six independent scalars at each P and t. These can be specified in several alternative ways, as follows:

(9) $\pi^{ij}(\xi, t)$ (the stress components in a coordinate system B);

(10) the principal values σ_i, the corresponding principal contravariant vectors $\boldsymbol{\theta}_i$, and their magnitudes $\|\boldsymbol{\theta}_i\|$;

(11) the values of $\boldsymbol{\pi} : \mathbf{v}^i \mathbf{v}^i$ (the normal components of traction) for six suitably oriented plane elements whose unit normals $\mathbf{v}^1, \mathbf{v}^2, \ldots, \mathbf{v}^6$ are given.

It is important to note in (10) that it is not sufficient to know the principal directions and principal values: one needs also to be given either the magnitudes $\|\boldsymbol{\theta}_i\|$ [for substitution in (6)(ii)] or, what is equivalent, the unit vectors $\boldsymbol{\theta}_i/\|\boldsymbol{\theta}_i\|$ along the principal directions. (11) follows from (9) on taking $\mathbf{v}^i B$ (for $i = 1, 2, \ldots, 6$) equal to the transposes of the matrices $\delta_1, \delta_2, \delta_3$, [0, 1, 1], [1, 0, 1], and [1, 1, 0], multiplied by the appropriate normalizing numbers.

(12) **Problem** If the normal components of traction across the six plane elements in (11) are each equal to $-p$, show that

(i) $\pi = -p\gamma^{-1}$;
(ii) the normal component of traction across every plane element is $-p$;
(iii) there is no tangential component of traction across any plane.

The state of stress in this case is called an *isotropic stress* or a *hydrostatic pressure p*. The result is a simple illustration of (11), which shows that knowledge of a sufficient number of *normal* components enables the complete state of stress, and therefore the tangential components of traction, to be determined.

(13) **Definition** The body tensor $\mathbf{\Pi} = \pi + p\gamma^{-1}$, where p is any scalar, is called the contravariant *extra-stress* tensor. The extra-stress tensor is said to be *deviatoric* if $\mathbf{\Pi} : \gamma = 0$, i.e., if $p = -\frac{1}{3}\pi : \gamma$.

For incompressible materials, it is often convenient to use the extra stress tensor instead of the stress tensor in constitutive equations. *In the particular case in which the incompressible material is a Newtonian liquid,* the extra stress is necessarily deviatoric. For other incompressible materials, there is usually no advantage in using a *deviatoric* extra stress tensor; the extra variable p (along with the other dependent variables) can in principle be determined in any well-defined isothermal problem by solving the simultaneous set of equations comprising the constitutive equations, the stress equations of motion, and the constant-volume condition, subject to an appropriate set of boundary and initial value conditions. In certain simple problems, the specification of a component of traction on some boundary is sufficient to determine p.

Space tensor fields for describing stress may be defined from the body stress tensor field by transfer at time t. Thus

(14) $$\boldsymbol{p}(\mathrm{Q}, t) = \mathbb{T}(t)\pi(\mathrm{P}, t) \qquad (\mathrm{P} \overset{t}{\leftrightarrow} \mathrm{Q})$$

is an absolute, second-rank, contravariant space stress tensor field, and

(15) $$\boldsymbol{g}(\mathrm{Q}) \cdot \boldsymbol{p}(\mathrm{Q}, t) \cdot \boldsymbol{g}(\mathrm{Q}) = \mathbb{T}(t)\gamma(\mathrm{P}, t) \cdot \pi(\mathrm{P}, t) \cdot \gamma(\mathrm{P}, t)$$

is the associated covariant space stress tensor field. Mixed space stress tensor fields can be defined in a similar fashion. In contrast, there is only one absolute Cartesian space stress tensor field $\mathsf{p} = \mathsf{p}(\mathrm{Q}, t)$ of second rank:

(16) $$\mathsf{p} = \mathbb{T}(t)\pi = \mathbb{T}(t)\pi \cdot \gamma = \mathbb{T}(t)\gamma \cdot \pi \cdot \gamma,$$

because of 5.4(59). By transferring (2) and (3) to space at time t, we have

(17) $$d\boldsymbol{f} = \boldsymbol{p} \cdot \boldsymbol{n}\, dA, \qquad \boldsymbol{p} = \tilde{\boldsymbol{p}},$$

where $\boldsymbol{df} = \mathbb{T}(t)\, \boldsymbol{d\varphi}$ and $\boldsymbol{n} = \mathbb{T}(t)\mathbf{v}$ are respectively the contravariant space vector for the contact force across dA and the covariant space vector for the unit normal to dA at time t. Equations of the same form evidently hold with Cartesian vectors and the Cartesian stress tensor used in place of general space vectors and the contravariant stress tensor.

The important features of the body stress tensor field $\pi(\mathrm{P}, t)$ may be summarized as follows:

(18) **Properties of $\pi(P, t)$**

- (i) π is independent of any choice of coordinate system;
- (ii) π is a one-state quantity;
- (iii) π gives a complete description of the state of stress at time t;
- (iv) π is unchanged by rigid rotation of the body when the state of stress is constant.

The associated body stress tensor fields (4) possess the same four properties. *The space stress tensor fields* defined in (14)–(16) possess properties (i)–(iii) but not (iv), and therefore *give a less simple description of stress than that given by any of the body stress tensors.* As far as we are aware, *there is no space field that possesses all four of the properties* (18). Certain space fields have been defined [such as $\mathbb{T}(t')\pi(\mathrm{P}, t)$] that possess properties (i), (iii), and (iv) but not (ii); since stress is a one-state quantity, such two-state tensors involve an extra and unnecessary complicating feature. In certain circumstances, however, the introduction of a second state t', (often called a reference state) may be useful; but the use of body tensor fields for describing stress always leaves one free to choose whether or not to use a reference state. The *exclusive* use of space fields, on the other hand, forces one to choose between property (ii) and property (iv).

In (18)(iv), we use the phrase "constant state of stress." The meaning to be assigned to this phrase has received less attention in the rheological literature than it deserves. We suggest that it is essential to consider separately the cases in which the body is moving rigidly or is deforming, i.e., in which $\gamma(\mathrm{P}, t)$ is, or is not, independent of t. For the rigid motion case, the following definition follows at once from (11) and the definition of the state of stress:

(19) **Definition** *The state of stress in a rigidly moving body is constant* if, and only if, for every infinitesimal material plane element dA, the normal component of traction is constant. In such a situation, $\gamma(\mathrm{P}, t)$ and $\pi(\mathrm{P}, t)$ are independent of t.

It is this definition which is used in (18)(iv).

(20) **Problem** A body is rotating rigidly in space; the state of stress is constant and is not isotropic. Prove that the space stress tensor $\mathbf{p}(Q, t)$ varies with t even if the same particle P remains at Q throughout.

(21) **Definition** A body is said to undergo *stress relaxation* during a given time interval if (i) $\gamma(P, t)$ is independent of t and (ii) $\pi(P, t)$ varies with t. *The term stress relaxation will not be used whenever the body is deforming.*

If the shape of a body is changing with time, then the term stress relaxation is inapplicable, according to this definition. More generally, we suggest that *no rheologically useful meaning can be assigned to the phrase, constant stress in a deforming body* (Lodge, 1964, Appendix 1), except possibly in the degenerate cases in which the stress is zero or isotropic, or the deformation is isotropic (i.e., a dilatation).

In certain experiments, e.g., in Couette flow between cylinders in relative rotation, after prolonged flow at constant shear rate, the traction on the cylinders usually becomes constant (i.e., independent of time), and in fact the traction across each shearing surface presumably also becomes constant. *We cannot then assert, however, that the state of stress has become constant*, for the traction (or the noımal component of traction) across *other* material surfaces varies with time. *Constancy of traction across each member of a one-parameter family of material surfaces is not sufficient to secure constancy of the state of stress*, for, according to any reasonable meaning which one might try to assign to such a term, constancy of traction across every infinitesimal material plane element would be required. When the material is deforming, this requirement is just too stringent to be of practical value.

Since variability is the antithesis of constancy, the preceding discussion suggests that we cannot assign a useful meaning to the phrase "variable state of stress" in a material which is deforming. We can, however, talk about the variation of traction across any one given material surface, and we believe that this should suffice for practical needs. These comments raise the further questions as to what behavior is implied by the constancy of the various stress tensors, since their dependence on time, or independence of time, are well-defined concepts.

When the body is moving rigidly, constancy of π implies constancy of $\pi \cdot \gamma$ and $\gamma \cdot \pi \cdot \gamma$, and means that the state of stress is constant; constancy of $\mathbf{p}$ is of no rheologically useful significance, since it requires that, in addition, the body shall not be rotating in space; the rheologically useful property is that the various convected derivatives of $\mathbf{p}$ are zero. *When the body is deforming*, constancy of one body stress tensor does not imply constancy of the others, nor that the state of stress is constant (since this is undefined);

constancy of $\boldsymbol{\pi} \cdot \boldsymbol{\gamma}$ implies constancy of principal values and principal axes of stress, according to (7); the implications of constancy of $\boldsymbol{\pi}$ or of $\boldsymbol{\gamma} \cdot \boldsymbol{\pi} \cdot \boldsymbol{\gamma}$ are given in the following two problems.

(22) **Problem** If $\partial\boldsymbol{\pi}(\mathrm{P}, t)/\partial t = \mathbf{0}$, prove that $f_n/(dh)^2$ is independent of time, where f_n denotes the component of traction normal to an arbitrary material surface $\sigma(\xi) = c$, and dh denotes the local separation of $\sigma(\xi) = c$ and $\sigma(\xi) = c + dc$.

(23) **Problem** If $\partial(\boldsymbol{\gamma} \cdot \boldsymbol{\pi} \cdot \boldsymbol{\gamma})/\partial t = \mathbf{0}$, prove that the polar diagram of $f_n^{-1/2}$ (assuming that $f_n > 0$) centered on P is represented by the same material surface at different times. (This polar diagram at time t is the set of particles $\{\mathrm{P}'\}$, where $\overrightarrow{\mathrm{PP}'}$ is normal to an arbitrary plane material element dA at P and $\mathrm{PP}' = \varepsilon f_n^{-1/2}$, where ε is an arbitrary, constant, infinitesimally small number and f_n is the normal component of traction at time t across dA; the set $\{\mathrm{P}'\}$ is generated by taking all possible elements dA through P.)

We shall see in Chapter 6 that constancy of the contravariant extra stress tensor $\boldsymbol{\Pi}$ occurs with a particular kind of incompressible rubberlike material whose properties are derivable from a molecular network model in the so-called Gaussian approximation.

5.7 Isotropic functions and orthogonal tensors

In this section, we consider relations of the form $\boldsymbol{\sigma} = F(\boldsymbol{\lambda})$, where $\boldsymbol{\sigma}$ and $\boldsymbol{\lambda}$ denote absolute right-covariant, mixed, second-rank body tensors at the same particle P. The $\boldsymbol{\lambda}$ *will be assumed to have real principal values* λ_i (which will often be taken to be all different, to simplify proofs), and real, linearly independent, contravariant principal axes $\boldsymbol{\alpha}_i$; the reciprocal set of vectors $\boldsymbol{\alpha}^i$, defined by the equations

(1) $$\boldsymbol{\alpha}_i \cdot \boldsymbol{\alpha}^j = \delta_i^j \qquad (i, \; = 1, 2, 3),$$

are then covariant principal axes, and from 2.8(2), we have the following equations:

(2) $$\boldsymbol{\delta} = \boldsymbol{\alpha}_i \boldsymbol{\alpha}^i;$$

(3) $$\boldsymbol{\lambda} = \lambda_i \boldsymbol{\alpha}_i \boldsymbol{\alpha}^i;$$

(4) $$\boldsymbol{\lambda}^2 = \lambda_i^2 \boldsymbol{\alpha}_i \boldsymbol{\alpha}^i \qquad [\text{from (1) and (3), with } \boldsymbol{\lambda}^2 \equiv \boldsymbol{\lambda} \cdot \boldsymbol{\lambda}].$$

In some applications, we shall take $\boldsymbol{\lambda} = \boldsymbol{\gamma}_0^{-1} \cdot \boldsymbol{\gamma}$, where $\boldsymbol{\gamma}_0$ and $\boldsymbol{\gamma}$ denote body metric tensors in states t_0 and t, so that the conditions of Theorem 2.8(2) are satisfied; we shall also have $\boldsymbol{\sigma} = \boldsymbol{\pi} \cdot \boldsymbol{\gamma}$ or $\boldsymbol{\Pi} \cdot \boldsymbol{\gamma}$, in which case the equation

$\boldsymbol{\sigma} = F(\boldsymbol{\lambda})$ represents a stress–strain relation, or constitutive equation, for a perfect elastic solid. For brevity, we derive only those results that we require in our present applications; further results in the theory of invariants and symmetry classes for tensor functions (usually expressed in terms of Cartesian space tensors), with particular reference to applications in continuum mechanics, have been derived by Rivlin and co-workers and are summarized by Truesdell and Noll (1965), who give the necessary references.

Equations (2)–(4), when the λ_i are all different, can be solved to give the equation

$$\boldsymbol{\alpha}_1\boldsymbol{\alpha}^1 = [\boldsymbol{\lambda}^2 - (\lambda_2 + \lambda_3)\boldsymbol{\lambda} + \lambda_2\lambda_3\boldsymbol{\delta}]/[(\lambda_1 - \lambda_2)(\lambda_1 - \lambda_3)] \tag{5}$$

and two similar equations for $\boldsymbol{\alpha}_2\boldsymbol{\alpha}^2$ and $\boldsymbol{\alpha}_3\boldsymbol{\alpha}^3$ obtained by cyclic interchange of indices 1, 2, and 3.

Comparison of Eqs. (3) and (4) suggests the following simple method of defining mixed tensor functions of $\boldsymbol{\lambda}$ having the same principal axes as $\boldsymbol{\lambda}$; if any function $F(x)$ of a single real variable x is given, we define a function $F(\boldsymbol{\lambda})$ (whose values are mixed tensors) by the equation

$$F(\boldsymbol{\lambda}) = F(\lambda_i)\boldsymbol{\alpha}_i\boldsymbol{\alpha}^i. \tag{6}$$

Equation (4) is the case $F(x) = x^2$; other important cases are

$$e^{\boldsymbol{\lambda}} = e^{\lambda_i}\boldsymbol{\alpha}_i\boldsymbol{\alpha}^i; \qquad \boldsymbol{\lambda}^{1/2} = \lambda_i^{1/2}\boldsymbol{\alpha}_i\boldsymbol{\alpha}^i \qquad (\lambda_i \geq 0). \tag{7}$$

It is easy to verify that $\boldsymbol{\lambda}^{1/2} \cdot \boldsymbol{\lambda}^{1/2} = \boldsymbol{\lambda}$, $e^{\boldsymbol{\lambda}} \cdot e^{\boldsymbol{\lambda}} = e^{2\boldsymbol{\lambda}}$, etc. More generally,

$$(\exp \boldsymbol{\lambda}) \cdot \exp \boldsymbol{\lambda}' = \exp(\boldsymbol{\lambda} + \boldsymbol{\lambda}') \tag{8}$$

if $\boldsymbol{\lambda}$ and $\boldsymbol{\lambda}'$ have the same principal axes;

$$F(\boldsymbol{\lambda}) \cdot F(\boldsymbol{\lambda}') = F(\boldsymbol{\lambda}') \cdot F(\boldsymbol{\lambda}) \tag{9}$$

if $\boldsymbol{\lambda}$ and $\boldsymbol{\lambda}'$ have the same principal axes. It follows from (5) and (6) that

$$F(\boldsymbol{\lambda}) = F_{\mathrm{I}}\boldsymbol{\lambda}^2 + F_{\mathrm{II}}\boldsymbol{\lambda} + F_{\mathrm{III}}\boldsymbol{\delta} \qquad (\lambda_i \text{ all different}), \tag{10}$$

where F_{I}, F_{II}, and F_{III} are symmetric functions of λ_i defined by the matrix equation

$$\begin{bmatrix} F_{\mathrm{I}} \\ F_{\mathrm{II}} \\ F_{\mathrm{III}} \end{bmatrix} = \sum \frac{F(\lambda_1)}{(\lambda_1 - \lambda_2)(\lambda_1 - \lambda_3)} \begin{bmatrix} 1 \\ -\lambda_2 - \lambda_3 \\ \lambda_2\lambda_3 \end{bmatrix}; \tag{11}$$

$\sum$ denotes the sum of three terms obtained by cyclic interchange of indices 1, 2, and 3. *Thus any four powers of* $\boldsymbol{\lambda}$ *are linearly dependent*; $\boldsymbol{\delta} \equiv \boldsymbol{\lambda}^0$, $\boldsymbol{\lambda}$, and $\boldsymbol{\lambda}^2$ form a *basis* for all functions of the type (6). On using (10) with $F(\boldsymbol{\lambda}) = \boldsymbol{\lambda}^3$ and taking the matrix representation in any body coordinate system B, we obtain an alternative proof of the Cayley–Hamilton theorem 1.3(24). Weissenberg (1935, p. 98) has used a definition of type (6) (for Cartesian tensors) in rheological applications.

In the discussion of second-rank tensor functions, we have used mixed tensors because the fact that $\boldsymbol{\lambda}^2$ (for example) is a mixed tensor, if $\boldsymbol{\lambda}$ is, makes the discussion a little simpler. Corresponding results for contravariant and for covariant tensors can be obtained from these results without difficulty. Corresponding results for left-covariant tensors can be obtained at once by taking the transpose of the right-covariant tensor equations; in particular, it follows from (6) or (10) that

(12) $$F(\tilde{\boldsymbol{\lambda}}) = \widetilde{F(\boldsymbol{\lambda})} = \lambda_i \boldsymbol{\alpha}^i \boldsymbol{\alpha}_i .$$

We now show that Eq. (6), or (10), represents the most general function $F(\boldsymbol{\lambda})$ (having values which are mixed tensors) that is *isotropic* in a sense now to be defined. For this purpose, we first define orthogonal body tensors.

(13) **Definition** A right-covariant, mixed, second-rank body tensor $\boldsymbol{\Omega}$ is *orthogonal* [or γ *orthogonal*, or t *orthogonal* if $\gamma = \gamma(\mathrm{P}, t)$] if $\tilde{\boldsymbol{\Omega}} \cdot \gamma \cdot \boldsymbol{\Omega} = \gamma$. Such a tensor is evidently necessarily *unimodular* in the sense that

(14) $$|\boldsymbol{\Omega}| = \pm 1; \qquad \text{also,} \quad \Omega \equiv \boldsymbol{\Omega} B \quad \text{is an orthogonal matrix if} \quad \gamma B = I.$$

A t-orthogonal tensor is thus represented by an orthogonal matrix $\Omega = \tilde{\Omega}^{-1}$ in any body coordinate system that is rectangular Cartesian in state t. In an arbitrary coordinate system (Weyl, 1939, p. 65), we have

(15) $$\tilde{\boldsymbol{\Omega}} \cdot \gamma = \gamma \cdot \boldsymbol{\Omega}^{-1}.$$

(16) **Theorem** The linear mapping $\boldsymbol{\Omega}\colon\{\boldsymbol{\theta}\} \to \{\boldsymbol{\theta}\}$ of contravariant vectors onto themselves leaves invariant the γ magnitudes of, and the angles between, all vectors if and only if $\boldsymbol{\Omega}$ is γ orthogonal.

(17) **Theorem** If $\boldsymbol{\alpha}_i$ ($i = 1, 2, 3$) are any three linearly independent contravariant vectors, and $\bar{\boldsymbol{\alpha}}_i$ are any three contravariant vectors such that $\boldsymbol{\alpha}_i \cdot \gamma \cdot \boldsymbol{\alpha}_j = \bar{\boldsymbol{\alpha}}_i \cdot \gamma \cdot \bar{\boldsymbol{\alpha}}_j$ for $i, j = 1, 2, 3$, then there exists a γ-orthogonal tensor $\boldsymbol{\Omega}$ such that $\bar{\boldsymbol{\alpha}}_i = \boldsymbol{\Omega} \cdot \boldsymbol{\alpha}_i$ ($i = 1, 2, 3$), and the $\bar{\boldsymbol{\alpha}}_i$ are linearly independent.

Proof of (16) If $\boldsymbol{\theta} \to \bar{\boldsymbol{\theta}} = \boldsymbol{\Omega} \cdot \boldsymbol{\theta}$ and $\boldsymbol{\theta}' \to \bar{\boldsymbol{\theta}}' = \boldsymbol{\Omega} \cdot \boldsymbol{\theta}'$, then $\bar{\boldsymbol{\theta}} \cdot \gamma \cdot \bar{\boldsymbol{\theta}}' = \boldsymbol{\theta} \cdot \tilde{\boldsymbol{\Omega}} \cdot \gamma \cdot \boldsymbol{\Omega} \cdot \boldsymbol{\theta}' = \boldsymbol{\theta} \cdot \gamma \cdot \boldsymbol{\theta}'$, by (13), if $\boldsymbol{\Omega}$ is γ orthogonal; thus magnitudes and angles (evaluated using γ) are unchanged by any γ-orthogonal transformation. The truth of the converse is obvious, and can readily be demonstrated from (17).

Proof of (17) Since the $\boldsymbol{\alpha}_i$ are linearly independent and three in number, there exists a set of three vectors $\boldsymbol{\alpha}^i$ reciprocal to them such that $\boldsymbol{\alpha}_i \boldsymbol{\alpha}^i = \boldsymbol{\delta}$. Clearly, $\boldsymbol{\Omega} \equiv \bar{\boldsymbol{\alpha}}_i \boldsymbol{\alpha}^i$ is a right covariant, mixed tensor which gives the equations $\boldsymbol{\Omega} \cdot \boldsymbol{\alpha}_i = \bar{\boldsymbol{\alpha}}_i$, as required; further, $\tilde{\boldsymbol{\Omega}} \cdot \boldsymbol{\gamma} \cdot \boldsymbol{\Omega} = \boldsymbol{\alpha}^i \bar{\boldsymbol{\alpha}}_i \cdot \boldsymbol{\gamma} \cdot \bar{\boldsymbol{\alpha}}_j \boldsymbol{\alpha}^j = \tilde{\boldsymbol{\delta}} \cdot \boldsymbol{\gamma} \cdot \boldsymbol{\delta} = \boldsymbol{\gamma}$, which proves that $\boldsymbol{\Omega}$ is $\boldsymbol{\gamma}$ orthogonal, as stated. Finally, from (14), it follows that $\boldsymbol{\Omega}^{-1}$ exists; therefore $\boldsymbol{\alpha}_i = \boldsymbol{\Omega}^{-1} \cdot \bar{\boldsymbol{\alpha}}_i$, and since the $\boldsymbol{\alpha}_i$ are linearly independent, so also are the $\bar{\boldsymbol{\alpha}}_i$.

(18) **Theorem** A $\boldsymbol{\gamma}$-orthogonal transformation $\boldsymbol{\psi} \to \bar{\boldsymbol{\psi}} = \boldsymbol{\psi} \cdot \boldsymbol{\Omega}^{-1}$ of covariant vectors $\{\boldsymbol{\psi}\}$ onto themselves leaves invariant all magnitudes and angles, and inner products $\boldsymbol{\psi} \cdot \boldsymbol{\theta}$ when contravariant vectors $\{\boldsymbol{\theta}\}$ are transformed according to the equation $\boldsymbol{\theta} \to \bar{\boldsymbol{\theta}} = \boldsymbol{\Omega} \cdot \bar{\boldsymbol{\theta}}$. [The proof is similar to the proof of (16).]

A right-covariant mixed tensor $\boldsymbol{\sigma}$ can be expressed in the form

(19) $$\boldsymbol{\sigma} = \sigma^i{}_j \boldsymbol{\beta}_i \boldsymbol{\beta}^j = \bar{\sigma}^i{}_j \bar{\boldsymbol{\beta}}_i \bar{\boldsymbol{\beta}}^j,$$

where $\boldsymbol{\beta}_i$ and $\boldsymbol{\beta}^i$ are base vectors for any body coordinate system B, and $\bar{\boldsymbol{\beta}}_i$ and $\bar{\boldsymbol{\beta}}^i$ are base vectors for any other body coordinate system $\bar{B}$. $\sigma^i{}_j$ and $\bar{\sigma}^i{}_j$ denote components of $\boldsymbol{\sigma}$ in B and $\bar{B}$.

(20) **Definition** A tensor $\boldsymbol{\sigma}$ is $\boldsymbol{\Omega}$ *invariant* if $\sigma^i{}_j = \bar{\sigma}^i{}_j$ when $\bar{\boldsymbol{\beta}}_i = \boldsymbol{\Omega} \cdot \boldsymbol{\beta}_i$ and $\bar{\boldsymbol{\beta}}^j = \boldsymbol{\beta}^j \cdot \boldsymbol{\Omega}^{-1}$, where $\boldsymbol{\Omega}$ is $\boldsymbol{\gamma}$ orthogonal. Similar definitions will be used for covariant and contravariant tensors.

Thus a tensor is $\boldsymbol{\Omega}$ invariant if it has the same descriptions when referred to any two "$\boldsymbol{\gamma}$-equivalent" body coordinate systems which are themselves related by the rigid rotation (in the body) represented by the $\boldsymbol{\gamma}$-orthogonal tensor $\boldsymbol{\Omega}$. In a similar way, we can define scalar and tensor *functions* of vectors and tensors to be $\boldsymbol{\Omega}$ invariant. To save space, we shall consider only the case of *isotropic* functions (i.e., functions which are $\boldsymbol{\Omega}$ invariant for *all* $\boldsymbol{\gamma}$-orthogonal $\boldsymbol{\Omega}$) because this is the case of main interest in our present applications. Again for brevity, we shall use only the "full $\boldsymbol{\gamma}$-orthogonal group" (i.e., all $\boldsymbol{\gamma}$-orthogonal $\boldsymbol{\Omega}$, whether $|\boldsymbol{\Omega}|$ is $+1$ or -1).

Associated with a given mixed tensor $\boldsymbol{\sigma}$, expressed as in (19), it is convenient to define a new tensor $\bar{\boldsymbol{\sigma}}$ by the equation

(21) $$\bar{\boldsymbol{\sigma}} = \sigma^i{}_j \bar{\boldsymbol{\beta}}_i \bar{\boldsymbol{\beta}}^j.$$

We can now express (20) in the form

(22) $$\boldsymbol{\sigma} = \bar{\boldsymbol{\sigma}} = \boldsymbol{\Omega} \cdot \boldsymbol{\sigma} \cdot \boldsymbol{\Omega}^{-1} \qquad \text{(for } \boldsymbol{\Omega}\text{-invariant } \boldsymbol{\sigma}\text{)}.$$

(23) **Definition** A function F relating mixed tensors $\boldsymbol{\sigma}$ and $\boldsymbol{\lambda}$ is *isotropic* (or γ *isotropic*) if $\boldsymbol{\sigma} = F(\boldsymbol{\lambda})$ implies that $\bar{\boldsymbol{\sigma}} = F(\bar{\boldsymbol{\lambda}})$ for all γ-orthogonal $\boldsymbol{\Omega}$, where $\bar{\boldsymbol{\sigma}} = \boldsymbol{\Omega} \cdot \boldsymbol{\sigma} \cdot \boldsymbol{\Omega}^{-1}$ and $\bar{\boldsymbol{\lambda}} = \boldsymbol{\Omega} \cdot \boldsymbol{\lambda} \cdot \boldsymbol{\Omega}^{-1}$, i.e., if

$$\boldsymbol{\Omega} \cdot F(\boldsymbol{\lambda}) \cdot \boldsymbol{\Omega}^{-1} = F(\boldsymbol{\Omega} \cdot \boldsymbol{\lambda} \cdot \boldsymbol{\Omega}^{-1}) \qquad \text{when} \quad \tilde{\boldsymbol{\Omega}} \cdot \boldsymbol{\gamma} \cdot \boldsymbol{\Omega} = \boldsymbol{\gamma}.$$

This is to be valid for all right-covariant mixed tensors $\boldsymbol{\lambda}$.

(24) **Definition** An absolute scalar function $s(\boldsymbol{\lambda}, \boldsymbol{\alpha}_i, \boldsymbol{\alpha}^j)$ of a right-covariant tensor $\boldsymbol{\lambda}$, contravariant vectors $\boldsymbol{\alpha}_i$, and covariant vectors $\boldsymbol{\alpha}^i$, is γ *isotropic* if it is $\boldsymbol{\Omega}$ invariant for all γ-orthogonal $\boldsymbol{\Omega}$, i.e., if

(25) $$s(\boldsymbol{\Omega} \cdot \boldsymbol{\lambda} \cdot \boldsymbol{\Omega}^{-1}, \boldsymbol{\Omega} \cdot \boldsymbol{\alpha}_i, \boldsymbol{\alpha}^j \cdot \boldsymbol{\Omega}^{-1}) = s(\boldsymbol{\lambda}, \boldsymbol{\alpha}_i, \boldsymbol{\alpha}^j) \qquad \text{when} \quad \tilde{\boldsymbol{\Omega}} \cdot \boldsymbol{\gamma} \cdot \boldsymbol{\Omega} = \boldsymbol{\gamma}.$$

Theorems for γ-isotropic scalar functions $s(\ )$

(26) $s(\boldsymbol{\lambda}) = h(\lambda_1, \lambda_2, \lambda_3)$ where h is a symmetric function of λ_i.

(27) $s(\boldsymbol{\alpha}_1, \boldsymbol{\alpha}_2, \boldsymbol{\alpha}_3) = h(\boldsymbol{\alpha}_i \cdot \boldsymbol{\gamma} \cdot \boldsymbol{\alpha}_j)$ if $\boldsymbol{\alpha}_i$ are linearly independent.

(28) $s(\boldsymbol{\alpha}_1, \boldsymbol{\alpha}_2, \boldsymbol{\alpha}_3, \boldsymbol{\gamma})$ is independent of $\boldsymbol{\alpha}_i$ and $\boldsymbol{\gamma}$ if $\boldsymbol{\alpha}_i$ are γ orthonormal.

Theorems for γ-isotropic mixed tensor functions $F(\boldsymbol{\lambda})$ (when $\boldsymbol{\gamma} \cdot \boldsymbol{\lambda}$ is symmetric)

(29) $F(\boldsymbol{\lambda})$ and $\boldsymbol{\lambda}$ have the same principal axes.

(30) $F(\boldsymbol{\lambda}) = F_{\text{I}}\boldsymbol{\lambda}^2 + F_{\text{II}}\boldsymbol{\lambda} + F_{\text{III}}\boldsymbol{\delta}$,

where

(31) F_{I}, F_{II}, and F_{III} are symmetric scalar functions of λ_i expressible in terms of λ_i and a single scalar function of λ_i.

(32) $F(\boldsymbol{\lambda})$ is γ_1 isotropic for all metric tensors $\boldsymbol{\gamma}_1$.

(33) $F(\boldsymbol{\lambda}) = a(\operatorname{tr} \boldsymbol{\lambda})\boldsymbol{\delta} + b\boldsymbol{\lambda}$, where a and b are constants, if $F(\boldsymbol{\lambda})$ is linear and homogeneous in $\boldsymbol{\lambda}$.

Proof of (27) In the set $\{\boldsymbol{\alpha}_i\}$ of all ordered triads of linearly independent contravariant vectors $\boldsymbol{\alpha}_i$, there is a proper subset $V(A)$, say, for which $[\boldsymbol{\alpha}_r \cdot \boldsymbol{\gamma} \cdot \boldsymbol{\alpha}_c] = A$, a given symmetric matrix. From (17), it follows that, given any two members $\boldsymbol{\alpha}_i$ and $\bar{\boldsymbol{\alpha}}_i$ of $V(A)$, we can find a γ-orthogonal $\boldsymbol{\Omega}$ such that $\bar{\boldsymbol{\alpha}}_i = \boldsymbol{\Omega} \cdot \boldsymbol{\alpha}_i$. But $s(\bar{\boldsymbol{\alpha}}_i) = s(\boldsymbol{\alpha}_i)$, since s is γ isotropic, and so $s(\boldsymbol{\alpha}_i)$ has the same value for every triad in $V(A)$. In other words, in order to determine the value of $s(\boldsymbol{\alpha}_i)$, it is sufficient to know in which subset $V(A)$ the triad $\boldsymbol{\alpha}_i$ belongs. Hence $s(\boldsymbol{\alpha}_i) = h(A)$, for some function h, which proves the theorem.

Proof of (28) Since the $\boldsymbol{\alpha}_i$ are γ orthonormal (i.e., $\boldsymbol{\alpha}_i \cdot \boldsymbol{\gamma} \cdot \boldsymbol{\alpha}_j = \delta_{ij}$), we can write $\boldsymbol{\gamma}^{-1} = \boldsymbol{\alpha}_i \boldsymbol{\alpha}_i$, by 2.7(16)(iii). Hence $s(\boldsymbol{\alpha}_i, \boldsymbol{\gamma})$ is an isotropic scalar function of $\boldsymbol{\alpha}_i$ alone, and therefore, by (27), depends on the values of $\boldsymbol{\alpha}_i \cdot \boldsymbol{\gamma} \cdot \boldsymbol{\alpha}_j = \delta_{ij}$ alone. This proves the result.

Proof of (26) From 2.8(2), any right-covariant tensor $\boldsymbol{\lambda}$, such that $\boldsymbol{\gamma} \cdot \boldsymbol{\lambda}$ is symmetric, can be written in the form (3), where $\boldsymbol{\alpha}_i$ can be chosen to be $\boldsymbol{\gamma}$ orthonormal; then $\boldsymbol{\alpha}^i = \boldsymbol{\gamma} \cdot \boldsymbol{\alpha}_i$, and thus $s(\boldsymbol{\lambda})$ can be expressed as a $\boldsymbol{\gamma}$-isotropic function of $\boldsymbol{\alpha}_i$, $\boldsymbol{\gamma}$, and the scalars λ_i, and is therefore, by (28), a function of λ_i alone, i.e., $s(\lambda_i \boldsymbol{\alpha}_i \boldsymbol{\alpha}^i) = h(\lambda_1, \lambda_2, \lambda_3)$, as stated. To show that h is symmetric, we first interchange λ_1 and λ_2 in this equation; we next interchange $\boldsymbol{\alpha}_1$ and $\boldsymbol{\alpha}_2$, which can be done by means of a $\boldsymbol{\gamma}$-orthogonal $\boldsymbol{\Omega}$ and which therefore leaves the value of $s(\boldsymbol{\lambda})$ unchanged. In this way, we see that $h(\lambda_1, \lambda_2, \lambda_3) = h(\lambda_2, \lambda_1, \lambda_3)$. Similar interchanges of λ_2 and λ_3, etc., complete the proof that h is symmetric. It is easy to verify that the $\boldsymbol{\Omega}$ for interchanging $\boldsymbol{\alpha}_1$ and $\boldsymbol{\alpha}_2$ is given by the equation

$$(34) \qquad \boldsymbol{\Omega} = \boldsymbol{\alpha}_2 \boldsymbol{\alpha}^1 + \boldsymbol{\alpha}_1 \boldsymbol{\alpha}^2 + \boldsymbol{\alpha}_3 \boldsymbol{\alpha}^3 \qquad (\boldsymbol{\alpha}_i \cdot \boldsymbol{\gamma} \cdot \boldsymbol{\alpha}_j = \delta_{ij};\ \boldsymbol{\alpha}_i \cdot \boldsymbol{\alpha}^j = \delta_i^j);$$

$$(35) \qquad \boldsymbol{\Omega} \cdot [\boldsymbol{\alpha}_1, \boldsymbol{\alpha}_2, \boldsymbol{\alpha}_3] = [\boldsymbol{\alpha}_2, \boldsymbol{\alpha}_1, \boldsymbol{\alpha}_3]; \qquad \tilde{\boldsymbol{\Omega}} \cdot \boldsymbol{\gamma} \cdot \boldsymbol{\Omega} = \boldsymbol{\gamma}.$$

Proof of (29)–(31) We may express $\boldsymbol{\lambda}$ in the form (3) with its principal axis basis $\boldsymbol{\alpha}_i$ taken to be $\boldsymbol{\gamma}$ orthonormal. Since the $\boldsymbol{\alpha}_i \boldsymbol{\alpha}^j$ form a basis for right-covariant mixed tensors, we may write

$$(36) \qquad F(\boldsymbol{\lambda}) = F(\lambda_k \boldsymbol{\alpha}_k \boldsymbol{\alpha}^k) = F_j^i(\lambda_1, \lambda_2, \lambda_3) \boldsymbol{\alpha}_i \boldsymbol{\alpha}^j,$$

where the $F_j^i = \boldsymbol{\alpha}^i \cdot F(\boldsymbol{\lambda}) \cdot \boldsymbol{\alpha}_j$ are absolute scalar functions of λ_k, $\boldsymbol{\alpha}_k$, and $\boldsymbol{\gamma}$; it is easy to verify that these scalar functions are also $\boldsymbol{\gamma}$ isotropic, and are therefore, from (28), functions of λ_k alone. It is easy to verify that the mixed tensor $\boldsymbol{\Omega}$ given by

$$(37) \qquad \boldsymbol{\Omega} = -\boldsymbol{\alpha}_1 \boldsymbol{\alpha}^1 + \boldsymbol{\alpha}_2 \boldsymbol{\alpha}^2 + \boldsymbol{\alpha}_3 \boldsymbol{\alpha}^3 = \boldsymbol{\Omega}^{-1}$$

is $\boldsymbol{\gamma}$ orthogonal and has the property that $\boldsymbol{\Omega} \cdot \boldsymbol{\lambda} \cdot \boldsymbol{\Omega}^{-1} = \boldsymbol{\lambda}$. Using this $\boldsymbol{\Omega}$ in (23) and (36), it is easy to see that $F_2^1 = F_3^1 = F_1^2 = F_1^3 = 0$. Similarly, $F_3^2 = F_2^3 = 0$. Hence (36) reduces to

$$(38) \qquad F(\boldsymbol{\lambda}) = F_i^i \boldsymbol{\alpha}_i \boldsymbol{\alpha}^i,$$

which proves (29). Using $\boldsymbol{\Omega}$ given by (34) (to interchange $\boldsymbol{\alpha}_1$ and $\boldsymbol{\alpha}_2$) and (38), with the isotropy of $F(\boldsymbol{\lambda})$, it is a straightforward matter to show that $F_1^1(\lambda_1, \lambda_2, \lambda_3) = F_2^2(\lambda_2, \lambda_1, \lambda_3)$ and $F_3^3(\lambda_1, \lambda_2, \lambda_3) = F_3^3(\lambda_2, \lambda_1, \lambda_3)$. Using these and similar results, it follows that

$$(39) \qquad F_1^1(\lambda_1, \lambda_2, \lambda_3) = F_2^2(\lambda_3, \lambda_1, \lambda_2) = F_3^3(\lambda_2, \lambda_3, \lambda_1),$$

showing that all F_j^i are expressible in terms of a single scalar function of λ_1, λ_2, and λ_3, and that F_2^2 and F_3^3 come from F_1^1 by cyclic interchanges of λ_1, λ_2, and λ_3. Using (5) and (38), we easily obtain (30), where the coefficients are given by (11) with $F(\lambda_1)$ replaced by $F_1^1(\lambda_1, \lambda_2, \lambda_3)$ and are therefore symmetric functions of λ_k expressible in terms of λ_k and a single scalar function of λ_k, as stated in (31).

Proof of (32) For any nonsingular, right-covariant, mixed tensor $\mathbf{\Omega}$, the tensor $\mathbf{\Omega} \cdot \boldsymbol{\lambda} \cdot \mathbf{\Omega}^{-1}$ has the same principal values as $\boldsymbol{\lambda}$ has; it therefore follows from (30) and (31) that $F(\mathbf{\Omega} \cdot \boldsymbol{\lambda} \cdot \mathbf{\Omega}^{-1}) = \mathbf{\Omega} \cdot F(\boldsymbol{\lambda}) \cdot \mathbf{\Omega}^{-1}$ for any such $\mathbf{\Omega}$, and hence, in particular, for any such $\mathbf{\Omega}$ satisfying the equation $\tilde{\mathbf{\Omega}} \cdot \boldsymbol{\gamma}_1 \cdot \mathbf{\Omega} = \boldsymbol{\gamma}_1$. Thus $F(\boldsymbol{\lambda})$ is $\boldsymbol{\gamma}_1$ isotropic, for any metric tensor $\boldsymbol{\gamma}_1$, as stated.

(40) **Theorem** A symmetric function $f(\lambda_1, \lambda_2, \lambda_3)$ of three variables λ_i can always be expressed in terms of λ_{I}, λ_{II}, and λ_{III} alone, where $\lambda_{\text{I}} = \lambda_1 + \lambda_2 + \lambda_3$, $\lambda_{\text{II}} = \lambda_2\lambda_3 + \lambda_3\lambda_1 + \lambda_1\lambda_2$, and $\lambda_{\text{III}} = \lambda_1\lambda_2\lambda_3$.

Proof of (40) λ_i are the roots in x of the cubic equation $x^3 - \lambda_{\text{I}}x^2 + \lambda_{\text{II}}x - \lambda_{\text{III}} = 0$, and are therefore determined, apart from order, by the coefficients λ_{I}, λ_{II}, and λ_{III}. Since f is symmetric, the order does not affect its value, which is therefore determined by the values of λ_{I}, λ_{II}, and λ_{III}. This proves the theorem.

Proof of (33) Since $F(\boldsymbol{\lambda})$ is linear and homogeneous in $\boldsymbol{\lambda}$, we have $F(a\boldsymbol{\lambda} + b\boldsymbol{\mu}) = aF(\boldsymbol{\lambda}) + bF(\boldsymbol{\mu})$, for arbitrary scalars a, b and mixed tensors $\boldsymbol{\lambda}$, $\boldsymbol{\mu}$. Since the principal values of $a\boldsymbol{\lambda} + b\boldsymbol{\mu}$ are $a\lambda_i + b\mu_i$, it follows from (30) and (31) that

$$\begin{aligned} &F_{\text{I}}(a\lambda_i + b\mu_i)(a\boldsymbol{\lambda} + b\boldsymbol{\mu})^2 + F_{\text{II}}(a\lambda_i + b\mu_i)(a\boldsymbol{\lambda} + b\boldsymbol{\mu}) + F_{\text{III}}(a\lambda_i + b\mu_i)\boldsymbol{\delta} \\ &\quad = a[F_{\text{I}}(\lambda_i)\boldsymbol{\lambda}^2 + F_{\text{II}}(\lambda_i)\boldsymbol{\lambda} + F_{\text{III}}(\lambda_i)\boldsymbol{\delta}] + b[F_{\text{I}}(\mu_i)\boldsymbol{\mu}^2 + F_{\text{II}}(\mu_i)\boldsymbol{\mu} + F_{\text{III}}(\mu_i)\boldsymbol{\delta}]. \end{aligned}$$

Since $\boldsymbol{\lambda}$ and $\boldsymbol{\mu}$ are arbitrary tensors, it is easy to show that we can equate coefficients of various powers. From the coefficient of $\boldsymbol{\lambda} \cdot \boldsymbol{\mu}$, we have $F_{\text{I}} = 0$. From the coefficients of $\boldsymbol{\lambda}$, we have $F_{\text{II}}(a\lambda_i + b\mu_i) = aF_{\text{II}}(\lambda_i)$, and hence $F_{\text{II}} = \text{const}$. From the coefficients of $\boldsymbol{\delta}$, it follows that F_{III} is linear and homogeneous, so that $F_{\text{III}}(\lambda_i) = a^i\lambda_i$, for some constants a^i; but $F_{\text{III}}(\lambda_i)$ is symmetric in λ_i, by (31), and so $a^1 = a^2 = a^3$ and $F_{\text{III}}(\lambda_i) = a^1 \operatorname{tr} \boldsymbol{\lambda}$. This completes the proof of (33), which is otherwise almost self-evident if one uses (30) and (40).

We state, without complete proof, the following result:

(41) **Theorem** If $\boldsymbol{\sigma} = F(\boldsymbol{\lambda}, \boldsymbol{\mu})$ is isotropic and bilinear in the mixed tensors $\boldsymbol{\lambda}$ and $\boldsymbol{\mu}$, $\boldsymbol{\sigma}$ being a mixed tensor, then F can be written in the form

$$F(\boldsymbol{\lambda}, \boldsymbol{\mu}) = a_1\boldsymbol{\lambda} \cdot \boldsymbol{\mu} + a_2\boldsymbol{\mu} \cdot \boldsymbol{\lambda} + b_1\lambda_{\text{I}}\boldsymbol{\mu} + b_2\mu_{\text{I}}\boldsymbol{\lambda} + c(\boldsymbol{\lambda} : \boldsymbol{\mu})\boldsymbol{\delta},$$

where a_i, b_i, and c are constants.

It is easy to verify that F, expressed in this form, is isotropic, and it is obviously bilinear. It is more difficult to prove that *all* isotropic, bilinear functions can be so expressed; we cannot use the representation (30), because $F(\boldsymbol{\lambda}, \boldsymbol{\mu})$ for constant $\boldsymbol{\mu}$, say, is not an isotropic function of $\boldsymbol{\lambda}$ alone. The presence of a second tensor $\boldsymbol{\mu}$ (different from the unit tensor $\boldsymbol{\delta}$) suffices to destroy isotropy. We can, however, argue heuristically that the preceding expression is general because it contains all mixed tensors "that we can think of," bilinear in $\boldsymbol{\lambda}$ and $\boldsymbol{\mu}$, which can be formed from $\boldsymbol{\lambda}$, $\boldsymbol{\mu}$, $\boldsymbol{\delta}$, and absolute scalars. In our applications, we shall take $a_1 = a_2$ and $b_1 = b_2$, since we shall have $\boldsymbol{\sigma} = \boldsymbol{\pi} \cdot \boldsymbol{\gamma}$ with $\tilde{\boldsymbol{\pi}} = \boldsymbol{\pi}$.

A further simplification may be noted here for applications to incompressible materials (Pipkin, 1964a). If $\boldsymbol{\lambda} = \boldsymbol{\gamma}^{-1} \cdot \dot{\boldsymbol{\gamma}}$, then $\lambda_{\rm I} = 0$, from the constant-volume condition 5.5(4), since the λ_i in this case are the roots in $\dot{s}$ of 2.8(28). The terms in $\lambda_{\rm I} = \operatorname{tr} \boldsymbol{\lambda}$ thus vanish from (30) and (41) when $\boldsymbol{\lambda}$ is a strain-rate tensor. Further, if $\boldsymbol{\lambda}' = \boldsymbol{\gamma}^{-1} \cdot \boldsymbol{\gamma}' - \boldsymbol{\delta}$ is a strain tensor for states t and t' (of metric tensors $\boldsymbol{\gamma}$ and $\boldsymbol{\gamma}'$), and the strain is small in the sense, say, that $(\boldsymbol{\lambda}' : \boldsymbol{\lambda}')^{1/2} = O(\varepsilon)$ is small, then $\lambda'_{\rm I}$ (although linear in $\boldsymbol{\lambda}'$ and hence apparently of order ε) is in fact of order ε^2; this, too, results in a simplification of representations such as (33) and (41) when one needs to retain terms up to a certain order only, for terms involving $\lambda'_{\rm I}$ are of higher order. In the bilinear case (41), we are taking $\boldsymbol{\mu} = \boldsymbol{\lambda}''$ to be a strain tensor of the same order as $\boldsymbol{\lambda}'$. The results may be summarized as follows, when we make the further simplification of omitting terms in the unit tensor $\boldsymbol{\delta}$ (because such terms can be absorbed, without loss of generality, into the extra stress tensor $\boldsymbol{\Sigma} = \boldsymbol{\Pi} \cdot \boldsymbol{\gamma}$):

Isotropic functions of small-strain tensors, constant volume:

(42) $$F(\boldsymbol{\lambda}') = a\boldsymbol{\lambda}' + O(\varepsilon^2) \qquad [F(\boldsymbol{\lambda}') \text{ linear}];$$

(43) $$F(\boldsymbol{\lambda}', \boldsymbol{\lambda}'') = a(\boldsymbol{\lambda}' \cdot \boldsymbol{\lambda}'' + \boldsymbol{\lambda}'' \cdot \boldsymbol{\lambda}') + O(\varepsilon^3) \qquad [F(\boldsymbol{\lambda}', \boldsymbol{\lambda}'') \text{ bilinear}];$$

(44) $$\boldsymbol{\lambda}' \equiv \boldsymbol{\gamma}^{-1} \cdot \boldsymbol{\gamma}' - \boldsymbol{\delta}, \qquad \boldsymbol{\lambda}'' \equiv \boldsymbol{\gamma}^{-1} \cdot \boldsymbol{\gamma}'' - \boldsymbol{\delta};$$
$$\varepsilon^2 = O(\boldsymbol{\lambda}' : \boldsymbol{\lambda}') = O(\boldsymbol{\lambda}'' : \boldsymbol{\lambda}'').$$

(45) **Problem** Prove the statement made above, namely, that $\lambda'_{\rm I} = O(\varepsilon^2)$ when the volume is constant for the strain $\boldsymbol{\gamma} \to \boldsymbol{\gamma}'$, where $\boldsymbol{\lambda}' = \boldsymbol{\gamma}^{-1} \cdot \boldsymbol{\gamma}' - \boldsymbol{\delta}$, and $\boldsymbol{\lambda}' : \boldsymbol{\lambda}' = O(\varepsilon^2)$.

(46) **Definition** A tensor is $\boldsymbol{\gamma}$ *isotropic* if it is $\boldsymbol{\Omega}$ invariant for all $\boldsymbol{\gamma}$-orthogonal $\boldsymbol{\Omega}$; thus second-rank tensors $\boldsymbol{\pi}$, $\boldsymbol{\sigma}$, and $\boldsymbol{\pi}^0$ are $\boldsymbol{\gamma}$ isotropic if, for all right-covariant tensors $\boldsymbol{\Omega}$ satisfying the equation $\tilde{\boldsymbol{\Omega}} \cdot \boldsymbol{\gamma} \cdot \boldsymbol{\Omega} = \boldsymbol{\gamma}$, we have:

$$\boldsymbol{\pi} = \bar{\boldsymbol{\pi}} = \boldsymbol{\Omega} \cdot \boldsymbol{\pi} \cdot \tilde{\boldsymbol{\Omega}} \qquad (\boldsymbol{\pi} \text{ contravariant});$$
$$\boldsymbol{\sigma} = \bar{\boldsymbol{\sigma}} = \boldsymbol{\Omega} \cdot \boldsymbol{\sigma} \cdot \boldsymbol{\Omega}^{-1} \qquad (\boldsymbol{\sigma} \text{ right-covariant mixed});$$
$$\boldsymbol{\pi}^0 = \bar{\boldsymbol{\pi}}^0 = \tilde{\boldsymbol{\Omega}}^{-1} \cdot \boldsymbol{\pi}^0 \cdot \boldsymbol{\Omega}^{-1} \qquad (\boldsymbol{\pi}^0 \text{ covariant}).$$

(47) **Theorem** If $\boldsymbol{\pi}$, $\boldsymbol{\sigma}$, and $\boldsymbol{\pi}^0$ are $\boldsymbol{\gamma}$ isotropic, and if $\boldsymbol{\pi}$ and $\boldsymbol{\pi}^0$ are symmetric and $\boldsymbol{\sigma}$ has three real principal values, then $\boldsymbol{\pi} = a\boldsymbol{\gamma}^{-1}$, $\boldsymbol{\sigma} = b\boldsymbol{\delta}$, and $\boldsymbol{\pi}^0 = c\boldsymbol{\gamma}$, for some scalars a, b, and c.

Proof Since $\boldsymbol{\pi}$ is symmetric, it has three real principal values, σ^i, say, and three $\boldsymbol{\gamma}$-orthonormal contravariant principal axes $\boldsymbol{\alpha}_i$, formed with respect to $\boldsymbol{\gamma}$; and we can write $\boldsymbol{\pi} = \sigma^i \boldsymbol{\alpha}_i \boldsymbol{\alpha}_i$. By (17), there exists a $\boldsymbol{\gamma}$-orthogonal $\boldsymbol{\Omega}$ that "interchanges $\boldsymbol{\alpha}_1$ and $\boldsymbol{\alpha}_2$," i.e., such that $\boldsymbol{\alpha}_2 = \boldsymbol{\Omega} \cdot \boldsymbol{\alpha}_1 = \boldsymbol{\alpha}_1 \cdot \tilde{\boldsymbol{\Omega}}$, $\boldsymbol{\alpha}_1 = \boldsymbol{\Omega} \cdot \boldsymbol{\alpha}_2$, $\boldsymbol{\alpha}_3 = \boldsymbol{\Omega} \cdot \boldsymbol{\alpha}_3$. [This $\boldsymbol{\Omega}$ is given in (34).] Since $\boldsymbol{\pi}$ is $\boldsymbol{\gamma}$ isotropic, it follows from the definition (46) that $\sigma^1 \boldsymbol{\alpha}_1 \boldsymbol{\alpha}_1 + \sigma^2 \boldsymbol{\alpha}_2 \boldsymbol{\alpha}_2 + \sigma^3 \boldsymbol{\alpha}_3 \boldsymbol{\alpha}_3 = \sigma^1 \boldsymbol{\alpha}_2 \boldsymbol{\alpha}_2 + \sigma^2 \boldsymbol{\alpha}_1 \boldsymbol{\alpha}_1 + \sigma^3 \boldsymbol{\alpha}_3 \boldsymbol{\alpha}_3$, and therefore $\sigma^1 = \sigma^2$. Similarly, $\sigma^2 = \sigma^3$. Hence $\boldsymbol{\pi} = \sigma^1 \boldsymbol{\alpha}_i \boldsymbol{\alpha}_i = \sigma^1 \boldsymbol{\gamma}^{-1}$, where σ^1 is a scalar, which proves the first result. The other two results can be proved in a similar manner by using the representations $\boldsymbol{\sigma} = \sigma^i \boldsymbol{\alpha}_i \boldsymbol{\alpha}^i$ and $\boldsymbol{\pi}^0 = \sigma^i \boldsymbol{\alpha}^i \boldsymbol{\alpha}^i$, where $\boldsymbol{\alpha}_i$ and $\boldsymbol{\alpha}^i$ are $\boldsymbol{\gamma}$-orthonormal principal axes of $\boldsymbol{\sigma}$ or of $\boldsymbol{\pi}^0$.

Theorem (47) evidently justifies calling a state of stress isotropic if $\boldsymbol{\pi} = -p\boldsymbol{\gamma}^{-1}$; it is easy to verify that in this case at any particle P the contact force $d\boldsymbol{\varphi}$ is a $\boldsymbol{\gamma}$-isotropic function of the unit normal $\mathbf{v}$. Theorem (47) shows that $\boldsymbol{\gamma}$-isotropic tensors, when of second rank, are scalar multiples of $\boldsymbol{\gamma}$, $\boldsymbol{\gamma}^{-1}$, or $\boldsymbol{\delta}$, showing that isotropy is a very restrictive condition on a tensor. It is even more restrictive on a vector, for the reader may readily verify (e.g., by considering the isotropy condition $\boldsymbol{\theta} = \bar{\boldsymbol{\theta}} = \boldsymbol{\Omega} \cdot \boldsymbol{\theta}$ for a contravariant vector $\boldsymbol{\theta}$, and using, say, $\boldsymbol{\Omega} = \boldsymbol{\alpha}_1 \boldsymbol{\alpha}^2 - \boldsymbol{\alpha}_2 \boldsymbol{\alpha}^1 - \boldsymbol{\alpha}_3 \boldsymbol{\alpha}^3$, where $\boldsymbol{\alpha}_i$ and their reciprocals $\boldsymbol{\alpha}^i$ are $\boldsymbol{\gamma}$ orthonormal) that *the only $\boldsymbol{\gamma}$-isotropic vector is the zero vector*. We state, without proof, the result that *the most general fourth-rank covariant tensor $\boldsymbol{\kappa}$ that is $\boldsymbol{\gamma}$ isotropic has components in any B of the form*

(48) $$\kappa_{ijkm} = a\gamma_{ij}\gamma_{km} + b\gamma_{ik}\gamma_{jm} + c\gamma_{im}\gamma_{jk},$$

where a, b, and c are scalars; this is evidently the covariant tensor analog of the well-known corresponding result [e.g., Eq. (20), Jeffreys (1931, p. 70)] for Cartesian tensors.

(49) **Definition** A tensor-valued function $F(\boldsymbol{\lambda})$ of a tensor $\boldsymbol{\lambda}$ is differentiable if there exists a fourth-rank tensor $(\partial/\partial\boldsymbol{\lambda})F(\boldsymbol{\lambda})$ (called the derivative of F at $\boldsymbol{\lambda}$), independent of $\boldsymbol{\varepsilon}$, such that $F(\boldsymbol{\lambda} + \boldsymbol{\varepsilon}) - F(\boldsymbol{\varepsilon}) = \boldsymbol{\varepsilon} : (\partial/\partial\boldsymbol{\lambda})F(\boldsymbol{\lambda}) + O(\boldsymbol{\varepsilon} : \boldsymbol{\varepsilon})$ for all $\boldsymbol{\varepsilon}$.

It is easy to see that, on referring this definition to any body coordinate system B and writing tensor components in a row (instead of the usual square matrix), the components of the derivative are just the elements of a Jacobian. Similar definitions can be given for tensors of other rank and for vector-valued functions of vectors, in particular.

The elementary definition of a tensor function $F(\boldsymbol{\lambda})$ given in (6) can be applied whenever $\boldsymbol{\lambda}$ has three real principal values (all different). For certain purposes (e.g., in integrating tensor differential equations and in discussing constant stretch history), we need an extended definition which applies when $\boldsymbol{\lambda}$ has some complex principal values. In fact, the only function that we need in such cases is the exponential function, and one can easily extend the definition of an exponential function of a matrix to that of a tensor. If $f(z)$ is a power series in the complex variable z, with radius of convergence R, and λ is a matrix such that, for each eigenvalue λ_i, $\|\lambda_i\| < R$, then the matrix series $f(\lambda)$ converges (in the sense that the series composed of the (i, j)th elements converges, for each i and j) to a matrix, which we denote by $f(\lambda)$. The exponential series has infinite radius of convergence, and so can be used to define $\exp \lambda$, for every square matrix λ, and to verify the expected properties ($e^{\lambda}e^{\mu} = e^{\lambda+\mu}$ when $\lambda\mu = \mu\lambda$; $de^{\lambda}/dt = \dot{\lambda}e^{\lambda} = e^{\lambda}\dot{\lambda}$) [cf., e.g., Bronson (1969, pp. 103–106), Finkbeiner (1960), Rinehart (1955)].

(50) **Definition** A sequence $\{\boldsymbol{\sigma}(n)\}$ $(n = 1, 2, \ldots)$ of right-covariant, mixed tensors at P converges to a limit $\boldsymbol{\sigma}$ (a right-covariant mixed tensor at P) if there exists a body coordinate system B such that the sequence $\{\sigma(n)\}$ of representative matrices in B converges to a matrix $\sigma = \boldsymbol{\sigma}B$. [From the transformation law 2.5(8), it is easy to show that in any other $\bar{B}$, the sequence $\{\bar{\sigma}(n)\}$ converges to $\bar{\sigma} = \boldsymbol{\sigma}\bar{B}$.]

(51) **Definition** The exponential function $e^{\boldsymbol{\lambda}} = \boldsymbol{\sigma}$, say, of a right-covariant mixed tensor $\boldsymbol{\lambda}$, is defined by the equation

$$e^{\boldsymbol{\lambda}} = \lim_{N \to \infty} \sum_{n=0}^{N} \boldsymbol{\lambda}^n / n!$$

5.8 Constant stretch history

The definition of steady flow used in classical hydrodynamics, namely $\partial v^i(y, t)/\partial t = 0$ (where v^i are velocity components of a particle at place y at time t), is not of direct interest in rheology, because the concept is not invariant under a rigid rotation. It is natural to inquire whether, in some sense to be determined, a "rheologically significant" definition of steady flow can be given. We have offered one such definition in the case of shear flow [namely, $\dot{s}$ and ζ independent of time, in 3.2(24)]. More generally, we must seek some condition to impose on the body metric tensor history $\{\boldsymbol{\gamma}(\mathrm{P}, t)\}$, for each given particle P, because any rheologically significant restriction on flow histories must be so expressible.

It seems natural, in the first instance, to consider histories which possess some property of invariance with respect to arbitrary translations along the time axis in the sense that for each value of t and τ, *the relation between* $\boldsymbol{\gamma}(\mathrm{P}, t)$ *and* $\boldsymbol{\gamma}(\mathrm{P}, t+\tau)$ *should be expressible in terms of an operator which can depend on* τ *and* P *but not on* t. The simplest such relation is one which is linear in both metric tensors, namely

$$(1) \qquad \boldsymbol{\gamma}(\mathrm{P}, t+\tau) = \boldsymbol{\Theta}(\mathrm{P}, \tau) : \boldsymbol{\gamma}(\mathrm{P}, t), \qquad \partial\boldsymbol{\Theta}/\partial t = \mathbf{0},$$

where $\boldsymbol{\Theta}(\mathrm{P}, \tau)$ is some fourth-rank tensor which is independent of t. We know of no investigation in the literature of flows satisfying a condition as general as this one, but considerable attention has been devoted to the case in which $\boldsymbol{\Theta}$ is a product of two (essentially equal) second-rank tensors, namely the case of *flows of constant stretch history*, which we shall define by the equations

$$(2) \qquad \boldsymbol{\gamma}(\mathrm{P}, t+\tau) = \tilde{\boldsymbol{\mu}}(\mathrm{P}, \tau) \cdot \boldsymbol{\gamma}(\mathrm{P}, t) \cdot \boldsymbol{\mu}(\mathrm{P}, \tau), \qquad \partial\boldsymbol{\mu}(\mathrm{P}, \tau)/\partial t = \mathbf{0},$$

$$(3) \qquad \boldsymbol{\mu}(\mathrm{P}, 0) = \boldsymbol{\delta} \qquad (\boldsymbol{\mu} \text{ a right-covariant mixed tensor}).$$

It follows that (2) must be expressible in the form

$$(4) \qquad \boldsymbol{\gamma}(\mathrm{P}, t+\tau) = \exp(\tilde{\boldsymbol{\mu}}_0 \tau) \cdot \boldsymbol{\gamma}(\mathrm{P}, t) \cdot \exp(\boldsymbol{\mu}_0 \tau), \qquad \boldsymbol{\mu}_0 = \left[\frac{\partial\boldsymbol{\mu}(\mathrm{P}, \tau)}{\partial\tau}\right]_{\tau=0}.$$

To see this, we differentiate (2) with respect to τ, and put τ equal to zero after the differentiation, obtaining the result $\dot{\boldsymbol{\gamma}} = \tilde{\boldsymbol{\mu}}_0 \cdot \boldsymbol{\gamma} + \boldsymbol{\gamma} \cdot \boldsymbol{\mu}_0$. This is a linear, first-order differential equation, which must possess a unique solution [when an initial value, $\boldsymbol{\gamma}(\mathrm{P}, 0)$ say, is given]. The solution is evidently given by (4) with $t = 0$ and $\tau = t$, and (4) itself is then readily obtained. It is also easy to derive the following further equations:

$$(5) \qquad \ddot{\boldsymbol{\gamma}} = \tilde{\boldsymbol{\mu}}_0 \cdot \dot{\boldsymbol{\gamma}} + \dot{\boldsymbol{\gamma}} \cdot \boldsymbol{\mu}_0, \quad \text{etc.};$$

$$(6) \qquad \boldsymbol{\gamma}^{-1}(\mathrm{P}, t+\tau) = \exp(-\tilde{\boldsymbol{\mu}}_0 \tau) \cdot \boldsymbol{\gamma}^{-1}(\mathrm{P}, t) \cdot \exp(-\boldsymbol{\mu}_0 \tau).$$

It is easy to verify from 3.8(6), 3.8(7), and 3.8(12) that *shear-free flow with constant strain rates* $\dot{\kappa}_i$ *is a flow of constant stretch history* in which $\boldsymbol{\mu}_0 = -\frac{1}{2}\dot{\kappa}_i \boldsymbol{\beta}_i \boldsymbol{\beta}^i$, where $\boldsymbol{\beta}_i$ and $\boldsymbol{\beta}^i$ are base vectors for any one of the body coordinate systems that are orthogonal throughout the motion. From 3.3(20), 3.3(23) (with $t' = t + \tau$), and 3.3(14), it is easy to see that *unidirectional shear flow at constant shear rate* $\dot{s}$ *is a flow of constant stretch history* in which $\boldsymbol{\mu}_0 = \dot{s}\boldsymbol{\alpha}_1\boldsymbol{\alpha}^2$ ($\boldsymbol{\alpha}_1$ tangential to a shear line; $\boldsymbol{\alpha}^2$ normal to a shear surface). If the strain rates vary with time, these flows are no longer flows of constant stretch history.

In order to relate our definition to others, we transfer (4) to space at time t, using Cartesian space tensors; from 5.4(14), we obtain the result

$$(7) \qquad \mathbf{C}_t(t+\tau) = \exp(\tilde{\mathbf{m}}\tau) \cdot \mathbf{U} \cdot \exp(\mathbf{m}\tau), \qquad \text{where } \mathbf{m} = \mathbf{m}(t) = \mathbb{T}(t)\boldsymbol{\mu}_0.$$

Using 11.3(11) and (7), it follows that the displacement gradient tensor $\mathbf{F}$ must be expressible in the form

(8) $$\mathbf{F}_t(t+\tau) = \mathbf{R}(t,\tau)\cdot\exp[\tau\mathbf{m}(t)], \qquad \tilde{\mathbf{R}} = \mathbf{R}^{-1},$$

for some orthogonal tensor function $\mathbf{R}(t,\tau)$. From 5.4(31)(iii) applied to the case $\mathbf{F}\equiv\mathbf{m}$, $\mathbf{\Phi}\equiv\boldsymbol{\mu}$, with arbitrary B and rectangular Cartesian C, we obtain from (7) the equations

(9) $$m^i_j(t) = \left(\frac{\partial y^i}{\partial\xi^r}\frac{\partial\xi^s}{\partial y^j}\right)_t \mu^r_{(0)s}(\xi);$$

eliminating the constant components $\mu^r_{(0)s}$ between these equations and the equations obtained from them by putting $t=0$, using 11.3(2) and the fact that $\mathbf{F}_0^{-1}(t) = \mathbf{F}_t(0)$, which is a particular case of

(10) $$\mathbf{F}_t(s)\cdot\mathbf{F}_{t'}(t) = \mathbf{F}_{t'}(s),$$

which follows from 11.3(2), we obtain the equation

(11) $$\mathbf{m}(t) = \mathbf{F}_0(t)\cdot\mathbf{m}(0)\cdot\mathbf{F}_t(0) \qquad (\xrightarrow{t}\partial\boldsymbol{\mu}/\partial t = \mathbf{0}),$$

which is the space tensor version of the body tensor condition $\partial\boldsymbol{\mu}/\partial t = \mathbf{0}$. From the definition of $e^{\boldsymbol{\mu}_0\tau}$ as a power series in $\boldsymbol{\mu}_0\tau$, it follows that

(12) $$\mathbb{T}(t)\exp(\boldsymbol{\mu}_0\tau) = \exp[\mathbb{T}(t)\boldsymbol{\mu}_0\tau] = \exp[\mathbf{m}(t)\tau],$$

and also, from (11), that

(13) $$\exp[\mathbf{m}(t)\tau] = \mathbf{F}_0(t)\cdot\exp[\tau\mathbf{m}(0)]\cdot\mathbf{F}_t(0).$$

Using (8), (10), and (13), we obtain the equation

(14) $$\mathbf{F}_0(t+\tau) = \mathbf{R}(t,\tau)\cdot\mathbf{F}_0(t)\cdot\exp[\tau\mathbf{m}(0)], \qquad (\tilde{\mathbf{R}} = \mathbf{R}^{-1}).$$

This is a consequence of the body tensor equation (2) and is in fact equivalent to (2), because it is easy to verify that each step taken in deriving (14) from (2) is reversible, and so (2) can be derived from (14). On putting $t=0$ in (14) and using the fact that $\mathbf{F}_0(0) = \mathbf{U}$, which follows from 11.3(2), we obtain the equation

(15) $$\mathbf{F}_0(\tau) = \mathbf{Q}(\tau)\cdot\exp[\tau\mathbf{m}(0)], \qquad \mathbf{Q} = \tilde{\mathbf{Q}}^{-1} \quad [\equiv\mathbf{R}(0,\tau)],$$

which is the same as Eq. (2.2) obtained by Noll (1962) and shown by him to be a necessary and sufficient condition for a motion to be at constant stretch history (≡ "substantially stagnant motion") as defined by Coleman (1961, 1962). We have thus proved that (2) *is the body tensor condition that a motion be at constant stretch history*, for it is a straightforward matter, starting from (15), to derive (14) [i.e., to prove that (14) must hold for some orthogonal tensor function $\mathbf{R}(t,\tau)$], and we have already shown that (2) can be derived from (14).

It follows from (4) that a motion of constant stretch history is completely determined when at each particle P, the constant mixed tensor $\boldsymbol{\mu}_0$ [or the Cartesian tensor $\mathbf{m}(0)$] is given; at most, the nine components of $\boldsymbol{\mu}_0$ in any B can be assigned arbitrary values. The symmetric tensors $\partial^n\gamma/\partial t^n$ ($n = 1, 2, 3, \ldots$) (each of which in B has at most six *a priori* independent components) cannot therefore be independent; Wang (1965) has shown that a knowledge of the corresponding space tensors $\mathbf{A}^n(t)$ for $n = 1, 2$, and 3 at any one t is sufficient to determine the strain history $\mathbf{C}_t(t-s)$ for $-\infty \leq s \leq 0$.

Huilgol (1969) has shown that the orthogonal rheometer flow is one of constant stretch history. Using 3.7(4), 3.7(6), and 3.7(7), it is possible to show that (2) is satisfied with

$$(16)\quad \boldsymbol{\mu}(\mathrm{P}, \tau) = \boldsymbol{\beta}_2\boldsymbol{\beta}^2 + (\boldsymbol{\beta}_1\boldsymbol{\beta}^1 + \boldsymbol{\beta}_3\boldsymbol{\beta}^3)\cos\Omega\tau - \boldsymbol{\beta}_1\boldsymbol{\beta}^3\xi^1\sin\Omega\tau - \boldsymbol{\beta}_3\boldsymbol{\beta}^1\frac{\cos\Omega\tau}{\xi^1},$$

where $\boldsymbol{\beta}_i(B, \mathrm{P})$ and $\boldsymbol{\beta}^i(B, \mathrm{P})$ are the base vectors for the body coordinate system B defined in Section 3.7. Since these vectors are independent of time, it follows that the flow is one of constant stretch history. One can derive (16) either by factorizing the matrix $\gamma(\xi, t+\tau) = \gamma(\mathrm{P}, t+\tau)B$ to obtain the matrix $\mu(\xi, \tau) = \boldsymbol{\mu}(\mathrm{P}, \tau)B$ and then writing $\boldsymbol{\mu} = \mu^i{}_j\boldsymbol{\beta}_i\boldsymbol{\beta}^j$ (I owe this factorization to K. Walters), or one can use the matrix $m(0)$ (which Huilgol shows is equal to the velocity gradient matrix in a rectangular Cartesian coordinate system), and derive from it the matrix $[\mu^r_{(0)c}]$ using (9) at $t = 0$, from which one obtains the very simple result

$$(17)\quad \boldsymbol{\mu}_0 = -\xi^1\Omega\boldsymbol{\beta}_1\boldsymbol{\beta}^3 + \frac{\Omega}{\xi^1}\boldsymbol{\beta}_3\boldsymbol{\beta}^1.$$

Hence $\boldsymbol{\mu}_0{}^2 \cdot \boldsymbol{\mu}_0{}^n = -\boldsymbol{\Omega}^2\boldsymbol{\mu}_0{}^n$, and the exponential series

$$(18)\quad \boldsymbol{\mu}(\mathrm{P}, \tau) = \exp(\boldsymbol{\mu}_0\tau) = \sum_{n=0}^{\infty}(\tau\boldsymbol{\mu}_0)^n/n!$$

separates into two series whose sums involve $\cos\Omega\tau$ and $\sin\Omega\tau$; using the fact that $\boldsymbol{\mu}_0{}^2 = -\Omega^2(\boldsymbol{\beta}_1\boldsymbol{\beta}^1 + \boldsymbol{\beta}_3\boldsymbol{\beta}^3)$, one finally obtains the stated result (16).

An alternative method of determining whether a given flow is one of constant stretch history, and which can always be applied when $\gamma(\mathrm{P}, t)$ is a given function of t, is illustrated in the next example.

If one is given a flow, it is sometimes difficult to decide whether it is one of constant stretch history or not, if one uses the definition (2): this equation is nonlinear in the unknown tensor function $\boldsymbol{\mu}(\mathrm{P}, \tau)$. It is always possible, however, to use the linear equation

$$(19)\quad \dot{\gamma} = \tilde{\boldsymbol{\mu}}_0 \cdot \gamma + \gamma \cdot \boldsymbol{\mu}_0,$$

which is equivalent to (2). Referred to any convenient body coordinate system B, (19) yields a matrix equation for the unknown constant matrix μ_0, and is thus equivalent to *six linear* scalar equations for the nine components of μ_0. Since these components must be independent of time, while components of $\dot{\gamma}$ and γ will depend on time, we usually have a set of conditions which are incompatible. For example, if one uses 3.8(5), 3.8(7), and 3.8(9) for the balance rheometer flow, one finds that no $\boldsymbol{\mu}_0$ exists that satisfies (19) and is independent of time. Thus *balance rheometer flow (with constant Ω) is not a flow at constant stretch history.*

The fact that the orthogonal rheometer flow is one of constant stretch history casts doubt on the value of considering constant-stretch-history flows as a class, for the class includes rheologically unsteady flows (as in the orthogonal rheometer, where the shear angle ζ is time dependent) and rheologically steady flows (e.g., steady unidirectional shear flows, in which shear rate and shear angle are independent of time). Since constant-stretch-history flows thus do not, in fact, achieve the aim (mentioned at the beginning of this section) of providing a definition for a class of rheologically steady flows, we are tempted to conjecture that *no universally applicable, rheologically significant, meaning can be given to the term steady flow.*

6 CONSTITUTIVE EQUATIONS FOR VISCOELASTIC MATERIALS

6.1 General forms for constitutive equations

For viscoelastic materials in general and polymeric solids and liquids in particular, most constitutive equations which have been proposed so far can be expressed in the form

$$\mathscr{G}\{\boldsymbol{\pi}', \boldsymbol{\gamma}', t' \,|\, -\infty < t' \leq t; \boldsymbol{\delta}, \boldsymbol{\kappa}_a \,|\, a = 1, 2, \ldots, M\} = \mathbf{0}, \tag{1}$$

where $\boldsymbol{\pi}' \equiv \boldsymbol{\pi}(\mathrm{P}, t')$ and $\boldsymbol{\gamma}' \equiv \boldsymbol{\gamma}(\mathrm{P}, t')$ are the only time-dependent (or state-dependent) body tensor variables, and $\boldsymbol{\kappa}_a \equiv \boldsymbol{\kappa}_a(\mathrm{P}, T)$ $(a = 1, 2, \ldots, M)$ are time-independent body tensors which will be called *material constant* body tensors. T denotes the temperature. In (1), $\mathscr{G}$ denotes an operator, formed from the various operators which can be defined for body tensor fields, which when applied to the tensors $\boldsymbol{\kappa}_a$ and the tensor histories $\{\boldsymbol{\pi}'\}$ and $\{\boldsymbol{\gamma}'\}$, yields a symmetric covariant (or contravariant) body tensor (of second rank).

Equation (1) is merely a body tensor version of the very general class of equations which Oldroyd (1950a) proposed in "convected components" of space tensor fields; it is shown in Section 11.1 that convected components of space tensor fields are numerically equal to B components of corresponding body tensor fields, when the body coordinate system B is chosen to be the same as that which generates the convected family of space coordinate systems. The relation between the two formalisms was noted without proof by Lodge (1951) and with proof by Lodge (1972).

The following fundamental points should be noted.

(i) Because (1) is a tensor equation, it has a significance independent of any particular choice of coordinate system from the set $\mathscr{B}$, and therefore is a possible equation for describing the rheological properties of a material. Equation (1) represents one unambiguous way of (automatically) satisfying the general requirement that equations used for describing properties of a material shall be expressible in a form which does not involve any unwanted dependence on a choice of coordinate system, for the choice of coordinate system involves an arbitrariness which has nothing to do with material properties.

(ii) Because (1) is a body tensor equation in which $\boldsymbol{\pi}'$ and $\boldsymbol{\gamma}'$ are the only tensor variables, it automatically satisfies [because of 2.6(25)(iv) and 5.6(18)(iv)] the general requirement that an admissible constitutive equation shall describe rheological properties of a material element that are independent of the position, orientation, and motion (relative to axes fixed in space) of that element. This requirement appears to be suitable in polymer rheology, and indeed in most of continuum mechanics, where no external fields act so as to affect the rheological properties of an element and where rotation (relative to axes fixed in space) is not so rapid that it could have a significant effect on molecular motions. The requirement would not be satisfied if other body tensor variables, such as the body vorticity tensor $\boldsymbol{\omega}$, defined in 11.3(22), were included, with $\boldsymbol{\pi}'$ and $\boldsymbol{\gamma}'$, in the arguments of $\mathscr{G}$. The essential assumption underlying (1) is that the stress tensor field $\boldsymbol{\pi}(\mathrm{P}, t')$ and the metric tensor field $\boldsymbol{\gamma}(\mathrm{P}, t')$ [with the temperature field $T(\mathrm{P}, t')$, if the flow history is not isothermal] form what we might call a complete set of rheological variables.

(iii) Within the class of constitutive equations of the form (1), different materials will be distinguished from one another by (a) different combinations of body tensor operators, denoted by $\mathscr{G}$, and (b) different values and kinds of the material constant tensors, denoted by $\boldsymbol{\kappa}_a$. In particular, if the material is a solid in the (Oldroyd) sense that its constitutive equation includes variables describing some unique state of "permanent rheological significance," then the material constants could include the metric tensor $\boldsymbol{\gamma}_0(\mathrm{P}, T)$ for such a state; the important examples of this type are the perfectly elastic solids (considered in Section 6.3), for which (1) can be solved to give $\boldsymbol{\pi}(\mathrm{P}, t)$ as a function of $\gamma_0(\text{P}, T)$, the metric tensor for the stress-free state at temperature T, and $\gamma(\text{P}, t)$, the metric tensor for a variable state, labeled t.

(iv) The material symmetry properties (e.g., isotropy, anisotropy, transverse isotropy, etc.) are described by the symmetry properties of the material constant tensors $\boldsymbol{\kappa}_a$. In particular, if all the material constant tensors $\boldsymbol{\kappa}_a$ ($a = 1, 2, \ldots, M$) are expressible in terms of absolute scalars, the unit tensor $\boldsymbol{\delta}$, and the metric tensor $\gamma_0(\text{P}, T)$ for some unique state (the stress-free state, for example, if there is a unique state, at constant temperature, which is attained either instantaneously or ultimately, when the stress is zero), then it is entirely in accordance with usage in physics and the classical theory of elasticity to call the material *isotropic*.

The distinction between material symmetry and admissibility of equations as constitutive equations was clearly brought out by Oldroyd (1950a), who used the term "inherently anisotropic" for materials for which "not all physical constants are combinations of scalar constants and universally constant tensors," and who postulated admissibility for all equations, form-invariant under arbitrary transformations from one convected coordinate system to another, in which the only tensor variables were convected components of stress and the metric.

(v) In most cases of interest for polymeric materials, $\mathscr{G}$ will be restricted so as to include only one-particle operators, so that (1) may be called a one-particle equation. This seems a reasonable restriction to adopt until such time as experimental (or other) evidence forces us to raise it. The restriction excludes, for example, the use of covariant derivatives, since these involve a differentiation which is a limiting process involving more than one particle; constitutive equations involving covariant derivatives are mentioned in Section 10.4 in connection with discrepancies in normal stress measurements. Equation (1), when $\mathscr{G}$ is a one-particle operator, includes all the so-called simple fluids (Lodge and Stark, 1972); these are thus included in the very general class of viscoelastic materials considered by Oldroyd (1950a).

We shall be mainly concerned with incompressible viscoelastic materials whose constitutive equations can be written in the form

$$\text{(2)}\quad \boldsymbol{\Pi}(\text{P}, t) \equiv \boldsymbol{\pi}(\text{P}, t) + p\gamma^{-1}(\text{P}, t) = \mathscr{F}\{\gamma(\text{P}, t') | -\infty < t' \leq t\} \equiv \mathscr{F}\{\gamma'\}$$

$$\text{(3)}\qquad |\gamma^{-1}(\text{P}, t) \cdot \gamma(\text{P}, t')| = 1,$$

where $\mathscr{F}$ is a one-particle functional, isotropic at time t in the sense that

$$\text{(4)}\qquad \mathscr{F}\{\tilde{\boldsymbol{\Omega}}^{-1} \cdot \gamma' \cdot \boldsymbol{\Omega}^{-1}\} = \boldsymbol{\Omega} \cdot \mathscr{F}\{\gamma'\} \cdot \tilde{\boldsymbol{\Omega}} \qquad [\gamma' \equiv \gamma(\text{P}, t')]$$

for all constant, right-covariant, mixed tensors $\boldsymbol{\Omega}$ that are t orthogonal, i.e., that satisfy Eq. 5.7(15), namely

(5) $$\tilde{\mathbf{\Omega}} \cdot \gamma \cdot \mathbf{\Omega} = \gamma \qquad [\gamma \equiv \gamma(\mathrm{P}, t)].$$

A material is said to be *incompressible* if at constant temperature the volume of every material element is constant, whatever stress is applied; (3) expresses the constancy of volume. It is assumed that for an incompressible material the *metric tensor history* $\{\gamma'\}$ $(-\infty < t' \leq t)$ determines the stress at time t only to within an unknown additive isotropic stress $p\gamma^{-1}$, where $p \equiv p(\mathrm{P}, t)$ is a scalar function to be determined (in a given problem, by the complete system of simultaneous equations and boundary conditions, such as those listed in Section 8.10).

The use of the term isotropic in connection with (4) is consistent with the usage in Section 5.7 and may be related to the usage given in connection with (1), as follows. It is a fundamental property of the various tensor operations that they always give matrix equations (when referred to an arbitrary coordinate system B) which are *form invariant* with respect to arbitrary changes of coordinate system $B \to \bar{B}$; thus, if $\mathscr{G}\{\boldsymbol{\theta}_i\}$ denotes any body tensor function of any set of body tensors $\{\boldsymbol{\theta}_i\}$ (of any kind), then for any B and $\bar{B}$ in $\mathscr{B}$, we have

(6) $$\mathscr{G}\{\boldsymbol{\theta}_i\}B = \mathscr{G}\{\theta_i\}, \qquad \mathscr{G}\{\boldsymbol{\theta}_i\}\bar{B} = \bar{\mathscr{G}}\{\bar{\theta}_i\},$$

where $\bar{\mathscr{G}}\{\theta_i\}$ and $\mathscr{G}\{\bar{\theta}_i\}$ are matrix functions of the matrix sets $\{\theta_i\}$ and $\{\bar{\theta}_i\}$. Form invariance is then expressed by the equation

(7) $$\mathscr{G} \equiv \bar{\mathscr{G}}.$$

The index i here can assume values in a discrete set and in a continuous set of numbers.

It is evident that (7) is still satisfied if we allow $\mathscr{G}$ to include absolute scalars and the unit tensor $\boldsymbol{\delta}$ (in addition to the standard tensor operations of addition, subtraction, multiplication, contraction, taking of inverse and transpose), for an absolute scalar is represented by the same number in every B and $\boldsymbol{\delta}$ is represented by the same matrix I in every B. However, (7) would not be satisfied if $\mathscr{G}$ included nonisotropic tensors, i.e., tensors that are not numerical tensors.

If we now consider (1) in the case in which all the material constant tensors $\boldsymbol{\kappa}_a$ can be expressed in terms of absolute scalars and the unit tensor $\boldsymbol{\delta}$, and we suppose that the implicit equation (1) can be solved so as to give $\boldsymbol{\pi}$ explicitly in terms of the history $\{\gamma'\}$, then, on replacing $\boldsymbol{\pi}$ by $\mathbf{\Pi}$ for incompressible materials, we obtain an equation of the form (2), in which $\mathscr{F}$ can include absolute scalars (such as t, t', T, and material constants) and $\boldsymbol{\delta}$, but no anisotropic tensors. It then follows from (7) that the representatives

$\mathscr{F}$ and $\bar{\mathscr{F}}$ of $\mathscr{F}$ in any B and $\bar{B}$ satisfy the equation

$$(8) \qquad \mathscr{F} = \bar{\mathscr{F}}.$$

From 2.5(8), we have $\bar{\Pi} = \Lambda\Pi\tilde{\Lambda}$ and $\bar{\gamma}' = \tilde{\Lambda}^{-1}\gamma'\Lambda^{-1}$; hence from (8),

$$(9) \qquad \Lambda\mathscr{F}\{\gamma'\}\tilde{\Lambda} = \mathscr{F}\{\tilde{\Lambda}^{-1}\gamma'\Lambda^{-1}\} \qquad [\Lambda \equiv \Lambda_{\mathrm{P}}(B, \bar{B})].$$

For any given B and P, we can choose a $\bar{B}$ so that the matrix Λ has any given nonsingular value. Hence (9) must hold for all time-independent, nonsingular matrices Λ, at each P.

It is easy to see that (9) includes (or implies) (4): For, given any B and any $\mathbf{\Omega}$ satisfying (5), we can choose a $\bar{B}$ such that (at any given P) $\Lambda_{\mathrm{P}}(B, \bar{B}) = \Omega \equiv \mathbf{\Omega}B$. Hence (9) gives the equation $\Omega\mathscr{F}\{\gamma'\}\tilde{\Omega} = \mathscr{F}\{\tilde{\Omega}^{-1}\gamma'\Omega^{-1}\}$, which is just the B representative of the required equation (4). Thus $\mathscr{F}\{\tilde{\mathbf{\Omega}}^{-1}\cdot\boldsymbol{\gamma}'\cdot\mathbf{\Omega}^{-1}\} - \mathbf{\Omega}\cdot\mathscr{F}\{\boldsymbol{\gamma}'\}\cdot\tilde{\mathbf{\Omega}}$ is a tensor that has zero representative matrix in one coordinate system (B) and is therefore the zero tensor. QED.

Thus the constitutive equation (2) holds, with $\mathscr{F}$ t isotropic, provided that $\mathscr{F}$ includes no anisotropic tensors. This means, in particular, that $\mathscr{F}$ includes no constant metric tensor $\boldsymbol{\gamma}_0(\mathrm{P}, T)$, for any such tensor is not t isotropic except in the particular case $\boldsymbol{\gamma}(\mathrm{P}, t) = \boldsymbol{\gamma}_0(\mathrm{P}, T)$; the possibility $\boldsymbol{\gamma}(\mathrm{P}, t) = x\boldsymbol{\gamma}_0(\mathrm{P}, T)$, where $x \neq 1$, is excluded by the constant-volume condition. Thus (2) describes a material which can be regarded as a liquid in the sense that its constitutive equation contains no reference to any one state of permanent rheological significance.

6.2 Constitutive equations from molecular theories

There are a few constitutive equations for polymeric materials which have been derived from well-defined molecular models on the basis of certain simplifying assumptions and approximations. Here, we shall simply list the equations with references to the literature where the details can be found. The equations are of considerable importance in their own right, because of the information which they give concerning the influence of molecular structure on rheological properties, and have also had a considerable influence on the choice of constitutive equations to be found in the literature on viscoelastic materials.

The materials considered here are all incompressible; the constitutive equations are of the general form 6.1(2), involving the extra stress tensor $\mathbf{\Pi}$, and are to be supplemented by the incompressibility condition 6.1(3).

The *rubberlike solid* constitutive equation

$$(1) \qquad \mathbf{\Pi}(\mathrm{P}, t) = ckTN_0\boldsymbol{\gamma}_0^{-1}(\mathrm{P}, T)$$

is derived from a molecular network model in the so-called Gaussian approximation. c is a number whose value is probably in the range 1 to 3; k is Boltzmann's constant; T is the absolute temperature; N_0 is the concentration (in number per ml) of junctions (i.e., chemical crosslinks) between polymer molecules in the "active" network; and γ_0 is the metric tensor in the stress-free state at temperature T. The theory is due to James and Guth (1947), James (1947), Wall (1942), Flory (1953), and Treloar (1958); the body tensor form of the equations has been given by Lodge (1968). A recent review has been given by Treloar (1973). The elastic properties are attributed to thermal motion of the polymer molecules; this accounts for the presence of the factor kT in the constitutive equation. *The molecular network mechanism in the Gaussian approximation is responsible for the fact that it is the contravariant body tensors which are simply related to one another*; further, the properties are governed by a single material constant cN_0, equal to the concentration of effective network segments (a *segment* is a length of polymer molecule joining two junctions with no intervening junction present).

The *rubberlike liquid* constitutive equation

$$\Pi(\mathrm{P}, t) = \int_{t'=-\infty}^{t} \mu(t-t', T)\gamma^{-1}(\mathrm{P}, t')\, dt' - \eta_s \frac{\partial \gamma^{-1}(\mathrm{P}, t)}{\partial t} \tag{2}$$

has been derived from a model developed from the network model for a rubberlike solid by assuming that the network junctions have finite lifetimes. The *memory function* μ is given by an equation of the form

$$\mu(\tau, T) = kT \sum_i L_i(T) \exp[-\tau/\tau_i(T)], \tag{3}$$

where $L_i(T)$ is the creation rate per unit volume for segments of type i having a loss probability per unit time equal to $\tau_i(T)$. The term $\mu(t - t', T)\, dt'/kT$ is equal to the concentration at time t of network segments which were created during the time interval $(t', t' + dt')$. The term in η_s represents a contribution from the solvent, assumed to be an incompressible, Newtonian liquid of viscosity η_s. The theory is due to Green and Tobolsky (1946) and Lodge (1954, 1956); the body tensor form of the equation has been given by Lodge (1968).

The *bead-spring model* for dilute polymer solutions leads to constitutive equations which are also of the form (2) and (3), but the constants in the memory function have a different interpretation and are all expressible in terms of three unknown constants: N, the number of Gaussian springs per polymer molecule; b, the root-mean-square length of a free spring in a solvent at rest; and f_0, the friction constant of each "bead," where the polymer molecule is represented as a set of $N + 1$ equal beads joined linearly by the

N identical springs. The memory function is of the form

(4) $$\mu(\tau, T) = nkT \sum_{i=1}^{N} \tau_i^{-1} \exp(-\tau/\tau_i),$$

where n denotes the number of polymer molecules per ml of solution,

(5) $$\tau_i = f_0 b^2/6kT\lambda_i \qquad (i = 1, 2, \ldots, N),$$

and $\lambda_i = \lambda_i(N, h^*)$ are characteristic values of a certain $N \times N$ symmetric matrix B; $h^* = (12\pi^3)^{-1/2} f_0/\eta_s b$. The theory is due to Rouse (1953) and Zimm (1956); the constitutive equations (2) and (4) were derived by Lodge and Wu (1971); the most recent computation of "exact" values for λ_i is due to Lodge and Wu (1972), who tabulate these values together with appropriate functions of use in calculating rheological properties for steady and oscillatory shear flows. Armstrong (1973) has extended the theory to slightly non-Gaussian springs for the case $N = 1$, with neglect of bead/bead hydrodynamic interaction, obtaining a constitutive equation of the form

(6)

$$\left(1 + \tau_{01}\frac{\partial}{\partial t}\right)\mathbf{\Pi}^* + (\eta_1 - a_1 \boldsymbol{\gamma} : \mathbf{\Pi}^*)\frac{\partial \boldsymbol{\gamma}^{-1}}{\partial t} = \tau_{02}\frac{\partial}{\partial t}\{(\boldsymbol{\gamma} : \mathbf{\Pi}^*)\mathbf{\Pi}^* + \mathbf{\Pi}^* \cdot \boldsymbol{\gamma} \cdot \mathbf{\Pi}^*\},$$

where τ_{01}, τ_{02}, η_1, and a_1 are constants, and $\mathbf{\Pi}^* \equiv \mathbf{\Pi} + \eta_s(\partial \boldsymbol{\gamma}^{-1}/\partial t)$ represents the polymer contribution to the extra stress tensor.

6.3 Perfectly elastic solids

In order to generalize the constitutive equation 6.2(1) (appropriate to a rubberlike "elastic solid"), it is simplest to consider a material for which *states of thermodynamic equilibrium can be identified and put into one-to-one correspondence with values of the body metric tensor* $\boldsymbol{\gamma}$ *and the temperature* T, so that A, the Helmholtz free energy per unit mass, is expressible as a function $A = A(\boldsymbol{\gamma}, T)$. The constants in A can include material constant body tensors $\boldsymbol{\kappa}_a(\mathrm{P}, T_0)$, evaluated for convenience at some arbitrary but definite temperature T_0, and the body metric tensor $\boldsymbol{\gamma}_0 = \boldsymbol{\gamma}_0(\mathrm{P}, T_0)$ in the stress-free state (supposed unique) at T_0. A is an absolute body scalar field whose change

(1) $$dA = dW - S\,dT$$

for an arbitrary infinitesimal change of equilibrium state from $(\boldsymbol{\gamma}, T)$ to $(\boldsymbol{\gamma} + d\boldsymbol{\gamma}, T + dT)$ is governed by the first and second laws of thermodynamics; S denotes the entropy per unit mass, and dW, the work done on unit mass of material by external tractions and body forces (minus the increase in kinetic energy, which can be taken to be zero in this context), is given by the equation

$$2\rho\, dW = \boldsymbol{\pi} : d\boldsymbol{\gamma}, \tag{2}$$

where ρ denotes the density of the material (an absolute scalar). A derivation of (2) is given in Section 8.5 [8.5(22)]. The analysis here can be thought of as applying, in the first instance, to an arbitrarily small material element which includes a particle P; the preceding variables are evaluated at P.

On referring (2) to an arbitrary body coordinate system B and using the fact that dT is arbitrary and the fact that $d\gamma_{ij} = d\gamma_{ji}$ are either arbitrary (if the material is compressible) or arbitrary subject only to the constant-volume condition (if the material is incompressible)

$$\gamma^{ij}\, d\gamma_{ji} = 0 \qquad \text{(constant volume)}, \tag{3}$$

it is a straightforward matter to derive the following general form of constitutive equation:

$$\rho\left(\frac{\partial}{\partial \gamma_{ij}} + \frac{\partial}{\partial \gamma_{ji}}\right) A([\gamma_{rc}], T) = \begin{cases} \pi^{ij} & \text{(compressible)}, \\ \Pi^{ij} & \text{(incompressible)}. \end{cases} \tag{4}$$

The hydrostatic pressure variable p in the extra stress $\boldsymbol{\Pi} = \boldsymbol{\pi} + p\boldsymbol{\gamma}^{-1}$ arises as a Lagrange multiplier associated with the scalar constraint (3). We also obtain the familiar equation $S = -\partial A/\partial T$. From the relation between body tensor components and convected components of space tensors (established in Section 11.1), it follows that (4) are the same as the equations given by Brillouin (1925, 1938) and Oldroyd (1950b), and used extensively in non-linear elasticity theory by Green and Zerna (1968) and Green and Adkins (1960). The equations apply to isotropic and to anisotropic perfectly elastic solids.

The term *perfectly elastic solid* is used in the literature to apply to materials having one, or more, of the following properties at any given temperature: (i) there is a unique stress-free shape; (ii) there is a one-to-one correspondence between stress and strain; (iii) there is a "strain-energy function" which can be expressed in the form $A(\boldsymbol{\gamma}, T)$ and which determines the stress in any state according to (4); (iv) following any change of shape, the new value of stress is attained instantaneously, and vice versa. Property (i) is used, for example, by Love (1927, p. 92): "The property of recovery of an original size and shape is the property that is termed elasticity." Oldroyd (1950b) uses the term "perfect" for property (iv). Prager (1961) discusses different uses of the term "elastic," and, in particular, uses a term "hyperelastic" for property (iii) (or for the corresponding property for adiabatic, instead of isothermal, strains). It is clear that (ii) includes (i), but the nature of the relations between other properties is complicated: for example, if a "(iii) body" is to possess property (ii), then Eqs. (4), which in general express a tensor $\boldsymbol{\pi}$ as a nonlinear function of a tensor $\boldsymbol{\gamma}$, must have a unique inverse; conditions for the existence of such an inverse are difficult to formulate.

(5) **Problem** For the rubberlike solid defined by 6.2(1), show that the Helmholtz free energy $A = (ckTN_0\boldsymbol{\gamma}:\boldsymbol{\gamma}_0^{-1}/2\rho) + \text{const.}$

(6) **Problem** If $\varphi(\boldsymbol{\varepsilon})$ denotes an absolute scalar function of a variable, second-rank, covariant body tensor $\boldsymbol{\varepsilon}$ of weight n, and the representative function $\varphi(\varepsilon_{ij})$ in an arbitrary body coordinate system B is differentiable, prove that $\partial\varphi/\partial\varepsilon_{ij}$ are the components in B of a second-rank, contravariant, body tensor of weight $-n$. This tensor may be denoted by $\partial\varphi/\partial\boldsymbol{\varepsilon}$.

For *isotropic* perfectly elastic solids, the material constant tensors must all be expressible in terms of absolute scalars, $\boldsymbol{\delta}$, and $\boldsymbol{\gamma}_0$, and hence the strain-energy function A (an absolute scalar) must be expressible as a function of absolute scalars formed from these quantities and the metric tensor $\boldsymbol{\gamma}$. From $\boldsymbol{\gamma}$ and $\boldsymbol{\gamma}_0$, we can form the following three independent absolute scalars:

(7)
$$I_1 = e_1^2 + e_2^2 + e_3^2,$$
$$I_2 = (e_2 e_3)^2 + (e_3 e_1)^2 + (e_1 e_2)^2,$$
$$I_3 = (e_1 e_2 e_3)^2,$$

where $e_i = e_i(t_0, t)$ $(= ds/ds_0$ for a principal axis) are the principal elongation ratios for the strain $t_0 \to t$ (where t_0 labels the stress-free state $\boldsymbol{\gamma}_0$). From the definition 2.8(21), we have

(8)
$$|\boldsymbol{\gamma} - e_i^2\boldsymbol{\gamma}_0| = 0.$$

From the Cayley–Hamilton theorem 5.7(10) applied to the mixed tensor $\boldsymbol{\lambda} \equiv \boldsymbol{\gamma}\cdot\boldsymbol{\gamma}_0^{-1}$, it is easy to show that the *strain invariants* I_i are given by the equations

(9)
$$I_1 = \boldsymbol{\gamma}:\boldsymbol{\gamma}_0^{-1}, \qquad I_2 = I_3\,\boldsymbol{\gamma}_0:\boldsymbol{\gamma}^{-1}, \qquad I_3 = |\boldsymbol{\gamma}\cdot\boldsymbol{\gamma}_0^{-1}|,$$

and, moreover, that the invariants formed from any power of the mixed strain tensor $\boldsymbol{\gamma}\cdot\boldsymbol{\gamma}^{-1}$ are expressible in terms of I_1, I_2, and I_3. It is reasonable, then, to consider the case in which A can be expressed as a function of I_1, I_2, I_3, and T, and of no other variables. On repeating the argument used to obtain (4) from (1) and (2), using the equations

(10)
$$dI_1 = \boldsymbol{\gamma}_0^{-1}:d\boldsymbol{\gamma}, \qquad dI_2 = -I_3(\boldsymbol{\gamma}^{-1}\cdot\boldsymbol{\gamma}_0\cdot\boldsymbol{\gamma}^{-1}):d\boldsymbol{\gamma} + \boldsymbol{\gamma}_0:\boldsymbol{\gamma}^{-1}\,dI_3,$$
$$dI_3 = I_3\boldsymbol{\gamma}^{-1}:d\boldsymbol{\gamma},$$

which are obtained from (9) for an arbitrary, infinitesimal, change $d\boldsymbol{\gamma} = \widetilde{d\boldsymbol{\gamma}}$

in the metric tensor in the stressed state, it is easy to obtain the following constitutive equation:

$$(11) \qquad \pi = 2\rho\{(I_2 A_2 + I_3 A_3)\gamma^{-1} + A_1\gamma_0^{-1} - A_2\gamma^{-1}\cdot\gamma_0\cdot\gamma^{-1}\},$$

where $A_i = \partial A(I_1, I_2, I_3, T)/\partial I_i\,(i = 1, 2, 3)$, and the material is compressible. If the material is incompressible, then the constitutive equation takes the form

$$(12) \quad \Pi = 2(W_1\gamma_0^{-1} - W_2\gamma^{-1}\cdot\gamma_0\cdot\gamma^{-1}), \qquad W_i = \rho\,\partial A(I_1, I_2, 1, T)/\partial I_i \quad (i = 1, 2).$$

The symbol W used here for the Helmholtz free energy per unit volume should not be confused with the symbol dW used for work done in (1). Equations (11) and (12) are body tensor versions of equations derived and used extensively by Rivlin for the solution of problems in large-strain elasticity; a full account of this and related work has been given by Green and Adkins (1960) and by Truesdell and Noll (1965). A summary and a comparison with experimental data on rubbers have been given by Treloar (1958, 1973).

(13) **Problem** Prove that for an isotropic, perfectly elastic solid having a constitutive equation of the form (11), the principal axes of stress coincide with the principal axes for the strain $\gamma_0 \to \gamma$, and that corresponding principal values σ_i and e_i of stress and strain are related by the equations

$$(14) \qquad \sigma_i = 2\rho\{(I_2 A_2 + I_3 A_3) + A_1 e_i^2 - A_2 e_i^{-2}\} \qquad (i = 1, 2, 3).$$

(15) **Problem** If A is expressed in terms of strain invariants J_i, where $J_1 = I_1$, $J_2 = e_1{}^4 + e_2{}^4 + e_3{}^4$, and $J_3 = I_3$, obtain the constitutive equation in the alternative form

$$\pi = 2\rho\{A_3'\gamma^{-1} + A_1'\gamma_0^{-1} + 2A_2'\gamma_0^{-1}\cdot\gamma\cdot\gamma_0^{-1}\},$$

where $A_i' = \partial A(J_1, J_2, J_3, T)/\partial J_i$.

It is evident that the Gaussian stress–strain equation 6.2(1) is that particular case of (12) in which $W_1 = ckTN_0/2$ and $W_2 = 0$. One equation which has been used extensively with rubbers is due to Mooney (1940), and has W_1 constant (as in the Gaussian equation) but includes the term with W_2 constant; the term in W_2 in (12) (when W_2 is constant) is sometimes called the Mooney term; the usual notation is $C_i = W_i$ in this context.

The strain-energy function A and its derivatives must satisfy certain conditions, probably involving inequalities, if it is to be used for describing actual materials. For example, a tensile stress must give rise to an increase (not a decrease) in length of an elastic filament. Progress in seeking necessary and sufficient conditions on A has been reviewed by Truesdell and Noll

(1965). In connection with certain integral constitutive equations for viscoelastic liquids which are obtainable from the Mooney term alone [by replacing t_0 in (12) by t', W_2 by a function $m(t - t')$, followed by integration over the range $-\infty \leq t' \leq t$], it is worth noting that *the elastic solid equation* (12) *with* $W_1 = 0$ *and* W_2 *independent of strain is unrealistic for application to rubberlike polymers*: It is easy to verify that for simple elongation, the tensile force has a finite upper bound (equal to one-third the value of the initial Young's modulus times the initial cross-sectional area).

The problem of obtaining the "incremental stress–strain relations" for an infinitesimal strain $d\boldsymbol{\gamma}$ following a finite strain $\boldsymbol{\gamma}_0 \to \boldsymbol{\gamma}$ has been treated by Green *et al.* (1952) and by Green and Adkins (1960). In body tensors, it is easy to write down the relations between $d\boldsymbol{\pi}$ and $d\boldsymbol{\gamma}$ by differentiating (11) (for compressible materials), or the relations between $d\boldsymbol{\Pi}$ and $d\boldsymbol{\gamma}$ by differentiating (12) (for incompressible materials). In the latter case, the equations are of the form

(16) $$d\boldsymbol{\Pi} = \boldsymbol{\Phi} : d\boldsymbol{\gamma},$$

where $\boldsymbol{\Phi}$ is a fourth-rank body tensor involving $\boldsymbol{\gamma}_0$, $\boldsymbol{\gamma}$, W_i, and $\partial^2 W/\partial I_i\, \partial I_j$.

6.4 Integral constitutive equations

Equation (2) is an example of what may be called an *integral constitutive equation*, since it expresses the extra stress at time t in terms of a time integral over the flow history. The first integral forms of constitutive equations were published by Boltzmann (1874, 1876) in order to describe what was called the "elastic aftereffect" observed in certain metals; the equations [which include equations equivalent to the linear viscoelasticity equations 6.1(5)] are admissible only in the case in which the displacement gradients are all infinitesimally small. Two of the unlimited number of possible generalizations of these equations, admissible without restriction, were given by Lodge (1954); one of these is included in an equation given by Oldroyd (1950a), which also included a time integral of stress. Expansions of constitutive equations which express the stress as a sum of terms involving multiple integrals of flow history have been considered in very general forms by Green and Rivlin (1957). Equations involving single and triple integrals have been used by Ward (1971) to describe certain data.

(1) The most extensive use of integral constitutive equations in the literature in connection with comparison with experimental data and with molecular theories for polymeric materials has involved the single integral equations listed in Table 6.4(1).

Table 6.4(1) Constitutive Equations of the Form[a] $\boldsymbol{\Pi} = \int_{-\infty}^{t} (\mu\boldsymbol{\gamma}'^{-1} + \nu\boldsymbol{\gamma}^{-1}\cdot\boldsymbol{\gamma}'\cdot\boldsymbol{\gamma}^{-1})\,dt'$

	$\mu(\tau)$	$\nu(\tau)$	Reference
1.	$m(\tau)$	0	Green and Tobolsky (1946), Lodge (1954, 1956, 1968), Walters (1960, 1962), Fredrickson (1962), Lodge and Wu (1971).
2.	0	$m(\tau)$	Oldroyd (1950a), Lodge (1954), Fredrickson (1962), Coleman and Noll (1961)
3.	$m_1(\tau)$	$m_2(\tau)$	Ward and Jenkins (1958)
4.	$2\,\partial W(\tau, I_1', I_2')/\partial I_1'$	$2\,\partial W(\tau, I_1', I_2')/\partial I_2'$	Kaye (1962), Bernstein *et al.* (1963)
5.	$(1+\frac{1}{2}\varepsilon)m(\tau, \dot{\gamma}_{\mathrm{II}}')$	$\frac{1}{2}\varepsilon m(\tau, \dot{\gamma}_{\mathrm{II}}')$	Spriggs *et al.* (1966), Bird and Carreau (1968), $\varepsilon = 0$: Yamamoto (1971, 1972)
6.	$(1+\frac{1}{2}\varepsilon)m\left(\tau, \underset{t'}{\overset{t}{\dot{\gamma}_{\mathrm{II}}''}}\right)$	$\frac{1}{2}\varepsilon m\left(\tau, \underset{t'}{\overset{t}{\dot{\gamma}_{\mathrm{II}}''}}\right)$	$\varepsilon = 0$ or -2: Fredrickson (1962); $\varepsilon = 0$: Bogue and Doughty (1966), Chen and Bogue (1972), Meister (1971), Carreau (1972)
7.	$(1+\frac{1}{2}\varepsilon)m(\tau)h(\tau - t_R)$	$\frac{1}{2}\varepsilon m(\tau)h(\tau - t_R)$	Tanner and Simmons (1967), Tanner (1969)
8.	$\int_0^{\tau_m} H(s)e^{-\tau/s}\,ds/s^2$	0	Leonov (1964), Booij (1970), τ_m assigned empirically
9.	$m\left(\underset{t'}{\overset{t}{\pi_{\mathrm{I}}''}}, \underset{t'}{\overset{t}{\pi_{\mathrm{II}}''}}, \underset{t'}{\overset{t}{\pi_{\mathrm{III}}''}}\right)$	0	Kaye (1966), Lodge (1968)

[a] *Notation*: $\tau = t - t'$, $\boldsymbol{\gamma} = \boldsymbol{\gamma}(\mathrm{P}, t)$, $\boldsymbol{\gamma}' = \boldsymbol{\gamma}(\mathrm{P}, t')$, $\dot{\boldsymbol{\gamma}}' = \partial\boldsymbol{\gamma}'/\partial t'$, etc. $\boldsymbol{\Pi} = \boldsymbol{\pi} + p\boldsymbol{\gamma}^{-1}$.
Invariants: strain: $I_1' = \boldsymbol{\gamma} : \boldsymbol{\gamma}'^{-1}$, $I_2' = \boldsymbol{\gamma}^{-1} : \boldsymbol{\gamma}'$;
strain rate: $\dot{\gamma}_{\mathrm{II}} = \mathrm{tr}(\dot{\boldsymbol{\gamma}}\cdot\boldsymbol{\gamma}^{-1}\cdot\dot{\boldsymbol{\gamma}}\cdot\boldsymbol{\gamma}^{-1})$;
stress: $\pi_{\mathrm{I}} = \boldsymbol{\pi} : \boldsymbol{\gamma}$, $\pi_{\mathrm{II}} = \mathrm{tr}(\boldsymbol{\pi}\cdot\boldsymbol{\gamma}\cdot\boldsymbol{\pi}\cdot\boldsymbol{\gamma})$, $\pi_{\mathrm{III}} = |\boldsymbol{\pi}\cdot\boldsymbol{\gamma}|$.

$\underset{t'}{\overset{t}{f}}(x'')$ denotes a functional of $x'' = x(t'')$ over the range $t' \le t'' \le t$. $h(u) = 0$ $(u < 0)$ or 1 $(u \ge 0)$.

For certain cases in 1, 5, 6, and 9, $m = \sum_{p=1}^{N} a_p \exp(\int_t^{t'} dt''/\tau_p'')$, where a_p and τ_p'' are given in Table 6.4(2).

In 7, $t_R = t'$ is chosen so that $[(1+\frac{1}{2}\varepsilon)I_1' + \frac{1}{2}\varepsilon I_2'] = b$, where b is a constant chosen to fit the data.

(2) The integrands are listed in Table 6.4(2). Where necessary, published equations cited in these tables have been transformed by techniques involving transfer from space to body and integration by parts (as illustrated in Section 6.6).

Table 6.4(2) Expressions Used for Memory Function Coefficients a_p and Exponents τ_p'' in Table 6.4(1)[a]

Number in Table 6.4(1)	a_p	τ_p''	Constants chosen to fit data	Reference
1	kTL_p	τ_p	L_p, τ_p ($p = 1, 2, \ldots, N$)	Network model Lodge (1968)
1	nkT/τ_p	$f_0 b^2/(6kT\lambda_p)$	N, b, f_0 (see Section 6.2)	Rouse/Zimm model: Lodge and Wu (1971)
9	a/τ_1'	$[b + 3c\pi_{II}'' - c(\pi_I'')^2]^{-1}$	a, b, c ($N = 1$)	Kaye (1966)
6	G_p/λ_p	$\{\lambda_p^{-1} + (\dot{\gamma}_{II}'')^{1/2}[b + \lambda_1(\dot{\gamma}_{II}'')^{1/2}]^n\}^{-1}$	b, λ_p, G_p ($p = 1, 2, \ldots, N$), n	Meister (1971)
5	$\eta_p \lambda_{2p}^{-2}(1 + \frac{1}{2}\lambda_{1p}^2 \dot{\gamma}_{II}')^{-1}$	λ_{2p}	N, ε, η_0, λ_1, λ_2, α_1, α_2, n_1, n_2	Bird and Carreau (1968)
6	$\eta_p \lambda_{1p}^{-2} f'$	$\lambda_{1p} g''$	N, ε, η_0, λ_1, λ_2, α, R, S, c	Carreau model A (1972)
6	$\eta_p \lambda_{1p}^{-2} f_p'$	$\lambda_{1p} g_p''$	N, ε, η_0, λ_1, λ_2, α, R, S, c	Carreau model B (1972)

[a] *Notation*: for 5 and 6: $\eta_p = \eta_0 \lambda_{1p} / \sum_{q=1}^{N} \lambda_{1q}$; for 5: $\lambda_{1p} = \lambda_1[(1 + n_1)/(p + n_1)]^{\alpha_1}$, $\lambda_{2p} = \lambda_2[(1 + n_2)/(p + n_2)]^{\alpha_2}$; for 6: $\lambda_{1p} = 2^{\alpha}\lambda_2(p + 1)^{-\alpha}$; for 6, model A: $f' = (1 + \frac{1}{2}c^2\lambda_1{}^2\dot{\gamma}_{II}')^{2R}(1 + \frac{1}{2}\lambda_1{}^2\dot{\gamma}_{II}')^{-3S}$; $g'' = (1 + \frac{1}{2}c^2\lambda_1{}^2\dot{\gamma}_{II}'')^{-R}(1 + \frac{1}{2}\lambda_1{}^2\dot{\gamma}_{II}'')^{S}$; for 6, model B: $f'_{p-1} = (1 + \frac{1}{2}c^2\lambda_2{}^2\dot{\gamma}_{II}')^{-2R}\{1 + [\frac{1}{2}(2^{\alpha}\lambda_1)^2\dot{\gamma}_{II}']^S/p^{2\alpha}\}$; $g''_{p-1} = (1 + \frac{1}{2}c^2\lambda_2{}^2\dot{\gamma}_{II}'')^R/f''_{p-1}$.

6.5 Differential constitutive equations

(1) In the literature, various constitutive equations have been considered which are expressible in the form of differential equations in which $\boldsymbol{\pi}$ and $\boldsymbol{\gamma}$ can be regarded as dependent variables and t as the independent variable. Those that are of first order and linear in $\boldsymbol{\pi}$ are listed in Table 6.5(1), to which should be added the equation for the incompressible Newtonian liquid, namely

(2) $$\boldsymbol{\Pi} = \eta - \partial\boldsymbol{\gamma}^{-1}/\partial t \qquad \text{or} \qquad \boldsymbol{\Pi}^0 = \eta\dot{\boldsymbol{\gamma}},$$

where η is a constant equal to the viscosity. Higher-order equations have been considered by Spriggs and Bird (1964) and Spriggs (1965).

There seems to be no rheologically very important distinction between integral and differential constitutive equations, nor any special reason for considering differential equations that are linear in the body stress tensor. It is perhaps of interest to note that such linear equations can be formally integrated, using a tensor version of the matrizant method, to give equations in which the stress tensor is expressed as an integral, or more generally as a series of multiple integrals, over the body metric tensor history. The method may be sketched (without rigor) as follows, using a simple extension of the matrix method [see, e.g., Frazer *et al.* (1952, pp. 53, 218)], which was first used for constitutive equations by Goddard and Miller (1966).

We consider the differential equation

(3) $$(\partial\boldsymbol{\pi}/\partial t) - \boldsymbol{\pi}\cdot\tilde{\boldsymbol{\theta}} - \boldsymbol{\theta}\cdot\boldsymbol{\pi} = \boldsymbol{\varphi},$$

where $\boldsymbol{\pi}$ and $\boldsymbol{\varphi}$ are symmetric, contravariant tensors and $\boldsymbol{\theta}$ is a right-covariant mixed tensor. All tensors in this section will be of second rank, evaluated at an arbitrary particle P which we shall omit from the arguments. The tensors can, for the present purposes, be regarded as functions of time t alone. The necessary changes can be made easily if analogous equations using covariant tensors (e.g., $\boldsymbol{\Pi}^0$ and $\boldsymbol{\theta}^0$) are to be integrated. We shall regard $\boldsymbol{\theta}$ and $\boldsymbol{\varphi}$ as given tensor functions of time and $\boldsymbol{\pi}$ as the unknown function of time.

We seek a tensor integrating factor $\boldsymbol{\omega}$. Taking $\boldsymbol{\omega}^{-1}\cdot(3)\cdot\tilde{\boldsymbol{\omega}}^{-1}$, we have $(\partial/\partial t)(\boldsymbol{\omega}^{-1}\cdot\boldsymbol{\pi}\cdot\tilde{\boldsymbol{\omega}}^{-1}) = \boldsymbol{\omega}^{-1}\cdot\boldsymbol{\varphi}\cdot\tilde{\boldsymbol{\omega}}^{-1}$, whose solution is

(4) $$(\boldsymbol{\omega}^{-1}\cdot\boldsymbol{\pi}\cdot\tilde{\boldsymbol{\omega}}^{-1})_t = (\boldsymbol{\omega}^{-1}\cdot\boldsymbol{\pi}\cdot\tilde{\boldsymbol{\omega}}^{-1})_{t_0} + \int_{t_0}^{t}(\boldsymbol{\omega}^{-1}\cdot\boldsymbol{\varphi}\cdot\tilde{\boldsymbol{\omega}}^{-1})_{t'}\,dt',$$

provided only that $\boldsymbol{\omega}$ satisfies the equation $\partial\boldsymbol{\omega}^{-1}/\partial t = -\boldsymbol{\omega}^{-1}\cdot\boldsymbol{\theta}$, or, what is equivalent,

(5) $$\dot{\boldsymbol{\omega}} = \boldsymbol{\theta}\cdot\boldsymbol{\omega}.$$

Table 6.5(1) Differential Constitutive Equations, Linear and of First Order in the Extra-Stress Tensor[a]:

$$\lambda_1(\partial\boldsymbol{\Pi}^0/\partial t) + \boldsymbol{\Pi}^0 - (\boldsymbol{\theta}\cdot\boldsymbol{\Pi}^0 + \boldsymbol{\Pi}^0\cdot\tilde{\boldsymbol{\theta}}) + (\boldsymbol{\Pi}^0\cdot\boldsymbol{\gamma}^{-1}) : \boldsymbol{\varphi} = \tfrac{1}{2}(\boldsymbol{\psi} + \tilde{\boldsymbol{\psi}})$$

	$\boldsymbol{\theta}$	$\boldsymbol{\varphi}$	$\boldsymbol{\psi}$	Reference
1	$\mathbf{0}$	$\mathbf{0}$	$\eta_0(\dot{\boldsymbol{\gamma}} + \lambda_2\ddot{\boldsymbol{\gamma}})$	Oldroyd (1950a)
2	$\lambda_1\dot{\boldsymbol{\gamma}}\cdot\boldsymbol{\gamma}^{-1}$	$\mathbf{0}$	$\eta_0\{\dot{\boldsymbol{\gamma}} + \lambda_2(\ddot{\boldsymbol{\gamma}} - 2\dot{\boldsymbol{\gamma}}\cdot\boldsymbol{\gamma}^{-1}\cdot\dot{\boldsymbol{\gamma}})\}$	Oldroyd (1950a)
3	$\lambda_1\dot{\boldsymbol{\gamma}}\cdot\boldsymbol{\gamma}^{-1}$	$\mathbf{0}$	$G_0\boldsymbol{\gamma} + \eta_s\{\dot{\boldsymbol{\gamma}} + \lambda_1(\ddot{\boldsymbol{\gamma}} - 2\dot{\boldsymbol{\gamma}}\cdot\boldsymbol{\gamma}^{-1}\cdot\dot{\boldsymbol{\gamma}})\}$	Giesekus (1962), Lodge (1970) (elastic dumbbell suspension)
4	$\tfrac{1}{2}\lambda_1\dot{\boldsymbol{\gamma}}\cdot\boldsymbol{\gamma}^{-1}$	$\mathbf{0}$	$\eta_0\dot{\boldsymbol{\gamma}}$	Zaremba (1903, 1937), Fromm (1933, 1948)
5	$\lambda_1\dot{\boldsymbol{\gamma}}\cdot\boldsymbol{\gamma}^{-1}$	$\mathbf{0}$	$c\lambda_1\dot{\boldsymbol{\gamma}}$	Tanner (1965), White and Metzner (1963) [with $\boldsymbol{\gamma}^{-1} : \boldsymbol{\Pi}^0 = 0$ and $\lambda_1 = \lambda_1(\dot{\gamma}_{\text{II}})$]
6	$\mu_1\dot{\boldsymbol{\gamma}}\cdot\boldsymbol{\gamma}^{-1}$	$\mu_0\,\boldsymbol{\delta}\dot{\boldsymbol{\gamma}} + \nu_1\dot{\boldsymbol{\gamma}}\cdot\boldsymbol{\gamma}^{-1}\boldsymbol{\gamma}$	$\eta_0\{\dot{\boldsymbol{\gamma}} + \lambda_2\ddot{\boldsymbol{\gamma}} - \mu_2\dot{\boldsymbol{\gamma}}\cdot\boldsymbol{\gamma}^{-1}\cdot\dot{\boldsymbol{\gamma}} + \nu_2\dot{\gamma}_{\text{II}}\boldsymbol{\gamma}\}$	Oldroyd (1958)
7	$(\mu_1\dot{\boldsymbol{\gamma}} + \nu_1\ddot{\boldsymbol{\gamma}})\cdot\boldsymbol{\gamma}^{-1}$	$\mu_0\,\boldsymbol{\delta}\dot{\boldsymbol{\gamma}}$	$\eta_0\{\dot{\boldsymbol{\gamma}} + \lambda_2\ddot{\boldsymbol{\gamma}} - \dot{\boldsymbol{\gamma}}\cdot\boldsymbol{\gamma}^{-1}\cdot(\mu_2\dot{\boldsymbol{\gamma}} + \nu_2\ddot{\boldsymbol{\gamma}})\}$	Bird (1972), Bird and Armstrong (1972), Armstrong and Bird (1973) (Rigid dumbbell suspension, approximate theory)

[a] *Notation*: Covariant extra-stress tensor: $\boldsymbol{\Pi}^0 = \boldsymbol{\gamma}(\mathrm{P}, t)\cdot\boldsymbol{\Pi}(\mathrm{P}, t)\cdot\boldsymbol{\gamma}(\mathrm{P}, t)$. $\dot{\boldsymbol{\gamma}} = \partial\boldsymbol{\gamma}(\mathrm{P}, t)/\partial t$; $\ddot{\boldsymbol{\gamma}} = \partial^2\boldsymbol{\gamma}(\mathrm{P}, t)/\partial t^2$; $\dot{\gamma}_{\text{II}} = \mathrm{Tr}(\dot{\boldsymbol{\gamma}}\cdot\boldsymbol{\gamma}^{-1}\cdot\dot{\boldsymbol{\gamma}}\cdot\boldsymbol{\gamma}^{-1})$.

Integrating this equation as it stands, we have

$$(6)\qquad \begin{aligned} \boldsymbol{\omega}(t) &= \boldsymbol{\omega}(t_0) + \int_{t_0}^{t} \boldsymbol{\theta}' \cdot \boldsymbol{\omega}(t')\, dt'; \\ \boldsymbol{\omega}(t') &= \boldsymbol{\omega}(t_0) + \int_{t_0}^{t'} \boldsymbol{\theta}'' \cdot \boldsymbol{\omega}(t'')\, dt'', \ldots \end{aligned}$$

and, substituting for $\boldsymbol{\omega}(t')$ from the second of these equations in the first, for $\boldsymbol{\omega}(t'')$ from the third in the second, and so on, we obtain a formal solution of (5) in the form

$$(7)\qquad \boldsymbol{\omega}(t) = \boldsymbol{\Omega}_{t_0}^{t} \cdot \boldsymbol{\omega}(t_0),$$

where

$$\boldsymbol{\Omega}_{t_0}^{t} \equiv \boldsymbol{\delta} + \int_{t_0}^{t} dt'\, \boldsymbol{\theta}' + \int_{t_0}^{t} dt'\, \boldsymbol{\theta}' \cdot \int_{t_0}^{t'} dt''\, \boldsymbol{\theta}'' + \cdots.$$

The "tensorzant" $\boldsymbol{\Omega}_{t_0}^{t} = \boldsymbol{\Omega}_{t_0}^{t}\{\boldsymbol{\theta}\}$ is a right-covariant mixed tensor whose value depends on the history of $\boldsymbol{\theta}$ throughout the interval (t_0, t). Since (7) must hold for arbitrarily chosen functions $\boldsymbol{\omega}(t_0)$, it is easy to show that

$$(8)\qquad \boldsymbol{\Omega}_{t_0}^{t} \cdot \boldsymbol{\Omega}_{t}^{t_0} = \boldsymbol{\delta}; \qquad \boldsymbol{\Omega}_{t_0}^{t} \cdot \boldsymbol{\Omega}_{t_1}^{t_0} = \boldsymbol{\Omega}_{t_1}^{t}.$$

Finally, from (4), (7), and (8), we obtain a formal solution of (3) in the form

$$(9)\qquad \boldsymbol{\pi}(t) = \boldsymbol{\Omega}_{t_0}^{t} \cdot \boldsymbol{\pi}(t_0) \cdot \tilde{\boldsymbol{\Omega}}_{t_0}^{t} + \int_{t_0}^{t} \boldsymbol{\Omega}_{t'}^{t} \cdot \boldsymbol{\varphi}(t') \cdot \tilde{\boldsymbol{\Omega}}_{t'}^{t}\, dt'.$$

In particular, if $\dot{\boldsymbol{\theta}} = \mathbf{0}$, then it is easy to show from (7) that

$$(10)\qquad \boldsymbol{\Omega}_{t'}^{t}\{\boldsymbol{\theta}\} = \exp[(t - t')\boldsymbol{\theta}] \qquad (\dot{\boldsymbol{\theta}} = \mathbf{0}).$$

If $\boldsymbol{\theta} = \boldsymbol{\gamma}^{-1} \cdot \dot{\boldsymbol{\gamma}}$ and the flow is one of constant stretch history so that, from 5.8(4), we have $\boldsymbol{\gamma}(t) = [\exp(\tilde{\boldsymbol{\mu}}_0 t)] \cdot \boldsymbol{\gamma}(0) \cdot \exp(\boldsymbol{\mu}_0 t)$, where $\boldsymbol{\mu}_0$ is constant, then it is easy to show that $\boldsymbol{\theta} = [\exp(-\boldsymbol{\mu}_0 t)] \cdot \boldsymbol{\varepsilon} \cdot \exp(\boldsymbol{\mu}_0 t)$, where

$$\boldsymbol{\varepsilon} \equiv \boldsymbol{\mu}_0 + \boldsymbol{\gamma}^{-1}(0) \cdot \tilde{\boldsymbol{\mu}}_0 \cdot \boldsymbol{\gamma}(0);$$

in this case Eq. (5) can be integrated as it stands [after left multiplication by $\exp(\boldsymbol{\mu}_0 t)$], with the result

$$(11)\qquad \boldsymbol{\Omega}_{t_0}^{t}\{\boldsymbol{\theta}\} = [\exp(-\boldsymbol{\mu}_0 t)] \cdot \exp[(\boldsymbol{\mu}_0 + \boldsymbol{\varepsilon})(t - t_0)] \cdot \exp(\boldsymbol{\mu}_0 t_0).$$

This cannot be simplified unless $\boldsymbol{\mu}_0$ and $\boldsymbol{\varepsilon}$ commute.

6.6 Alternative forms for constitutive equations

By obvious processes involving integration by parts, differentiation, and rearrangement, many constitutive equations can be expressed in various alternative forms. These processes are trivial when body tensor equations are used, but the equivalence of the corresponding space tensor equations is often far from obvious. We illustrate these processes for the case of the rubberlike liquid, 6.2(2), with the solvent term (in η_s) omitted for brevity. Under suitable conditions, the following are some of the alternative forms for these constitutive equations:

(1) $$\boldsymbol{\Pi} = \int_{-\infty}^{t} \mu(t-t')\boldsymbol{\gamma}'^{-1}\,dt';$$

(2) $$\boldsymbol{\Pi} = -\int_{-\infty}^{t} G(t-t')\frac{\partial \boldsymbol{\gamma}'^{-1}}{\partial t'}\,dt', \qquad \text{where} \quad G(t) = \int_{t}^{\infty}\mu(s)\,ds;$$

(3) $$\boldsymbol{\Pi}^0 = \int_{-\infty}^{t} G(t-t')\boldsymbol{\gamma}\cdot\boldsymbol{\gamma}'^{-1}\cdot\dot{\boldsymbol{\gamma}}'\cdot\boldsymbol{\gamma}'^{-1}\cdot\boldsymbol{\gamma}\,dt' \qquad (\boldsymbol{\Pi}^0 \equiv \boldsymbol{\gamma}\cdot\boldsymbol{\Pi}\cdot\boldsymbol{\gamma});$$

(4) $$\left(1+\tau_1\frac{\partial}{\partial t}\right)\boldsymbol{\Pi} = a_1\tau_1\boldsymbol{\gamma}^{-1};$$

(5) $$\left(1+\tau_1\frac{\partial}{\partial t}\right)\boldsymbol{\Pi}^0 = a_1\tau_1\boldsymbol{\gamma} + \tau_1(\dot{\boldsymbol{\gamma}}\cdot\boldsymbol{\gamma}^{-1}\cdot\boldsymbol{\Pi}^0 + \boldsymbol{\Pi}^0\cdot\boldsymbol{\gamma}^{-1}\cdot\dot{\boldsymbol{\gamma}});$$

(6) $$\left(1+\tau_1\frac{\partial}{\partial t}\right)\boldsymbol{\Pi} = -a_1\tau_1^2\frac{\partial\boldsymbol{\gamma}^{-1}}{\partial t};$$

(7) $$\left(1+\tau_1\frac{\partial}{\partial t}\right)\left(1+\tau_2\frac{\partial}{\partial t}\right)\boldsymbol{\Pi} = (a_1\tau_1 + a_2\tau_2)\boldsymbol{\gamma}^{-1} + (a_1+a_2)\tau_1\tau_2\frac{\partial\boldsymbol{\gamma}^{-1}}{\partial t};$$

(8) $$\left(1+\tau_1\frac{\partial}{\partial t}\right)\left(1+\tau_2\frac{\partial}{\partial t}\right)\boldsymbol{\Pi}$$
$$= -(a_1\tau_1^2 + a_2\tau_2^2)\frac{\partial\boldsymbol{\gamma}^{-1}}{\partial t} - (a_1\tau_1 + a_2\tau_2)\tau_1\tau_2\frac{\partial^2\boldsymbol{\gamma}^{-1}}{\partial t^2}.$$

For (7) and (8), we have taken $\mu(s) = a_1 e^{-s/\tau_1} + a_2 e^{-s/\tau_2}$, where a_i and τ_i are constants; for (4)–(6), $\mu(s) = a_1 e^{-s/\tau_1}$.

Equation (2) is obtained from (1) on integration by parts; the function $G(t)$ is called the *relaxation modulus*. Equation (3) is obtained from (2) on using the definition $\boldsymbol{\Pi}^0 = \boldsymbol{\gamma}\cdot\boldsymbol{\Pi}\cdot\boldsymbol{\gamma}$, and the derivative of $\boldsymbol{\gamma}'\cdot\boldsymbol{\gamma}'^{-1} = \boldsymbol{\delta}$. Equation (4) is obtained from (1) on using the exponential form for $\mu(s)$ and differentiating under the integral sign. Equation (5) is obtained from (4) on using the definition of $\boldsymbol{\Pi}^0$. Equation (6) is obtained from (2) as (4) is obtained from (1). Equations (7) and (8) are obtained from (1) and (2), respectively, in a similar way on using a second differentiation under the integral sign.

6.7 Memory-integral expansions

Constitutive equations expressed as a series in which the nth term involves an n-fold integration over the flow history have been considered by Green and Rivlin (1957, 1960), Green *et al.* (1959), Coleman and Noll (1961), Chacon and Rivlin (1964), Pipkin (1964a, 1966), and Walters (1970a). We shall here outline the main results in a form based mainly on Walters' treatment, but expressed in terms of body tensors. The outline will be heuristic, rather than rigorous, to the extent that we shall not attempt to offer necessary and sufficient conditions either for convergence of the series or for a cutoff series to represent a valid approximation to a given constitutive equation. We shall also confine our attention to incompressible materials whose constitutive equation can be expressed in the form

(1) $\boldsymbol{\Sigma} = \mathscr{F}\{\boldsymbol{\lambda}'\} \qquad [\boldsymbol{\Sigma} \equiv \boldsymbol{\Pi}\cdot\boldsymbol{\gamma};\ \ \boldsymbol{\lambda}' + \boldsymbol{\delta} = \{\text{(a)}\ \ \boldsymbol{\gamma}^{-1}\cdot\boldsymbol{\gamma}' \ \ \text{or} \ \ \text{(b)}\ \ \boldsymbol{\gamma}'^{-1}\cdot\boldsymbol{\gamma}\}],$

where $\mathscr{F}$ is a one-particle, t-isotropic function of the history $\{\boldsymbol{\lambda}'\}$, i.e.,

(2) $$\boldsymbol{\Omega}\cdot\mathscr{F}\{\boldsymbol{\lambda}'\}\cdot\boldsymbol{\Omega}^{-1} = \mathscr{F}\{\boldsymbol{\Omega}\cdot\boldsymbol{\lambda}'\cdot\boldsymbol{\Omega}^{-1}\}$$

for all constant $\boldsymbol{\Omega}$ such that $\tilde{\boldsymbol{\Omega}}\cdot\boldsymbol{\gamma}\cdot\boldsymbol{\Omega} = \boldsymbol{\gamma}$. The tensors $\boldsymbol{\Sigma}$, $\boldsymbol{\lambda}'$, and $\boldsymbol{\Omega}$ are all right covariant, mixed; $\boldsymbol{\Sigma}$ and $\boldsymbol{\gamma}$ as usual denote values at time t. The range of t' is $(-\infty, t)$.

Guided by the expansion 1.4(4) for a functional, it is reasonable to suppose that there is a wide class of constitutive equations (1) for which $\mathscr{F}$ can be expressed in the form of a series

(3) $$\mathscr{F}\{\boldsymbol{\lambda}'\} - \mathscr{F}\{\mathbf{0}\} = \int_{-\infty}^{t} \boldsymbol{\kappa}_4(t-t') : \boldsymbol{\lambda}'\, dt' + \int_{-\infty}^{t}\int_{-\infty}^{t} \boldsymbol{\lambda}'' : \boldsymbol{\kappa}_6(t-t', t-t'') : \boldsymbol{\lambda}'\, dt'\, dt'' + \cdots,$$

where $\boldsymbol{\kappa}_4$, $\boldsymbol{\kappa}_6$, ... denote tensors of rank 4, 6, ..., which are independent of $\boldsymbol{\lambda}'$. The series (3) represents an expansion about the "rest history" $\boldsymbol{\lambda}' = \mathbf{0}$, and is formally similar to a Maclaurin expansion in the sense in which 1.4(4) is similar to a Taylor expansion. Considerable simplification can be made when we use the isotropy of $\mathscr{F}$, the fact that terms in $\boldsymbol{\delta}$ can be absorbed into the extra stress tensor $\boldsymbol{\Sigma}$, and the fact that terms in $\operatorname{Tr}\boldsymbol{\lambda}'$ are of second order of smallness when $\boldsymbol{\lambda}'$ is of first order [cf. 5.7(45)].

Since $\mathscr{F}$ is t isotropic, it follows from (2) that $\mathscr{F}\{\mathbf{0}\}$ is a $\boldsymbol{\gamma}$-orthogonal mixed tensor, and is therefore, from 5.7(47), a scalar multiple of $\boldsymbol{\delta}$, which can be absorbed into $\boldsymbol{\Sigma}$ without loss of generality, since

(4) $$\boldsymbol{\Sigma} = \boldsymbol{\pi}\cdot\boldsymbol{\gamma} + p\boldsymbol{\delta} \qquad (p \text{ an arbitrary scalar}).$$

Since (3) is an identity in $\boldsymbol{\lambda}'$, valid for all histories $\{\boldsymbol{\lambda}'\}$ in some suitably chosen set, it is reasonable to assume that each term on the right is individually t isotropic. The single-integral term can then easily be dealt with by considering the particular history for which $\boldsymbol{\lambda}' = \boldsymbol{\lambda}_1\, \delta(t - t' - \tau)$, where $\boldsymbol{\lambda}_1$ is a constant tensor and τ a constant scalar; the integral is then equal to $\boldsymbol{\kappa}_4(\tau)\!:\!\boldsymbol{\lambda}_1$, which, as a $\boldsymbol{\gamma}$-isotropic, linear function of $\boldsymbol{\lambda}_1$, must be expressible in the form $a(\text{Tr}\, \boldsymbol{\lambda}_1)\boldsymbol{\delta} + m_1(\tau)\boldsymbol{\lambda}_1$, by 5.7(33). Absorbing the term in $\boldsymbol{\delta}$ into $\boldsymbol{\Sigma}$, it follows that, without loss of generality, we may write $\boldsymbol{\kappa}_4(\tau)\!:\!\boldsymbol{\lambda}_1 = m_1(\tau)\boldsymbol{\lambda}_1$, for some scalar function m_1, and hence we have

$$\text{(5)} \qquad \boldsymbol{\Sigma} \equiv \boldsymbol{\pi} \cdot \boldsymbol{\gamma} - p\boldsymbol{\delta} = \int_{-\infty}^{t} m_1(t - t')\boldsymbol{\lambda}'\, dt' + O(\boldsymbol{\lambda}'\!:\!\boldsymbol{\lambda}'),$$

for the first approximation to the constitutive equation.

Similarly, the double-integral term can be simplified by using the history $\boldsymbol{\lambda}' = \boldsymbol{\lambda}_1\, \delta(t - t' - \tau_1) + \boldsymbol{\lambda}_2\, \delta(t - t' - \tau_2)$ and making use of 5.7(43); the result is

$$\text{(6)} \quad \boldsymbol{\Sigma} = \int_{-\infty}^{t} m_1(t - t')\boldsymbol{\lambda}'\, dt' + \int_{-\infty}^{t} \int_{-\infty}^{t} m_2(t - t', t - t'')\boldsymbol{\lambda}' \cdot \boldsymbol{\lambda}''\, dt'\, dt'' + \cdots,$$

where $m_2(r, s) = m_2(s, r)$ is a scalar function, symmetric in r and s.

If one takes $\boldsymbol{\lambda}' = \boldsymbol{\gamma}^{-1} \cdot \boldsymbol{\gamma}' - \boldsymbol{\delta}$ and transfers (3) to space at time t, one obtains Eq. (6.4) [with the tr $J(s_1)$ term omitted as being of higher order than those retained] of Coleman and Noll (1961), or Eq. (11) of Walters (1970a). If one takes $\boldsymbol{\lambda}' = \boldsymbol{\gamma}'^{-1} \cdot \boldsymbol{\gamma} - \boldsymbol{\delta}$, the first term in (6) is of the same form as that given by the molecular network theory and the bead/spring theory for polymer solutions, namely, Eq. 6.2(2) with the η_s term omitted. Armstrong and Bird (1973) have used (6) with the triple-integral terms added, with $\boldsymbol{\lambda}' = \boldsymbol{\gamma}^{-1} \cdot \dot{\boldsymbol{\gamma}}'$, and with transfer to space at time t, to obtain constitutive equations (up to the triple-integral terms) for dilute suspensions of rigid, rodlike particles with Brownian motion; they also offer an argument in support of their contention that the unknown kernels $m_1, m_2, \ldots$ can be determined from a consideration of *irrotational* flows (of arbitrary time dependence) alone.

From (6), one can readily obtain an alternative series when the flow history is smooth enough to justify use of the Taylor series

$$\text{(7)} \qquad \boldsymbol{\lambda}' \equiv \boldsymbol{\gamma}^{-1} \cdot \boldsymbol{\gamma}' - \boldsymbol{\delta} = (t' - t)\dot{\boldsymbol{\lambda}} + \tfrac{1}{2}(t' - t)^2\ddot{\boldsymbol{\lambda}} + \cdots,$$

where

$$\dot{\boldsymbol{\lambda}} \equiv (\partial \boldsymbol{\lambda}'/\partial t')_{t'=t} = \boldsymbol{\gamma}^{-1} \cdot \dot{\boldsymbol{\gamma}}, \qquad \ddot{\boldsymbol{\lambda}} \equiv (\partial^2 \boldsymbol{\lambda}'/\partial t'^2)_{t'=t} = \boldsymbol{\gamma}^{-1} \cdot \ddot{\boldsymbol{\gamma}}, \quad \text{etc.},$$

and $(\boldsymbol{\lambda}')_{t'=t} = \mathbf{0}$. If we suppose that the motion is slow, or "retarded," in the sense that $\dot{\lambda}$ is small, $\ddot{\lambda}$ is smaller, and so on, then we obtain the equation

$$\Sigma = a_1\dot{\lambda} + b_1\ddot{\lambda} + b_2\dot{\lambda}^2 + \cdots, \tag{8}$$

where

$$a_1 = -\int_0^\infty m_1(r)r\,dr, \tag{9}$$

$$b_1 = \tfrac{1}{2}\int_0^\infty m_1(r)r^2\,dr, \qquad b_2 = \int_0^\infty\int_0^\infty m_2(r, s)rs\,dr\,ds, \ldots.$$

6.8 Boltzmann's viscoelasticity theory: small displacements

On transferring the rubberlike-liquid equation 6.6(2) to space at time t, using 5.4(53), 5.4(55), and 5.4(60), we obtain the following equations referred to an arbitrary rectangular Cartesian coordinate system $C\colon \mathrm{Q} \to y$:

$$P^{ij}(y, t) = \int_{-\infty}^{t} G(t - t')\frac{\partial y^i}{\partial y'^r}\frac{\partial y^j}{\partial y'^s} A_{rs}^{(1)}(y', t')\,dt', \tag{1}$$

where particle P moves from y' at t' to y at t, and the strain-rate tensor components are given by $A_{rs}^{(1)}(y', t') = (\partial v'^r/\partial y'^s + \partial v'^s/\partial y'^r)$. For small values of $u^i \equiv y^i - y'^i$ (the displacement components), two approximations can be made, namely (i) $A_{rs}^{(1)}(y', t')$ can be replaced by $A_{rs}^{(1)}(y, t')$, and (ii) $\partial y^i/\partial y'^r$ can be replaced by δ_r^i. Hence

$$P^{ij}(y, t) \doteqdot \int_{-\infty}^{t} G(t - t')A_{ij}^{(1)}(y, t')\,dt' \qquad (u^i \text{ small}). \tag{2}$$

These are one form of Boltzmann's equations for an isotropic, incompressible, viscoelastic material (Boltzmann, 1874, 1876). The appropriate form for a compressible, isotropic, viscoelastic material, expressed in terms of Cartesian tensors, is as follows:

$$\mathbf{p}(\mathrm{Q}, t) \doteqdot \int_{-\infty}^{t} \{G(t - t')\mathbf{U}\cdot + \lambda(t - t')\mathbf{UU}:\}\mathbf{A}^1(\mathrm{Q}, t')\,dt' \qquad (u^i \text{ small}), \tag{3}$$

where $\lambda(t - t')$ is a second material function, associated with volume changes. Boltzmann proposed this class of equations for describing the "elastic aftereffect" (a dependence of elastic recovery on shear history) which Kohlrausch (1866) discovered in certain metals.

Lodge (1954) used body tensors to propose two possible generalizations of Boltzmann's equations which could be admissible constitutive equations without restriction on the displacement components or on the flow history; one generalization was the rubberlike liquid (an integral equation involving contravariant stress and metric body tensors); the other was an equation of similar form in which the contravariant body tensors were replaced by the corresponding covariant body tensors.

The Boltzmann equation (3) has been widely used for the analysis of small-strain data for polymeric materials (Ferry, 1970) and in various engineering calculations (Christensen, 1971).

The equation has occasionally been used [e.g., by Sips (1951) for steady Couette flow between cylinders in relative rotation] with apparently sensible results under conditions in which the restriction to small displacements is not made use of. Presumably in such circumstances the two approximations (i) and (ii) in some sense cancel one another, yielding a final equation (e.g., for viscosity, angular relative velocity, and cylinder radii) which is acceptable. But we suggest that Eq. (3) in such circumstances is physically unacceptable, because it states that the stress $\mathbf{p}(\mathrm{Q}, t)$ at place Q and time t depends on the strain-rate history of all material elements which pass through place Q during the interval $(-\infty, t)$; such a material is therefore not a simple fluid, since its constitutive function is not a *one*-particle function of strain (or strain-rate) history.

It is perhaps of interest to consider in this connection the equation

$$(4) \qquad \mathbf{p}(\mathrm{Q}, t) = \int_{-\infty}^{t} G(t - t')\mathbf{A}^1(\mathrm{Q}', t')\,dt' \qquad (\partial u^i/\partial y^i \text{ small}),$$

which is obtainable from the (admissible) rubberlike liquid equation (1) by using approximation (ii) only; the displacement gradients are taken to be small, but the displacement components are not otherwise restricted. In particular, in contrast to (2) and (3), the integrand is evaluated at Q′ instead of at Q. Equation (4) is thus a valid Cartesian tensor equation (since tensors at different places belong to the same linear space) which expresses $\mathbf{p}(\mathrm{Q}, t)$ as a *one*-particle functional of the strain-rate tensor. The equation is not, however, admissible as a constitutive equation, and therefore furnishes an illustration of the fact that when space tensor equations are used, an additional condition (over and above tensor invariance) must be imposed if inadmissible equations are to be excluded.

If one applies the inadmissible equation (4) to steady Couette flow in a narrow-gap concentric cylinder apparatus [taking the shear rate $\dot{s} \doteq r_1(\Omega_1 - \Omega_0)/(r_1 - r_0)$ to be small, and hence also the strain-rate tensor $\mathbf{A}^1$ to be small], it is a straightforward matter to show that at the wall, $r = r_1$, the viscosity $\eta = G_0\,\tau_1/(1 + 4\tau_1^{\,2}\Omega_1^{\,2})$. Thus the viscosity depends on the angular velocity Ω_1 of the wall $r = r_1$ and not on the difference of angular velocities of the two walls; this is physically unacceptable. [We have taken $G(\tau) = G_0\exp(-\tau/\tau_1)$, where G_0 and τ_1 are constants. Ω_0 denotes the angular velocity of the wall $r = r_0$.]

6.9 Classical elasticity and hydrodynamics

Let us consider an isotropic, perfectly elastic solid; the right-covariant stress tensor $\boldsymbol{\pi}\cdot\boldsymbol{\gamma}$ (in an arbitrary state t) can be expressed as an isotropic function of the right-covariant strain tensor $(\boldsymbol{\gamma}_0^{-1}-\boldsymbol{\gamma}^{-1})\cdot\boldsymbol{\gamma}$, where $\boldsymbol{\gamma}_0$ is the covariant metric tensor in the (unique) stress-free state t_0. For strains $t_0 \to t$ which are small in the sense that

(1) $$(\boldsymbol{\Delta}\cdot\boldsymbol{\gamma}\cdot\boldsymbol{\Delta}):\boldsymbol{\gamma} \ll 1, \qquad \text{where} \quad \boldsymbol{\Delta} \equiv \boldsymbol{\gamma}_0^{-1} - \boldsymbol{\gamma}^{-1},$$

the linear approximation to the stress–strain equation may, from 5.7(33), be written in the form

(2) $$\boldsymbol{\pi} = \tfrac{1}{2}\lambda(\boldsymbol{\Delta}:\boldsymbol{\gamma})\boldsymbol{\gamma}^{-1} + \mu\boldsymbol{\Delta},$$

where λ and μ are material constants (absolute scalars). On transferring these equations to space in state t, using 5.4(14) and 5.6(14), we get

(3) $$\boldsymbol{p}(\mathrm{Q}, t) = \tfrac{1}{2}\lambda(\boldsymbol{D}:\boldsymbol{g})\boldsymbol{g}^{-1}(\mathrm{Q}) + \mu\boldsymbol{D}, \qquad \text{where} \quad \boldsymbol{D} \equiv \boldsymbol{B}_t(\mathrm{Q}, t_0) - \boldsymbol{g}^{-1}(\mathrm{Q});$$

(4) $$(\boldsymbol{D}\cdot\boldsymbol{g}\cdot\boldsymbol{D}):\boldsymbol{g} \ll 1.$$

Let us now express the space strain tensor $\boldsymbol{D}$ in terms of the displacement components

(5) $$u^i = y^i - y_0^i,$$

where y^i and y_0^i denote rectangular Cartesian coordinates of the places Q and Q_0 which a typical particle P occupies in states t and t_0, respectively. We shall consider the case

(6) $$\|\partial u^i/\partial y^j\| \ll 1,$$

i.e., in which the displacement gradients are so small that their products may be neglected. This condition is more stringent than (4) [which is equivalent to (1)]: from 5.4(57), referred to $C:\mathrm{Q}\to y$, and (5), we have

(7) $$\boldsymbol{B}_t^{-1}(\mathrm{Q}, t_0)C = \left[\delta_{rc} - \frac{\partial u^r}{\partial y^c} - \frac{\partial u^c}{\partial y^r} + \frac{\partial u^k}{\partial y^r}\frac{\partial u^k}{\partial y^c}\right].$$

The "classical approximation" (6) enables one to omit the quadratic terms in (7), and then the inverse matrix is obtained, to this order of approximation, by changing the sign of the remaining (small) displacement gradient terms, with the result

(8) $$\mathbf{B}_t(\mathrm{Q}, t_0) \doteq 2\mathbf{E} + \mathbf{U},$$

where

$$\mathbf{E}C = \left[\frac{\partial u^r}{\partial y^c} + \frac{\partial u^c}{\partial y^r}\right] \qquad \text{(small displacement gradients).}$$

The Cartesian tensor $\mathbf{E}$ is called the *infinitesimal strain tensor*. It is evident that (6) implies (4), because $DC \doteq 2\mathbf{E}C$, and $\mathbf{E}$ is small when (6) is valid; the converse is not true, however, for if we consider the particular example $u^1 = 2y^1 + \varepsilon y^2$, $u^2 = \varepsilon y^1 + 2y^2$, $u^3 = 0$, we find that all D^{ij} are $O(\varepsilon)$, where ε can be arbitrarily small, while some $\partial u^i/\partial y^j$ are equal to 2. Thus (4) is not sufficient to ensure (6). Finally, from (3) and (8) referred to C, we obtain the small-displacement gradient equations

$$(9) \qquad p^{ij}(y, t) = \frac{1}{2}\lambda\left(\frac{\partial u^1}{\partial y^1} + \frac{\partial u^2}{\partial y^2} + \frac{\partial u^3}{\partial y^3}\right)\delta_{ij} + \mu\left(\frac{\partial u^i}{\partial y^j} + \frac{\partial u^j}{\partial y^i}\right),$$

which form the basis of most of the classical theory of elasticity as presented, for example, by Love (1927). Here, λ and μ are the Lamé constants, in the usual notation.

The constitutive equation for an incompressible Newtonian liquid of viscosity η can be written in the form

$$(10) \qquad \mathbf{\Pi}^0 = \eta\dot{\gamma},$$

which is evidently admissible as a constitutive equation without restriction on the flow history (provided only that the time derivative exists). Using 5.4(60), the corresponding Cartesian space tensor equation obtained by transfer at time t is

$$(11) \qquad \mathbf{P}(\mathrm{Q}, t) = \eta\mathbf{A}^1(\mathrm{Q}, t), \qquad P^{ij}(y, t) = \eta\left(\frac{\partial v^i}{\partial y^j} + \frac{\partial v^j}{\partial y^i}\right),$$

where the $v^i(y, t)$ denote the velocity components in $C : \mathrm{Q} \to y$, and the $P^{ij} = p^{ij} + p\delta_{ij}$ denote the components in C of the extra-stress tensor. The constant-volume condition 8.5(7) and the stress equations of motion take the form

$$(12) \qquad \frac{\partial v^i}{\partial y^i} = 0,$$

$$(13) \qquad -\frac{\partial p}{\partial y^i} + \frac{\partial P^{ij}}{\partial y^j} = \rho(a^i - X^i), \qquad a^i = \frac{\partial v^i}{\partial t} + v^k\frac{\partial v^i}{\partial y^k},$$

where the X^i denote the components per unit mass of the external body force.

Eliminating P^{ij} between (11) and (13) with use of (12), we obtain the Navier–Stokes equations

$$(14) \qquad -\frac{\partial p}{\partial y^i} + \eta \frac{\partial^2 v^i}{\partial y^k \, \partial y^k} = \rho\left(\frac{\partial v^i}{\partial t} + v^k \frac{\partial v^i}{\partial y^k} - X^i\right) \qquad (i = 1, 2, 3),$$

which form the basis of much of classical hydrodynamics. In contrast to the equations of classical elasticity theory, which are admissible only when the displacement gradients are small, the constitutive equations (11), and hence also the Navier–Stokes equations, are admissible without restriction on the state of flow. Equations (13) come from C(5),(6) with $h_i = 1$, $G^i_{jk} = 0$, and $x^i = y^i$.

7 REDUCED CONSTITUTIVE EQUATIONS FOR SHEAR FLOW AND SHEAR-FREE FLOW

7.1 Incompressible viscoelastic liquids in unidirectional shear flow

Let us consider a unidirectional shear flow, as defined in 3.2(2) and 3.2(19). We do not restrict the time dependence of the shear rate $\dot{s}(\mathrm{P}, t)$; the angle of shear $\zeta(\mathrm{P}, t)$, however, is everywhere independent of t, because the flow is unidirectional. At each particle P, we introduce shear flow bases $\boldsymbol{\alpha}_i(\mathrm{P}, t)$ and $\boldsymbol{\alpha}^i(\mathrm{P}, t)$, as defined in 3.3(1)–3.3(5) and illustrated in Fig. 3.3(6). Their essential features are as follows:

(1) $\boldsymbol{\alpha}^2$ is normal to a shearing surface; $\boldsymbol{\alpha}_1$ is tangential to a line of shear;

$$\boldsymbol{\gamma}'^{-1} - \boldsymbol{\gamma}^{-1} = -s(\boldsymbol{\alpha}_2 \boldsymbol{\alpha}_1 + \boldsymbol{\alpha}_1 \boldsymbol{\alpha}_2) + s^2 \boldsymbol{\alpha}_2 \boldsymbol{\alpha}_2, \qquad s = \int_t^{t'} \dot{s}(\mathrm{P}, t'')\, dt''.$$

Each set is orthonormal at time t, and the sets are reciprocal to one another. We have used 3.3(17) and 3.3(21) (with $s \equiv s_1$, $s_3 = 0$).

The outer products $\boldsymbol{\alpha}_i \boldsymbol{\alpha}_j$ form a basis for contravariant, second-rank body tensors at P, and so we may write [as in 2.7(17)]

(2) $$\boldsymbol{\pi}(\mathrm{P}, t) = \widehat{p^{ij}} \boldsymbol{\alpha}_i \boldsymbol{\alpha}_j = \widehat{\pi^{ij}} \boldsymbol{\alpha}_i \boldsymbol{\alpha}_j,$$

where $\widehat{p^{ij}}$ are absolute scalars, called *physical components of stress* referred to the basis $\boldsymbol{\alpha}_i$. (This agrees with a more familiar definition according to which physical components are equal to tensor components referred to a coordinate system which is locally rectangular Cartesian: We can always choose such a coordinate system which, at time t, has $\boldsymbol{\alpha}_i$ tangential to its coordinate curves at P.)

It is convenient to define absolute scalars σ, N_1, and N_2 as follows:

(3) $\sigma = \widehat{p^{12}} = \boldsymbol{\alpha}^1 \boldsymbol{\alpha}^2 : \boldsymbol{\pi}$ (shear stress),

$N_1 = \widehat{p^{11}} - \widehat{p^{22}} = (\boldsymbol{\alpha}^1 \boldsymbol{\alpha}^1 - \boldsymbol{\alpha}^2 \boldsymbol{\alpha}^2) : \boldsymbol{\pi}$ (primary normal stress difference),

$N_2 = \widehat{p^{22}} - \widehat{p^{33}} = (\boldsymbol{\alpha}^2 \boldsymbol{\alpha}^2 - \boldsymbol{\alpha}^3 \boldsymbol{\alpha}^3) : \boldsymbol{\pi}$ (secondary normal stress difference).

An argument to be given shows that for a wide class of incompressible materials undergoing an arbitrary unidirectional shear flow throughout the time interval $(-\infty, t)$, the state of stress at time t necessarily has certain simplifying features which make σ, N_1, and N_2 the only stress quantities of rheological significance. Measurements of these three quantities provide fundamental information which can be used to test the applicability of proposed constitutive equations for given materials. For the important types 3.3(33) and 3.3(34) of unidirectional shear flow, the terms viscosity, dynamic viscosity, etc., can be defined as follows.

(4) **Definitions for undirectional shear flow**

(i) *Steady flow*: When σ and $\dot{s}$ are independent of t, $\sigma/\dot{s} = \eta$, the *viscosity*.
(ii) *Oscillatory flow*: If $\dot{s} = \alpha\omega \cos \omega t$ and $\sigma = \alpha\omega(\eta' \cos \omega t + \eta'' \sin \omega t) + O(\alpha^2)$, then η' and $G' = \omega\eta''$ are called the *dynamic viscosity* and *dynamic rigidity*, respectively.

The following additional terms are also used:

complex viscosity:	$\eta^* = \eta' - i\eta''$,
complex modulus:	$G^* = G' + iG''$,
complex compliance:	$J^* = G^{*-1} = J' + iJ''$,
loss modulus:	$G'' = \omega\eta'$,
loss angle:	$\delta = \tan^{-1}(G''/G')$.

When a polymeric liquid previously at rest is made to undergo shear flow, a finite time is usually required before the shear stress attains a constant value (or a steady oscillatory form) and the preceding definitions are applicable. In oscillatory shear, the amplitude α is usually kept small when the dynamic viscosity and modulus are measured (Ferry, 1970); a few measurements with larger amplitudes have been made (Shen, 1971), and the conditions under which nonlinear effects first appear may give useful information about possible forms for the appropriate constitutive equations (Astarita, 1972).

(5) **Theorem** If an incompressible material having a constitutive equation of the form $\mathbf{\Pi} = \mathscr{F}\{\boldsymbol{\gamma}'^{-1}\}$, where $\mathscr{F}$ is a t-isotropic, one-particle functional, undergoes unidirectional shear flow throughout the interval $-\infty \leq t' \leq t$, then $\widehat{p^{13}} = \widehat{p^{23}} = 0$ and σ, N_1, and N_2 are functionals of the shear rate $\dot{s}$.

Proof From 3.3(17), 3.3(21) (with $s = s_1$, $s_3 = 0$), (2), and (5), we have, at time t, the equations

(6)

$$\widehat{p^{ij}}\boldsymbol{\alpha}_i\boldsymbol{\alpha}_j + p\boldsymbol{\gamma}^{-1} = \mathscr{F}\{\boldsymbol{\gamma}^{-1} - s(\boldsymbol{\alpha}_2\boldsymbol{\alpha}_1 + \boldsymbol{\alpha}_1\boldsymbol{\alpha}_2) + s^2\boldsymbol{\alpha}_1\boldsymbol{\alpha}_1\}, \qquad s = \int_t^{t'} \dot{s}(\mathrm{P}, t'')\, dt'',$$

which are valid for any unidirectional shear flow history and for any shear flow basis $\boldsymbol{\alpha}_i = \boldsymbol{\alpha}_i(\mathrm{P}, t)$ chosen in accordance with (1). If $\boldsymbol{\alpha}_i$ is one such shear flow basis, then $(-\boldsymbol{\alpha}_1, -\boldsymbol{\alpha}_2, \boldsymbol{\alpha}_3)$ is another. Moreover, the substitution

(7) $$(\boldsymbol{\alpha}_1, \boldsymbol{\alpha}_2, \boldsymbol{\alpha}_3) \to (-\boldsymbol{\alpha}_1, -\boldsymbol{\alpha}_2, \boldsymbol{\alpha}_3)$$

leaves $\boldsymbol{\gamma}^{-1}$ and $\boldsymbol{\gamma}'^{-1}$ unchanged, and therefore, from (6), leaves $\widehat{p^{ij}}\boldsymbol{\alpha}_i\boldsymbol{\alpha}_j + p\boldsymbol{\gamma}^{-1}$ unchanged. But from (6) we see that $\widehat{p^{ij}} + p\delta^{ij} = \boldsymbol{\alpha}^i \cdot \mathscr{F}\{\boldsymbol{\gamma}'^{-1}\} \cdot \boldsymbol{\alpha}^j$ are absolute scalar functions of (at most) $\boldsymbol{\alpha}_i$ and $\boldsymbol{\gamma}$, and of no other vectors or tensors (except $\boldsymbol{\delta}$). The $\boldsymbol{\alpha}_i$ are $\boldsymbol{\gamma}$ orthonormal. Since $\mathscr{F}$ is given to be t isotropic, it is easy to verify that these scalar functions are also $\boldsymbol{\gamma}$ isotropic. The conditions of Theorem 5.7(28) thus apply, and therefore the $\widehat{p^{ij}} + p\delta^{ij}$ are absolute scalar functionals of the shear rate $\dot{s}$ alone; in particular, they are invariant under the substitution (7). It follows that $\widehat{p^{13}} = \widehat{p^{23}} = 0$; hence $\widehat{p^{31}} = \widehat{p^{32}} = 0$, because $p^{ij} = p^{ji}$. It also follows that σ, N_1, and N_2 are absolute scalar functionals of $\dot{s}$ alone. Q.E.D.

Similarly, by using the substitution

(8) $$(\dot{s}, \boldsymbol{\alpha}_1, \boldsymbol{\alpha}_2, \boldsymbol{\alpha}_3) \to (-\dot{s}, -\boldsymbol{\alpha}_1, \boldsymbol{\alpha}_2, \boldsymbol{\alpha}_3),$$

it is easy to show that

(9) $$\sigma(\mathrm{P}, \dot{s}') = -\sigma(\mathrm{P}, -\dot{s}'), \qquad N_k(\mathrm{P}, \dot{s}') = N_k(\mathrm{P}, -\dot{s}') \qquad (k = 1, 2).$$

For steady shear flow, it follows that

$$\eta(\dot{s}) = \eta(-\dot{s}). \tag{10}$$

It is clear from the proof of Theorem (5) that its validity depends on a certain symmetry property of unidirectional shear flow [expressed by the invariance of γ'^{-1} under (7)] and on the isotropy of the material [which leads to the conclusion that the $\widehat{p^{ij}} + p\delta^{ij}$ are invariant under (7)]. If the material functional $\mathscr{F}$ included anisotropic physical constant vectors or tensors, these could be combined with $\boldsymbol{\alpha}_i$ to form scalars whose value would change under (7) and could change the value of $\widehat{p^{ij}}$. A similar statement is true if the functional $\mathscr{F}$ is not a one-particle functional: Covariant derivatives of γ' (formed, say, with the metric tensor γ or γ'') could introduce vectors other than $\boldsymbol{\alpha}_i$, and these could combine with $\boldsymbol{\alpha}_i$ to form scalars not invariant under (7). Weissenberg (1947) [see also Russell (1946) and Roberts (1952)] was the first to recognize that for a wide class of viscoelastic materials, the symmetry of unidirectional shear flow must lead to the simplification in the state of stress expressed by Theorem (5). An elementary proof for the case of rectilinear shear flow has been given by Lodge (1964, p. 62). The preceding proof is a body-tensor version of a method of proof due to Pipkin and Owen (1967, §3). A proof for time-dependent unidirectional shear flow (unidirectional shear flow ≡ "viscometric" flow) has been given by Coleman (1968).

Under the conditions of Theorem (5), the constitutive equation can be written in a simple, closed form in terms of time derivatives of the metric tensor; the coefficients involve the functionals σ, N_1, and N_2:

(11) **Corollary to Theorem (5)** If $\dot{s} \neq 0$ at time t, then

$$\boldsymbol{\Pi}(\mathrm{P}, t) = -\mathscr{k}_0 \frac{\partial \boldsymbol{\gamma}^{-1}}{\partial t} + \mathscr{k}_1 \frac{\partial^2 \boldsymbol{\gamma}^{-1}}{\partial t^2} + \mathscr{k}_2 \frac{\partial \boldsymbol{\gamma}^{-1}}{\partial t} \cdot \boldsymbol{\gamma} \cdot \frac{\partial \boldsymbol{\gamma}^{-1}}{\partial t},$$

where

$$\mathscr{k}_0 = \sigma\left(\mathrm{P}, \mathop{\dot{s}'}_{-\infty}^{t}\right)/\dot{s} + \ddot{s} N_1\left(\mathrm{P}, \mathop{\dot{s}'}_{-\infty}^{t}\right)/2\dot{s}^3; \tag{12}$$

$$\mathscr{k}_m = m N_m\left(\mathrm{P}, \mathop{\dot{s}'}_{-\infty}^{t}\right)/2\dot{s}^2 \qquad (m = 1, 2).$$

For *stress relaxation* after cessation at time t_1 of unidirectional shear flow, a similar equation gives $\boldsymbol{\Pi}(\mathrm{P}, t)$ for $t > t_1$ provided only that $\dot{s}$ and $\partial^m \boldsymbol{\gamma}^{-1}/\partial t^m$ are evaluated at time t_1 (the latest time at which $\dot{s}$ is not zero).

Proof From (2), (3), and 3.3(19), it follows that

$$\pi - \widehat{p^{22}\gamma^{-1}} = \sigma(\boldsymbol{\alpha}_1\boldsymbol{\alpha}_2 + \boldsymbol{\alpha}_2\boldsymbol{\alpha}_1) + N_1\boldsymbol{\alpha}_1\boldsymbol{\alpha}_1 - N_2\boldsymbol{\alpha}_3\boldsymbol{\alpha}_3. \tag{13}$$

From 3.3(28) (with $\dot{s}_3 = 0$, $\dot{s}_1 = \dot{s}$) and 3.3(32), we can express $\boldsymbol{\alpha}_1\boldsymbol{\alpha}_2 + \boldsymbol{\alpha}_2\boldsymbol{\alpha}_1$, etc., in terms of $\dot{s}$ and $\partial^m\gamma^{-1}/\partial t^m$. On writing p for $-(\widehat{p^{22}} + N_2)$, one obtains from (13) the result stated in the corollary for the case in which $\dot{s} \neq 0$ at time t. This condition is used in dividing by $\dot{s}$. For the case of stress relaxation following cessation at time t_1 of unidirectional shear flow, we can apply the result (13) at time $t > t_1$, because the flow history is a unidirectional shear flow up to time t (although there is no flow from t_1 to t). Because $\dot{s}$ is zero from t_1 to t, it follows from 3.3(14) (with $\dot{s}_1 = \dot{s}$ and $\dot{s}_3 = 0$) that $\boldsymbol{\alpha}_i(t) = \boldsymbol{\alpha}_i(t_1)$. But $\boldsymbol{\alpha}_i(t_1)$ can be expressed in terms of $\dot{s}$ and $\partial^m\gamma^{-1}/\partial t^m$ at $t = t_1$ as before, since $\dot{s} \neq 0$ at t_1. The upper "limit" in the functionals σ, N_1, and N_2, however, is t. The stated result then follows. This completes the proof of Corollary (11).

(14) **Problem** Show that the following is an equivalent form of (11):

$$\boldsymbol{\Pi}^0 = \left(\frac{\sigma}{\dot{s}} + \frac{N_1\ddot{s}}{2\dot{s}^3}\right)\dot{\gamma} - \frac{N_1}{2\dot{s}^2}\ddot{\gamma} + \frac{N_1 + N_2}{\dot{s}^2}\dot{\gamma}\cdot\gamma^{-1}\cdot\dot{\gamma}.$$

(15) **Problem** For the incompressible Newtonian liquid defined by the equation

$$\boldsymbol{\Pi} = -\eta_0\,\partial\gamma^{-1}/\partial t,$$

prove that in unidirectional shear flow, $\sigma = \eta_0\,\dot{s}$ and that $N_1 = N_2 = 0$.

(16) **Problem** For the rubberlike liquid 6.2(2) with $\eta_s = 0$, show that in an arbitrary unidirectional shear flow, $N_2 = 0$, and

$$\begin{pmatrix}\sigma\\ N_1\end{pmatrix} = \int_{-\infty}^{t} \mu(t-t')\begin{pmatrix}s\\ s^2\end{pmatrix} dt', \qquad \text{where} \quad s = \int_{t}^{t'} \dot{s}(\mathrm{P}, t'')\,dt''.$$

For the particular case in which the unidirectional shear flow is steady (i.e., when $\dot{s}$ is independent of time), the reduced constitutive equation (15), on transfer to space at time t, gives an equation published by Ericksen (1960). For time-dependent shear rate $\dot{s}$, Truesdell and Noll [1965, Eq. (106.6) and (106.23)] have given reduced constitutive equations involving a Cartesian tensor (**N**, in their notation), related to the displacement gradient, in place of the Cartesian second-strain-rate tensor $\mathbf{A}^2$, which would be obtained from the term $\ddot{\gamma}$ in our Eq. (14). Equation (11), with constant coefficients, describes a "second-order fluid."

For future convenience, we note the following equation, valid for an arbitrary unidirectional shear flow:

$$\pi = \widehat{\pi^{ii}}\boldsymbol{\alpha}_i\boldsymbol{\alpha}_i + \sigma(\boldsymbol{\alpha}_1\boldsymbol{\alpha}_2 + \boldsymbol{\alpha}_2\boldsymbol{\alpha}_1). \tag{17}$$

The equation

$$\widehat{\pi^{ij}} = \widehat{p^{ij}} \qquad (B \overset{t}{\equiv} S) \tag{18}$$

holds whenever B and S are congruent. We also have the following alternative forms of (17):

$$\pi - \sigma(\boldsymbol{\alpha}_1\boldsymbol{\alpha}_2 + \boldsymbol{\alpha}_2\boldsymbol{\alpha}_1) = \widehat{\pi^{11}}\boldsymbol{\alpha}_i\boldsymbol{\alpha}_i - N_1\boldsymbol{\alpha}_2\boldsymbol{\alpha}_2 - (N_1 + N_2)\boldsymbol{\alpha}_3\boldsymbol{\alpha}_3 \tag{19}$$

$$= \widehat{\pi^{22}}\boldsymbol{\alpha}_i\boldsymbol{\alpha}_i + N_1\boldsymbol{\alpha}_1\boldsymbol{\alpha}_1 - N_2\boldsymbol{\alpha}_3\boldsymbol{\alpha}_3 \tag{20}$$

$$= \widehat{\pi^{33}}\boldsymbol{\alpha}_i\boldsymbol{\alpha}_i + (N_1 + N_2)\boldsymbol{\alpha}_1\boldsymbol{\alpha}_1 + N_2\boldsymbol{\alpha}_2\boldsymbol{\alpha}_2 . \tag{21}$$

We have shown that 3.3(20), with $s = s_1$ and $s_3 = 0$, characterizes unidirectional shear flows. On transferring this equation to space at time t, we obtain the equation

$$\mathbf{C}_t(Q, t') - \mathbf{U} = s(\mathbf{a}^1\mathbf{a}^2 + \mathbf{a}^2\mathbf{a}^1) + s^2\mathbf{a}^2\mathbf{a}^2, \tag{22}$$

where

$$\mathbf{a}^i = \mathbb{T}(t)\boldsymbol{\alpha}^i, \tag{23}$$

and we have used Cartesian vectors and tensors. Since the $\boldsymbol{\alpha}^i$ are t orthonormal, so also are the $\mathbf{a}^i$. If the flow is steady, then $s = (t' - t)\dot{s}_0$, say, and (22) is equivalent to Eq. (2.6) of Yin and Pipkin (1970), who have shown that it characterizes the class of flows called viscometric [e.g., by Coleman *et al.* (1966, p. 29)]. It follows that viscometric flow is synonymous with steady, unidirectional shear flow. Truesdell and Noll (1965, p. 433) use the term viscometric even when the flow is not steady; in this usage, the term *viscometric* flow is synonomous with *unidirectional shear* flow. Caswell (1967) has given a general treatment of shear flow near a rigid surface.

7.2 Oscillatory and steady shear flow: low-frequency relations

For steady shear flow (shear rate $\dot{s}_0$) and small-amplitude oscillatory shear (angular frequency ω), taken either separately [as in 3.3(33), 3.3(34), and 7.1(4)] or in parallel or orthogonal superposition [as in 3.3(35) and 3.3(36)], the following limiting relations between material properties can be derived theoretically, making only formal assumptions about the constitutive equation, and tested experimentally:

$$\left.\begin{array}{ll}(1) & \lim_{\omega\to 0}\eta'(\omega) = \lim_{\dot{s}_0\to 0}\eta(\dot{s}_0)\\ (2) & \lim_{\omega\to 0}\omega^{-2}G'(\omega) = \lim_{\dot{s}_0\to 0}\tfrac{1}{2}\dot{s}_0^{-2}N_1(\dot{s}_0)\end{array}\right\}\quad\begin{pmatrix}\text{separate oscillatory and}\\ \text{steady shear flows}\end{pmatrix};$$

$$\left.\begin{array}{ll}(3) & \lim_{\omega\to 0}\eta'_{\|}(\dot{s}_0, \omega) = \dfrac{d}{d\dot{s}_0}[\dot{s}_0\,\eta(\dot{s}_0)]\\ (4) & \lim_{\omega\to 0}\theta'_{i,\|}(\dot{s}_0, \omega) = \dfrac{d}{d\dot{s}_0}N_i(\dot{s}_0)\quad (i = 1, 2)\end{array}\right\}\quad\text{(parallel superposition).}$$

$$(5)\quad \lim_{\omega\to 0}\eta'_{\perp}(\dot{s}_0, \omega) = \eta(\dot{s}_0)\qquad\text{(orthogonal superposition).}$$

The symbols $\eta'_{\|}$, $\theta'_{i,\|}$, and $\eta'_{\perp}$, to be defined, refer to material functions associated with small-amplitude oscillatory shear superposed on steady shear flow. In all cases considered here, *the family of shearing surfaces is the same for both steady and oscillatory contributions to the shear flow*. In three cases (steady, oscillatory, and parallel superposition), the shear flow is unidirectional and so the reduced constitutive equation 7.1(11) and other results of Section 7.1 are applicable; we shall show that from these results one can prove (1), (3), and (4) very simply. In the case of orthogonal superposition, on the other hand, the (resultant) shear flow is not unidirectional, and a different method will be used to prove (5).

Derivations of these relations, valid for general incompressible viscoelastic liquids, have been given by Coleman and Markovitz (1964) [for (1) and (2)], Pipkin (1968) [for (3) and (5)], and Bernstein and Fosdick (1970) [for (4)]. Booij (1970) has reviewed the subject and reported results of measurements taken with various viscoelastic materials. Parallel superposition measurements have been reported by Osaki *et al.* (1963), Booij (1966), Walters and Jones (1968), and Kataoka and Ueda (1969). Orthogonal superposition measurements have been reported by Simmons (1966, 1968) and Tanner and Simmons (1967). When special forms of constitutive equations are used, further predictions can be made concerning the behavior at higher frequencies; Booij (1970) has shown that the comparison of predictions with parallel superposition data is a particularly valuable method for assessing the merits of various constitutive hypotheses. Weissenberg (1935) was the first to discuss (under the heading, "anisotropy of virtual work") the importance of small strains superposed on steady flows of viscoelastic materials.

Since the limiting relations (1)–(5) should apply to a wide range of viscoelastic materials, we shall give a derivation of these relations here. This will also furnish a useful illustration of the present formalism. With the constitutive assumptions of 7.1(5), the relations (1)–(5) can be derived from the assumption that $\mathscr{F}\{\gamma'^{-1}\}$ is Fréchet differentiable at the rest history [in the case of (1) and (2)] and at the steady shear flow history [in the case of (3)–(5)]. *Each relation is then a consequence of the equality of (Gateaux) derivatives taken along the appropriate two directions in shear-rate function space*, one derivative being taken along a steady shear flow direction and the other along an oscillatory shear direction (Section 1.4).

Proof of (3) In the case of *parallel superposition*, the flow is a unidirectional shear flow with shear rate $\dot{s}' = \dot{s}_0 + \alpha\omega \cos \omega t'$, and hence, by 7.1(5), the shear stress σ at time t can be expressed as a functional $\not{f}$ of $\dot{s}'$ alone, which, by 1.4(2), can be expanded, for small oscillatory amplitude α, in the form

$$(6) \quad \sigma = \not{f}\{\dot{s}_0 + \alpha\omega \cos \omega t'\} = \not{f}(\dot{s}_0) + \int_{-\infty}^{t} k(t - t', \dot{s}_0)\alpha\omega \cos \omega t' \, dt' + O(\alpha^2),$$

where $k(t - t', \dot{s}_0)$, the Fréchet derivative, is independent of the incremental shear rate history $\alpha\omega \cos \omega t'$ [$\equiv h(t')$, in the notation of 1.4(2)]. The leading term $\not{f}(\dot{s}_0)$ is a *function* of $\dot{s}_0$, since $\dot{s}_0$ is independent of time throughout the interval $(-\infty, t)$, and is in fact equal to $\dot{s}_0 \eta(\dot{s}_0)$, according to the definition 7.1(4) of viscosity η. Equation (6) can be rewritten in the form

$$(7) \qquad \sigma = \dot{s}_0 \eta(\dot{s}_0) + \alpha\omega[\eta'_{\parallel}(\dot{s}_0, \omega) \cos \omega t + \eta''_{\parallel}(\dot{s}_0, \omega) \sin \omega t] + O(\alpha^2),$$

where

$$(8) \qquad \eta'_{\parallel} = \int_0^{\infty} k(\tau, \dot{s}_0) \cos \omega\tau \, d\tau, \qquad \eta''_{\parallel} = \int_0^{\infty} k(\tau, \dot{s}_0) \sin \omega\tau \, d\tau.$$

Equation (7) can be taken as an operational definition of $\eta'_{\parallel}$ and $\eta''_{\parallel}$, since the shear stress σ can be measured as a function of t and analyzed so as to separate the fundamentals in $\cos \omega t$ and $\sin \omega t$; a change of oscillatory amplitude α can be used to test whether the terms $O(\alpha^2)$ (which will, in any case, all involve higher-frequency components) are negligible. To prove (3), we consider the case $\dot{s}' = \dot{s}_0 + \dot{s}_1$, where $\dot{s}_1$ is small and independent of time; in place of (6), we have

$$(9) \quad \sigma = (\dot{s}_0 + \dot{s}_1)\eta(\dot{s}_0 + \dot{s}_1) = \dot{s}_0 \eta(\dot{s}_0) + \dot{s}_1 \int_{-\infty}^{t} k(t - t', \dot{s}_0) \, dt' + O(\dot{s}_1^2).$$

The first equality follows because the combined flow is steady. Taking the limit $\dot{s}_1 \to 0$ in (9), we obtain the equation

$$(10) \qquad \frac{d}{d\dot{s}_0} [\dot{s}_0 \eta(\dot{s}_0)] = \int_0^{\infty} k(\tau, \dot{s}_0) \, d\tau.$$

Taking the limit $\omega \to 0$ in (8), we obtain from (10) the required result (3).

Proof of (1) If one could justify changing the order of taking limits, (1) could be obtained at once from (3) by letting $\dot{s}_0 \to 0$. Alternatively, one can prove (1) directly by expanding $\not{f}\{\dot{s}'\}$ about the rest history $\dot{s}' = 0$; the proof follows exactly similar lines to that given for (3). For oscillatory shear of small amplitude α, we have [using 1.4(2) with $x = 0$ and $h(t') = \alpha\omega \cos \omega t'$]

(11) $$\sigma = \not{f}\{\alpha\omega \cos \omega t'\} = \not{f}(0) + \int_{-\infty}^{t} k_0(t - t')\alpha\omega \cos \omega t' \, dt' + O(\alpha^2),$$

where $\not{f}(0) = 0$ and k_0, the Fréchet derivative at the rest history $\dot{s}' = 0$ differs from k [in (6)], the Fréchet derivative at the steady shear flow history $\dot{s}' = \dot{s}_0$. Hence

(12) $$\sigma = \alpha\omega[\eta'(\omega) \cos \omega t + \eta''(\omega) \sin \omega t] + O(\alpha^2)$$

[in agreement with the definitions of η' and η'' given in 7.1(4)], where

(13) $$\eta' = \int_0^{\infty} k_0(\tau) \cos \omega\tau \, d\tau, \qquad \eta'' = \int_0^{\infty} k_0(\tau) \sin \omega\tau \, d\tau.$$

On the other hand, for steady shear flow of small shear rate $\dot{s}_0$, we have

(14) $$\sigma = \dot{s}_0 \eta(\dot{s}_0) = \not{f}\{0 + \dot{s}_0\} = \not{f}\{0\} + \dot{s}_0 \int_{-\infty}^{t} k_0(t - t') \, dt' + O(\dot{s}_0^{\,2}),$$

and hence

(15) $$\eta(\dot{s}_0) = \int_0^{\infty} k_0(\tau) \, d\tau + O(\dot{s}_0^{\,2}).$$

The required result (1) follows at once from (13) and (15).

Proof of (4) This follows exactly the same lines as the proof of (3) on replacing σ, $\eta'_{\parallel}$, nnd $\dot{s}_0 \eta(\dot{s}_0)$ by N_i, $\theta'_{i,\parallel}$, and $N_i(\dot{s}_0)$, respectively. The flow is unidirectional, and the result (4) follows from the Fréchet differentiability of the normal stress difference functionals 7.1(5) of shear rate. The equations (for parallel superposition) analogous to (7) are

(16) $$N_i = N_i(\dot{s}_0) + \alpha\omega[\theta'_{i,\parallel}(\dot{s}_0, \omega) \cos \omega t + \theta''_{i,\parallel}(\dot{s}_0, \omega) \sin \omega t] + O(\alpha^2),$$

for $i = 1, 2$, and *these serve for operational definitions of* $\theta'_{i,\parallel}$ *and* $\theta''_{i,\parallel}$, *the normal stress difference dynamic coefficients for small-amplitude oscillatory shear in parallel superposition on steady shear flow.* From 7.1(9), the $N_i(\dot{s}_0)$ are even functions of $\dot{s}_0$ and so $dN_i(\dot{s}_0)/d\dot{s}_0 \to 0$ as $\dot{s}_0 \to 0$; it follows from (4) that $\lim_{\dot{s}\to 0} \lim_{\omega \to 0} \theta'_{i,\parallel} = 0$, which is consistent with the fact that *for oscillatory shear by itself, the fundamental terms for* N_i *are of double frequency* (i.e., involve $\cos 2\omega t$ and $\sin 2\omega t$) because of 7.1(9).

If we rewrite (3) in the form

$$\eta'_{\|}(\dot{s}_0, 0) = \eta(\dot{s}_0)\left\{1 + \frac{d \log_e \eta(\dot{s}_0)}{d \log_e \dot{s}_0}\right\} \tag{17}$$

and use (1), then it follows that *when* $\eta(\dot{s}_0)$ *is a decreasing function* (as it is for most polymeric liquids), *we have*

$$\eta'_{\|}(\dot{s}_0, 0) < \eta(\dot{s}_0) < \eta(0) = \eta'(0). \tag{18}$$

Proof of (2) Unlike (1), (3), and (4), Eq. (2) relates different stress components, namely the shear stress σ (through G') and the primary normal stress difference N_1. The use of Fréchet differentiability of the individual (scalar) functionals σ and N_i is no longer sufficient for our needs, and we must have recourse to Eq. 6.7(6)—a tensor equation whose validity rests on the Fréchet differentiability about the rest history, and the isotropy, of $\mathscr{F}$. Equation (2) evidently involves small departures from the rest history.

Taking $\boldsymbol{\lambda} = (\boldsymbol{\gamma}'^{-1} - \boldsymbol{\gamma}^{-1})\cdot\boldsymbol{\gamma}$ and using 3.3(21) with $s' = \int_t^{t'} \dot{s}_1'' \, dt''$ and $\dot{s}_3 = 0$, Eq. 6.7(6), with 7.1(3), yields the results [to $O(s'^3)$]

$$\sigma = -\int_{-\infty}^{t} m_1(t - t')s' \, dt'; \tag{19}$$

$$N_1 = \int_{-\infty}^{t} m_1(t - t')s'^2 \, dt'; \qquad N_2 = \int_{-\infty}^{t} m_2(t - t', t - t'')s's'' \, dt' \, dt'', \tag{20}$$

which show that σ *and* N_1 (*but not* N_2) *are determined* (*at the lowest order of approximation*) *by a single scalar function* m_1. This result does not depend on the foregoing choice for the strain tensor $\boldsymbol{\lambda}'$, since the same form 6.7(6) is obtained whenever $\mathscr{F}$ is t isotropic; replacing the $\boldsymbol{\lambda}'$ here by any $\boldsymbol{\gamma}$-isotropic function of $\boldsymbol{\lambda}'$ (which vanishes when $\boldsymbol{\lambda}'$ vanishes) will lead also to a t-isotropic $\mathscr{F}$, and hence to an equation of the same form as 6.7(6).

For steady shear flow, we have $s' = (t' - t)\dot{s}$ ($\dot{s} = \text{const}$), and (19) and (20) give the results

$$\eta(\dot{s}) = \int_0^{\infty} m_1(u)u \, du + O(\dot{s}^2); \tag{21}$$

$$\dot{s}^{-2}N_1(\dot{s}) = \int_0^{\infty} m_1(u)u^2 \, du + O(\dot{s}^2); \tag{22}$$

$$\dot{s}^{-2}N_2(\dot{s}) = \int_0^{\infty} m_2(u, v)uv \, du \, dv + O(\dot{s}^2). \tag{23}$$

Only the second of these is required for the present purpose; the other two are included for completeness. We have used 7.1(9) and 7.1(10) to justify writing $O(\dot{s}^2)$ [instead of $O(\dot{s})$] in these three equations.

For oscillatory shear, we have $\dot{s}' = \alpha\omega \cos \omega t'$, $s' = \alpha(\sin \omega t' - \sin \omega t)$, and hence from (19) and 7.1(4), we obtain the results

$$(24)\quad \eta'(\omega) = \omega^{-1} \int_0^\infty m_1(u) \sin \omega u \, du; \qquad G'(\omega) = \int_0^\infty m_1(u)(1 - \cos \omega u) \, du.$$

On taking the limit $\dot{s} \to 0$ in (22) and the limit $\omega \to 0$ in (24), the required result (2) follows immediately. It is also easy to see from (19) and (20) that the lowest-order terms in N_1 and N_2 involve $\cos 2\omega t$ and $\sin 2\omega t$.

Proof of (5) As in 3.3(36), we consider the orthogonal superposition of two unidirectional shear flows, as follows:

(25) constant shear rate $\dot{s}_1$; shear line tangent $\boldsymbol{\alpha}_1$ (unperturbed flow); oscillatory shear $\dot{s}_3' = \alpha\omega \cos \omega t'$, shear line tangent $\boldsymbol{\alpha}_3$ (perturbation, α small).

Taking the t' derivative of 3.3(20), we obtain the result $\dot{\boldsymbol{\gamma}}' = \dot{\boldsymbol{\gamma}}_1' + \delta\dot{\boldsymbol{\gamma}}'$, where

$$(26)\qquad \dot{\boldsymbol{\gamma}}_1' = \dot{s}_1(\boldsymbol{\alpha}^1\boldsymbol{\alpha}^2 + \boldsymbol{\alpha}^2\boldsymbol{\alpha}^1) + 2\dot{s}_1^2(t' - t)\boldsymbol{\alpha}^2\boldsymbol{\alpha}^2;$$
$$\delta\dot{\boldsymbol{\gamma}}' = \dot{s}_3'(\boldsymbol{\alpha}^3\boldsymbol{\alpha}^2 + \boldsymbol{\alpha}^2\boldsymbol{\alpha}^3) + O(\dot{s}_3'^2).$$

We use δ to denote an increment generated by the perturbation $\dot{s}_3'$; it is sufficient to work to first order in $\dot{s}_3'$ only.

Since $\boldsymbol{\gamma}' = \boldsymbol{\gamma} + \int_t^{t'} \dot{\boldsymbol{\gamma}}'' \, dt''$, we can write $\boldsymbol{\Pi} = \mathscr{F}\{\boldsymbol{\gamma}'^{-1}\} = \mathscr{H}\{\boldsymbol{\gamma}, \dot{\boldsymbol{\gamma}}'\}$; using a tensor analog of the Fréchet derivative formula 1.4(2) (with x and h replaced by $\dot{\boldsymbol{\gamma}}_1'$ and $\delta\dot{\boldsymbol{\gamma}}'$, respectively), we have $\boldsymbol{\Pi} = \boldsymbol{\Pi}_1 + \delta\boldsymbol{\Pi}_1 + O(\alpha^2)$, where

$$(27)\qquad \boldsymbol{\Pi}_1 = \sigma(\dot{s}_1)(\boldsymbol{\alpha}_1\boldsymbol{\alpha}_2 + \boldsymbol{\alpha}_2\boldsymbol{\alpha}_1) + N_1(\dot{s}_1)\boldsymbol{\alpha}_1\boldsymbol{\alpha}_1 - N_2(\dot{s}_1)\boldsymbol{\alpha}_3\boldsymbol{\alpha}_3;$$

$$(28)\qquad \delta\boldsymbol{\Pi}_1 = \int_{-\infty}^t \boldsymbol{\kappa}_4(t - t', \dot{s}_1, \boldsymbol{\alpha}^i) : (\boldsymbol{\alpha}^3\boldsymbol{\alpha}^2 + \boldsymbol{\alpha}^2\boldsymbol{\alpha}^3)\dot{s}_3' \, dt'.$$

The kernel $\boldsymbol{\kappa}_4$ (representing a tensor analog of a Fréchet derivative at the unperturbed flow history) is a fourth-rank tensor which can depend on the unperturbed flow parameters but not on the perturbed flow parameters; this tensor possesses anisotropy generated by the unperturbed flow, and may be contrasted with the isotropic kernel $\boldsymbol{\kappa}_4$ in 6.7(3) associated with the derivative at the rest history. Equation (27) comes directly from 7.1(13), which applies to any unidirectional shear flow and therefore, in particular, to the present unperturbed flow. For the oscillatory shear perturbation (25), it follows from (28) that

$$(29)\qquad \boldsymbol{\alpha}^2\boldsymbol{\alpha}^3 : \delta\boldsymbol{\Pi}_1 = \alpha\omega(\eta_\perp' \cos \omega t + \eta_\perp'' \sin \omega t),$$

where

$$(30)\qquad \begin{pmatrix} \eta_\perp'(\dot{s}_1, \omega) \\ \eta_\perp''(\dot{s}_1, \omega) \end{pmatrix} = \int_0^\infty \boldsymbol{\alpha}^2\boldsymbol{\alpha}^3 : \boldsymbol{\kappa}_4(u, \dot{s}_1, \boldsymbol{\alpha}^i) : (\boldsymbol{\alpha}^3\boldsymbol{\alpha}^2 + \boldsymbol{\alpha}^2\boldsymbol{\alpha}^3) \begin{pmatrix} \cos \omega u \\ \sin \omega u \end{pmatrix} du.$$

Equation (29) can be taken as an operational definition of the dynamic coefficients $\eta'_\perp$ and $\eta''_\perp$ associated with the small-amplitude oscillatory shear whose lines of shear are orthogonal to those of the steady shear flow of shear rate $\dot{s}_1$.

To relate $\eta'_\perp(\dot{s}_1, 0)$ to the steady flow function $\eta(\dot{s}_1)$, we can (i) apply (26)–(28) with $\dot{s}'_3 = \text{const}$, and (ii) obtain $\boldsymbol{\Pi}$ (to first order in $\dot{s}_3$) for the superposed flow (25) with $\dot{s}'_3 = \text{const}$, by an alternative route, as follows: This superposed flow is unidirectional, with shear line tangent $\boldsymbol{\alpha}_4$ and normal $\boldsymbol{\alpha}_5$ [as in Fig. 3.3(6)] and shear rate $\dot{s} = (\dot{s}_1^2 + \dot{s}_3^2)^{1/2}$, and so 7.1(13) gives the result

$$\text{(31)} \qquad \boldsymbol{\Pi} = \boldsymbol{\Pi}_1 + \delta\boldsymbol{\Pi}_1 = \sigma(\dot{s})(\boldsymbol{\alpha}_4\boldsymbol{\alpha}_2 + \boldsymbol{\alpha}_2\boldsymbol{\alpha}_4) + N_1(\dot{s})\boldsymbol{\alpha}_4\boldsymbol{\alpha}_4 - N_2(\dot{s})\boldsymbol{\alpha}_5\boldsymbol{\alpha}_5,$$

where, from 3.3(13) and the fact that $\dot{s} = \dot{s}_1 + O(\dot{s}_3^2)$, it follows that

$$\text{(32)} \qquad \boldsymbol{\alpha}_4 = \boldsymbol{\alpha}_1 + \frac{\dot{s}_3}{\dot{s}_1}\boldsymbol{\alpha}_3 + O(\dot{s}_3^2); \qquad \boldsymbol{\alpha}_5 = \boldsymbol{\alpha}_3 - \frac{\dot{s}_3}{\dot{s}_1}\boldsymbol{\alpha}_1 + O(\dot{s}_3^2).$$

Since $\delta\boldsymbol{\Pi}_1$ calculated by these two methods (i) and (ii) must give the same result, we obtain the following body tensor version of some of Pipkin's consistency relations:

$$\text{(33)} \qquad \sigma(\dot{s}_1)(\boldsymbol{\alpha}_3\boldsymbol{\alpha}_2 + \boldsymbol{\alpha}_2\boldsymbol{\alpha}_3) + (N_1 + N_2)_{\dot{s}_1}(\boldsymbol{\alpha}_1\boldsymbol{\alpha}_3 + \boldsymbol{\alpha}_3\boldsymbol{\alpha}_1)$$
$$= \dot{s}_1 \int_0^\infty \boldsymbol{\kappa}_4(u, \dot{s}_1, \boldsymbol{\alpha}^i) : (\boldsymbol{\alpha}^3\boldsymbol{\alpha}^2 + \boldsymbol{\alpha}^2\boldsymbol{\alpha}^3)\, du.$$

On contracting this equation from the left twice with $\boldsymbol{\alpha}^2\boldsymbol{\alpha}^3$ and comparing the result with the limit $\omega \to 0$ of (30), the required result (5) is obtained at once.

7.3 Orthogonal rheometer: small-strain limit

Using the notation of Section 3.7 and Fig. 3.7(1), we shall show that the components F_1 and F_3 of the tangential total force exerted by an incompressible viscoelastic liquid on the lower plate (of radius R) when the angular velocity Ω is constant and the displacement angle $\tan^{-1} a$ is small are given by the equation

$$\text{(1)} \qquad F_3 + iF_1 = \pi R^2 a[G'(\Omega) - i\Omega\eta'(\Omega)] + O(a^2) \qquad (i = \sqrt{-1}),$$

where η' and G' denote the dynamic viscosity and dynamic modulus of the liquid. F_1 and F_3 have the directions of Oy^1 and Oy^3, respectively. The result depends on having the assumed state of flow considered in Section 3.7, for which, from 3.3(17), 3.3(21), and 3.7(11), we have

(2) $$\gamma'^{-1} - \gamma^{-1} = -s_1(\boldsymbol{\alpha}_2\boldsymbol{\alpha}_1 + \boldsymbol{\alpha}_1\boldsymbol{\alpha}_2) - s_3(\boldsymbol{\alpha}_2\boldsymbol{\alpha}_3 + \boldsymbol{\alpha}_3\boldsymbol{\alpha}_2) + O(a^2),$$

(3) $$s_1 + is_3 = a[\exp\{-i(\xi^3 + \Omega t')\} - \exp\{-i(\xi^3 + \Omega t)\}].$$

The ambiguity of sign of s_1 [arising from 3.3(13)(i) and 3.7(11)] has been resolved in (2) and (3) by considering the sign of the rate of change of length of a material line OP [using 3.3(26), which goes with 3.3(21) and hence with (2)], where P is a particle on a plane above Oy^1y^3 in Fig. 3.7(1) located instantaneously above the axis Oy^1 not far from the vertical through O.

The required force components at time t are given by the equation

(4) $$F_3 + iF_1 = \iint \boldsymbol{\alpha}^2 \cdot \boldsymbol{\pi} \cdot (\boldsymbol{\alpha}^1 + i\boldsymbol{\alpha}^3)e^{i\phi}\xi^1\, d\xi^1\, d\xi^3, \qquad \phi = \xi^3 + \Omega t,$$

where the integration extends over the disk $r = R$. Since a is small, the flow involves a small perturbation on the rest history, and so we can use 6.7(6); the kernel m_1 is related to η' and G' by 7.2(24) provided that we take $\boldsymbol{\lambda}' = (\gamma'^{-1} - \gamma^{-1})\cdot\gamma$ in 6.7(6). The isotropic stress term contributes nothing to the tangential components F_1 and F_3; since $\boldsymbol{\Pi} = \boldsymbol{\Sigma}\cdot\gamma^{-1}$, we obtain the equation

(5)
$$F_3 + iF_1 = -\tfrac{1}{2}R^2 \int_0^{2\pi} d\xi^3 \int_{-\infty}^0 m_1(t - t')(s_1 + is_3)\exp i(\xi^3 + \Omega t)\, dt' + O(a^2),$$

on using (2), (4), and integrating with respect to ξ^1 over the range $(0, R)$. On using (3), integrating with respect to ξ^3, making the substitution $t - t' = u$, and comparing the result with 7.2(24), it is a straightforward matter to derive the required result (1).

It is interesting to see from (1) that although the dynamic functions η' and G' are defined, in the first instance, for a unidirectional shear flow with time-dependent shear rate [as in 7.1(4)], these functions can nevertheless be determined from (steady) force measurements in the orthogonal rheometer operating with time-independent angular velocity Ω, i.e., under conditions in which the shear rate is independent of time, as shown in 3.7(12). The reason is that the flow is not "rheologically steady," because the shear lines rotate relative to the material.

Instruments of this type are available commercially (Maxwell and Chartoff, 1965; Macosko and Starita, 1971) and are used for measuring η' and G', particularly for molten polymers. The axial component F_2 of force can also be measured, but can be less universally interpreted: it is obvious from (2) that no terms of first order in a survive the process of contraction with $\boldsymbol{\alpha}^2\boldsymbol{\alpha}^2$, and so the leading terms in F_2 are $O(a^2)$ (or higher), and at least two terms in the memory integral expansion must be involved.

7.4 Shear-free flow

We consider a material having a constitutive equation of the form $\boldsymbol{\pi}^0 = \mathscr{F}(\boldsymbol{\gamma}')$, where $\mathscr{F}$ is a t isotropic, one-particle functional, subjected to an arbitrary shear-free flow throughout a time interval $(-\infty, t)$. $\boldsymbol{\pi}^0 \equiv \boldsymbol{\gamma}\cdot\boldsymbol{\pi}\cdot\boldsymbol{\gamma}$ denotes the covariant stress tensor.

The shear-free flow base vectors $\boldsymbol{\alpha}^i = \boldsymbol{\alpha}^i(\mathrm{P}, t)$ (orthonormal at t) have been expressed in terms of $\boldsymbol{\gamma}$ and $\dot{\boldsymbol{\gamma}}$ in Section 3.9, and $\boldsymbol{\gamma}'$ is expressed in terms of $\boldsymbol{\alpha}^i$ in 3.9(12). It follows that we can write

$$\boldsymbol{\pi}^0 = \widehat{p_{ij}}\,\boldsymbol{\alpha}^i\boldsymbol{\alpha}^j = \mathscr{F}(e_i^2\boldsymbol{\alpha}^i\boldsymbol{\alpha}^i), \tag{1}$$

where $\widehat{p_{ij}}$ (the "physical components" of stress referred to the orthonormal basis $\boldsymbol{\alpha}^i$) are absolute scalar functions of $\boldsymbol{\alpha}^i$, $\boldsymbol{\gamma}$, and $\boldsymbol{\delta}$, and functionals of $e_i(\mathrm{P}, t, t')$. Since $\mathscr{F}$ is t isotropic, it is easy to verify that $\widehat{p^{ij}}$ is $\boldsymbol{\gamma}$ isotropic; the $\boldsymbol{\alpha}^i$ are $\boldsymbol{\gamma}$ orthonormal; therefore, from 5.7(28) (or, rather, its analog for covariant vectors $\boldsymbol{\alpha}^i$), the $\widehat{p^{ij}}$ are independent of $\boldsymbol{\gamma}$ and $\boldsymbol{\alpha}^i$, and are scalar functionals of e_1, e_2, and e_3 alone. It follows, in particular, that $\widehat{p_{ij}}$ and the expression $\boldsymbol{\gamma}' = e_i^2\boldsymbol{\alpha}^i\boldsymbol{\alpha}^i$ are unaltered by the substitution

$$(\boldsymbol{\alpha}^1, \boldsymbol{\alpha}^2, \boldsymbol{\alpha}^3) \to (\boldsymbol{\alpha}^1, -\boldsymbol{\alpha}^2, -\boldsymbol{\alpha}^3), \qquad e_i \to e_i, \tag{2}$$

and by the substitution

$$(\boldsymbol{\alpha}^1, \boldsymbol{\alpha}^2, \boldsymbol{\alpha}^3) \to (\boldsymbol{\alpha}^2, \boldsymbol{\alpha}^3, \boldsymbol{\alpha}^1) \qquad (e_1, e_2, e_3) \to (e_2, e_3, e_1). \tag{3}$$

Using (2) with (1), we see that $\widehat{p_{12}} = \widehat{p_{13}} = 0$; similarly, $\widehat{p_{23}} = 0$. It follows that the $\boldsymbol{\alpha}^i$ are the principal directions of stress, and that *the principal directions of stress are everywhere orthogonal to the surfaces of the shear-free flow body coordinate system B* (i.e., B is a body coordinate system which is always orthogonal).

The substitution (3) with (1) shows that $\widehat{p_{11}}$, $\widehat{p_{22}}$, and $\widehat{p_{33}}$ are all expressible in terms of a single material functional $\widehat{p_{11}}(e_1, e_2, e_3)$ and hence we have

$$\boldsymbol{\pi}^0 = \boldsymbol{\alpha}^1\boldsymbol{\alpha}^1\widehat{p_{11}}(e_1, e_2, e_3) + \boldsymbol{\alpha}^2\boldsymbol{\alpha}^2\widehat{p_{11}}(e_2, e_3, e_1) + \boldsymbol{\alpha}^3\boldsymbol{\alpha}^3\widehat{p_{11}}(e_3, e_1, e_2). \tag{4}$$

The fact that the principal stresses can be expressed in terms of a single material functional in these circumstances has been noted by Coleman (1968). Using 3.9(13), we can rewrite (4) in the form of a "reduced" constitutive equation:

$$\boldsymbol{\pi}^0 = \hbar_1\boldsymbol{\gamma} + \hbar_2\dot{\boldsymbol{\gamma}} + \hbar_3\dot{\boldsymbol{\gamma}}\cdot\boldsymbol{\gamma}^{-1}\cdot\dot{\boldsymbol{\gamma}} \qquad (\dot{\kappa}_i \text{ unequal at } t); \tag{5}$$

$$\begin{bmatrix}\hbar_1\\ \hbar_2\\ \hbar_3\end{bmatrix} = \sum \frac{\widehat{p_{11}}(e_1, e_2, e_3)}{(\dot{\kappa}_1 - \dot{\kappa}_2)(\dot{\kappa}_1 - \dot{\kappa}_3)}\begin{bmatrix}\dot{\kappa}_2\dot{\kappa}_3\\ -\dot{\kappa}_2 - \dot{\kappa}_3\\ 1\end{bmatrix}; \tag{6}$$

$\sum$ in (6) means a sum of three terms obtained by cyclic permutation of $\dot{\kappa}_1$, $\dot{\kappa}_2$, $\dot{\kappa}_3$ simultaneously with e_1, e_2, e_3. In view of 3.9(5), the coefficients $\hbar_i$ in (5) can be regarded as functionals of the elongation rates $\dot{\kappa}_i''$ over the interval $-\infty \le t'' \le t$; by (6), the three functionals $\hbar_i$ are expressible in terms of a single material functional of the three variables $\dot{\kappa}_i''$ $(i = 1, 2, 3)$. For the validity of (5), which gives $\boldsymbol{\pi}^0(\mathrm{P}, t)$, it is necessary that the elongation rates be all different at t, whether or not they are all different at previous times.

In the case of steady, shear-free flow, the principal elongation rates are independent of time throughout the interval $(-\infty, t)$, and so the coefficients $\hbar_i$ in (5) reduce to functions (instead of functionals) of $\dot{\kappa}_i$. These functions are symmetric, and can be reexpressed in terms of three strain-rate invariants. The constitutive equation (5) (for a viscoelastic liquid in steady, shear-free flow) is then of the same form as the constitutive equation for a purely viscous liquid (valid for any flow) (Weissenberg, 1935, p. 163; Reiner, 1945; Rivlin, 1948; Lodge, 1973a). In this sense, we can argue qualitatively that *elastic effects cannot show up in prolonged steady shear-free flow.* Elastic effects do show up in steady *shear* flow, however; the appropriate reduced constitutive equation for this case is 7.1(14) (for incompressible liquids) which contains an extra term in $\ddot{\boldsymbol{\gamma}}$ [not present in (5)] which does not vanish in steady shear flow. It should be noted that the functionals σ, N_1, and N_2 occurring in the shear flow equation 7.1(14) are functionals of a single strain rate $\dot{s}$ and are not the same as the functionals $\hbar_i$ (which involve three strain rates $\dot{\kappa}_i$) occurring in the shear-free-flow equation (5). The importance of the distinction between shear flow and shear-free flow was appreciated by Weissenberg (1931), who considered the forms of rather general classes of constitutive equations for viscoelastic materials subjected to arbitrary shear-free flows.

It is clear that for incompressible materials, a similar route can be taken if $\boldsymbol{\pi}^0$ is replaced by $\boldsymbol{\Pi}^0$ throughout the preceding analysis.

By using the substitutions $(\boldsymbol{\alpha}^1, \boldsymbol{\alpha}^2, \boldsymbol{\alpha}^3) \to (\boldsymbol{\alpha}^1, \boldsymbol{\alpha}^3, \boldsymbol{\alpha}^2)$ with $(e_1, e_2, e_3) \to (e_1, e_3, e_2)$, it is easy to show from (4) (since $\boldsymbol{\pi}^0$ is unaltered) that

$$\widehat{p_{11}}(e_1, e_2, e_3) = \widehat{p_{11}}(e_1, e_3, e_2). \tag{7}$$

If $\dot{\kappa}_2 = \dot{\kappa}_3 \ne \dot{\kappa}_1$, it follows from (5) and 3.9(15) that $\boldsymbol{\pi}^0 = \bar{\hbar}_1\boldsymbol{\gamma} + \bar{\hbar}_2\dot{\boldsymbol{\gamma}}$, where $\bar{\hbar}_1$ and $\bar{\hbar}_2$ depend on a single functional of $\dot{\kappa}_1$ and $\dot{\kappa}_2$.

8 COVARIANT DIFFERENTIATION AND THE STRESS EQUATIONS OF MOTION

8.1 Divergence and curl

Most of the tensor operations considered so far have been "one-point" operations whose coordinate representations have not involved differentiation or integration with respect to coordinates. In this chapter, we consider operations involving differentiation with respect to coordinates. We develop the analysis in three stages: (i) operations with general tensors, not requiring covariant derivatives; (ii) covariant differentiation in a Euclidean manifold; and (iii) covariant differentiation in an affinely connected Riemannian manifold (Euclidean or non-Euclidean).

As an example to illustrate the problems involved in differentiation of general tensor fields, let us consider the transformation law for an absolute, covariant, vector field $\boldsymbol{u}$(Q), namely

(1) $$\bar{u}_i(\bar{x}) = \frac{\partial x^k}{\partial \bar{x}^i} u_k(x),$$

which comes from 2.3(31) and 2.4(2) (applied to a space field). On differentiating this equation, we obtain the result

(2) $$\frac{\partial \bar{u}_i(\bar{x})}{\partial \bar{x}^j} = \frac{\partial x^k}{\partial \bar{x}^i}\frac{\partial x^m}{\partial \bar{x}^j}\frac{\partial u_k(x)}{\partial x^m} + \frac{\partial^2 x^k}{\partial \bar{x}^i\,\partial \bar{x}^j}\,u_k(x).$$

The first two terms in this equation are just those which would occur in the transformation law for an absolute, covariant, second-rank tensor field (of components $\partial u_i/\partial x^j$), but *the third term is not a term which occurs in any type of tensor transformation law*, because all such laws involve first (not second) derivatives of coordinates. The second derivatives $\partial^2 x^k/\partial \bar{x}^i\,\partial \bar{x}^j$ are all zero if the coordinate systems are rectangular Cartesian but are not zero for arbitrary coordinate systems. We can therefore define a second-rank *Cartesian* tensor $\nabla \mathbf{u}$ by the equation

(3)
$$\nabla \mathbf{u}\, C = [\partial u_c/\partial y^r], \qquad \text{where} \quad \mathbf{u}C = [u_r], \quad C: \mathrm{Q} \rightarrow y, \quad \text{for all } C \text{ in } \mathscr{C},$$

by using the set $\mathscr{C}$ of all rectangular Cartesian space coordinate systems; similar definitions can be given if **u** is replaced by an absolute Cartesian tensor field of any rank.

For general vector and tensor fields, on the other hand, it is necessary to select combinations of derivatives (such as $\partial u_i/\partial x^j$) and other terms in such a way that the unwanted second derivatives $\partial^2 x^k/\partial \bar{x}^i\,\partial \bar{x}^j$ do not appear in the final transformation equations. The convected derivative $\mathscr{D}/\mathscr{D}t$, defined in 5.4(27), furnishes one example of such a process which yields, from a given space field $\boldsymbol{F}(\mathrm{Q}, t)$ of arbitrary type, another space field $\mathring{\boldsymbol{F}}(\mathrm{Q}, t)$ of the same type whose components include derivatives taken with respect to x^i; in this example, the velocity field $\boldsymbol{v}(\mathrm{Q}, t)$ of the flowing material is used in order to eliminate the unwanted second derivatives. The essential features of this process of elimination are contained in Problem 5.4(39), where the velocity $\boldsymbol{v}$ is replaced by an arbitrary absolute contravariant vector field $\boldsymbol{X}$, and the given field $\boldsymbol{F}$ is also taken to be an absolute, contravariant vector field $\boldsymbol{Y}$.

(4) In Table 8.1(4), we have listed this example and similar examples in which general tensors or scalars can be constructed from derivatives. In most cases, it will be seen that the given tensor fields have to be of a special type. Methods of defining derivatives of general tensor fields of any type will be considered later in this chapter.

The components, in an an arbitrary coordinate system $S: \mathrm{Q} \rightarrow x$, of the various quantities listed in Table 8.1(4) and not already given are as stated in the following equations:

(5) $(\mathrm{Curl}\ \boldsymbol{u})S = [\partial u_c/\partial x^r - \partial u_r/\partial x^c];$

Table 8.1(4). Multipoint General Tensor Operations Not Requiring Covariant Differentiation

Operator	Operand	Result	Reference
Volume integral	Relative scalar, weight − 1	Integral invariant	Section 5.5
Convected derivative	Space tensor field	Field of same type	5.4(27), 5.4(39)
grad	Absolute scalar field	Absolute covariant vector field	2.4(12)
Curl	Absolute covariant vector field	Absolute antisymmetric second-rank tensor field	8.1(5)
curl	Absolute covariant vector field	Relative contravariant vector field of weight one	8.1(6)
div	Relative contravariant vector field of weight 1	Relative scalar field of weight one	8.1(7)
∇^2	Absolute scalar field	Absolute scalar field	8.1(9)
Div	Absolute contravariant vector field	Absolute scalar field	8.1(8)

(6) $(\text{curl } \boldsymbol{u})S = -[e^{rij}\, \partial u_i/\partial x^j];$

(7) $(\text{div } \boldsymbol{w})S = \partial w^i/\partial x^i;$

(8) $(\text{Div } \boldsymbol{v})S = (\sqrt{|g|}\,)^{-1}\, \partial(\sqrt{|g|}\; v^i)/\partial x^i;$

(9) $\nabla^2 s = (\sqrt{|g|}\,)^{-1}(\partial/\partial x^i)(\sqrt{|g|}\; g^{ij}\, \partial s/\partial x^j).$

The given fields are of the following types: $\boldsymbol{u}$ is an absolute covariant vector; $\boldsymbol{v}$ is an abolute contravariant vector; $\boldsymbol{w}$ is a relative contravariant vector of weight 1; s is an absolute scalar. The proof of the tensor properties stated in Table 8.1(4) is straightforward. In (2), the second derivative is symmetric in i and j, and therefore the antisymmetric part of $\partial u_i/\partial x^j$ satisfies the transformation law for an absolute covariant tensor of second rank, as stated in (5). The curl $\boldsymbol{u}$ [in (6)] can be obtained from Curl $\boldsymbol{u}$ by double contraction with the contravariant alternating tensor, whose weight is 1, by 5.2(6), and hence curl $\boldsymbol{u}$ is a relative contravariant vector of weight 1, as stated. The transformation law for $\boldsymbol{w}$ is of the form

(10) $$\bar{w} = |L|^{-1} L w, \qquad L = [\partial \bar{x}^r/\partial x^c],$$

from Table 5.1(4) (with $v = w$ and $n = 1$); on differentiating and contracting, we obtain the equation

$$(11) \qquad \frac{\partial \bar{w}^i}{\partial \bar{x}^i} = |L|^{-1}\left\{\left(-\frac{\partial \log_e |L|}{\partial x^j} + \frac{\partial^2 \bar{x}^i}{\partial x^s \, \partial x^j} \frac{\partial x^s}{\partial \bar{x}^i}\right) w^j + \frac{\partial w^j}{\partial x^j}\right\}.$$

From 1.3(33) (with $A = L$ and $t = x^j$) and the fact that $L^{-1} = [\partial x^r/\partial \bar{x}^c]$, it follows that

$$(12) \qquad \frac{\partial \log_e |L|}{\partial x^j} = \operatorname{tr}\left(L^{-1} \frac{\partial L}{\partial x^j}\right) = \frac{\partial x^s}{\partial \bar{x}^i} \frac{\partial^2 \bar{x}^i}{\partial x^j \, \partial x^s}.$$

Equation (11) therefore simplifies to give the transformation law for a relative scalar of weight one, as stated in (7). Equation (8) then follows from (7) on taking $\boldsymbol{w} = \sqrt{|\boldsymbol{g}|}\ \boldsymbol{v}$, and using the fact that $\sqrt{|\boldsymbol{g}|}$ is a relative scalar of weight 1 [see 5.5(2)]. Finally, (9) follows from (8) on taking $v^i = g^{ij}\, \partial s/\partial x^j$, which is permissible since $\partial s/\partial x^j$ are components of an absolute covariant vector field.

(13) **Problem** Let $\boldsymbol{W}$ be an *antisymmetric* relative contravariant tensor field of weight 1 and rank 2, with components $W^{ij}(x)$ in any S. By contracting with an arbitrary absolute covariant vector and using a quotient theorem (or otherwise), prove that $\partial W^{ij}/\partial x^j$ are components of a contravariant relative vector of weight 1.

(14) **Problem** Let S be an *orthogonal* coordinate system, and write $g = \operatorname{diag}[h_1^2, h_2^2, h_3^2]$ $(h_i > 0)$. Show that (8) and (9) can be written in the form

$$\operatorname{Div} \boldsymbol{v} = \frac{1}{h_1 h_2 h_3} \frac{\partial}{\partial x^i}\left(\frac{h_1 h_2 h_3}{h_i} \widehat{v^i}\right), \qquad \nabla^2 s = \frac{1}{h_1 h_2 h_3} \frac{\partial}{\partial x^i}\left(\frac{h_1 h_2 h_3}{h_i^2} \frac{\partial s}{\partial x^i}\right),$$

where the $\widehat{v^i}$ are the physical components in S of the absolute contravariant vector field $\boldsymbol{v}$ [as defined by the spatial analog of 2.7(21)].

8.2 Covariant differentiation in a Euclidean manifold

We now wish to obtain from any given general space tensor field a general space tensor field which is, in some suitable sense, a spatial derivative of the given field. The difficulty lies in the fact that general space tensors associated with two distinct places Q and Q′ belong to disjoint linear spaces and cannot therefore be subtracted from one another. If the space manifold is Euclidean, however, then any given general space tensor field gives rise to a unique Cartesian space tensor field (of the same rank), as shown in Section 4.4; one can form the difference of the Cartesian tensors at Q and Q′, and hence define a Cartesian tensor spatial derivative; to this tensor field, there corresponds a general space tensor field (of rank that is one higher than that of the original given field) that is a spatial derivative of the given field and is unique if one agrees to take it to be of the same type as the given field with an extra "left-covariant rank" associated with the spatial derivative operator ∇. This field is called the *covariant derivative* of the given field. We shall illustrate this procedure in this section, using vectors and second-rank tensors, and will apply the results in Section 8.4 to obtain the stress equations of motion, which involve spatial derivatives of the (second-rank) stress tensor. For a non-Euclidean manifold, Cartesian tensor fields cannot be defined, the preceding method cannot be used, and it is customary to add extra postulates (to generate an "affine connection") which enable one to define covariant differentiation (Section 8.5).

We now show how to define the covariant derivative $\nabla \boldsymbol{v}$ of any given absolute contravariant space vector field $\boldsymbol{v}(\mathrm{Q})$. Let $\mathbf{v}(\mathrm{Q})$ be the corresponding Cartesian vector field (i.e., the field that has, in any rectangular Cartesian coordinate system C, the same representative matrix), and let Q and Q′ be any two neighboring places. Then the equation

(1) $$\mathbf{v}(\mathrm{Q}') - \mathbf{v}(\mathrm{Q}) = \mathbf{dr} \cdot \nabla \mathbf{v}(\mathrm{Q}) \qquad (\mathbf{dr} = \overrightarrow{\mathrm{QQ}'})$$

defines a second-rank Cartesian tensor field $\nabla \mathbf{v}$, since the Cartesian vector $\mathbf{dr}$ is arbitrary. It is a straightforward matter to show, from (1) and the equation $\mathbf{v}(\mathrm{Q}) = v^i(y)\mathbf{b}_i(C)$, that

(2) $$\nabla \mathbf{v} = \frac{\partial v^i}{\partial y^j}\, \mathbf{b}^j(C)\mathbf{b}_i(C),$$

where $\mathbf{b}_i(C) = \mathbf{b}^i(C)$ are the base vectors for an arbitrary rectangular Cartesian coordinate system $C : \mathrm{Q} \to y$. These base vectors are independent of position.

In order to obtain a general tensor equation corresponding to (1) (and hence to define a general tensor $\nabla \boldsymbol{v}$), it is necessary to associate all quantities in (1) with the same place, Q say, and to express the equation in terms of Cartesian base vectors $\mathbf{b}_i(S, \mathrm{Q})$ and $\mathbf{b}^i(S, \mathrm{Q})$ associated with an *arbitrary* space coordinate system $S : \mathrm{Q} \to x$. Since the vectors are Cartesian, we can simply transfer $\mathbf{v}(\mathrm{Q}')$ parallel to itself to obtain a vector, $\mathbf{v}(\mathrm{Q}' \mid \mathrm{Q})$ say, which we associate with Q:

(3) $$\mathbf{v}(\mathrm{Q}' \mid \mathrm{Q}) = \mathbf{v}(\mathrm{Q}') \qquad \text{(Cartesian vectors).}$$

(4) See Fig. 8.2(4).

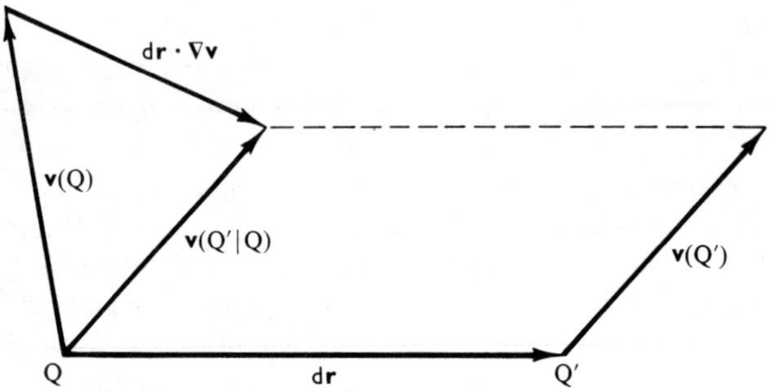

Figure 8.2(4) Definition of the covariant derivative $\nabla\mathbf{v}$ of a vector field $\mathbf{v}(Q)$. Under parallel transfer, the field vector $\mathbf{v}(Q')$ (at Q′) gives rise to a vector $\mathbf{v}(Q'|Q)$ (at Q); the difference between this vector and $\mathbf{v}(Q)$, the field vector at Q, is obtained from $\mathbf{dr} = \overrightarrow{QQ'}$ by contraction with $\nabla\mathbf{v}$.

Corresponding to Eq. (1), with (3), there is now a unique *general tensor* equation which we may write in the form

(5) $$\boldsymbol{v}(Q'|Q) - \boldsymbol{v}(Q) = \boldsymbol{dx} \cdot \nabla\boldsymbol{v}(Q),$$

where $\boldsymbol{v}(Q'|Q)$ is the absolute contravariant vector (at Q) that corresponds to the Cartesian vector $\mathbf{v}(Q'|Q)$. The left-hand side of (5) is thus an absolute contravariant vector at Q; $\boldsymbol{dx}$ $(= \overrightarrow{QQ'})$ is an arbitrary (infinitesimal) absolute contravariant vector at Q; hence, from a quotient theorem (see Section 5.3), it follows that $\nabla\boldsymbol{v}$ is an absolute, left-covariant, mixed, second-rank tensor at Q. Equation (5) therefore defines $\nabla\boldsymbol{v}$, which we call the covariant derivative of $\boldsymbol{v}$.

To obtain a useful expression for $\nabla\boldsymbol{v}$ in terms of general base vectors $\boldsymbol{b}_i(S, Q)$, we need to express $\boldsymbol{v}(Q'|Q)$ in terms of $\boldsymbol{b}_i(S, Q)$. From 4.5(8), we have

(6) $$\boldsymbol{v}(Q) = v^i(x)\boldsymbol{b}_i(S, Q) \qquad \text{(contravariant vectors)},$$

(7) $$\mathbf{v}(Q) = v^i(x)\mathbf{b}_i(S, Q) \qquad \text{(Cartesian vectors)},$$

with the same coefficients $v^i(x)$ in each representation, where $\mathbf{v}$ is the Cartesian vector corresponding to the given contravariant vector $\boldsymbol{v}$. The *Cartesian* tangent base vectors $\mathbf{b}_i(S, Q')$ can be expressed in terms of $\mathbf{b}_i(S, Q)$ by equations

(8) $$\mathbf{b}_i(S, Q') = \mathbf{b}_i(S, Q) + \frac{\partial \mathbf{b}_i(S, Q)}{\partial x^j}\, dx^j = \{\delta_i^k + G_{ij}^k(x)\, dx^j\}\mathbf{b}_k(S, Q),$$

where

(9) $$G_{ij}^k(x) = \mathbf{b}^k(S, Q) \cdot \frac{\partial \mathbf{b}_i(S, Q)}{\partial x^j}.$$

From 4.5(3), it follows that

(10) $$G^k_{ij} = G^k_{ji}.$$

From (3) and (7), we have

(11) $$\mathbf{v}(Q'\,|\,Q) = v^i(x + dx)\mathbf{b}_i(S, Q') = \left(v^i + \frac{\partial v^i}{\partial x^r}\,dx^r\right)(\delta^k_i + G^k_{ij}\,dx^j)\mathbf{b}_k(S, Q),$$

on using (8). Since $\mathbf{dr} = dx^i\ \mathbf{b}_i(S, Q)$, it now follows from (1) and (11) that

(12) $$\mathbf{dr} \cdot \nabla\mathbf{v} = v^k{}_{,j}(x)\mathbf{b}_k(S, Q),$$

where

(13) $$v^k{}_{,j} = \frac{\partial v^k}{\partial x^j} + v^i G^k_{ij},$$

and hence that

(14) $$\nabla\mathbf{v}(Q) = v^k{}_{,j}(x)\mathbf{b}^j(S, Q)\mathbf{b}_k(S, Q) \qquad \text{(Cartesian tensors)},$$

on using the fact that the dx^i are arbitrary infinitesimals. Equation (14) involves Cartesian tensors associated with the same place Q, and scalar coefficients $v^k{}_{,j}$, and so it gives rise to a unique general tensor equation

(15) $$\nabla\boldsymbol{v}(Q) = v^k{}_{,j}(x)\boldsymbol{b}^j(S, Q)\boldsymbol{b}_k(S, Q) \qquad \text{(general tensors)},$$

which, as required, expresses *the covariant derivative* $\nabla\boldsymbol{v}$ *of an absolute contravariant vector field* $\boldsymbol{v}(Q)$ *in terms of base vectors for an arbitrary space coordinate system S*. The coefficients $v^k{}_{,j}$ are given by (13) in terms of G^k_{ij}; the G^k_{ij} are given by (9) in terms of derivatives of Cartesian base vectors which have no counterpart in terms of general base vectors. However, by differentiating equations 4.5(10) and using 4.5(13), it is easy to derive the equations

(16) $$G^k_{ij} = g^{kr}G_{rij}, \qquad \text{where} \quad G_{kij} = \frac{1}{2}\left(\frac{\partial g_{ki}}{\partial x^j} + \frac{\partial g_{kj}}{\partial x^i} - \frac{\partial g_{ij}}{\partial x^k}\right),$$

(17) $$G_{kij} + G_{ikj} = \frac{\partial g_{ki}}{\partial x^j}.$$

(18) **Problem** Prove that $|g|^{-1/2}\,\partial|g|^{1/2}/\partial x^i = G^k_{ki}$.

The quantities G_{kij} and G^k_{ij}, when defined by (16), are called Christoffel symbols of the first and second kind, respectively, and are usually denoted by Γ_{kij} and Γ^k_{ij} or by $[k, ij]$ and $\{^{\,k}_{ij}\}$. We shall retain the symbols Γ_{kij} and Γ^k_{ij} to denote the corresponding quantities associated with the body manifold.

(19) **Problem** For an orthogonal coordinate system, with

$$g = \text{diag}[h_1{}^2, h_2{}^2, h_3{}^2],$$

derive Eq. (2) in Appendix C.

As an important example involving covariant differentiation of an absolute contravariant vector field, we consider the *acceleration* $\boldsymbol{a}(\text{Q}, t)$ defined, for a given material with contravariant velocity field $\boldsymbol{v}(\text{Q}, t)$, by the equation

$$\text{(20)} \qquad \boldsymbol{a}(\text{Q}, t) = \lim_{t' \to t} \frac{\boldsymbol{v}(\text{Q}' \,|\, \text{Q}, t') - \boldsymbol{v}(\text{Q}, t)}{t' - t},$$

where $\boldsymbol{v}(\text{Q}' \,|\, \text{Q}, t')$ is the vector at Q obtained by parallel transfer of $\boldsymbol{v}(\text{Q}', t')$ and Q and Q′ denote the places occupied at times t and t' by a particle **P**, so that $\boldsymbol{dx} = \overrightarrow{\text{QQ}'} = \boldsymbol{v}(\text{Q}, t)\, dt$ $(dt = t' - t)$. By adding and subtracting $\boldsymbol{v}(\text{Q}' \,|\, \text{Q}, t)$ in the numerator, it is easy to show that

$$\text{(21)} \qquad \boldsymbol{a}(\text{Q}, t) = \left(\frac{\partial}{\partial t} + \boldsymbol{v}(\text{Q}, t) \cdot \boldsymbol{\nabla}\right) \boldsymbol{v}(\text{Q}, t) = a^i(x, t)\boldsymbol{b}_i(S, \text{Q}),$$

$$\text{(22)} \qquad a^i(x, t) = \frac{\partial v^i}{\partial t} + v^k\, \partial v^i/\partial x^k + v^j v^k G^i_{jk} \qquad [v^i = v^i(x, t)].$$

The acceleration field of a moving material therefore involves terms quadratic in the velocity field, even when referred to a rectangular Cartesian coordinate system (i.e., when $G^i_{jk} = 0$). The differential operator

$$\text{(23)} \qquad \frac{D}{Dt} = \frac{\partial}{\partial t} + \boldsymbol{v}(\text{Q}, t) \cdot \boldsymbol{\nabla}$$

occurring in (21) is called the *hydrodynamic derivative* or *material derivative*, and represents a rate of change of a space field "at constant particle," when applied to a function of Q and t.

It should be noted that it is immaterial whether, in defining $\boldsymbol{\nabla v}$, we transfer $\boldsymbol{v}(\text{Q}')$ to Q, as in (5), or $\boldsymbol{v}(\text{Q})$ to Q′, as in the equation

$$\boldsymbol{v}(\text{Q}') - \boldsymbol{v}(\text{Q} \,|\, \text{Q}') = \boldsymbol{dx} \cdot \boldsymbol{\nabla v}(\text{Q}').$$

(24) We have used the example of an absolute contravariant space vector field $\boldsymbol{v}(\text{Q})$ to illustrate the method used to define its covariant derivative $\boldsymbol{\nabla v}$ when the underlying manifold is Euclidean. We can immediately extend this procedure so as to apply to any type of space tensor field $\boldsymbol{v}(\text{Q})$ (scalar, vector, or tensor of any rank; absolute or relative) by simply retaining Eq. (5) (for arbitrary $\boldsymbol{dx}$) as the definition of $\boldsymbol{\nabla v}$. Only the equations giving the components

of ∇v in any S will change as the type of tensor field is changed. The main consequences of this definition are easy to derive, and are presented in Table 8.2(24) for convenience.

Before proving the results in Table 8.2(24), we note from the definition (5) that *the covariant derivative operator* ∇ *has the usual properties of a differential operator*: it is a linear operator, and when applied to a product such as uv, gives the result embodied in the equation

(25) $$dx \cdot \nabla(uv) = (dx \cdot \nabla u)v + u(dx \cdot \nabla v).$$

This follows from (5) on adding and subtracting $u(\mathrm{Q})v(\mathrm{Q}' \mid \mathrm{Q})$. It is a weakness of the present tensor notation that we cannot remove the arbitrary factor $dx\cdot$ from (25) as it stands, because the right-hand side would be a third-rank tensor if the factor $dx\cdot$ were removed, and the ordering of the tensor factors would then present a notational problem. The corresponding problem does not arise, however, when we refer the equation to an arbitrary coordinate

Table 8.2(24) Covariant Derivatives of Absolute Space Tensor Fields[a]

(i)	Scalar s	$\nabla s = s_{,k} b^k,$	$s_{,k} = \dfrac{\partial s}{\partial x^k}$
(ii)	Contravariant vector v	$\nabla v = v^i{}_{,k} b^k b_i,$	$v^i{}_{,k} = \dfrac{\partial v^i}{\partial x^k} + v^r G^i_{rk}$
(iii)	Covariant vector u	$\nabla u = u_{i,k} b^k b^i,$	$u_{i,k} = \dfrac{\partial u_i}{\partial x^k} - u_r G^r_{ik}$
(iv)	Contravariant second-rank tensor p	$\nabla p = p^{ij}{}_{,k} b^k b_i b_j,$	$p^{ij}{}_{,k} = \dfrac{\partial p^{ij}}{\partial x^k} + p^{rj} G^i_{rk} + p^{ir} G^j_{rk}$
(v)	Covariant second-rank tensor q	$\nabla q = q_{ij,k} b^k b^i b^j,$	$q_{ij,k} = \dfrac{\partial q_{ij}}{\partial x^k} - q_{rj} G^r_{ik} - q_{ir} G^r_{jk}$
(vi)	Right-covariant mixed tensor m	$\nabla m = m^i{}_{j,k} b^k b_i b^j,$	$m^i{}_{j,k} = \dfrac{\partial m^i{}_j}{\partial x^k} + m^r{}_j G^i_{rk} - m^i{}_r G^r_{jk}$
(vii)	Covariant base vectors b^i	$\nabla b^i = -G^i_{jk} b^k b^j$	
(viii)	Contravariant base vectors b_i	$\nabla b_i = G^j_{ik} b^k b_j$	
(ix)	Metric tensors g, g^{-1}	$\nabla g = 0, \quad \nabla g^{-1} = 0$	
(x)	Unit tensor I	$\nabla I = 0.$	

[a] $b_i = b_i(S, \mathrm{Q})$ and $b^i = b^i(S, \mathrm{Q})$ are base vectors for an arbitrary coordinate system $S : \mathrm{Q} \rightarrow x$.

system $S:\mathrm{Q}\to x$; if, for example, $\boldsymbol{u}$ is a covariant vector and $\boldsymbol{v}$ is a contravariant vector, then it is easy to verify that

$$(26)\qquad (u_i v^j)_{,k} = u_{i,k} v^j + u_i v^j{}_{,k}.$$

There are similar results whatever types of tensor are chosen for $\boldsymbol{u}$ and $\boldsymbol{v}$, and it is obvious that similar equations hold if the factors $\boldsymbol{u}$ and $\boldsymbol{v}$ are contracted. If one of the two factors is a scalar s, then clearly

$$(27)\qquad \nabla(s\boldsymbol{T}) = (\nabla s)\boldsymbol{T} + s\,\nabla\boldsymbol{T},$$

where $\boldsymbol{T}$ is a tensor field of any type; in this case, the ordering of factors on the right-hand side presents no problem.

Proof of Results in Table 8.2(24) (i) follows from (5) (with $\boldsymbol{v}$ replaced by s) and the fact that $s(\mathrm{Q}'\,|\,\mathrm{Q}) = s(\mathrm{Q}')$. (ii) has been proved already in (13) and (15). (viii) follows from (ii) and the fact that $b^j_{(i)} = \delta^j_i$. (iii) follows on taking the covariant derivative of the equation $\boldsymbol{b}_i \cdot \boldsymbol{u} = u_i$ (where $\boldsymbol{u}_i$ is a scalar) and using (i), (viii), and 2.5(19). (vii) follows from (iii) and the fact that $b^{(i)}_j = \delta^i_j$. To prove (iv), we operate with $\boldsymbol{dx}\cdot\nabla$ from the left on the equation $\boldsymbol{p} = p^{ij}\boldsymbol{b}_i\boldsymbol{b}_j$ (where p^{ij} are scalars), and use (i), (27), and (viii); on moving a scalar factor $\boldsymbol{dx}\cdot\boldsymbol{b}^k$ to the left and using the fact that $\boldsymbol{dx}$ is arbitrary, the validity of (iv) can be established. (v) and (vi) can be proved in a similar manner, using (i), (vii), and (viii). (x) follows from (vi) and the fact that $\boldsymbol{IS} = [\delta^r_c]$. (ix) follows from (v) and (17), and then from taking the covariant derivative of the equation $\boldsymbol{g}\cdot\boldsymbol{g}^{-1} = \boldsymbol{I}$, using (x). This completes the proof of Table 8.2(24).

(28) **Definition** A tensor field $\boldsymbol{T}(\mathrm{Q})$ of any type is *homogeneous* if $\nabla\boldsymbol{T} = \boldsymbol{0}$. For example, a state of stress is homogeneous if $\nabla\boldsymbol{p} = \boldsymbol{0}$, and a strain $t'\to t$ is homogeneous if $\nabla\boldsymbol{C}_t(\mathrm{Q}, t') = \boldsymbol{0}$.

(29) **Problem** Prove that if $\boldsymbol{P} = \boldsymbol{p} + p\boldsymbol{g}^{-1}$ is the contravariant space extra stress tensor (where p is a scalar), then $P^{ij}{}_{,k} = p^{ij}{}_{,k} + g^{ij}\,\partial p/\partial x^k$.

8.3 Covariant derivatives of body tensor fields

Since the body manifold is Euclidean (i.e., there exists a body coordinate system which is rectangular Cartesian at any given instant t), the method of Section 8.2 can be used to define covariant derivatives of all types of body tensor field. Instead of repeating the method of Section 8.2, we shall transfer the space field results to the body by means of the operator $\mathbb{T}(t)$; since *covariant differentiation is a one-state operation*, it commutes with $\mathbb{T}(t)$ and the results of the transfer operation can be written down immediately: the body tensor version of Table 8.2(24) involves the body Christoffel symbols $\Gamma^i_{jk}(t)$, where

$$\text{(1)} \qquad \Gamma^i_{jk} = \gamma^{ir}\Gamma_{rjk}, \qquad \Gamma_{ijk} = \frac{1}{2}\left(\frac{\partial\gamma_{ij}}{\partial\xi^k} + \frac{\partial\gamma_{ik}}{\partial\xi^j} - \frac{\partial\gamma_{jk}}{\partial\xi^i}\right),$$

which depend on time because the body metric tensor does. Accordingly, we use $\mathbf{\nabla}_t$ for the covariant derivative operator at time t in the body manifold. This operator may be defined in terms of the space field operator $\mathbf{\nabla}$ as follows.

Let $\mathbf{\Phi}$ be any given body tensor field, and let $\boldsymbol{F}$ denote the space field obtained by transfer at any given time t. Then the body tensor field $\mathbf{\nabla}_t\mathbf{\Phi}$ is defined by transfer of $\mathbf{\nabla}\boldsymbol{F}$ at time t, i.e.,

$$\text{(2)} \qquad \mathbf{\nabla}_t\mathbf{\Phi} = \mathbb{T}^{-1}(t)\,\mathbf{\nabla}\boldsymbol{F}, \qquad \text{where} \quad \boldsymbol{F} = \mathbb{T}(t)\mathbf{\Phi}.$$

Since $\mathbf{\Phi}$ is an arbitrary body tensor field, we can rewrite (2) symbolically in the form

$$\text{(3)} \qquad \mathbf{\nabla}_t = \mathbb{T}^{-1}(t)\,\mathbf{\nabla}\mathbb{T}(t), \qquad \text{or} \qquad \mathbb{T}(t)\mathbf{\nabla}_t = \mathbf{\nabla}\mathbb{T}(t).$$

To obtain useful expressions for $\mathbf{\nabla}_t$ for an arbitrary body coordinate system $B: \mathrm{P} \to \xi$, we use the space coordinate system $S: \mathrm{Q} \to x$ that is congruent to B at time t; the base vectors for the two systems are therefore related by the transfer operator $\mathbb{T}(t)$:

$$\text{(4)} \qquad \boldsymbol{\beta}^k(B, \mathrm{P}) = \mathbb{T}(t)\boldsymbol{b}^k(S, \mathrm{Q}); \qquad \boldsymbol{\beta}_i(B, \mathrm{P}) = \mathbb{T}(t)\boldsymbol{b}_i(S, \mathrm{Q}) \qquad (B \overset{t}{\equiv} S).$$

Further, we have at time t the relations

$$\text{(5)} \qquad \xi^i = x^i, \qquad \gamma_{ij}(\xi, t) = g_{ij}(x), \qquad \Gamma^i_{jk}(\xi, t) = G^i_{jk}(x) \qquad (B \overset{t}{\equiv} S).$$

It will be sufficient illustration to consider an absolute contravariant body vector field $\boldsymbol{\theta}(\mathrm{P})$. From (2), we have, with $\boldsymbol{v} \equiv \mathbb{T}(t)\boldsymbol{\theta}$,

$$\mathbf{\nabla}_t\boldsymbol{\theta} = \mathbb{T}^{-1}(t)\,\mathbf{\nabla}\boldsymbol{v} = \mathbb{T}^{-1}(t)\left(\frac{\partial v^i}{\partial x^k} + G^i_{jk}\,v^j\right)\boldsymbol{b}^k\boldsymbol{b}_i \qquad \text{[by 8.2(24)]}$$

$$= \left(\frac{\partial v^i}{\partial x^k} + G^i_{jk}\,v^j\right)\boldsymbol{\beta}^k\boldsymbol{\beta}_i \qquad \text{[by (4)]}.$$

Using (5) and the fact that $\theta^i = v^i$ (since $B \overset{t}{\equiv} S$), it follows that

$$\text{(6)} \qquad \mathbf{\nabla}_t\boldsymbol{\theta} = \theta^i_{\ ,k}\,\boldsymbol{\beta}^k\boldsymbol{\beta}_i, \qquad \text{where} \quad \theta^i_{\ ,k} \equiv \frac{\partial\theta^i}{\partial\xi^k} + \theta^j\Gamma^i_{jk}(t).$$

It is clear that a similar argument will yield body field expressions of the same form as the other space field expressions in Table 8.2(24). In particular, we have the following results:

(7) $\nabla_t \sigma = \boldsymbol{\beta}^k \, \partial\sigma/\partial\xi^k$ (σ an absolute scalar field);

(8) $\nabla_t \boldsymbol{\beta}^i = -\Gamma^i_{jk}(t)\boldsymbol{\beta}^k\boldsymbol{\beta}^j$; $\nabla_t \boldsymbol{\beta}_i = \Gamma^j_{ik}(t)\boldsymbol{\beta}^k\boldsymbol{\beta}_j$;

(9) $\nabla_t \boldsymbol{\delta} = \mathbf{0}$; $\nabla_t \boldsymbol{\gamma}(\mathrm{P}, t) = \mathbf{0}$;

(10) $$\nabla_t \boldsymbol{\pi} = \pi^{ij}{}_{,k}\,\boldsymbol{\beta}^k\boldsymbol{\beta}_i\boldsymbol{\beta}_j, \qquad \pi^{ij}{}_{,k} \equiv \frac{\partial\pi^{ij}}{\partial\xi^k} + \pi^{rj}\Gamma^i_{rk} + \pi^{ir}\Gamma^j_{rk}.$$

In (10), $\boldsymbol{\pi}$ is an absolute, second-rank, contravariant body tensor.

8.4 Curvature of surfaces

In this section, we give a brief treatment of the curvature of a surface in a three-dimensional Euclidean manifold in order to derive formulas for the principal radii of curvature in a form convenient for use with the stress equations of motion. Fuller treatments are given in many standard books on differential geometry [see, e.g., Weatherburn (1939, p. 66), Eisenhart (1909, especially pp. 447–449)]. We shall use general space vectors for the analysis in order that the analogous results for body vectors will be obvious.

(1) Let $\boldsymbol{n}(\mathrm{Q}_0)$, $\boldsymbol{n}(\mathrm{Q})$, ... be absolute, covariant unit vectors which are normals at Q_0, Q, ..., respectively, to any given surface. Let $\boldsymbol{n}(\mathrm{Q}_0 | \mathrm{Q})$ denote the vector (at Q) obtained by parallel transfer of $\boldsymbol{n}(\mathrm{Q}_0)$ from Q_0 to Q [Fig. 8.4(1)], where Q and Q_0 are neighboring points, and $\boldsymbol{dx} = \overrightarrow{\mathrm{Q}_0\mathrm{Q}}$.

(2) **Definition** If, given Q_0, a point Q can be found such that

$$\boldsymbol{n}(\mathrm{Q}) - \boldsymbol{n}(\mathrm{Q}_0 | \mathrm{Q}) = \boldsymbol{g}(\mathrm{Q}) \cdot \boldsymbol{dx}/\rho, \qquad \|\boldsymbol{n}(\mathrm{Q})\| = 1$$

for some scalar ρ, then $\boldsymbol{dx}$ is said to define a *principal direction of curvature* of the surface at Q_0, and ρ is the corresponding *principal radius of curvature*. A *line of curvature* is everywhere tangential to a principal direction of curvature.

Expressed rather loosely, this equation is the condition that consecutive normals intersect; for points Q_0 and Q not on a line of curvature, the (Cartesian vector) normals at Q_0 and Q, when produced, give skew lines whose shortest separation is of order $\mathrm{Q}_0\mathrm{Q}$; the separation is of a higher order when $\overrightarrow{\mathrm{Q}_0\mathrm{Q}}$ defines a principal direction of curvature. The main results which we require are stated in the following theorems.

(3) **Theorem** Through any point Q_0 on a given surface, there pass two lines of curvature; these lines are orthogonal at Q_0.

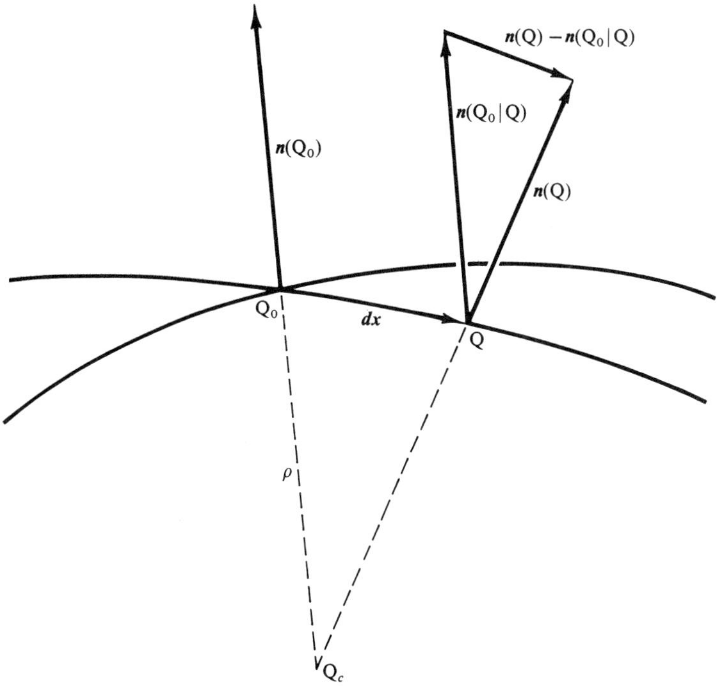

Figure 8.4(1) Principal directions of curvature of a surface. $\boldsymbol{dx} = \overrightarrow{Q_0 Q}$ defines a principal direction of curvature at Q_0 if, as $Q \to Q_0$, the surface normals $\boldsymbol{n}(Q)$ and $\boldsymbol{n}(Q_0)$ "intersect," at Q_c, say. $\boldsymbol{n}(Q)$, $\boldsymbol{n}(Q_0 \mid Q)$, and $\boldsymbol{dx}$ are then coplanar; Q_c is a center of curvature, and $Q_0 Q_c$ is a principal radius of curvature.

(4) **Dupin's Theorem** The coordinate lines of an orthogonal coordinate system in a three-dimensional manifold are lines of curvature on the coordinate surfaces.

(5) **Theorem** For an orthogonal coordinate system $S: Q \to x$, the principal radii of curvature ρ_{31} and ρ_{32} for the x^2 curve and the x^1 curve, respectively, through the point Q, on the surface $x^3 = \text{const}$ are given by the equations

$$\rho_{31} = h_2 h_3 \bigg/ \frac{\partial h_2}{\partial x^3}, \qquad \rho_{32} = h_1 h_3 \bigg/ \frac{\partial h_1}{\partial x^3},$$

where $[g_{ij}] = \text{diag}[h_1{}^2, h_2{}^2, h_3{}^2]$

Proof of (3) Using 8.2(5), we can rewrite (2) in the form

(6) $$\boldsymbol{dx} \cdot (\nabla \boldsymbol{n} - \boldsymbol{g}/\rho) = \boldsymbol{0},$$

which has the form of a principal axis (or eigenvalue) equation for the unknowns $\boldsymbol{dx}$ and ρ. The scalar equation obtained by contracting (6) with $\boldsymbol{n}$ is satisfied because $\|\boldsymbol{n}\| = 1$ and $\boldsymbol{n} \cdot \boldsymbol{dx} = 0$. Thus (6) is equivalent to two scalar equations $\boldsymbol{dx} \cdot (\boldsymbol{\nabla n} - \boldsymbol{g}/\rho) \cdot \boldsymbol{b}_k = 0$ $(k = 1, 2)$, or

(7) $$(\rho n_{k,\,m} - g_{mk})\, dx^m = 0 \qquad (k, m = 1, 2).$$

Without loss of generality, we have chosen a coordinate system $S:\mathrm{Q} \to x$ which has the given surface as a coordinate surface $x^3 = \text{const}$; $\boldsymbol{b}_k = \boldsymbol{b}_k(S, \mathrm{Q})$ $(k = 1, 2)$ are the contravariant base vectors tangential at Q_0 to $x^3 = \text{const}$. Equation (7) is of standard eigenvalue form, and there are therefore two values for ρ for each of which there exist nonzero solutions for (dx^1, dx^2). Since $n_i = (g^{33})^{-1/2}\delta_i^3$, it follows from 8.2(24)(iii) that

(8) $$n_{k,\,m} = -(g^{33})^{-1/2} G_{km}^3 \qquad (k, m = 1, 2).$$

Hence $n_{k,\,m} = n_{m,\,k}$, and therefore the roots in ρ are both real, and to each root corresponds a real solution for (dx^1, dx^2) (undetermined to the extent of an arbitrary factor). This proves that there are two principal directions of curvature at Q_0. The usual method of proof [see, e.g., 2.8(19)] can be used to show that these directions are orthogonal if the principal radii of curvature are unequal, and can be chosen to be orthogonal if the principal radii or curvature are equal; in the latter case, it is easy to show that every direction at Q_0 is a principal direction of curvature. This completes the proof of (3).

Proof of (4) and (5) We now use the preceding argument with the coordinate system $S:\mathrm{Q} \to x$ orthogonal. Using 8.2(19) to evaluate G_{km}^3, it is found from (8) that the matrix $[\rho n_{k,\,m} - g_{mk}]$ is diagonal, which proves that the x^1 and x^2 curves are lines of curvature on $x^3 = \text{const}$, and that the roots in ρ of the equation $|\rho n_{k,\,m} - g_{mk}| = 0$ are given by (5). Since similar arguments evidently apply to the other two coordinate surfaces, the proof of (4) and (5) is complete.

In connection with *helical flow* (Section 3.6), it is of interest to note that by applying Dupin's Theorem (4) to a cylindrical polar (orthogonal) coordinate system, the lines of curvature on a cylindrical shearing surface are the generators of the cylinder (straight lines parallel to the cylinder axis) and the circles orthogonal to them; *there cannot be an orthogonal coordinate system whose coordinate lines include the shearing lines*, since these are helices.

8.5 Stress equations of motion

The stress equations of motion relate the inertial force on a material element to the resultant contact force (exerted by neighboring material) and the external "body force" (usually that due to gravity). The stress equations of motion, when combined with the constitutive equations (appropriate to the material under consideration) and equations expressing, say, a strain tensor or the body metric tensor in terms of displacement variables, form a set of equations sufficient (in principle) to determine the flow of a material when suitable boundary values and initial conditions are specified (Section 8.10). For incompressible materials, the constant-volume condition provides an additional equation to "go with" the additional isotropic pressure variable p. For compressible materials, it is convenient to include the density ρ as a variable, and to add an extra equation expressing conservation of mass. If the temperature is not constant and uniform, additional equations are required, but we shall not consider these here.

It is assumed that any three-dimensional region $\mathscr{P}_1$ of the body possesses a *mass* $m(\mathscr{P}_1)$ which is a positive number, independent of time, given by an integral of the form

$$(1) \qquad m(\mathscr{P}_1) = \int_{\mathscr{P}_1} \mu(\xi)\, d\xi^1\, d\xi^2\, d\xi^3,$$

when referred to an arbitrary body coordinate system $B: \mathrm{P} \to \xi$. Thus $m(\mathscr{P}_1)$ is an integral invariant, and so the integrand $\mu(\xi)$ is the component in B of a relative body scalar of weight 1, since $d\xi^1\, d\xi^2\, d\xi^3$ has the transformation properties of a relative body scalar of weight -1, according to the body analog of 5.5(1). We define the density $\rho(\mathrm{P}, t)$ of the material at particle P and time t by the equation

$$(2) \qquad \rho\, dV = dm = \mu(\xi)\, d\xi^1\, d\xi^2\, d\xi^3,$$

where dm denotes the mass of the element $d\xi^1\, d\xi^2\, d\xi^3$ and the volume dV is given by 5.5(4). Thus ρ is an absolute scalar.

Conservation of mass may be expressed by the following alternative forms of equations:

$$(3) \quad 0 = \frac{\partial}{\partial t}\mu(\xi) = \frac{\partial}{\partial t}\{\rho(\xi, t)\sqrt{|\gamma(\xi, t)|}\};$$

$$(4) \quad -\frac{\partial}{\partial t}\rho(x, t) = \frac{1}{\sqrt{|g|}}\frac{\partial}{\partial x^k}(\rho v^k\sqrt{|g|}\,)_{x,t} = \mathrm{Div}(\rho \boldsymbol{v});$$

$$(5) \quad \frac{D}{Dt}\rho(\mathrm{Q}, t) = -\rho\, \mathrm{Div}\, \boldsymbol{v}.$$

Equations (4) and (5) are usually called equations of continuity. Equation (3) follows at once from (2) and 5.5(4), which give

$$(6) \qquad \mu = \rho\sqrt{|\gamma|}\,.$$

Equation (4) follows from 5.4(27) applied to the relative scalar $F = \rho\sqrt{|g|}$ (weight $n = 1$): since $\mu \xrightarrow{t} \rho\sqrt{|g|}$, from (6), we have $\partial\mu(\xi)/\partial t \xrightarrow{t} \mathring{F} = 0$, from (3); the first of equations (4) follows immediately, and the second from 8.1(8). Equation (5) follows at once from (4) and 8.2(23) applied to the absolute scalar ρ.

Constancy of volume (or the incompressibility condition) is expressed by the equation

$$\text{(7)} \qquad \operatorname{Div} \boldsymbol{v} = 0,$$

which follows from (5) with $D\rho/Dt = 0$.

The stress equations of motion in general space tensors, or in Cartesian tensors expressed in a form applicable to an arbitrary space coordinate system, can be derived by considering the forces acting on an infinitesimally small element of material, usually chosen to be of a particular shape. We shall use an alternative method which involves integration over an arbitrary region of material, and so it is convenient to use Cartesian tensor fields in the derivation; once the required equations are obtained in the form of differential equations, it is easy to write down the corresponding equations for general space (or body) tensor fields.

We consider an arbitrary three-dimensional region $\mathscr{P}_1$ of the moving body, occupying a region $\mathscr{Q}_1$ of space at any given time t. Application of Newton's second law of motion to the material $\mathscr{P}_1$ yields the equation

$$\text{(8)} \qquad \int_{\partial\mathscr{Q}_1} \mathbf{p}\cdot\mathbf{n}\, dA + \int_{\mathscr{Q}_1} \mathbf{X}\rho\, dV = \frac{d}{dt}\int_{\mathscr{P}_1} \mathbf{v}\, dm = \int_{\mathscr{P}_1} \frac{D\mathbf{v}}{Dt}\, dm,$$

where the first integral comes from the contact forces exerted by material across the boundary $\partial\mathscr{Q}_1$, the second comes from external forces having a resultant $\mathbf{X}$ per unit mass, and the third integral is the rate of change of momentum of the material $\mathscr{P}_1$. Here, $\mathbf{p}$, $\mathbf{n}$, $\mathbf{X}$, and $\mathbf{v}$ are Cartesian tensors, $\mathbf{v}$ being the velocity of the element dm. The right-hand expression for the rate of change of momentum can be obtained immediately by recognizing that the integral can be regarded as a body integral either with dm as an integration variable or [using (2)] with $d\xi^1\, d\xi^2\, d\xi^3$ as integration variable; the range of integration is then independent of t; the operator d/dt can be taken inside the integral, where it becomes the hydrodynamic derivative D/Dt, since in this context $\mathbf{v}$ must be expressed in terms of ξ^i and t (as independent variables). Assuming there are no singularities in the region of integration, we may apply Green's theorem to convert the surface integral into a volume integral. (If necessary, a rectangular Cartesian coordinate system can be used at this stage; the operator $\boldsymbol{\nabla}$ defined in Section 8.2 then has components d/dx^i, which occur in the familiar form of Green's theorem.) Finally, writing $dm = \rho\, dV$, and using the fact that the region of volume integration is arbitrary, it follows that the integrand must vanish, i.e., that

$$\nabla \cdot \tilde{\mathbf{p}} + \rho \mathbf{X} = \rho \mathbf{a} \qquad \text{(Cartesian tensors),} \tag{9}$$

where $\mathbf{a} = D\mathbf{v}/Dt$ is the Cartesian acceleration vector. We have not used the symmetry of the stress tensor in this derivation. Equation (9) is one form of the required stress equations of motion. Another form can be written down immediately:

$$\nabla \cdot \tilde{\boldsymbol{p}} + \rho \boldsymbol{X} = \rho \boldsymbol{a} \qquad \text{(general tensors);} \tag{10}$$

here, $\boldsymbol{p}$ is the second-rank contravariant stress tensor, $\boldsymbol{X}$ is a contravariant vector (corresponding to the Cartesian vector $\mathbf{X}$), and $\boldsymbol{a}$ is the contravariant acceleration vector; interpreting $\nabla \cdot \boldsymbol{p}$ as the covariant derivative of $\boldsymbol{p}$, contracted once, it follows that $\nabla \cdot \boldsymbol{p}$ is a contravariant vector. Thus $\nabla \cdot \tilde{\boldsymbol{p}} + \rho(\boldsymbol{X} - \boldsymbol{a})$ is a contravariant vector whose representative matrix in any rectangular Cartesian coordinate system is zero, by (9), and is therefore zero, which proves (10).

We can also immediately write down the corresponding stress equations of motion in terms of body tensors, by simply transferring (10) to the body at time t, and noting that the covariant derivative operator ∇ is a one-state operator. We thus obtain the equation

$$\nabla_t \cdot \boldsymbol{\pi} + \rho \boldsymbol{\Xi} = \rho \boldsymbol{\alpha} \qquad \text{(body tensors),} \tag{11}$$

where

$$\nabla_t \equiv \mathbb{T}^{-1}(t)\nabla, \qquad \boldsymbol{\Xi} = \mathbb{T}^{-1}(t)\boldsymbol{X}, \qquad \boldsymbol{\alpha} = \mathbb{T}^{-1}(t)\boldsymbol{a}. \tag{12}$$

As noted in Section 8.3, the covariant derivative operator ∇_t (for body tensor fields) depends on time, because it involves Christoffel symbols Γ^i_{jk} constructed from components $\gamma_{ij}(\xi, t)$ of the body metric tensor, which depends on time.

The stress equations of motion may easily be expressed in various alternative forms, some of which are as follows:

$$p^{ki}{}_{,k} = \frac{1}{\sqrt{|g|}} \frac{\partial}{\partial x^k}\left(p^{ki}\sqrt{|g|}\right) + G^i_{jk}\, p^{jk} = \rho(a^i - X^i) \qquad (\text{any } S); \tag{13}$$

$$-\frac{\partial p}{\partial x^i} + P^k{}_{i,k} = \rho(a_i - X_i) \qquad (\text{any } S); \tag{14}$$

$$(15)\quad \widehat{p^{ki}_{,k}} = \frac{1}{h_1h_2h_3}\frac{\partial}{\partial x^k}\left(\frac{h_1h_2h_3}{h_k}\widehat{p^{ki}}\right) + \frac{1}{h_{\hat{i}}h_j}\left(\widehat{p^{\hat{i}j}}\frac{\partial h_{\hat{i}}}{\partial x^j} - \widehat{p^{jj}}\frac{\partial h_j}{\partial x^{\hat{i}}}\right)$$

$$= \rho(\widehat{a^i} - \widehat{X^i}) \qquad \text{(orthogonal } S\text{, physical components);}$$

$$(16)\quad -\frac{\partial p}{\partial x^i} + \frac{h_{\hat{i}}}{h_1h_2h_3}\frac{\partial}{\partial x^k}\left(\frac{h_1h_2h_3}{h_k}\widehat{P^{ki}}\right) + \frac{1}{h_k}\left(\widehat{P^{\hat{i}k}}\frac{\partial h_{\hat{i}}}{\partial x^k} - \widehat{P^{kk}}\frac{\partial h_k}{\partial x^{\hat{i}}}\right)$$

$$= \rho h_{\hat{i}}(\widehat{a^{\hat{i}}} - \widehat{X^{\hat{i}}}) \qquad \text{(orthogonal } S\text{, physical components).}$$

Similar equations can be written down for the corresponding body tensor components. Equation (13) comes from (10) on using 8.2(24)(iv) and 8.2(18). Equation (14) comes from (13) on using the definition

$$(17)\qquad \boldsymbol{P} = \boldsymbol{p} + p\boldsymbol{g}^{-1}, \qquad P^k{}_i = g_{ij}P^{kj}, \qquad \text{etc.,}$$

of the space extra stress tensor $\boldsymbol{P}$, together with 8.2(24)(i), (ix). Equation (15) follows from (13) on using the equations

$$(18)\qquad \widehat{p^{ki}} = h_{\hat{k}}h_{\hat{i}}\,p^{\hat{k}\hat{i}}, \qquad \widehat{a^i} = h_{\hat{i}}\,a^{\hat{i}}, \qquad \widehat{X^i} = h_{\hat{i}}\,X^{\hat{i}}, \qquad \sqrt{|g|} = h_1h_2h_3,$$

appropriate to an orthogonal coordinate system S with $g = \mathrm{diag}[{h_1}^2, {h_2}^2, {h_3}^2]$; these follow from the space analog of 2.7(18) and 2.7(21). Equation (16) follows from (15) on using the equation

$$(19)\qquad \widehat{p^{ki}_{,k}} = \widehat{P^{ki}_{,k}} - \widehat{\frac{\partial p}{\partial x^i}} = \widehat{P^{ki}_{,k}} - \frac{1}{h_{\hat{i}}}\frac{\partial p}{\partial x^{\hat{i}}}.$$

In Appendices A and B, Eq. (15) is written out for cylindrical and spherical polar coordinate systems. Other forms for the stress equations are given in Section 8.10.

We now consider the rate at which work is being done by the external tractions acting across the boundary $\partial\mathcal{Q}_1$ of material instantaneously contained within an arbitrary three-dimensional region $\mathcal{Q}_1$ [as in the context of (8)]. Using Cartesian vectors and tensors, the required rate of working is equal to

$$\int_{\partial Q_1} \mathbf{v}\cdot\mathbf{p}\cdot\mathbf{n}\,dA = \int_{Q_1} \nabla\cdot(\tilde{\mathbf{p}}\cdot\mathbf{v})\,dV = \int\{(\nabla\cdot\tilde{\mathbf{p}})\cdot\mathbf{v} + \tfrac{1}{2}\mathbf{p}:\mathbf{A}^1\}\,dV$$

$$= \int\{(\mathbf{a}-\mathbf{X})\cdot\mathbf{v}\rho + \tfrac{1}{2}\mathbf{p}:\mathbf{A}^1\}\,dV.$$

We have here used Green's theorem, 5.4(60), and (9). But

$$\int \rho\mathbf{a}\cdot\mathbf{v}\,dV = \int(D\mathbf{v}/Dt)\cdot\mathbf{v}\,dm = \tfrac{1}{2}\int(D\mathbf{v}^2/Dt)\,dm = \tfrac{1}{2}(d/dt)\int\mathbf{v}^2\,dm,$$

which is equal to the rate of change of kinetic energy of the material in $\mathcal{Q}_1$. Here $\int \mathbf{X} \cdot \mathbf{v} \, dV$ is the rate at which external body forces do work on this material. Since the region of integration is arbitrary and every term in the equation is now in the form of a volume integral, it follows that the integrands must be equal, i.e., that

$$\dot{W} = (1/2\rho)\mathbf{p} : \mathbf{A}^1, \tag{20}$$

where $\dot{W}$ = (rate of working of external surface and body forces on unit mass of material) − (rate of increase of kinetic energy per unit mass). It follows at once that, in general space tensors and in body tensors, we have

$$\dot{W} = (1/2\rho)\boldsymbol{p} : \boldsymbol{A}^1 = (1/2\rho)\boldsymbol{\pi} : \dot{\boldsymbol{\gamma}}. \tag{21}$$

For an arbitrary infinitesimal strain $\boldsymbol{\gamma} \to \boldsymbol{\gamma} + \boldsymbol{d\gamma}$, it follows from (21) and the equations $\dot{\boldsymbol{\gamma}}\, dt = \boldsymbol{d\gamma}$ and $\dot{W}\, dt = dW$ that

$$dW = (1/2\rho)\boldsymbol{\pi} : \boldsymbol{d\gamma}. \tag{22}$$

8.6 Covariant differentiation in an affinely connected manifold

Although the definition of covariant differentiation given in Section 8.2 for a Euclidean manifold is sufficient for our present needs (i.e., for rheological applications in the nonrelativistic approximation), it is worthwhile presenting an alternative definition (Weyl, 1922) which is applicable whether the manifold is Euclidean or not and which is equivalent to the previous definition when the manifold is Euclidean. In addition to the extra generality, the present definition affords added insight into the process of covariant differentiation. In this section, we define covariant differentiation using an affine connection for a three-dimensional (space) manifold *without using the metric tensor field*; in the next section, we add an extra postulate in order to relate the affine connection to the metric tensor field, assuming that the space manifold is Riemannian. It will be apparent that the results are applicable also to the body manifold (since general tensor fields are used) and are immediately extendable to manifolds of arbitrary finite dimension.

We have seen that $\{\boldsymbol{v}(\mathrm{Q})\}$ (the set of all contravariant vectors at Q) and $\{\boldsymbol{v}(\mathrm{Q}')\}$ (the set of all contravariant vectors at Q′) form *disjoint* linear spaces when Q and Q′ denote two distinct places. We cannot give meaning to the statement that a vector at Q is equal to a vector at Q′. We can, however, add an extra property to the manifold which enables us to set up a *one-to-one correspondence between vectors at* Q *and vectors at* Q′, when Q and Q′ are any two *neighboring* places; the manifold is then said to be *linearly*, or *affinely, connected*, if the mapping is linear in both members and in the displacement vector $\boldsymbol{dx} = \overrightarrow{\mathrm{QQ}'}$. This seems to be the next best thing to having a vector at Q equal to a vector at Q′. The fact that we start with contravariant vectors is immaterial, for an affine relation between sets of contravariant vectors induces a similar relation between sets of covariant vectors, and also between sets of higher-rank tensors of any given type. Vectors or tensors related under an affine connection are also said to be obtained from one another by *parallel transfer* (the word "transfer" is misleading, because the vector stays where it is, but is conventional and need cause no trouble). By successive application to arbitrarily small displacements, the process of parallel transfer can obviously be used along any given curve. By transferring a vector round a closed curve, we are led to define the fourth-rank Riemann–Christoffel curvature tensor, whose vanishing is a necessary and sufficient condition that the original vector be recovered and that the manifold, if Riemannian, be Euclidean. To the extent that the following arguments involve infinitesimal triangles and quadrilaterals, they are perhaps heuristic, but readily visualizable; rigorous proofs can be found in many textbooks [e.g., Synge and Schild (1949)].

(1) **Definition** A manifold $\mathcal{Q} \equiv \{\mathrm{Q}\}$ is *affinely connected* if for every two neighboring points Q and Q′, there exists a mapping $\boldsymbol{G}(\mathrm{Q})\colon \{\boldsymbol{v}(\mathrm{Q})\} \to \{\boldsymbol{v}(\mathrm{Q}')\}$ which is one-to-one, linear in $\boldsymbol{v}(\mathrm{Q})$, $\boldsymbol{v}(\mathrm{Q}')$, and $\boldsymbol{dx} = \overrightarrow{\mathrm{QQ}'}$, and reduces to the identity mapping when $\boldsymbol{dx} = \boldsymbol{0}$. Here, $\{\boldsymbol{v}(\mathrm{Q})\}$ denotes the set of all absolute contravariant vectors at Q. We shall write $\boldsymbol{G}(\mathrm{Q})\colon \boldsymbol{v}(\mathrm{Q}) \to \boldsymbol{v}(\mathrm{Q}|\mathrm{Q}')$, so that $\boldsymbol{v}(\mathrm{Q}|\mathrm{Q}')$ denotes the vector (at Q′) that corresponds to a given vector $\boldsymbol{v}(\mathrm{Q})$ (at Q), or, equivalently, $\boldsymbol{v}(\mathrm{Q}) \xrightarrow{\boldsymbol{G}(\mathrm{Q})} \boldsymbol{v}(\mathrm{Q}|\mathrm{Q}')$.

Referred to an arbitrary space coordinate system $S\colon \mathrm{Q} \to x$, it follows that the equations for the correspondence must be expressible in the form

$$[v^r(x)] \xrightarrow{G(Q)} \boldsymbol{v}(\mathrm{Q}|\mathrm{Q}')S = [v^r(x) - G^r_{ij}(x)\, dx^j\, v^i(x)], \tag{2}$$

where $\boldsymbol{v}(\mathrm{Q})S = [v^r(x)]$ and $G^r_{ij}(x)$ denote the "components" of the affine connection $\boldsymbol{G}(\mathrm{Q})$. The $G^r_{ij}(x)$ can depend on x^k but not on dx^k or on v^r.

(3) **Problem** Prove that (2) follows from (1).

(4) **Problem** Prove that G^m_{qp} and $\bar{G}^i_{rs}$, the components of $\boldsymbol{G}(\mathrm{Q})$ in any two coordinate systems $S\colon \mathrm{Q} \to x$ and $\bar{S}\colon \mathrm{Q} \to \bar{x}$, satisfy the equations

$$\frac{\partial^2 \bar{x}^i}{\partial x^q\, \partial x^p} - \frac{\partial \bar{x}^i}{\partial x^m} G^m_{qp} = -\frac{\partial \bar{x}^r}{\partial x^q}\frac{\partial \bar{x}^s}{\partial x^p} \bar{G}^i_{rs}.$$

Deduce that $\bar{G}^i_{rs} = \bar{G}^i_{sr}$ if $G^m_{pq} = G^m_{qp}$; in this case, $\boldsymbol{G}$(Q) is said to be *symmetric*. It is clear from this transformation law that $\boldsymbol{G}$(Q) *is not a tensor.*

(5) **Problem** Let Q, Q′, and Q″ be any three noncollinear neighboring points. Points Q_1 and Q_2 are defined by transfer of the contravariant vectors $\overrightarrow{QQ'}$ and $\overrightarrow{QQ''}$:

$$\overrightarrow{QQ'} \xrightarrow[Q\to Q'']{G(Q)} \overrightarrow{Q''Q_1}; \qquad \overrightarrow{QQ''} \xrightarrow[Q\to Q']{G(Q)} \overrightarrow{Q'Q}$$

[see Fig. 8.6(6)]. Prove that a necessary and sufficient condition that $Q_1 = Q_2$ for any given Q and arbitrary neighboring Q′ and Q″ is that $\boldsymbol{G}$(Q) be symmetric.

(6) Figure 8.6(6).

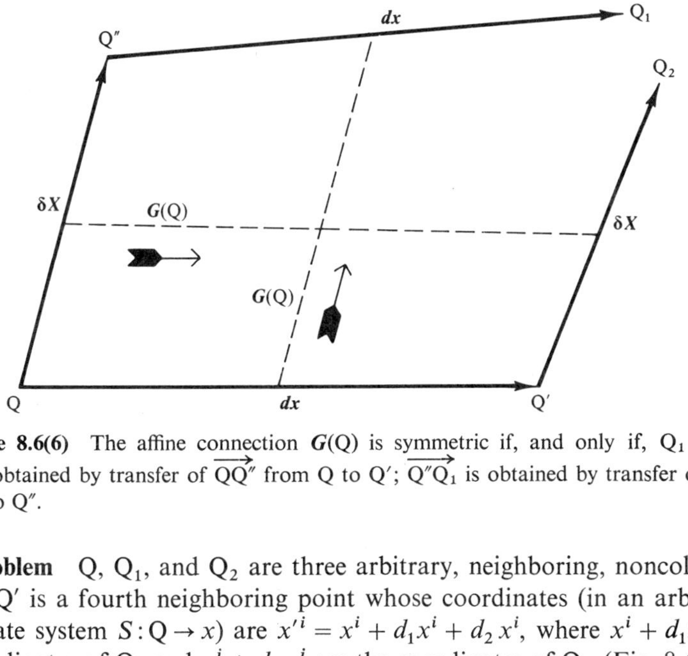

Figure 8.6(6) The affine connection $\boldsymbol{G}$(Q) is symmetric if, and only if, $Q_1 = Q_2$. $\overrightarrow{Q'Q_2}$ is obtained by transfer of $\overrightarrow{QQ''}$ from Q to Q′; $\overrightarrow{Q''Q_1}$ is obtained by transfer of $\overrightarrow{QQ'}$ from Q to Q″.

(7) **Problem** Q, Q_1, and Q_2 are three arbitrary, neighboring, noncollinear points. Q′ is a fourth neighboring point whose coordinates (in an arbitrary coordinate system $S: Q \to x$) are $x'^i = x^i + d_1x^i + d_2x^i$, where $x^i + d_1x^i$ are the coordinates of Q_1 and $x^i + d_2x^i$ are the coordinates of Q_2 (Fig. 8.6(8)).

(8) See Fig. 8.6(8).

An arbitrary contravariant vector $\boldsymbol{v}$(Q) is transferred via Q_1 to Q′, yielding V_1(Q′), and via Q_2 to Q′, yielding V_2(Q′). Prove that a necessary and sufficient

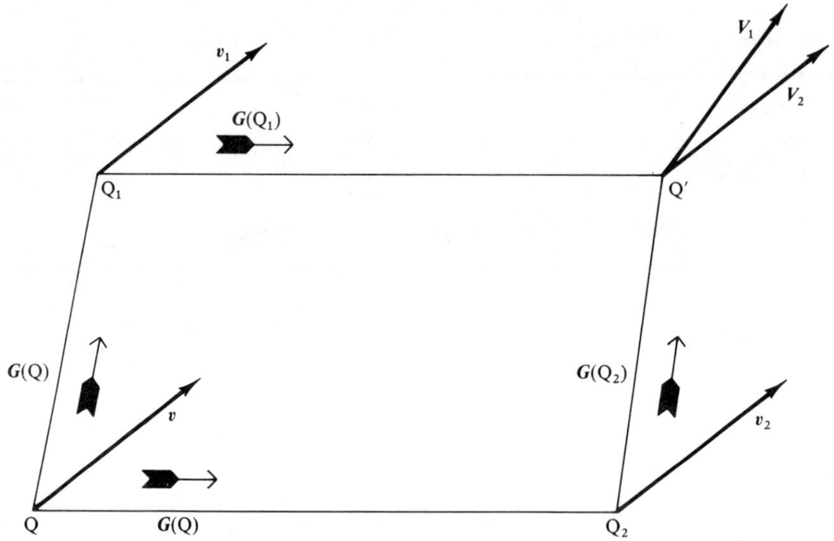

Figure 8.6(8) Definition of the Riemann–Christoffel curvature tensor $\boldsymbol{R}$. A vector $\boldsymbol{v}(Q)$ is transferred from Q to Q′ by two routes: via Q_1, yielding V_1, and via Q_2, yielding V_2. $V_1 - V_2 = = \boldsymbol{v} \cdot \boldsymbol{R} : (\boldsymbol{d}_1\boldsymbol{x}\, \boldsymbol{d}_2\, \boldsymbol{x})$, where $\boldsymbol{d}_1\boldsymbol{x} = \overrightarrow{QQ_1}$ and $\boldsymbol{d}_2\,\boldsymbol{x} = \overrightarrow{QQ_2}$. $V_1 = V_2$ if and only if $\boldsymbol{R} = 0$.

condition that $V_1(Q') = V_2(Q')$ is that the Riemann–Christoffel fourth-rank tensor $\boldsymbol{R}$ be zero, where

$$\text{(9)} \qquad \boldsymbol{R}(Q)S = [R^i_{jsk}], \qquad R^i_{jsk} = G^i_{rs}\, G^r_{jk} - G^i_{rk}\, G^r_{js} + \frac{\partial G^i_{jk}}{\partial x^s} - \frac{\partial G^i_{js}}{\partial x^k}.$$

Show further that

$$\text{(10)} \qquad R^i_{jsk} + R^i_{jks} = 0;$$

$$\text{(11)} \qquad R^i_{jsk} + R^i_{skj} + R^i_{jks} = 0 \qquad \text{when} \quad G^i_{jk} = G^i_{kj}.$$

An alternative way of expressing (7) is to say that if and only if $\boldsymbol{R}$ is zero does the transfer of an arbitrary contravariant vector round an arbitrary quadrilateral yield the original vector. By approximating an arbitrary closed curve by the edges of an arbitrarily fine mesh of quadrilaterals, it is easy to see that the result can be extended *for transfer round an arbitrary closed curve in the manifold: One obtains the original vector if and only if* $\boldsymbol{R}$ *vanishes throughout the manifold.* Such a manifold is called *flat*. It follows that on transferring a contravariant vector from any given point to any other, the resulting vector is independent of the path. In other words, a given contravariant vector $\boldsymbol{v}(Q)$ at any point Q in a flat space gives rise to a vector field over the whole space: The vector $\boldsymbol{v}(\mathrm{Q}')$ at any other point Q′ is obtained by parallel transfer along any path and is unique. We shall see that such a field is *homogeneous* in the sense that its covariant derivative (formed using the affine connection $\boldsymbol{G}$ to be described later) is everywhere zero.

The covariant derivative of a contravariant vector field can now be defined as before by 8.2(5), provided that we interpret $\boldsymbol{v}(\mathrm{Q}' \mid \mathrm{Q})$ as the vector at Q obtained by parallel transfer (using $\boldsymbol{G}$) from Q′:

(12) **Definition** The *covariant derivative* $\nabla \boldsymbol{v}$ of an absolute contravariant vector field $\boldsymbol{v}(\mathrm{Q})$ is defined by the equation

$$\boldsymbol{dx} \cdot \nabla \boldsymbol{v}(\mathrm{Q}) = \boldsymbol{v}(\mathrm{Q}' \mid \mathrm{Q}) - \boldsymbol{v}(\mathrm{Q}) \qquad [\boldsymbol{dx} = \overrightarrow{\mathrm{QQ}'}; \quad \boldsymbol{v}(\mathrm{Q}') \xrightarrow{\boldsymbol{G}(\mathrm{Q}')} \boldsymbol{v}(\mathrm{Q}' \mid \mathrm{Q})],$$

in which Q′ is an arbitrary point in the neighborhood of Q.

The right-hand side is a vector at Q; $\boldsymbol{dx}$ is an arbitrary vector at Q; from a quotient theorem, it follows (as in Section 8.2) that $\nabla \boldsymbol{v}$ is a left-covariant mixed tensor. In an arbitrary coordinate system $S : \mathrm{Q} \to x$, this equation takes the component form

$$dx^k \, v^i{}_{,k} = v^i(x') + G^i_{rk}(x') \, dx^k \, v^r(x') - v^i(x),$$

where $v^i{}_{,k}$ denotes the components of $\nabla \boldsymbol{v}$ as in 8.2(15). We have used (2) with Q and Q′ interchanged. In the term in dx^k on the right-hand side, we can replace x' by x in the arguments of G^i_{rk} and v^r, because the error involved is of order $(dx)^2$ and is unimportant in the present context. [In (7), on the other hand, the variation of $\boldsymbol{G}$ with position is essential.] Since dx^k is arbitrary, it then follows that $v^i{}_{,k}$ is given by 8.2(13) with G^i_{rk} representing the components of the affine connection $\boldsymbol{G}(\mathrm{Q})$.

The preceding definitions of affine connection and covariant differentiation have been given for absolute contravariant vectors. It is simplest to define the corresponding operations for absolute covariant vectors in such a way that *absolute scalars are unchanged by parallel transfer*. By applying this principle to the absolute scalar formed from the inner product $\boldsymbol{u} \cdot \boldsymbol{v}$ of a covariant vector $\boldsymbol{u}$ and a contravariant vector $\boldsymbol{v}$, so that

(13) $$\boldsymbol{u}(\mathrm{Q}) \cdot \boldsymbol{v}(\mathrm{Q}) = \boldsymbol{u}(\mathrm{Q} \mid \mathrm{Q}') \cdot \boldsymbol{v}(\mathrm{Q} \mid \mathrm{Q}'),$$

it is easy to show that

(14) $$\boldsymbol{u}(\mathrm{Q} \mid \mathrm{Q}')S = [u_r(x) + G^i_{rj}(x) \, dx^j \, u_i(x)] \qquad (\boldsymbol{dx} = \overrightarrow{\mathrm{QQ}'}),$$

to the first order in dx^j. Further, defining the covariant derivative $\nabla \boldsymbol{u}$ by replacing $\boldsymbol{v}$ by $\boldsymbol{u}$ in (12), it is easy to show that $\nabla \boldsymbol{u}$ has components $u_{i,k}$ given by 8.2(24)(iii). The other results in Table 8.2(24) can be derived readily.

(15) **Problem** Using (4), show that it is possible to choose a coordinate

system $\bar{S}$ in such a way that $\bar{G}^i_{rs}(\bar{x}) = 0$ at any one given point (of coordinates $\bar{x}^i$) if and only if $\boldsymbol{G}$ is symmetric.

(16) **Problem** Prove that the vector field obtained by parallel transfer from a given contravariant vector $\boldsymbol{v}(\mathrm{Q})$ at a point Q in a flat space is homogeneous [as defined in 8.2(28)].

(17) **Problem** Prove that any set of linearly independent vectors at Q yields a set of vectors at Q′, obtained by parallel transfer along a given curve, which are themselves linearly independent.

(18) **Problem** In a space with a symmetric affine connection, prove that every *homogeneous* absolute covariant vector field $\boldsymbol{B}(\mathrm{Q})$ satisfies the equation Curl $\boldsymbol{B} = \boldsymbol{0}$ [Curl is defined in 8.1(5)].

8.7 The affine connection for a Riemannian manifold

The affine connection has been defined, and covariant differentiation developed, without using the metric tensor field $\boldsymbol{g}(\mathrm{Q})$ in Section 8.6. We now consider the case of greatest interest: an affinely connected Riemannian manifold. The stipulation made in Section 8.6 that parallel transfer should leave all absolute scalars unchanged in value means, in particular, that the magnitudes of all vectors shall be unchanged in value. This leads to a relation between the affine connection $\boldsymbol{G}$ and the metric tensor $\boldsymbol{g}$, which can be completed if we also restrict $\boldsymbol{G}$ to be symmetric, as the following result shows.

(1) **Problem** It is given that in a Riemannian manifold of metric tensor field $\boldsymbol{g}(\mathrm{Q})$ and affine connection $\boldsymbol{G}(\mathrm{Q})$, the magnitude of every absolute *contravariant* vector $\boldsymbol{v}(\mathrm{Q})$ is unchanged for every parallel transfer (i.e., $\|\boldsymbol{v}(\mathrm{Q})\| = \|\boldsymbol{v}(\mathrm{Q}|\mathrm{Q}')\|$). Show that the angle between every pair of absolute contravariant vectors at Q is unchanged on parallel transfer; and that, in any $S:\mathrm{Q} \to x$,

(2) $$\partial g_{ij}/\partial x^k = G_{ijk} + G_{jik}, \qquad \text{where} \quad G_{ijk} = g_{ir}G^r_{jk}, \quad G^i_{jk} = g^{ir}G_{rjk}.$$

If, further, $\boldsymbol{G}$ is symmetric, show that it is completely determined by $\boldsymbol{g}(\mathrm{Q})$ according to the equations

(3) $$2G_{kij} = \partial g_{ki}/\partial x^j + \partial g_{kj}/\partial x^i - \partial g_{ij}/\partial x^k;$$

that the magnitude of every absolute *covariant* vector is unchanged on parallel transfer; and that the angle between any two absolute covariant vectors is unchanged on parallel transfer.

It is evident that (2) and (3) are the same as 8.2(16) and 8.2(17). This completes the rederivation of the equations (for general vectors and tensors) of Section 8.2 by the present method, which is evidently applicable whether the Riemannian manifold be Euclidean or not. From now on, *we shall consider a manifold which is Riemannian with a symmetric affine connection satisfying the conditions (and therefore the equations) of Problem (1).* We summarize the main results related to the conditions for a Riemannian manifold to be Euclidean [see 2.7(15) for definitions]. The idea underlying the following ordering of results is to put down conditions which enable one to construct a Cartesian coordinate system (throughout the manifold) by a process involving parallel transfer, to all points of the manifold, of a given set of linearly independent base vectors $\boldsymbol{B}^i(\mathrm{Q})$ at an arbitrary point Q. The base vector fields so constructed are to be the base vectors of a Cartesian coordinate system.

(4) **Problem** Show that a necessary and sufficient condition that a coordinate system $S:\mathrm{Q} \to x$ shall exist having as normal base vectors a given set $\boldsymbol{B}^i(\mathrm{Q})$ of three linearly independent absolute covariant vector fields is that $\text{Curl } \boldsymbol{B}^i = \boldsymbol{0}$ $(i = 1, 2, 3)$.

(5) **Problem** Show that a necessary and sufficient condition that a Cartesian coordinate system shall exist for a given space is that the space be flat (i.e., that $\boldsymbol{R} = \boldsymbol{0}$ everywhere) [use 8.6(17), 8.6(18), and (2)–(4)].

(6) **Problem** The *coefficients of rotation* γ^{nhk} for a set of three linearly independent, absolute, covariant vector fields $\boldsymbol{B}^i(\mathrm{Q})$ are absolute scalars defined by the equation $\gamma^{nhk} = B_{i,j}^{(n)} B_r^{(h)} B_s^{(k)} g^{ir} g^{js}$, where $[B_r^{(n)}] = \boldsymbol{B}^n S$. [See, e.g., Eisenhart (1926, p. 97, Eq. (30.1)), for the case in which $\boldsymbol{B}^i$ are orthogonal.] Prove that (i) $\gamma^{nhk} - \gamma^{nkh} = \boldsymbol{B}_k \cdot (\text{Curl } \boldsymbol{B}^n) \cdot \boldsymbol{B}_h$ (where $\boldsymbol{B}_k = \boldsymbol{g}^{-1} \cdot \boldsymbol{B}^k$), and that (ii) $\gamma^{nhk} = \gamma^{nkh}$ is a necessary and sufficient condition that the field $\boldsymbol{B}^n(\mathrm{Q})$ be *normal*, i.e., that there shall exist a one-parameter family of *surfaces* everywhere normal to $\boldsymbol{B}^n$, assuming that $\boldsymbol{G}$ is symmetric.

8.8 Compatibility conditions

We have seen that in any given body coordinate system $B:\mathrm{P} \to \xi$, the body metric tensor field $\boldsymbol{\gamma}(\mathrm{P}, t)$ is represented by a symmetric, positive-definite matrix $[\gamma_{rc}(\xi, t)]$ whose elements are functions of variables ξ^1, ξ^2, ξ^3, and t. We now consider whether any such matrix can be the representative of a body metric tensor field in some state t. The essential question concerns the dependence of the elements on $\xi \equiv \xi^1, \xi^2, \xi^3$ at fixed t.

The body manifold is Euclidean; at any instant t, there exists therefore a body coordinate system $\bar{B}:\mathrm{P} \to \bar{\xi}$ which is rectangular Cartesian, and so

$\boldsymbol{\gamma}(\mathrm{P}, t)\bar{B} = I$. The transformation law 2.6(10) takes the form $\gamma = \tilde{\Lambda}\Lambda$, i.e.,

(1) $$\gamma_{ij}(\xi, t) = \frac{\partial \bar{\xi}^k}{\partial \xi^i}\frac{\partial \bar{\xi}^k}{\partial \xi^j} \qquad (\bar{B} \text{ rectangular Cartesian at } t).$$

Because of the symmetry of the matrix $[\gamma_{rc}]$, not more than six of its elements can be independent at any one particle. If the matrix is to represent a metric tensor field, Eq. (1) shows that not more than three of its elements can be independent functions of ξ, for all the elements must be expressible in terms of three functions $\bar{\xi}^k(\xi^1, \xi^2, \xi^3)$ $(k = 1, 2, 3)$. Elimination of these functions between (1) and partial derivatives of (1) yields a set of equations, called *compatibility conditions*, which given functions $\gamma_{ij}(\xi, t)$ must satisfy if they are to represent a metric tensor field for a Euclidean manifold. A form of these conditions is given in the following theorems.

(2) **Theorem** The Riemann–Christoffel tensor $\boldsymbol{R}[\boldsymbol{\gamma}(\mathrm{P}, t)]$ formed from the body metric tensor $\boldsymbol{\gamma}(\mathrm{P}, t)$ in any state t is zero.

Proof. By hypothesis, the space manifold is Euclidean with a symmetric affine connection $\boldsymbol{G}[\boldsymbol{g}(\mathrm{Q})]$ given by 8.7(2) and 8.7(3). The operations involved in the definitions of the affine connection and the Riemann–Christoffel tensor $\boldsymbol{R}[\boldsymbol{g}(\mathrm{Q})]$ are evidently all one-state operations which, in any given state t, induce corresponding quantities in the body:

(3) $$\boldsymbol{G}[\boldsymbol{g}(\mathrm{Q})] \xrightarrow{t} \boldsymbol{\Gamma}[\boldsymbol{\gamma}(\mathrm{P}, t)], \qquad \boldsymbol{\Gamma} \text{ symmetric};$$

(4) $$\boldsymbol{R}[\boldsymbol{g}(\mathrm{Q})] \xrightarrow{t} \boldsymbol{R}[\boldsymbol{\gamma}(\mathrm{P}, t)] = \mathbb{T}^{-1}(t)\boldsymbol{R}[\boldsymbol{g}(\mathrm{Q})].$$

Since a zero space tensor induces a zero body tensor under the correspondence $\mathbb{T}(t)$, and $\boldsymbol{R}[\boldsymbol{g}(\mathrm{Q})] = \boldsymbol{0}$, it follows that $\boldsymbol{R}[\boldsymbol{\gamma}(\mathrm{P}, t)] = \boldsymbol{0}$, as stated. (We use $\boldsymbol{R}[\boldsymbol{\gamma}]$ for the body tensor and $\boldsymbol{R}[\boldsymbol{g}]$ for the space tensor, in this case.)

(5) **Theorem** Given any state t, with body metric tensor field $\boldsymbol{\gamma}(\mathrm{P}, t)$, and any body coordinate system $B : \mathrm{P} \to \xi$, there exists a continuous deformation $t \to t_1$ of the body such that B is rectangular Cartesian in the new state t_1.

Proof Choose any space coordinate system $S : \mathrm{Q} \to x$. In any state t_1, attainable from t by continuous deformation of the body, let P be at Q_1 of coordinates x_1^i. Since $t \to t_1$ is continuous, there exist three equations $x_1^i = f^i(\xi, t_1)$ involving sufficiently differentiable functions f^i such that the transformation $x_1 \to \xi$ is one-to-one. From 2.2(13) with $t = t_1$, we have

(6) $$\gamma_{ij}(\xi, t_1) = g_{rs}(x_1)\frac{\partial x_1^r}{\partial \xi^i}\frac{\partial x_1^s}{\partial \xi^j}.$$

We have $R[g(x_1)] = \boldsymbol{R}[\boldsymbol{g}(\mathrm{Q}_1)]S = 0$; from 8.7(5), it follows that functions

f^i of the stated type exist such that (4) is satisfied with $\gamma_{ij}(\xi, t_1) = \delta_{ij}$, i.e., such that B is rectangular Cartesian in state t_1, which proves the theorem.

(7) **Theorem** In order that a given positive-definite (symmetric), covariant, second-rank body tensor field $\boldsymbol{\psi}(\mathrm{P})$ shall be the body metric tensor field in some state t_1, it is necessary and sufficient that $\boldsymbol{R}(\boldsymbol{\psi}) = \mathbf{0}$.

Proof. The necessity follows at once from (2). Since $\boldsymbol{R}(\boldsymbol{\psi}) = \mathbf{0}$, it follows from 8.7(5) that given any $B: \mathrm{P} \to \xi$, there exists a one-to-one, sufficiently differentiable transformation $\xi \to \bar{\xi}$ such that $\psi_{ij}\, d\xi^i\, d\xi^j = d\bar{\xi}^r\, d\bar{\xi}^r$, where $[\psi_{rc}] = \boldsymbol{\psi}B$. Hence $\mathrm{P} \to \bar{\xi}$ is a coordinate system, $\bar{B}$ say, and $\boldsymbol{\psi}\bar{B} = I$. From (5), there exists a state t_1 in which $\bar{B}$ is rectangular Cartesian, i.e., $\boldsymbol{\gamma}(\mathrm{P}, t_1)\bar{B} = I = \boldsymbol{\psi}(\mathrm{P})\bar{B}$. Thus $\boldsymbol{\gamma}(\mathrm{P}, t_1)$ and $\boldsymbol{\psi}(\mathrm{P})$ are tensors of the same type at P that have equal representative matrices in one coordinate system, $\bar{B}$; hence $\boldsymbol{\psi}(\mathrm{P}) = \boldsymbol{\gamma}(\mathrm{P}, t_1)$, which proves the theorem.

It follows from (3), (4), 8.7(2), and 8.7(3) that the Riemann–Christoffel tensor $\boldsymbol{R}[\boldsymbol{\gamma}(\mathrm{P}, t)]$ and the affine connection $\boldsymbol{\Gamma}[\boldsymbol{\gamma}(\mathrm{P}, t)]$ for the body manifold can be expressed as follows, in terms of the base vectors $\boldsymbol{\beta}_i(B, \mathrm{P})$ and $\boldsymbol{\beta}^i(B, \mathrm{P})$ for an arbitrary body coordinate system $B: \mathrm{P} \to \xi$:

(8) $$\boldsymbol{R}[\boldsymbol{\gamma}(\mathrm{P}, t)] = R^i_{jsk}\boldsymbol{\beta}_i\boldsymbol{\beta}^j\boldsymbol{\beta}^s\boldsymbol{\beta}^k;$$

(9) $$R^i_{jsk} = \Gamma^i_{rs}\Gamma^r_{jk} - \Gamma^i_{rk}\Gamma^r_{js} + \frac{\partial \Gamma^i_{jk}}{\partial \xi^s} - \frac{\partial \Gamma^i_{js}}{\partial \xi^k};$$

(10) $$\Gamma^i_{jk} = \gamma^{ir}\Gamma_{rjk}; \qquad \Gamma_{kij} = \frac{1}{2}\left(\frac{\partial \gamma_{ki}}{\partial \xi^j} + \frac{\partial \gamma_{kj}}{\partial \xi^i} - \frac{\partial \gamma_{ij}}{\partial \xi^k}\right).$$

The equation $\boldsymbol{R}(\boldsymbol{\gamma}) = \mathbf{0}$ which a body metric tensor has to satisfy is thus represented in any B by a set of *nonlinear* partial differential equations of second order, in which the γ_{ij} are the dependent variables and the ξ^i are the independent variables; t is a constant. It follows that, in general, *a linear combination of body metric tensor fields for states t, t', ... is not a body metric tensor field for any state attainable by continuous deformation from states t, t', ...* An exception occurs when all the states t, t', ... are related by *homogeneous* deformations [defined in 11.2(1)].

(11) **Problem** If the deformation $t' \to t$ is homogeneous for $t_0 \le t' < t$ and $\boldsymbol{\theta}(\mathrm{P}) \equiv \int_{t_0}^{t} K(t, t')\boldsymbol{\gamma}^{-1}(\mathrm{P}, t')\, dt'$, where $K(t, t')$ is a nonnegative, absolute scalar function of t and t', prove that $\boldsymbol{\theta}^{-1}(\mathrm{P}) = \boldsymbol{\gamma}(\mathrm{P}, t^*)$, the body metric tensor field for some state t^* attainable from t by continuous (and homogeneous) deformation. (A similar result can be proved wth covariant tensors in place of contravariant tensors.)

These results are important in connection with the phenomenon known as *die swell*, jet swell, or extrusion expansion (Lodge, 1958): Many polymeric liquids exhibit a very large increase in diameter on flowing out of a tube of circular cross section, and a very large increase in thickness on flowing out of a slit. Diameter increases of as much as 200% are readily attainable with suitable liquids flowing under conditions of substantially constant volume. The effect has been widely investigated and is probably due, at least in part, to elastic recovery effects arising from (i) memory of converging flow at the tube entry, and (ii) lateral expansion (Lodge, 1958; 1964, p. 131) following shear flow in the slit. As a first approximation, it is reasonable to assume that as soon as liquid leaves the tube, the stress becomes isotropic (or zero) at the free liquid boundary. Considering the rubberlike liquid 6.2(2) (with $\eta_s = 0$), for purposes of illustration, let us suppose that we have some given flow throughout $(-\infty, t)$ and that from t onward, $\boldsymbol{\pi}(\mathrm{P}, t') = \mathbf{0}$; assuming that the memory function μ is continuous, it follows at once on putting $\boldsymbol{\pi} = \mathbf{0}$ in the constitutive equation that there is an instantaneous change of shape $\boldsymbol{\gamma}(\mathrm{P}, t) \to \boldsymbol{\gamma}^*(\mathrm{P})$, say, where the body metric tensor $\boldsymbol{\gamma}^*(\mathrm{P})$ for the "instantaneously recoverable state" is given by the equations

$$(12) \qquad p\boldsymbol{\gamma}^*(\mathrm{P})^{-1} = \int_{-\infty}^{t} \mu(t-t')\boldsymbol{\gamma}^{-1}(\mathrm{P}, t')\, dt', \qquad |\boldsymbol{\gamma}^* \cdot \boldsymbol{\gamma}^{-1}| = 1.$$

The second equation comes from the constant-volume condition and, with the first, serves also to determine the value of the scalar p, which turns out to be positive [Lodge (1964, Eq. (7.15)) with $p = \mu_0^*$]. Equation (12) thus furnishes an example of (11). We may conclude that since the flow history for flow through a tube or slit involves *inhomogeneous* deformations, the conditions leading to the derivation of Eq. (12) cannot apply over the whole cross section at the exit plane of the tube or slit because *the tensor field* $\boldsymbol{\gamma}^*(\mathrm{P})$ *describing the instantaneously recoverable state for a rubberlike liquid does not satisfy the compatibility condition* $\boldsymbol{R}(\boldsymbol{\gamma}^*) = \mathbf{0}$. Thus even if the stress becomes instantaneously zero (or isotropic) for liquid elements at the free boundary "immediately" after the exit plane, *the stress cannot be zero or isotropic over the whole exit plane*; in particular, therefore, it seems unreasonable to assume (as some have assumed) that the stress on the tube centerline at the exit plane is that due to atmospheric pressure. These features are otherwise evident if one considers that in the liquid outside the exit plane, liquid elements at different distances from the axis of the tube, having undergone very different flow histories, will be "trying" to execute elastic recovery in varying amounts, and therefore, by acting against each other, can be expected to generate nonisotropic stress distributions which may persist for some distance downstream of the exit plane. They may well also lead to "anticipatory" flow patterns upstream of the exit plane; the common assumption that "fully developed" shear flow persists up to the exit plane may therefore also be seriously in error.

8.9 Boundary conditions

(1) The contravariant space traction vector $\boldsymbol{Y}$ acting on a surface of covariant unit normal $\boldsymbol{n}$ satisfies the equation $\boldsymbol{Y} = \boldsymbol{p} \cdot \boldsymbol{n}$, where $\boldsymbol{p}$ is the contravariant space stress tensor. From this equation, together with 2.7(6), 2.7(21), and 5.4(31), it is easy to derive the forms of the boundary conditions shown in Table 8.9(1) when the traction is specified on a material surface whose equation is of the form $\sigma(\xi) = c$ referred to an arbitrary body coordinate system $B : \mathrm{P} \to \xi$.

Table 8.9(1) Boundary Conditions Expressed in Terms of Contravariant Body Stress Tensor Components π^{ij} Referred to an Arbitrary Body Coordinate System $B : \mathrm{P} \to \xi$

	Boundary condition[a]	Surface	Traction
(i)	$Y^k = \dfrac{\partial x^k}{\partial \xi^i} \dfrac{\pi^{ij}\sigma_j}{(\gamma^{rs}\sigma_r\sigma_s)^{1/2}}$	$\sigma(\xi) = c$	$\boldsymbol{Y} = Y^k \boldsymbol{b}_k(S, \mathrm{Q})$, arbitrary S
(ii)	$\widehat{Y^k} = \left(\dfrac{g_{\hat{k}\hat{k}}}{\gamma^{11}}\right)^{1/2} \dfrac{\partial x^k}{\partial \xi^i} \pi^{i1}$	$\xi^1 = c$	Physical components $\widehat{Y^k}$ in an orthogonal $S : \mathrm{Q} \to x$
(iii)	$(\pi^{ij} + p_a \gamma^{ij})\sigma_j = 0$	$\sigma(\xi) = c$	$\boldsymbol{Y} = -p_a \boldsymbol{n}$
(iv)	$\pi^{i1} = -p_a \gamma^{i1}$	$\xi^1 = c$	$\boldsymbol{Y} = -p_a \boldsymbol{n}$

[a] $\sigma_i = \partial\sigma/\partial\xi^i$. $\partial x^k/\partial\xi^i = \partial f^k(\xi, t)/\partial\xi^i$. $\boldsymbol{n}$ is the covariant space unit normal to the surface $\sigma(\xi) = c$.

When the motion of a boundary, or part of a boundary, is given, then the functions $f^i(\xi, t)$ are specified functions of t for those values of ξ satisfying the equation $\sigma(\xi) = c$. The f^i are the functions in the equations $x^i = f^i(\xi, t)$ [2.2(4)], which describe the motion of the body referred to arbitrary body and space coordinate systems B and S. If the velocity is specified, then $\partial f^i(\xi, t)/\partial t$ are given functions of t on the boundary.

8.10 Simultaneous equations for isothermal flow problems

To obtain a solution to a given flow problem, one can, in principle, solve the simultaneous system of integrodifferential equations composed of the given constitutive equations for the material of interest and the universal stress equations of motion, subject to boundary conditions and initial conditions appropriate to the given flow problem. For all but a few problems, the equations are formidable. The equations can be expressed in many different (but equivalent) ways. Here we shall merely collect two sets of equations, for purposes of illustration, to show, in particular, that the number of independent equations is equal to the number of dependent variables. For extra simplicity, we consider incompressible materials, which at present constitute the class of greatest interest in polymer rheology.

I $B : \mathrm{P} \to \xi$ Rectangular Cartesian at t; Arbitrary $S : Q \to x$

(1) 16 *dependent variables*: $\pi^{ij}\ (= \pi^{ji})$, $\gamma'_{ij}(= \gamma'_{ji})$, x'^i, p.

(2) 4 *independent variables*: ξ^i, t.

(3) 16 *simultaneous equations*:

(i) $\pi^{ij} + p\delta^{ij} = \mathscr{F}^{ij}\{\gamma'_{rs}, t, t'\}$ ($\mathscr{F}^{ij}$ are given functionals);

(ii) $\gamma'_{ij} = \dfrac{\partial x'^r}{\partial \xi^i}\dfrac{\partial x'^s}{\partial \xi^j} g_{rs}(x')$ (g_{rs} are given functions);

(iii) $\partial|\gamma'|/\partial t' = 0$.

(iv) $$\frac{\partial \pi^{sr}}{\partial \xi^s} = \rho\left(\frac{\partial^2 x^i}{\partial t^2} + G^i_{jk}\frac{\partial x^j}{\partial t}\frac{\partial x^k}{\partial t} - X^i\right) g_{iu}(x)\frac{\partial x^u}{\partial \xi^r}.$$

(4) *Conditions on a boundary* $\sigma(\xi) = c$

(i) x^i or $\partial x^i/\partial t$ are given functions of t and ξ^i; or

(ii) $\pi^{ij}\sigma_j(\sigma_r\sigma_r)^{-1/2}\ \partial x^k/\partial \xi^i = Y^k$ (given functions of t and ξ^i).

(5) *Condition at* t: $\gamma_{ij} = \delta_{ij}$.

II Arbitrary $B : \mathrm{P} \to \xi$; Rectangular Cartesian $C : \mathrm{Q} \to y$

(6) 16 *dependent variables*: $\pi^{ij}\ (= \pi^{ji})$, $\gamma'_{ij}\ (= \gamma'_{ji})$, y'^i, p.

(7) 4 *independent variables*: ξ^i, t.

(8) 16 *simultaneous equations*:

(i) $\pi^{ij} + p\gamma^{ij} = \mathscr{F}^{ij}\{\gamma'_{rs}, t, t'\}$;

(ii) $\gamma'_{ij} = \dfrac{\partial y'^r}{\partial \xi^i}\dfrac{\partial y'^r}{\partial \xi^j}$;

(iii) $\dfrac{\partial|\gamma'|}{\partial t'} = 0$;

(iv) $$\frac{\partial y^k}{\partial \xi^i}\left\{|\gamma|^{-1/2}\frac{\partial}{\partial \xi^s}(|\gamma|^{1/2}\pi^{si}) + \Gamma^i_{rs}\pi^{rs}\right\} = \rho\left(\frac{\partial^2 y^k}{\partial t^2} - X^k\right).$$

(9) *Boundary conditions on* $\xi^1 = c$

(i) y^i, or $\partial y^i/\partial t$, are given functions of t and ξ^i, or

(ii) $$\pi^{i1}\frac{\partial y^k}{\partial \xi^i}(\gamma^{11})^{-1/2} = Y^k \text{ (given functions of } t \text{ and } \xi^i).$$

For convenience, we repeat the following notations:

(10) $[\gamma^{rc}] = [\gamma_{rc}]^{-1}$; so $\gamma^{ij}\gamma_{jk} = \delta^i_k$; γ_{ij} and γ'_{ij} denote values at t and t'.

(11) $|\gamma| = \det \gamma_{rc}$.

(12) $$G^i_{jk} = g^{ir}\frac{1}{2}\left\{\frac{\partial g_{rj}}{\partial x^k} + \frac{\partial g_{rk}}{\partial x^j} - \frac{\partial g_{jk}}{\partial x^r}\right\}.$$

(13) $$\Gamma^i_{rs} = \gamma^{ik}\frac{1}{2}\left\{\frac{\partial \gamma_{kr}}{\partial \xi^s} + \frac{\partial \gamma_{ks}}{\partial \xi^r} - \frac{\partial \gamma_{rs}}{\partial \xi^k}\right\}.$$

(14) Y^k = components of surface traction referred to S (or C).

(15) X^k = components of external body force per unit mass, referred to S (or C); e.g., referred to C with the y^1 axis vertically upward, $X^k = -g\delta^k_1$, where g here denotes the acceleration due to gravity.

(16) ρ = density, a constant.

9 STRESS MEASUREMENTS IN UNIDIRECTIONAL SHEAR FLOW; THEORY

9.1 Stress equations of motion for unidirectional shear flow

In Section 7.1, we saw that for any incompressible, viscoelastic material describable by an isotropic, one-particle constitutive functional, the rheological behavior in an arbitrary unidirectional shear flow is governed by three functionals (N_1, N_2, and σ) of shear rate $\dot{s}$. One of the most important current tasks is to develop reliable experimental methods for determining these functionals for given polymeric liquids. For this purpose, rectilinear shear flow, though apparently the simplest unidirectional shear flow, is of no use: It cannot be generated in the laboratory, and even if it could, one could measure the traction on the shearing surfaces only; this would give σ, but not N_1 and N_2, for which two other normal components of traction must be determined.

Weissenberg (1947) and Russell (1946) were the first to note that *curvilinear shear flow could be used to determine N_1 and N_2 from measured values of spatial distributions of the component of traction normal to shearing surfaces.* The curvature of the shear surfaces and shear lines in the presence of nonzero N_1 and/or N_2 can give rise to a spatial variation of an additive isotropic contribution to the stress tensor field which shows up as a nonuniform spatial distribution of the normal component of traction on shearing surfaces, even when the flow history is essentially the same at all particles (as in shear flow between a cone and a touching plate). For this to be true, the principal directions of curvature must be suitably oriented; in the case of flow through a tube of circular cross section, they are not, and the axial gradient of normal traction on the wall is related to σ instead of to N_1 or N_2.

(1) The validity of these statements can be checked using the equations of Table 9.1(1), which gives the stress equations of motion in simplified form for the types of apparatus which have been used most in the determination of N_1 and N_2.

Methods of using the simplified stress equations of motion listed in Table 9.1(1) are illustrated in some detail in Sections 9.3 and 9.4. It is perhaps illuminating to note that the common principle of these methods of determining N_1 and N_2 can be made clear by examining the number 3 stress equation of motion, when expressed in the following form:

(2)

$$(\gamma_{33})^{-1/2}\frac{\partial \widehat{\pi^{33}}}{\partial \xi^3} - \frac{N_2}{\rho_{31}} - \frac{N_1 + N_2}{\rho_{32}} = \widehat{\pi^{j3}}_{,j} = \rho(\alpha^3 - \Xi^3) \qquad \begin{pmatrix} B \text{ orthogonal;} \\ \pi^{13} = \pi^{23} = 0 \end{pmatrix}.$$

This equation is referred to a shear-flow body coordinate system B that [for each of the flows in Table 9.1(1)] is orthogonal instantaneously at time t, and has the ξ^2 surfaces as shear surfaces and the ξ^1 curves as shear lines at all times. ρ_{31} and ρ_{32} denote the principal radii of curvature for a surface $\xi^3 =$ const, associated with those principal directions of curvature tangential, respectively, to the ξ^2 curve and the ξ^1 curve [cf. 8.4(4) and 8.4(5)]. Equation (2) is readily obtained from the body tensor version of 8.5(15) for the case $i = 3$, on putting $\pi^{13} = \pi^{23} = 0$, which is justified since the flow is a unidirectional shear flow. Equation (2) shows how the variation of the physical component of stress $\widehat{\pi^{33}}$ along a ξ^3 curve is governed by N_1, N_2, and the principal radii of curvature; α^3 and Ξ^3 here denote physical components of acceleration and body force and are usually negligible. Since one usually measures, not $\widehat{\pi^{33}}$, but $\widehat{\pi^{22}}$ (the physical component of traction normal to a shear surface $\xi^2 =$ const), it is more convenient to rewrite (2) in the following form:

(3)
$$\frac{\partial \widehat{\pi^{22}}}{\partial \xi^3} = \frac{\partial N_2}{\partial \xi^3} + (\gamma_{33})^{1/2}\left(\frac{N_1 + N_2}{\rho_{32}} + \frac{N_2}{\rho_{31}}\right) + \rho(\alpha^3 - \Xi^3),$$

which relates N_1 and N_2 to the pressure gradient $-\partial \widehat{\pi^{22}}/\partial \xi^3$.

Table 9.1(1) Simplified Forms of the Stress Equations of Motion in Some Unidirectional Shear Flows.[a]

Type of flow and coordinate system	Stress equations of motion	Stress components	Shear rate and velocity components
Rectilinear flow through tube of wide-slit cross section; rectilinear Cartesian coordinates: Ox along tube axis, Oy normal to wider wall. $(\xi^1, \xi^2, \xi^3) \overset{t}{\equiv} (x, y, z)$	$\frac{\partial \widehat{xx}}{\partial x} + \frac{\partial \sigma}{\partial y} = \rho\left(\frac{\partial v_x}{\partial t} + v_x \frac{\partial v_x}{\partial x} - X_x\right)$ $\frac{\partial \sigma}{\partial x} + \frac{\partial \widehat{yy}}{\partial y} = -\rho X_y$	$N_1 = \widehat{xx} - \widehat{yy}$ $N_2 = \widehat{yy} - \widehat{zz}$ $\sigma = \widehat{xy} = \widehat{yx}$ $0 = \widehat{zx} = \widehat{zy}$	$\dot{s} = \pm \frac{\partial v_x(y, t)}{\partial y}$ $v_y = v_z = 0$ $\frac{\partial}{\partial z}() = X_z = 0$
Rectilinear flow through tube or annulus of circular cross section. Cylindrical polars: Oz along tube axis. $(\xi^1, \xi^2, \xi^3) \overset{t}{\equiv} (z, r, \phi)$	$\frac{\partial \widehat{rr}}{\partial r} + \frac{N_2}{r} + \frac{\partial \sigma}{\partial z} = -\rho X_r$ $\frac{1}{r}\frac{\partial (r\sigma)}{\partial r} + \frac{\partial \widehat{zz}}{\partial z} = \rho\left(\frac{\partial v_z}{\partial t} + v_z \frac{\partial v_z}{\partial z} - X_z\right)$	$N_1 = \widehat{zz} - \widehat{rr}$ $N_2 = \widehat{rr} - \widehat{\phi\phi}$ $\sigma = \widehat{zr} = \widehat{rz}$ $0 = \widehat{\phi r} = \widehat{\phi z}$	$\dot{s} = \pm \frac{\partial v_z(r, t)}{\partial r}$ $v_r = v_\phi = 0$ $\frac{\partial}{\partial \phi}() = X_\phi = 0$
Couette flow between concentric circular cylinders in relative rotation. Cylindrical polar coordinates with Oz as cylinder axis. $(\xi^1, \xi^2, \xi^3) \overset{t}{\equiv} (\phi, r, z)$	$\frac{\partial \widehat{rr}}{\partial r} - \frac{N_1}{r} = -\rho(r\Omega^2 + X_r)$ $\frac{1}{r^2}\frac{\partial (r^2\sigma)}{\partial r} = \rho r\dot{\Omega}$ $\frac{\partial \widehat{zz}}{\partial z} = -\rho X_z$	$N_1 = \widehat{\phi\phi} - \widehat{rr}$ $N_2 = \widehat{rr} - \widehat{zz}$ $\sigma = \widehat{\phi r} = \widehat{r\phi}$ $0 = \widehat{z\phi} = \widehat{zr}$	$\dot{s} = \pm \frac{r\, \partial \Omega(r, t)}{\partial r}$ $v_\phi = r\Omega(r, t)$ $v_r = v_z = 0$ $\frac{\partial}{\partial \phi}() = X_\phi = 0$

Torsional flow between parallel circular plates in relative rotation about axis Oz of a cylindrical polar coordinate system. $(\xi^1, \xi^2, \xi^3) \overset{t}{\equiv} (\phi, z, r)$	$\dfrac{\partial \widehat{rr}}{\partial r} - \dfrac{N_1 + N_2}{r} = -\rho(r\Omega^2 + X_r)$ $\dfrac{\partial \sigma}{\partial z} = \rho r \dot{\Omega}$ $\dfrac{\partial \widehat{zz}}{\partial z} = -\rho X_z$	$N_1 = \widehat{\phi\phi} - \widehat{zz}$ $N_2 = \widehat{zz} - \widehat{rr}$ $\sigma = \widehat{\phi z} = \widehat{z\phi}$ $0 = \widehat{r\phi} = \widehat{rz}$	$\dot{s} = \pm \dfrac{r\,\partial\Omega(z,t)}{\partial z}$ $v_\phi = r\Omega(z,t)$ $v_r = v_z = 0$ $\dfrac{\partial}{\partial \phi}(\,) = X_\phi = 0$ $\dfrac{\partial^2 \Omega}{\partial z^2} = 0$
Torsional flow between a cone and touching plate in relative rotation about a common axis $\theta = 0$ of a spherical polar coordinate system. $(\xi^1, \xi^2, \xi^3) \overset{t}{\equiv} (\phi, \theta, r)$	$\dfrac{\partial \widehat{rr}}{\partial r} - \dfrac{N_1 + 2N_2}{r} = -\rho(r\Omega^2 \sin^2\theta + X_r)$ $\dfrac{1}{r}\dfrac{\partial \widehat{\theta\theta}}{\partial \theta} - \dfrac{N_1}{r}\cot\theta = -\rho\left(\dfrac{\Omega^2}{2r}\sin 2\theta + X_\theta\right)$ $\dfrac{\partial}{\partial \theta}(\sigma \sin^2\theta) = \rho r^2 \sin^3\theta\, \dot{\Omega}$	$N_1 = \widehat{\phi\phi} - \widehat{\theta\theta}$ $N_2 = \widehat{\theta\theta} - \widehat{rr}$ $\sigma = \widehat{\phi\theta} - \widehat{\theta\phi}$ $0 = \widehat{r\phi} = \widehat{r\theta}$	$\dot{s} = \pm \dfrac{\sin\theta\,\partial\Omega(\sigma,t)}{\partial \theta}$ $v_\phi = r\Omega(\theta,t)$ $v_r = v_\theta = 0$ $\dfrac{\partial}{\partial \phi}(\,) = X_\phi = 0$

[a] $\widehat{rr}$, $\widehat{r\phi}$, etc. denote physical components of stress for the appropriate orthogonal coordinate systems; $\widehat{rr} > 0$ for a tensile component of traction. Inertial terms are included for completeness, but give rise to secondary flows which may be incompatible with the unidirectional shear flow results $\pi^{13} = \pi^{23} = 0$ used in the simplifications. The flows admit a body coordinate system $B: \mathrm{P} \to \xi$ (with ξ^2 surfaces as shear surfaces and ξ^1 curves as shear lines) orthogonal at time t. X_r, X_z, etc. denote components of the acceleration due to gravity. Shear stress σ and normal stress differences N_1 and N_2 are functionals of shear rate $\dot{s}$. v_r, v_θ, etc. denote physical components of velocity.

9.2 The importance of N_1 and N_2

During the past thirty years, increasing efforts have been devoted to the study of N_1 and N_2, the primary and secondary normal stress differences generated by polymeric liquids in unidirectional shear flow and by polymeric solids, in the rubberlike state, in shear. The following points furnish some justification for these efforts.

(1) N_1 can be large [e.g., $N_1 = 5\sigma = 5 \times 10^5$ dyn cm^{-2} for Melt 1 at 1 sec^{-1}, Fig. 10.1(4)(iii)].

(2) N_1 is associated with finite strains and "elastic properties."

(3) N_1 and N_2 are new fundamental quantities which can be measured in an accurately known state of flow; in suitable conditions (Lipson and Lodge, 1968), $N_1 + 2N_2$ can be measured to an accuracy of about 1%.

(4) Measured values of N_1 and N_2 can be used to test the applicability of proposed constitutive equations.

(5) The sign of N_2 may influence the stability of processes used to coat wire with plastic (Jones, 1964, 1965; Tadmor and Bird, 1974).

(6) Values of N_1 and N_2 determine the direction in which a liquid flows through a "conical pump" (Maxwell and Scalora, 1959).

(7) Values of N_1, N_2, and σ give important correlations with quantities measured in flow birefringence (Janeschitz-Kriegl, 1969; Wales and Philippoff, 1973). These correlations lay the foundations for the stress analysis of flowing polymeric liquids from flow birefringence measurements by techniques which may be regarded as the liquid analog of the widely used photoelastic techniques for stress analysis in solids.

(8) N_1 is more sensitive than η to changes in molecular structure of polymeric liquids.

(9) A striking example of (8) is shown in Fig. 9.2(9); an aqueous solution of polyethylene oxide was pumped round a closed loop; samples were periodically removed, and the values of N_1 and η measured, using a Weissenberg rheogoniometer (Higashitani, 1973). During an hour of pumping, N_1 decreased by a factor of 5.3, while η decreased by a factor of only 1.3. It is likely that the high shear rates, near the teeth of the gear pump used, caused some breakdown of associations between different polymer molecules and possibly also some breakage of the longer molecules.

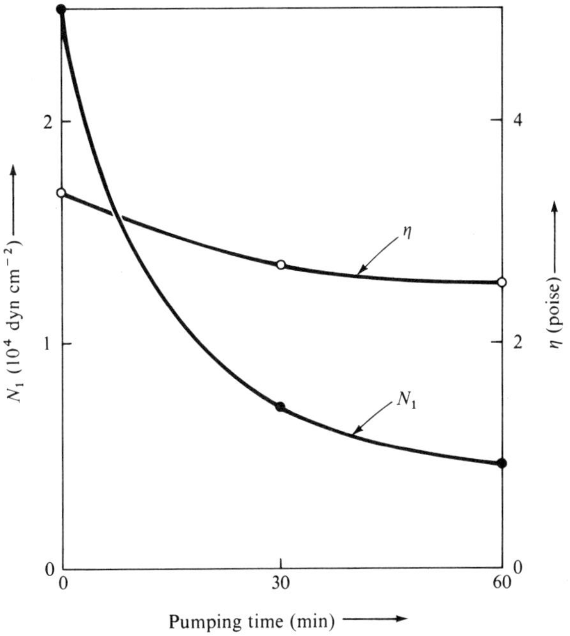

Figure 9.2(9) Demonstration that N_1 is more sensitive than η to changes in molecular structure of a polymer solution. A gear pump is used to pump 300 ml at 14.5 ml sec^{-1} of a 2% solution of polyethylene oxide (Polyox WSR 301 from Union Carbide Corp.) in water around a closed loop. A Weissenberg rheogoniometer is used to measure N_1 and η on samples withdrawn after different times of pumping (Higashitani, 1973).

9.3 Torsional flow, parallel plates

We consider an apparatus consisting of two parallel, rigid, circular plates, each of radius R, in relative rotation about a common vertical axis fixed in space. We use a cylindrical polar space coordinate system $S: \mathrm{Q} \to x$ with the axis $r = 0$ along the axis of rotation, and the plates located at $z = 0$ and $z = z_1$; the rim of either plate is $r = R$. In order to conform to our use of indexes for unidirectional shear flow, we take

$$(x^1, x^2, x^3) = (\phi, z, r). \tag{1}$$

In order to obtain a convenient form for the stress equations of motion at an arbitrary but definite time t, we shall choose a body coordinate system $B \overset{t}{\equiv} S$.

As a trial solution for the flow of an incompressible viscoelastic liquid contained between the plates and adhering to them, we shall take

$$\dot{\phi} = \Omega(z, t), \qquad \dot{z} = \dot{r} = 0, \tag{2}$$

where $\Omega(z, t)$ is at our disposal, and the dots, as usual, denote time derivatives at constant particle. This solution can evidently be made to satisfy the boundary conditions

$$\text{(3)} \qquad \dot{\phi} = 0 \quad \text{when} \quad z = 0; \qquad \dot{\phi} = \Omega_1(t) \quad \text{when} \quad z = z_1.$$

For this flow, Eqs. 2.2(4) evidently take the form

$$\text{(4)} \qquad \phi' = f^1(\xi, t') = \xi^1 + \int_t^{t'} \Omega(\xi^2, t'')\, dt'',$$

$$z' = f^2(\xi, t') = \xi^2, \qquad r' = f^3(\xi, t') = \xi^3.$$

For any two neighboring particles, it follows that

$$\text{(5)} \quad (ds')^2 = (r'\, d\phi')^2 + (dz')^2 + (dr')^2 = r^2(d\xi^1 + w\, d\xi^2)^2 + (d\xi^2)^2 + (d\xi^3)^2,$$

where

$$\text{(6)} \qquad w \equiv \int_t^{t'} \frac{\partial \Omega(\xi^2, t'')}{\partial \xi^2}\, dt'', \qquad \dot{w} \equiv \frac{\partial w}{\partial t'} = \frac{\partial \Omega(z, t')}{\partial z}.$$

Comparing (5) with 2.2(12), we see that

$$\text{(7)} \quad [\gamma_{rc}(\xi, t')] = \begin{bmatrix} r^2 & r^2 w & 0 \\ r^2 w & 1 + r^2 w^2 & 0 \\ 0 & 0 & 1 \end{bmatrix}, \qquad [\dot{\gamma}_{rc}(\xi, t')] = \begin{bmatrix} 0 & r^2 \dot{w} & 0 \\ r^2 \dot{w} & 2r^2 w \dot{w} & 0 \\ 0 & 0 & 0 \end{bmatrix}$$

Hence $|\gamma| = r^2$, and the conditions 3.2(4) and 3.2(6) for a shear flow are satisfied. Moreover, 3.2(5) is satisfied, so that B is a shear-flow coordinate system with $\xi^2 = c$ as a shearing surface. From 3.2(15), since $\dot{\gamma}_{32} = 0$, it follows that $\zeta = 0$, and so the shear flow is unidirectional with the ξ^1 curves as shear lines. From 3.2(13), we have (taking the positive sign)

$$\text{(8)} \qquad \dot{s}(\text{P}, t') = r\, \partial \Omega(z, t')/\partial z.$$

From 7.1(5), it follows that for any isotropic, incompressible viscoelastic material described by a one-particle constitutive functional, the physical components of stress at time t referred to the orthogonal shear-flow coordinate system S ($\overset{t}{\equiv} B$) can be written in the form

$$\text{(9)} \qquad [\widehat{p^{ij}}] = -pI + \begin{bmatrix} N_1 + N_2 & \sigma & 0 \\ \sigma & N_2 & 0 \\ 0 & 0 & 0 \end{bmatrix} = [\widehat{\pi^{ij}}],$$

where N_1, N_2, and σ are functionals of $\dot{s}$ alone. The stress equations of motion are given in Appendix A. With $v^i = (\Omega, 0, 0)$, $X_r = X_\phi = 0$, $X_z = -g$ (the acceleration due to gravity), the equations reduce to the following, when we set $\partial(\,)/\partial\phi = 0$ as a trial solution rendered plausible by the cylindrical symmetry of the problem:

$$\partial\sigma/\partial z = \rho r\dot{\Omega} \qquad (\phi \text{ equation}); \tag{10}$$

$$\partial(N_2 - p)/\partial z = \rho g \qquad (z \text{ equation}); \tag{11}$$

$$\partial p/\partial r + (N_1 + N_2)/r = \rho r\Omega^2 \qquad (r \text{ equation}). \tag{12}$$

For a given material, these are three equations for two dependent variables, p and Ω.

We shall consider the solution in the case in which *the inertial terms* (i.e., those containing Ω) *are negligible*. Then (10) is satisfied if $\dot{s}$ is independent of z, i.e., from (8), if Ω is linear in z:

$$\Omega(z, t) = z\Omega_1(t)/z_1. \tag{13}$$

Then N_2 is independent of z, and hence (11) has the solution $p = -\rho gz + p_0(r, t)$, for some function $p_0(r, t)$ which is given by the integral of (12), namely,

$$p_0(r, t) = p_0(R, t) + \int_R^r (N_1 + N_2)\frac{dr'}{r'}. \tag{14}$$

Thus the differential equations are compatible and possess a solution. The component of traction normal to the lower plate, $\widehat{zz}$, is given by the equation

$$\widehat{zz}(r, t) = N_2 - p = N_2\left(\dot{s}\left(r, \underset{-\infty}{\overset{t}{t'}}\right)\right) - p_0(R, t) + \int_R^r (N_1 + N_2)\bigg|_{r'}\frac{dr'}{r'}, \tag{15}$$

while the tangential component, $\widehat{z\phi}$, is given by the equation

$$\widehat{z\phi}(r, t) = \sigma\left(\dot{s}\left(r, \underset{-\infty}{\overset{t}{t'}}\right)\right), \qquad \dot{s} = r\Omega_1(t')/z_1. \tag{16}$$

There is thus a nonuniform radial distribution of normal and tangential traction over the lower (and therefore also over the upper) plate; this distribution depends on the material through the functionals N_1, N_2, and σ. The boundary conditions at the plates having been satisfied, it remains only to consider the conditions at the free liquid boundary, which will be taken to be cylindrical (i.e., with the equation $r = R$) when gravity is neglected. The tractions on this boundary are $\widehat{r\phi} = \widehat{rz} = 0$ and $\widehat{rr} = -p$, from (9), and can therefore be matched by atmospheric pressure $-p_a$ alone, so that

$$p(R, t) = p_0(R, t) = p_a \qquad \text{(gravity neglected)}. \tag{17}$$

It then follows from (15) that

(18) $$\widehat{zz}(R, t) = -p_a + N_2\left(\dot{s}\left(R, \overset{t}{\underset{-\infty}{t'}}\right)\right),$$

showing that *the value of normal traction at the rim is determined by the secondary normal stress difference*, whatever the form of the constitutive equation.

9.4 Torsional flow, cone and plate

We consider an apparatus consisting of a circular horizontal plate of radius R and a wide-angled cone rotating at angular velocity $\Omega_1(t)$ about its axis, which is vertical; the cone apex touches the center of the plate. We use a spherical polar coordinate system $S : \mathrm{Q} \to x$ with origin $r = 0$ at the center of the plate and axis $\theta = 0$ along the cone axis. We consider the flow of an incompressible viscoelastic liquid filling the gap between cone and plate and adhering to both surfaces. We choose the order of coordinates as follows:

(1) $$(x^1, x^2, x^3) = (\phi, \theta, r),$$

and we shall choose a body coordinate system $B : \mathrm{P} \to \xi$ that is congruent to S at an arbitrary but definite time t: $B \overset{t}{\equiv} S$.

As a trial solution for the motion, we shall take

(2) $$\dot{\phi} = \Omega(\theta, t), \qquad \dot{\theta} = \dot{r} = 0,$$

where the angular velocity function $\Omega(\theta, t)$ is at our disposal and can be chosen to satisfy the boundary conditions

(3) $$\dot{\phi}(\tfrac{1}{2}\pi, t) = 0, \qquad \dot{\phi}(\theta_1, t) = \Omega_1(t)$$

at the plate and cone surfaces. The gap angle $\Delta\theta \equiv \frac{1}{2}\pi - \theta_1$ in practice is usually small (about 1° to 8°). For this flow, Eqs. 2.2(4) evidently take the form

(4) $$\phi' = f^1(\xi, t') = \xi^1 + \int_t^{t'} \Omega(\xi^2, t'')\, dt'',$$

$$\theta' = f^2(\xi, t') = \xi^2, \qquad r' = f^3(\xi, t') = \xi^3.$$

For any two neighboring particles, it follows that

(5) $$\begin{aligned}(ds')^2 &= (r' \sin\theta'\, d\phi')^2 + (r'\, d\theta')^2 + (dr')^2 \\ &= (r \sin\theta)^2\, (d\xi^1 + w\, d\xi^2)^2 + (r\, d\xi^2)^2 + (d\xi^3)^2,\end{aligned}$$

where

$$(6) \qquad w \equiv \int_t^{t'} \frac{\partial\Omega(\xi^2, t'')}{\partial\xi^2}\, dt'', \qquad \dot{w} \equiv \frac{\partial w}{\partial t'} = \frac{\partial\Omega(\theta, t')}{\partial\theta}.$$

Comparing (5) with 2.2(12), we see that

$$(7) \qquad [\gamma_{rc}(\xi, t')] = \begin{bmatrix} r^2 \sin^2\theta & r^2 w \sin^2\theta & 0 \\ r^2 w \sin^2\theta & r^2(1 + w^2 \sin^2\theta) & 0 \\ 0 & 0 & 1 \end{bmatrix}, \qquad |\gamma| = r^4 \sin^2\theta,$$

and hence

$$(8) \qquad [\dot{\gamma}_{rc}(\xi, t')] = \begin{bmatrix} 0 & r^2 \dot{w} \sin^2\theta & 0 \\ r^2 \dot{w} \sin^2\theta & 2r^2 w\dot{w} \sin^2\theta & 0 \\ 0 & 0 & 0 \end{bmatrix}.$$

On comparing these results with 3.2(4)–(6), it follows that the flow is a shear flow with the ξ^2 surfaces (or θ surfaces) as shearing surfaces. From 3.2(15), since $\dot{\gamma}_{32} = 0$, it follows that $\zeta = 0$, so that the shear flow is unidirectional, with the ξ^1 curves as shear lines. From 3.2(13), taking the positive sign, *the shear rate is given by the equation*

$$(9) \qquad \dot{s}(\mathrm{P}, t) = \sin\theta\, \partial\Omega(\theta, t)/\partial\theta,$$

and is therefore independent of r. This is in contrast to the case of torsional flow between parallel plates, where the shear rate depends on r (not z) [see 9.3(8)], and makes it easier to use the cone-and-plate apparatus for determining σ, N_1, and N_2 from torque and pressure distribution measurements, because the flow history is uniform throughout the gap [provided that (2) holds].

Since the flow is a unidirectional shear flow, we can apply Theorem 7.1(5) and express the physical components of stress in the form 9.3(9) with N_1, N_2, and σ as functionals of the shear rate $\dot{s}$, given by (9). Neglecting inertial forces and gravity, and taking, as a trial solution, $\partial(\;)/\partial\phi = 0$ because of the complete rotational symmetry of the problem, the stress equations of motion [Appendix A, (4)–(6)] reduce to the following:

$$(10) \qquad \partial\sigma/\partial\theta + 2\sigma \cot\theta = 0 \qquad (\phi \text{ equation});$$

$$(11) \qquad \partial(N_2 - p)/\partial\theta - N_1 \cot\theta = 0 \qquad (\theta \text{ equation});$$

$$(12) \qquad \partial p/\partial r + (N_1 + 2N_2)/r = 0 \qquad (r \text{ equation}).$$

Since $\dot{s}$ is independent of r, it follows that N_1 and N_2 are independent of r, and hence, from (11), that $\partial^2 p/\partial r\, \partial\theta = 0$. This implies, from (12), that

$$(13) \qquad \partial(N_1 + 2N_2)/\partial\theta = 0.$$

For most polymeric liquids investigated so far, $N_1 + 2N_2 \neq 0$ (see later discussion); Eq. (10) implies that $\partial\dot{s}/\partial\theta \neq 0$ throughout the gap; hence (13) cannot be satisfied. The stress equations of motion are therefore incompatible, even when gravity and inertia are neglected, and the trial solution (3) must be inadmissible for incompressible viscoelastic liquids generally (Oldroyd, 1958; Ericksen, 1960). This is again in contrast to the case of torsional flow between parallel plates, for which the trial flow 9.3(2) (also a unidirectional shear flow) is a solution when gravity and inertia are neglected.

There is, however, good reason to believe that for gap angles not greater than, say, 10°, the condition (13) can be satisfied to an approximation sufficient to ensure that the "primary flow" (2) will predominate over such "secondary flow" that must occur whenever (13) is not satisfied exactly. We offer three reasons for this belief. First, Eq. (10) can be integrated to give the result

$$\sigma(\theta, t) = \sigma(\tfrac{1}{2}\pi, t)\operatorname{cosec}^2\theta = (3M/2\pi R^3)\operatorname{cosec}^2\theta, \tag{14}$$

where M denotes the torque which the liquid exerts on the plate $\theta = \frac{1}{2}\pi$. This equation governs the variation of shear stress, and hence also of shear rate $\dot{s}$, with θ, and shows that this variation is very small near the plate where $\operatorname{cosec}^2\theta = 1 + (\frac{1}{2}\pi - \theta)^2 + \cdots$. Since N_1 and N_2 are functionals of $\dot{s}$ alone, it is therefore reasonable to expect that the variation of $N_1 + 2N_2$ with θ will by very small near the plate (Lodge, 1964, p. 204). Second, the results of an approximate calculation of secondary flow (Adams and Lodge, 1964, pp. 168–169) show that the secondary flow velocity is about $(\Delta\theta_1)^4/320$ times the greatest value of primary flow velocity, and that the corresponding contribution $p_{(v)}$ to the pressure on the plate is given by the equation

$$p_{(v)}(r) = -0.05(\Delta\theta_1)^2(N_1 + 2N_2)\log_e(r/R) + p_{(v)}(R), \tag{15}$$

for the case of steady primary flow. Although the pressure perturbation $p_{(v)}$ has the same r dependence as has the primary pressure (see later discussion) (so that experimental data giving pressure varying linearly with $\log_e r$ do *not* prove that secondary flow is unimportant), the coefficient $0.05(\Delta\theta_1)^2$ is only 0.002 for a 10° gap angle ($\Delta\theta_1$ being in radians), and is even smaller for smaller gap angles; the secondary flow pressure perturbation should therefore be negligible. Third, the coefficient varies with gap angle $\Delta\theta_1$, whereas the corresponding coefficient for the primary flow pressure does not; the limited data published show that the slope of the line representing the variation of pressure with $\log_e r$ is independent of the gap angle at constant shear rate (Adams and Lodge, 1964; Kaye *et al.*, 1968, Fig. 5) and strongly suggest that the *effects of secondary flow associated with the impossibility of satisfying the "viscoelastic compatibility condition"* (13) *are negligible* for the polymeric liquids investigated. The occurrence of secondary flow velocity fields has been investigated experimentally and theoretically in cone-and-plate systems by various workers listed in Appendix F. It is quite possible that a secondary flow velocity field, too small to have a detectable influence on the measured pressure distribution, can be readily photographed during *prolonged* shear flow when dyed liquid is introduced locally at suitable places, and it would be dangerous to conclude from such information alone that the use of the primary flow analysis of pressure data is unreliable; the preceding discussion suggests that it is safer to *repeat the pressure measurements with different gap angles using the same shear rate history in order to assess the reliability of the use of the cone-and-plate apparatus for determining* $N_1 + 2N_2$. The results of these and other measurements (Kaye *et al.*, 1968) strongly suggest that the cone-and-plate apparatus is indeed most useful for measuring N_1 and N_2, at least when the shear rate is not too high. Attempts to produce high shear rates often result in drastic disturbance to the state of flow (Hutton, 1963) so that the apparatus is unusable. We shall proceed, then, with the analysis of the primary flow (2) alone.

Assuming that the free liquid boundary is part of the surface of a sphere $r = R$ and that the effects of gravity and surface tension are negligible, it follows from 9.3(9) that the boundary conditions

(16) $$-\widehat{rr}(R, t) = p(R, t) = p_a, \qquad \widehat{r\phi} = \widehat{r\theta} = 0,$$

are satisfied to the approximation that $\partial N_2/\partial\theta = 0$ [see (11)]; this is the same approximation as that already introduced and discussed. The failure in practice to produce a free boundary of spherical shape may not be too serious, because Kaye *et al.* (1968, p. 373) found that for the one solution tested, quite large changes in shape of the free boundary resulted in a change of only 6% in total thrust.

Finally, integration of (12) using the fact that N_1 and N_2 are independent of r gives the result

(17) $$p(r, t) = p(R, t) - (N_1 + 2N_2)\log_e(r/R).$$

From 9.3(9) and (1), we have $\widehat{\theta\theta} = -p + N_2$, and hence, usng (16) and (17), we obtain the following equations for the component $\widehat{\theta\theta}$ of traction normal to the plate:

(18) $$\widehat{\theta\theta}(r, t) = -p_a + N_2 + (N_1 + 2N_2)\log_e(r/R);$$

(19) $$\widehat{\theta\theta}(R, t) = -p_a + N_2.$$

In these equations, N_1 and N_2 are functionals of shear rate $\dot{s}$ alone, and $\dot{s}$ is given by (9) or, to a good approximation [see Adams and Lodge (1964, p. 183, Table 2)], by the equation

(20) $$\dot{s} = \Omega_1(t)/\Delta\theta_1.$$

Thus if the pressure $-\widehat{\theta\theta}$ on the plate is plotted as a function of $\log_e(r/R)$, one can obtain the values of N_1 and N_2 for any given unidirectional shear flow history (provided that the primary flow (2) is dominant) from the slope and intercept at the rim of the straight line which should be obtained. The total thrust F due to the shear flow is obtained by integrating $(\widehat{\theta\theta} - p_a)\, d(\pi r^2)$ and is given by the equation

(21) $$N_1 = 2F/(\pi R^2).$$

No assumption about the value of N_2 is involved in the derivation of (21).

9.5 Steady helical flow

We consider helical flow, as defined in Section 3.6, which is steady in the sense that the angular velocity Ω and the axial velocity v are independent of time. According to 3.6(8), the shear angle ζ is then independent of time and so also is the shear rate $\dot{s}$, given by 3.6(10). The shear flow is therefore unidirectional, and the simplified expression for the stress tensor given in Section 7.1 can be used. To make helical flow unidirectional, it would be sufficient to have the ratio v_2/Ω_2 given in 3.6(8) independent of time throughout the gap between the cylinders, rather than require both v and Ω to be independent of time, but such a wider possibility would be difficult to realise in practice, so we need not consider it. Helical flow was first considered by Rivlin (1956), who proposed to determine N_1 and N_2 from measured values of the difference of pressure on the cylinder walls obtained with different values for the ratio $v_2 : \Omega_2$.

We use the cylindrical polar space coordinate system $S: (x^1, x^2, x^3) = (\phi, r, z)$ and the results of Section 3.6. The shear lines are helices making a constant angle $\zeta(r)$ on a shear surface $r = \text{const}$. We use a body coordinate system $B \overset{t}{\equiv} S$, as in Section 3.6. At any particle P, the t-orthonormal shear-flow base vectors $(\boldsymbol{\alpha}_4, \boldsymbol{\alpha}_2, \boldsymbol{\alpha}_5)$ are related to the t-orthonormal base vectors $(\boldsymbol{\alpha}_1, \boldsymbol{\alpha}_2, \boldsymbol{\alpha}_3)$, tangential to the coordinate curves of B, by the equations

(1) $$\boldsymbol{\alpha}_4 = \boldsymbol{\alpha}_1 \cos\zeta + \boldsymbol{\alpha}_3 \sin\zeta, \qquad \boldsymbol{\alpha}_5 = -\boldsymbol{\alpha}_1 \sin\zeta + \boldsymbol{\alpha}_3 \cos\zeta,$$

according to 3.3(6) and 3.3(13).

(2) These vectors are illustrated in Fig. 9.5(2).

Since the flow is a unidirectional shear flow, we can use 7.1(22) with $\boldsymbol{\alpha}_1$ replaced by $\boldsymbol{\alpha}_4$ (the tangent to a shear line), so that

(3) $$\boldsymbol{\pi} = -p\boldsymbol{\alpha}_i\boldsymbol{\alpha}_i + \sigma(\boldsymbol{\alpha}_4\boldsymbol{\alpha}_2 + \boldsymbol{\alpha}_2\boldsymbol{\alpha}_4) + (N_1 + N_2)\boldsymbol{\alpha}_4\boldsymbol{\alpha}_4 + N_2\boldsymbol{\alpha}_2\boldsymbol{\alpha}_2.$$

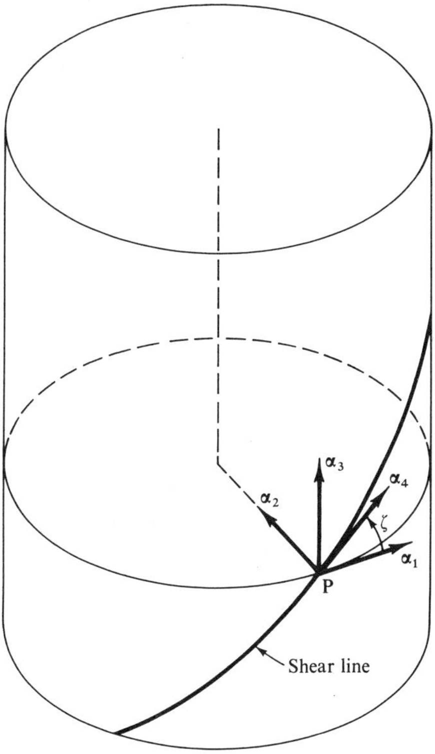

Figure 9.5(2) In helical flow, shear lines are helices on shear surfaces, which are cylinders. $(\boldsymbol{\alpha}_1, \boldsymbol{\alpha}_2, \boldsymbol{\alpha}_3)$ is a t-orthonormal shear-flow basis at P, the vectors $\boldsymbol{\alpha}_i$ being tangential to the coordinate curves of a shear-flow body coordinate system B which is instantaneously cylindrical polar at time t. $\boldsymbol{\alpha}_4$ is tangential to a shear line and makes an angle ζ with $\boldsymbol{\alpha}_1$.

Substituting for $\boldsymbol{\alpha}_4$, from (1), using 7.1(2) and the fact that $(\xi^1, \xi^2, \xi^3) \stackrel{t}{\equiv} (\phi, r, z)$, so that $\widehat{\pi^{11}} = \widehat{p^{11}} = \widehat{\phi\phi}$, etc., we have

$$(4) \qquad \widehat{\phi\phi} = -p + (N_1 + N_2)\cos^2\zeta, \qquad \widehat{rr} = -p + N_2,$$

$$\widehat{zz} = -p + (N_1 + N_2)\sin^2\zeta,$$

$$\widehat{rz} = \sigma \sin\zeta, \qquad \widehat{z\phi} = (N_1 + N_2)\cos\zeta\sin\zeta, \qquad \widehat{\phi r} = \sigma\cos\zeta.$$

Since σ, N_1, N_2, and ζ are functions of r alone ($\dot{s}$ being a function of r alone), the stress equations of motion [(4)–(6)] of Appendix A, with neglect of the inertia terms, reduce to the following:

$$(5)\qquad -\frac{\partial p}{\partial r} + \frac{\partial N_2}{\partial r} + r^{-1}(N_2 \sin^2 \zeta - N_1 \cos^2 \zeta) = 0;$$

$$(6)\qquad \frac{\partial}{\partial r}(r^2 \sigma \cos \zeta) - r\frac{\partial p}{\partial \phi} = 0;$$

$$(7)\qquad \frac{1}{r}\frac{\partial}{\partial r}(r\sigma \sin \zeta) - \frac{\partial p}{\partial z} = \rho g;$$

where g is the acceleration due to gravity, and the z axis is taken to point vertically upward. For a given material, we can regard σ, N_1, and N_2 as given functions of $\dot{s}(r)$; there are therefore three simultaneous equations in three unknown variables: p, $\dot{s}(r)$, and $\zeta(r)$, where $\dot{s}(r)$ occurs implicitly. We shall be content here to show that the equations are compatible.

Equations (5)–(7) show that $\partial p/\partial r$, $\partial p/\partial \phi$, and $\partial p/\partial z$ are functions of r alone; therefore $p = a\phi + bz + c(r)$, where a and b are constants. But $\widehat{rr} = -p + N_2$ must be a one-valued function of the coordinates; therefore $a = 0 = \partial p/\partial \phi$, and $p = bz + c(r)$, where $c(r)$ is a function to be determined. From (4), we see that $\partial \widehat{zz}/\partial z = -\partial p/\partial z = -b$, so b is equal to the axial pressure gradient and will be determined by the externally applied pressure gradient which generates flow in the axial direction. Now (6) and (7) integrate to give

$$(8)\qquad r^2 \sigma \cos \zeta = A; \qquad r\sigma \sin \zeta = \tfrac{1}{2}(b + \rho g)r^2 + B.$$

The constants of integration, A and B, must be independent of r and cannot depend on ϕ or z because the other terms depend on r alone. A is related to the torque per unit length applied to a cylinder. Equations (8) determine $\dot{s}$ and ζ as functions of r, and can therefore be satisfied. The remaining equation (5), since $\partial p/\partial r = dc(r)/dr$, then suffices to determine the unknown function $c(r)$, and hence also the radial variation of $\widehat{\phi\phi}$, $\widehat{rr}$, and $\widehat{zz}$, according to (4). The three stress equations of motion (when inertial terms are omitted) are therefore compatible, and so the assumed state of steady helical flow is one which can be maintained by the application of surface tractions alone to the external fluid boundaries. This analysis applies to all incompressible liquids for which the extra stress can be expressed as an isotropic, one-particle function of the strain history. The further analysis necessary to relate the various integration constants to measurable quantities and to determine the functions σ, N_1, and N_2 from measured quantities can be readily performed (Coleman *et al.*, 1966, pp. 38–41). Unsteady helical flow has been considered by Markovitz and Coleman (1964).

10 CONSTITUTIVE PREDICTIONS AND EXPERIMENTAL DATA

10.1 Shear and elongation of low-density polyethylene

There is a large and rapidly increasing amount of published material which presents the results of measurements of rheological properties of visco-elastic materials and which compares these results with the predictions of various forms of constitutive equations. In this chapter, we select a few of these comparisons in an attempt to bring out some of the more important features and to give some indication of the current state of knowledge.

Our selection is influenced by two requirements: The need to use data whose reliability has been established by suitable experimental tests; and the need to have data obtained, for a given material, *from a variety of rheological experiments* featuring flow histories which differ from one another both geometrically and temporally. These two requirements eliminate a substantial amount of published material from our consideration.

The importance of experimental reliability of data should be obvious, but it has received less attention in the rheological literature than it deserves. In the case of shear flow of very viscous molten plastics in the Weissenberg rheogoniometer (a commercial cone-and-plate apparatus) involving a step-function increase of shear rate, Meissner (1972) has recently shown that incorrect transient N_1 data can be obtained unless a suitable gap angle (6° to 10°) and diameter are used; it is also necessary to stiffen the commercial apparatus and to alter the method of measuring axial thrust. In elongational flow of molten plastics, Meissner (1971) has emphasized the importance of checking the homogeneity of the elongation obtained; molten polystyrene appears to be particularly prone to generate inhomogeneous elongation (Meissner, 1973).

The importance of using different types of strain when testing a constitutive hypothesis for materials subjected to large elastic strains has been emphasized by Treloar (1958, Fig. 5.10); stress–strain data for a rubber subjected to two different types of strain are consistent with the form of the strain-energy function predicted by the molecular network theory of rubber elasticity in the Gaussian approximation, but data for a third type of strain are not. In spite of this warning, some papers have appeared in which data for only one type of strain (usually simple elongation) have been used to test a proposed form of strain-energy function. It appears that the point is of particular importance where large-strain effects are significant and of little importance in the fields of classical elasticity (with infinitesimally small strains) and hydrodynamics where, after all, the form of constitutive equations are known and the only function of simple experiments is to assign values to material constants. For this reason, *experience in the classical fields can be misleading in polymer science and engineering*. It is reasonable to expect that similar considerations will apply not only to the deformation of rubberlike solids, but also to the flow of viscoelastic liquids. Tension–extension data for simple elongation of a rubberlike solid give very limited information about the form of the strain-energy function, no matter how large a range of elongational strain is used. We can expect, therefore, that (for example) *viscosity/shear-rate data for a polymer solution or molten plastic by themselves will give only very limited information about the constitutive equations, no matter how large the range of shear rate may be.*

(1) Meissner has published stress/time data for both shear flow and elongational flow for a molten low-density polyethylene, and has also published elastic recovery data following elongational flow. The polymer specification, types of measurement, references, and comparisons with theory are listed in Table 10.1(1). M_w and M_n denote the weight- and number-average molecular weights. $CH_3/1000C$ means the average number of methyl side groups per 1000 main-chain carbon atoms. "Melt flow index" is obtained from a capillary flow measurement according to a standard procedure. Additional calculations using the same constitutive equations to assess the stability of

Table 10.1(1) Summary of Experiments and Calculations for Melt 1

Specification of melt 1 (a low-density polyethylene)		Meissner (1971)
Density at 20°C	0.920 g ml^{-1}	
Melt flow index (190/2.16)	1.33	
Viscosity at 150°C ($\dot{s} \to 0$)	500,000 P	
M_w/M_n	28.1	
M_n	482,000	
CH_3/1000C	31	
Rheological measurements at 150°C		
Elongational flow, step function elongation rates		Meissner (1971)
Tensile stress versus time		
Ultimate free recovery versus previous strain		
Shear flow, step-function shear rates		Lodge and Meissner (1973)
N_1 versus t		
σ versus t		
Comparison of rubber-like liquid predictions (set B constants) with data		
Choice of memory function constants		Chang (1973), Lodge and Meissner (1973)
Stress/time data, elongation		Chang (1973), Lodge and Meissner (1973)
Recovery data, elongation		Chang (1973)
N_1 versus t, shear flow		Lodge and Meissner (1973)
σ versus t, shear flow		Lodge and Meissner (1973)

elongational flow and of thin-film extension have been made by Chang (1973) and Chang and Lodge (1972), but no data are available for comparison. Meissner (1973) has obtained some data in elongational flow giving the tensile stress versus time when an oscillatory elongation rate of small amplitude is superposed on a constant elongation rate; the oscillatory part of the tensile stress has a very large amplitude; calculations for this experiment have also been made by Chang (1973). We shall here give the results of experiments and calculations listed in Table 10.1(1).

(2) The stress/time data for elongational flows at various shear rates are given in Fig. 10.1(2); for clarity, some of the elongation rates and data points given in Meissner's paper (1971) have been omitted. The smooth curves represent the predictions of the rubberlike-liquid constitutive equation 6.2(2) with $\eta_s = 0$ and the constants a_r and τ_r in the memory function

(3) $$\mu(\tau) = \sum_{r=1}^{5} a_r e^{-\tau/\tau_r}$$

chosen to fit the data at the lowest elongation rate ($\dot{\varepsilon} = 0.001$ sec^{-1}). There is,

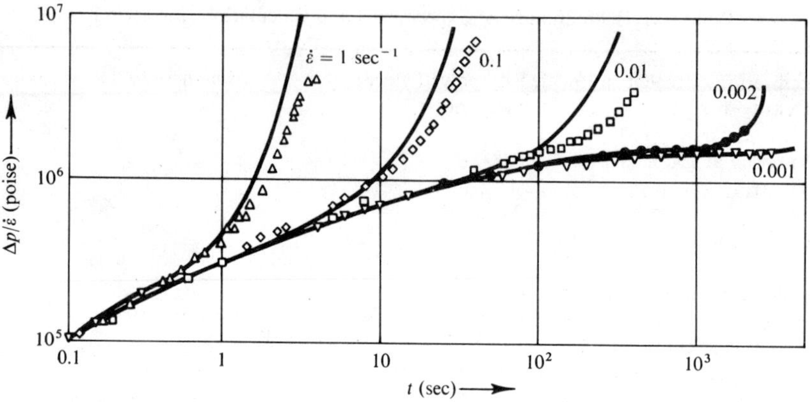

Figure 10.1(2) Variation of tensile stress Δp with time t during elongation at constant elongation rate $\dot{\varepsilon}$ started at $t = 0$. Points: values for a low-density polyethylene (melt 1 at 150°C) measured by Meissner (1971). Curves: calculated for a rubberlike liquid [having a constitutive equation 6.2(2) with $\eta_s = 0$ and $\mu(\tau) = \sum_{r=1}^{5} a_r \exp(-\tau/\tau_r)$] by Chang (1973); the memory function constants, chosen to fit the data at the two lowest elongation rates (0.001 and 0.002 sec^{-1}), are as follows (set B) [Quoted from Chang (1973), p. 33. Table I of Lodge and Meissner (1973) contains two misprints.]:

τ_r (sec):	1000	100	10	1	0.1
a_r (dyn cm^{-2} sec^{-1}):	1.601×10^{-3}	19.26	1723	6.64×10^4	3.972×10^6

The comparison between curves and points at the higher elongation rates (0.01, 0.1, and 1 sec^{-1}) gives a test of the molecular network theory used to derive the constitutive equation.

of course, much arbitrariness in the choice of constants and in the choice of the number of terms to include in the memory function; one could alternatively use an integral, instead of a sum of exponential terms, for the memory function. Many approximate methods of choosing the memory function (or, what is equivalent, the relaxation spectrum) so as to fit given data, usually η' and G' data in the so-called linear viscoelastic region, have been investigated (Ferry, 1970). In the present case, Chang and Lodge (1972) chose five terms, with time constants 0.01, 0.1, 1, 10, and 100 sec, because the elongation data covered nearly five units on the $\log_{10} t$ scale; a smooth curve was drawn through the data (for $\dot{\varepsilon} = 0.001\ \text{sec}^{-1}$), and values of the ordinates at five times were taken. Substitution of these values into the equations obtained from the rubberlike-liquid theory yielded five simultaneous equations which were solved by a computer in order to obtain values for the coefficients a_r ($r = 1, 2, \ldots, 5$) in (3). In some cases, negative values for one or more coefficients were obtained. By making small changes in the smoothed curve drawn through the data, it proved possible to obtain positive values for all the coefficients a_r.

Subsequently, Chang (1973) chose a second set of constants (set B) with $\tau_r = 10^{r-2}$ sec ($r = 1, 2, \ldots, 5$), with the coefficients a_r chosen to fit the stress-growth data in elongation at the two lowest elongation rates (0.001 and 0.002 sec^{-1}). The constitutive equation being thereby completely determined, the comparison of its predictions with other data furnishes a test of the molecular network theory from which the constitutive equation has been derived.

One such test is furnished in Fig. 10.1(2) by the comparison between the smooth curves at elongation rates 0.01, 0.1, and 1 sec^{-1} and the points, which represent results of measurements. It is seen that there is agreement at the smaller values of t, the time elapsed since the start of elongational flow, but disagreement at higher times.

(4) The remaining tests are represented in Fig. 10.1(4). The theoretical curves for N_1 and σ (as functions of time elapsed from the start of elongation) and ε_R, the ultimate free recovery (as a function of total strain x/x_0 imposed before recovery), were obtained by Chang (1973); the free recovery calculation involved a numerical solution of a nonlinear Volterra integral equation. It is noteworthy that very large values of elastic recovery are obtained, with values as high as 10 for the ratio x/x_R being measured following elongation at a rate $\dot{\varepsilon} = 1\ \text{sec}^{-1}$. Here, x denotes the length of a segment immediately before cutting and x_R denotes the length of the same segment when recovery is complete. $\varepsilon_R = \log_e(x/x_R)$. It is evident that the theory agrees well with measured recovery values at the highest elongation rate, $\dot{\varepsilon} = 1\ \text{sec}^{-1}$, but not at lower elongation rates. As far as we are aware, this is the first published comparison between large-strain free recovery data and predictions of an admissible constitutive equation for a viscoelastic liquid. The recovery is free in the sense (Lodge, 1964, p. 124) that it occurs with the stress zero or isotropic, and it is a feature of the elongation experiment that free recovery can be measured. Recovery measured following shear flow in a rotational apparatus, on the other hand, is usually constrained in the sense that some, but not all, stress components are zero.

(5) It is also seen from Fig. 10.1(4) that the theory gives a good description of the growth of stress (both σ and N_1) with time after the start of shear flow, at least for the earlier times. At longer times, however, the measured values decrease while the calculated values continue to increase, reaching ultimately constant values at each shear rate. With N_1, the calculated values are as much as 100 times greater than the measured values. Such a large discrepancy is not evident with the elongation results shown in Fig. 10.1(2); but these experiments involve smaller total strains measured from rest, as shown in Fig. 10.1(5).

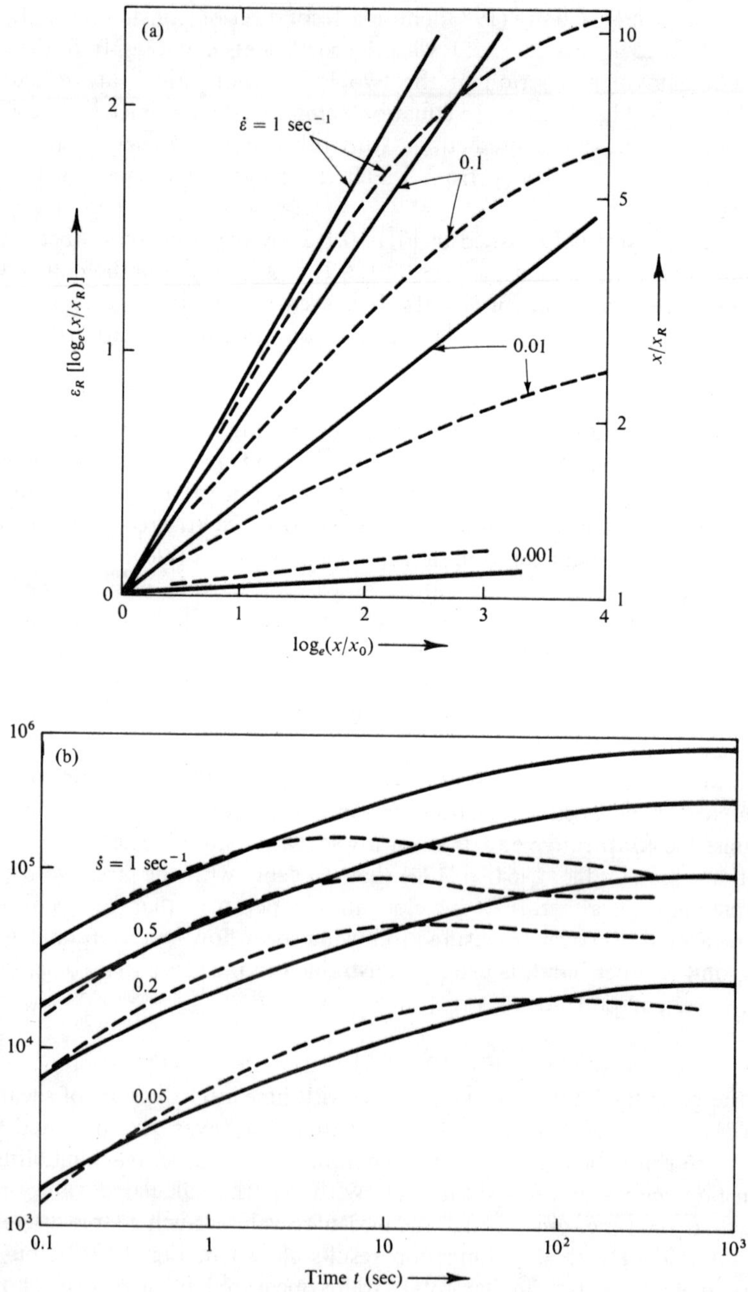

Figure 10.1(4)a and b

In Fig. 10.1(5), we have chosen the axes so as to combine in one graph the comparison between theory and experiment for all the stress-growth data obtained in both elongational and shear flows. The ordinate represents values of the ratio of the theoretical and measured values of Δp, a difference of principal values of stress. For elongation, Δp is simply the tensile force per unit area of cross section of the filament. For shear, the principal values of stress are, in general, all different; we have (arbitrarily) taken Δp to be the difference of those principal values that belong to the two principal axes that are tangential to a surface $\xi^3 = \text{const}$, so that we have

$$(6) \qquad \Delta p = (N_1{}^2 + 4\sigma^2)^{1/2} \qquad \text{(shear flow)}.$$

The abscissa represents values of $I_{IV}\,t$, where t is the time elapsed after the

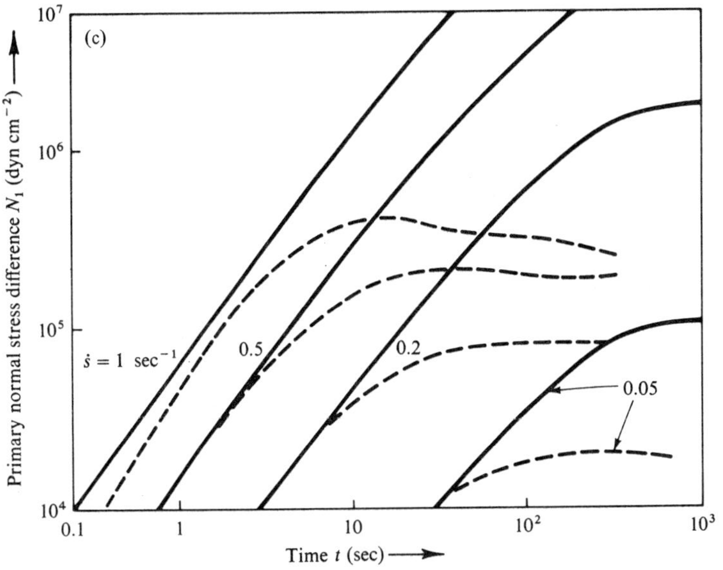

Figure 10.1(4)c

Figure 10.1(4) Tests of the applicability of the molecular network theory to a molten low-density polyethylene [melt 1 as in Table 10.1(1) and Fig. 10.1(2)]. *Full lines:* from the rubberlike-liquid constitutive equation calculated by Chang (1973) using the set B memory function constants [legend to Fig. 10.1(2)]. *Dashed curves:* values measured by Meissner (1971) and Lodge and Meissner (1973). (a) Free elastic recovery $\varepsilon_R \equiv \log_e(x/x_R)$ as a function of total elongation preceding recovery; a sample of initial length x_0 is elongated at a constant elongation rate $\dot{\varepsilon}$ to a length x; the stress is suddenly removed, and the sample contracts freely to a final length x_R. (b) Variation of shear stress σ with time t during shear flow at constant shear rate $\dot{s}$ started at $t = 0$. (c) Variation of primary normal stress difference N_1 with time t during shear flow at constant shear rate $\dot{s}$ started at $t = 0$.

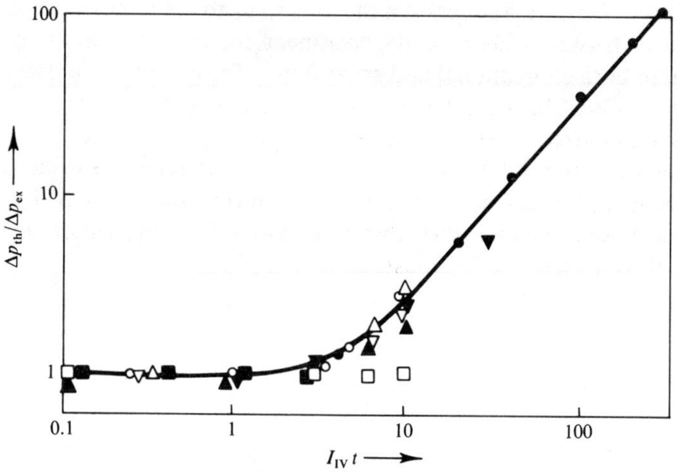

Figure 10.1(5) Comparison of network theory predictions for stress growth in shear and elongation with measured data for a low-density polyethylene melt: alternative representation of the results shown in Figs. 10.1(2) and 10.1(4)(b), (c). Δp_{th} and Δp_{ex} denote calculated and measured values of a difference of principal stresses; agreement between theory and experiment ($1 \leq \Delta p_{\text{th}}/\Delta p_{\text{ex}} \leq 1.1$) is obtained when $I_{\text{IV}}\, t \leq 3$. I_{IV} is a strain-rate invariant, defined in 10.1(7), having values equal to $3.01\dot{\varepsilon}$ in elongational flow and $\dot{s}$ in shear flow.

$\dot{\varepsilon}$	$\dot{s}$	sec^{-1}
□	■	0.001
△	▲	0.01
▽	▼	0.1
○	●	1

start of flow and I_{IV} denotes a strain-rate invariant given by the equation

$$(7) \qquad I_{\text{IV}} = \frac{1}{\sqrt{2}} \left\{ \left(\sum_{i=1}^{3} \kappa_i^2 \right)^{1/2} + \left| \sum_{i=1}^{3} \kappa_i^3 \right|^{1/3} \right\} = \begin{cases} 3.01\dot{\varepsilon} & \text{(elongation)}, \\ \dot{s} & \text{(shear)}. \end{cases}$$

This was chosen somewhat arbitrarily after a trial-and-error attempt to find an abscissa variable that would bring the shear and elongation points as close as possible to a common line. The preceding results are due to Lodge and Meissner (1973).

From this discussion, we can reach the following conclusions.

(i) The rubberlike-liquid theory describes stress-growth data in elongation and shear provided that the total strain (measured by $I_{\text{IV}}\, t$) from rest does not exceed a value of about three; discrepancies between measured and calculated values of Δp do not exceed 10%.

(ii) At higher values of $I_{IV}\,t$, the discrepancies become very large.
(iii) The extent of agreement or disagreement between theory and experiment is the same for shear flow and for elongational flow.
(iv) The theory successfully describes the large free elastic recovery measured at the highest elongation rate ($\dot{\varepsilon} = 1\ \text{sec}^{-1}$), but not at lower elongation rates.

It is well known that the rubberlike-liquid theory gives a qualitatively successful description of some of the main rheological properties (e.g. $N_1 > 0$, large elastic recovery, N_1 relaxes more slowly than σ after cessation of shear flow) of polymeric liquids, but is unsatisfactory in its inability to give any shear-rate dependence of viscosity and stress relaxation and any stress "overshoot" (e.g., the maximum through which σ passes as t increases after sudden start of shear flow). Various modifications listed in Tables 6.4(1) and 6.4(2) have been made in order to describe more data. One of the more successful of these is described in Section 10.3. The recent results of Meissner on low-density polyethylene and their comparison with the (unmodified) rubberlike-liquid theory given here may be important in helping to locate the origin of the failure of the theory and in offering some guidance as to which of the many possible modifications of the theory is the more likely to be appropriate to molten plastics. In the next section, we describe a recent novel proposal designed to furnish experimental guidance in this task.

10.2 Fast-strain tests of the Gaussian network hypothesis

The network theory leading to the rubberlike-liquid constitutive equation is based on certain assumptions (Lodge, 1960, 1968) which, for the present purpose, we shall separate into two groups called "good" and "bad":

Network theory, good assumptions

(1) The liquid is a viscoelastic, incompressible continuum whose extra-stress tensor $\boldsymbol{\Pi}(\text{P}, t)$ at any time t is determined by the values of the metric tensor $\boldsymbol{\gamma}(\text{P}, t,)$ for $-\infty < t' \leqslant t$ (at constant temperature) through a one-particle, t-isotropic functional: $\boldsymbol{\Pi}(\text{P}, t) = \mathscr{F}\{\boldsymbol{\gamma}(\text{P}, t')\}$.

(2) The molecular network "connectivity" is continually changing due to the loss and creation of temporary junctions which connect linear polymer molecules.

(3) At any instant t (following any flow history), the equilibrium network theory for a rubberlike solid (having permanent junctions and the same

connectivity) in the Gaussian approximation is applicable and yields an expression of the form

$$\Pi(P, t) = \mu_0^* \gamma^{-1}(P, t^*), \qquad \mu_0^* = ckTN_0^*(t),$$

where $N_0^*(t)$ denotes the segment concentration at time t and $\gamma(P, t^*)$ denotes the metric tensor for the state which would be obtained by instantaneous free recovery at time t [see 6.2(1)]; i.e., if the stress were made zero at time t, then there would be a strain $\gamma(P, t) \to \gamma(P, t^*)$ which would occur so rapidly (i.e., "fast," or "macroscopically instantaneously") that the network connectivity would not change.

For convenience, we shall call a material *Gaussian* if it satisfies these three assumptions.

These assumptions are plausible for polymer solutions of sufficient concentration and polymer melts of sufficiently high molecular weight. Qualitatively, they yield a material which is a liquid and which can exhibit large, rapid, elastic recovery. When a further plausible assumption concerning birefringence is added, one can derive stress/birefringence relations for arbitrary flow histories (Lodge, 1960) which are quantitatively successful in describing data for a variety of polymeric liquids (Janeschitz-Kriegl, 1969). There is reason, therefore, for calling these assumptions "good." They are not sufficient, however, to enable one to calculate $N_0^*(t)$ and $\gamma(P, t^*)$ for a given flow history or to calculate the form of the functional $\mathscr{F}$ in the constitutive equation. The aim of this present section is *to propose an experimental test of these assumptions* which could, in principle, be performed with a cone-and-plate rheogoniometer. Therefore, we should be able to investigate the validity of these assumptions [and evaluate the junction concentration $cN_0^*(t)$] in a manner which depends in no way on the following remaining assumptions which have been used, in conjunction with the good assumptions, to obtain the particular form for $\mathscr{F}$ represented by the rubberlike-liquid constitutive equation:

Network theory, bad assumptions

(4) The "short-time-average" positions of junctions move "affinely."

(5) The creation and loss rates of junctions are constant for a given material and temperature.

(6) On creation, new segments have short-time-average end-to-end vectors which have an isotropic distribution.

The various terms used here are defined, for example, by Lodge (1968).

The experimental test now proposed (Lodge and Meissner, 1972) depends on the possibility of making stress measurements in states t^- and t^+, where

$$\Delta\gamma = \gamma(\mathrm{P}, t^+) - \gamma(\mathrm{P}, t^-) \tag{7}$$

is a fast strain of abitrary magnitude. In this section, we use Δ to denote an increment associated with a fast strain $t^- \to t^+$.

(8) **Definition** For a polymeric material whose rheological behavior is governed by a molecular network, a strain is *fast* if no change of network connectivity occurs (i.e., if there is no loss or creation of junctions). In the applications considered here under assumption (3), a fast strain must be slow on the time scale governing thermal fluctuations of network segments in order that the network deformation can be regarded as quasistatic so that the equilibrium theory can be applied, at least at each state t^- and t^+.

It follows at once from (3) that

$$\Delta\mathbf{\Pi} = \mathbf{0} \qquad \text{(for any fast strain in a Gaussian material).} \tag{9}$$

This is merely a restatement of the fact that according to the equilibrium network theory for a rubberlike solid in the Gaussian approximation, the extra-stress tensor is constant in an arbitrary isothermal deformation.

An important particular example of a fast strain is furnished by the instantaneous free recovery $\gamma(\mathrm{P}, t) \to \gamma(\mathrm{P}, t^*)$; in this case, we have $t^- = t$, $t^+ = t^*$, and $\boldsymbol{\pi}(\mathrm{P}, t^+) = \mathbf{0}$. It is important to note that, in general, when the flow through $(-\infty, t)$ is a unidirectional shear flow, the recovery $t \to t^*$ is *not* a shear; in the case of a rubberlike liquid, for example, the recovery involves stretching and increase of separation of the material surfaces which were shearing surfaces (Lodge, 1964, p. 134), except for the case in which the memory function has the very special (and unrealistic) form $\mu(\tau) = \mu_0\,\delta(\tau - \tau_0)$. Associated with this fact is the result that in unidirectional shear flow, *a Gaussian material can generate a nonzero secondary normal stress difference* N_2 (Lodge and Meissner, 1972).

The Gaussian network hypothesis expressed in (3) involves the unknown quantities μ_0^* and $\gamma^* \equiv \gamma(\mathrm{P}, t^*)$; expressed in (9), these quantities are absent. Equation (9) therefore represents a test of the Gaussian network hypothesis. To see what measurements are involved, we consider an arbitrary unidirectional shear flow throughout the interval $(-\infty, t^+)$, incorporating a fast shear (t^-, t^+) of any magnitude Δs.

From (1) and 7.1(5), it follows that at any t in $(-\infty, t^+)$, we have

$$\mathbf{\Pi} = N_1 \boldsymbol{\alpha}_1 \boldsymbol{\alpha}_1 - N_2 \boldsymbol{\alpha}_3 \boldsymbol{\alpha}_3 + \sigma(\boldsymbol{\alpha}_1 \boldsymbol{\alpha}_2 + \boldsymbol{\alpha}_2 \boldsymbol{\alpha}_1) + p\gamma^{-1}, \tag{10}$$

where N_1, N_2, and σ are defined in 7.1(3), and $\boldsymbol{\alpha}_i$ is a shear flow basis, orthonormal at time t, with $\boldsymbol{\alpha}_1$ tangential to a shear line; the reciprocal basis $\boldsymbol{\alpha}^i$ has $\boldsymbol{\alpha}^2$ normal to a shearing surface. For the fast shear $t^- \to t^+$, the changes in the base vectors are given by the equations

(11) $\Delta\boldsymbol{\alpha}_1 = \mathbf{0}, \qquad \Delta\boldsymbol{\alpha}_2 = -\boldsymbol{\alpha}_1^- \Delta s, \qquad \Delta\boldsymbol{\alpha}_3 = \mathbf{0}$ (contravariant);

(12) $\Delta\boldsymbol{\alpha}^1 = \boldsymbol{\alpha}_-^2\, \Delta s, \qquad \Delta\boldsymbol{\alpha}^2 = \mathbf{0}, \qquad \Delta\boldsymbol{\alpha}^3 = \mathbf{0}$ (covariant).

These follow from 3.3(16) on putting $t = t^-$, $t' = t^+$, $s_3 = 0$, and $s_1 = \Delta s$. It follows that

(13) $$\Delta\boldsymbol{\gamma}^{-1} = \Delta(\boldsymbol{\alpha}_2\boldsymbol{\alpha}_2) = -(\boldsymbol{\alpha}_1^-\boldsymbol{\alpha}_2^- + \boldsymbol{\alpha}_2^-\boldsymbol{\alpha}_1^-)\Delta s + \boldsymbol{\alpha}_1^-\boldsymbol{\alpha}_1^-(\Delta s)^2;$$

(14) $$\Delta\boldsymbol{\gamma} = \Delta(\boldsymbol{\alpha}^1\boldsymbol{\alpha}^1) = (\boldsymbol{\alpha}_-^1\boldsymbol{\alpha}_-^2 + \boldsymbol{\alpha}_-^2\boldsymbol{\alpha}_-^1)\Delta s + \boldsymbol{\alpha}_-^2\boldsymbol{\alpha}_-^2(\Delta s)^2.$$

Hence, from (10), we have

(15) $$\Delta\boldsymbol{\Pi} = [\Delta N_1 - (\sigma^+ + \sigma^-)\,\Delta s - (\Delta\sigma - p^+\,\Delta s)\,\Delta s]\boldsymbol{\alpha}_1^-\boldsymbol{\alpha}_1^- + \boldsymbol{\alpha}_2^-\boldsymbol{\alpha}_2^-\,\Delta p - \boldsymbol{\alpha}_3^-\boldsymbol{\alpha}_3^-\,\Delta N_2 + (\Delta\sigma - p^+\,\Delta s)(\boldsymbol{\alpha}_1^-\boldsymbol{\alpha}_2^- + \boldsymbol{\alpha}_2^-\boldsymbol{\alpha}_1^-).$$

From the Gaussian condition (9), it now follows that

(16)
$$\Delta N_1 = (\sigma^+ + \sigma^-)\,\Delta s, \quad \Delta N_2 = 0, \quad \text{and} \quad G \equiv \Delta\sigma/\Delta s \quad \text{is independent of } \Delta s.$$

The last result follows from the fact that $\Delta p = \Delta\sigma - p^+\,\Delta s = 0$: For then clearly $\Delta\sigma/\Delta s = p^-$ $(\equiv G)$, which is independent of Δs, since p^- depends only on the flow (and possibly stress) history up to t^-.

The three conditions (16) *represent fast-shear tests of the Gaussian network hypothesis* [(3), with (1) and (2)] *which, in principle, can be investigated experimentally.*

Furthermore, since the constant-volume condition requires that

(17) $$|\boldsymbol{\gamma}^{-1}(\mathrm{P}, t^*) \cdot \boldsymbol{\gamma}(\mathrm{P}, t)| = 1,$$

it follows from (3) that

(18) $$(\mu_0^*)^3 = |\boldsymbol{\Pi} \cdot \boldsymbol{\gamma}| \qquad \text{(Gaussian material).}$$

Applying this result to (10), with $p^+ = p^- = G$, we see that

(19) $$(\mu_0^*)^3 = (G - N_2^-)(G^2 + GN_1^- - \sigma_-^2) \qquad \text{(Gaussian material; shear).}$$

It is reasonable to call G the *incremental shear modulus*, and μ_0^* the *network modulus*, or *shear modulus*.

Equation (19) shows that for a Gaussian liquid, *the network modulus* μ_0^* *can be determined from measured values of* N_1^-, N_2^-, σ^-, *and* σ^+. It is easy to show, since $p^+ = p^-$ and (10) can be used with $t = t^+$ or t^-, that N_1^-, N_2^-, and σ^- in (19) can be replaced by N_1^+, N_2^+, and σ^+, respectively. It is also clear that μ_0^* is independent of the magnitude of Δs.

(20) **Problem** A Gaussian liquid undergoes simple elongational flow throughout an interval $(-\infty, t^+)$, including a fast elongation $t^- \to t^+$ of principal elongation ratio λ. Prove that

$$(\mu_0^*)^3 = (x^+ - x^-/\lambda)(x^+ - x^-\lambda^2)^2/(\lambda^2 - \lambda^{-1})^3,$$

and that the value of μ_0^* is independent of λ, where $x \equiv \widehat{p^{11}} - \widehat{p^{22}}$, with the 1-direction along the elongation direction, and the 2-direction transverse.

The result (20) and the first two of Eqs. (16) have been given by Lodge and Meissner (1972). *The fact that* μ_0^*, *as given by* (20), *is independent of* λ *furnishes a further experimental test of the Gaussian hypothesis.* In general, it is to be expected that μ_0^* will vary with time, even during shear or elongation at constant strain rate, and we can make no prediction about the relation between values to be expected for shear flow and elongational flow, unless additional information [such as the "bad assumptions" (4)–(6)] is used. If these assumptions are used, it follows from 6.2(1), 6.2(3), 10.2(3), and Eqs. [5.9] and [5.11] of Lodge (1968) that

(21) $$\mu_0^* = ckT \sum_i L_i \tau_i$$

is independent of time and strain rate; but μ_0^* calculated from the rubberlike-liquid constitutive equations (Lodge, 1964; Fig. 7.2; H. Chang, 1973) varies with both. This contradiction suggests that *the affine deformation assumption* (4) *may be seriously in error for networks whose connectivity changes during flow of the polymeric material.* Furthermore, the rubberlike-liquid calculation (Chang, 1973) for elongation with a step-function elongation rate shows that μ_0^* is constant at first but later increases sharply; this (as is to be expected) is compatible with the stress-growth comparison between theory and data for low-density polyethylene as shown in Figs. 10.1(4) and 10.1(5). It is possible that a substantial improvement in the description of the data could be obtained if the affine deformation assumption were abandoned and the appropriate theory substituted, even without modifying the other "bad assumptions" (5) and (6). Modifications to (5) have been made by Kaye (1966), Bird and Carreau (1968), and Carreau (1972) [see Table 6.4(1)].

This discussion of "fast-strain" tests is closely related to Weissenberg's postulate that the stress tensor at any time t is an isotropic function of the "recoverable strain tensor" (Weissenberg, 1935, p. 159, b). The Gaussian relation (3) is one particular kind of isotropic relation between the mixed stress tensor $\boldsymbol{\Pi} \cdot \gamma$ and the mixed strain tensor $\gamma^{*-1} \cdot \gamma$; this differs from Weissenberg's class of relations in that we have the extra-stress tensor in place of a deviatoric stress tensor, and we take $\gamma \to \gamma^*$ to be the instantaneous (instead of the ultimate) free recovery (Lodge, 1973a). We do retain, however, the essential novel features of Weissenberg's postulate (novel, that is, when applied to liquids), namely, that stress shall be isotropically related to a finite-strain (instead of, for example, a strain rate) tensor which is itself related to the elastic recovery properties of the material.

The proposed fast-strain tests are also of interest in that they furnish an example in which useful measurements and analysis can be considered even when the complete constitutive equation is not known. The proposed tests are an extension (to finite strains, with use of N_1 as well as σ) of tests first performed by Miller *et al.* (1951). Coleman (1964) has considered thermodynamic aspects of fast strains.

10.3 Polystyrene/Aroclors data and Carreau's model B

(1) Carreau (1972) has subjected a constitutive equation [model B: number 6 in Tables 6.4(1) and 6.4(2)] to more tests than anyone else to date, as far as we are aware. For illustration, we select only one of the viscoelastic liquids used: a 4% solution of a polystyrene in Aroclors 1248 [Table 10.3(1)].

Table 10.3(1) Values of Constants in the Constitutive Equation Number 6 of Table 6.4(1) and Table 6.4(2) (Carreau "Model B") Assigned Empirically to Fit All the Data[a] Shown in Fig. 10.3(2) for a Solution Containing 4% by Weight of Polystyrene[b] in Aroclors 1248[c].

Constant	Value	Assigned from
η_0	2500 P	η at low $\dot{s}$
α	2.7	$d\eta'/d\omega$ at high ω
λ_2	4.078 sec	A value of η' at high ω
R	0.1605	N_1 values at two high $\dot{s}$ values
S	0.863	$d\eta/d\dot{s}$ at high $\dot{s}$
λ_1	2.923 sec	A value of η at high $\dot{s}$

[a] Data obtained by Ashare (1967).
[b] $M_w = 1.8 \times 10^6$; $M_w/M_n = 1.20$.
[c] Viscosity 3.0 P at 25°C.

(2) Of the nine constants listed in Table 6.4(2), Carreau took $N = \infty$ (an upper limit in certain summations), $\varepsilon = 0$ (because no data for N_2 were available), and $c = 1$. The remaining six constants were then determined empirically so as to obtain the best overall fit to constant steady-shear-flow values of η and N_1 at various shear rates $\dot{s}$, and oscillatory-shear values of η' and G' at various frequencies ω [Fig. 10.3(2)]. The values of the constants are given in Table 10.3(1). Since only six constants were used to fit four functions

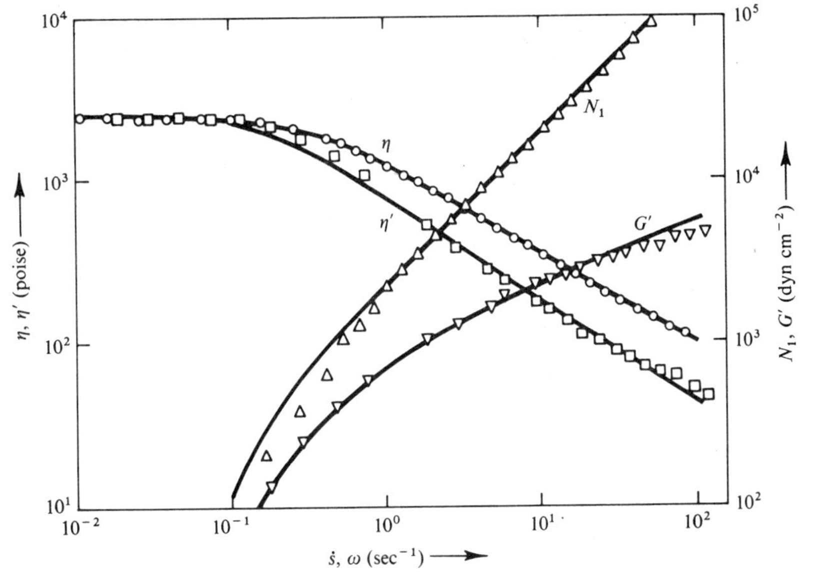

Figure 10.3(2) Steady shear flow and oscillatory shear properties of a 4% solution of polystyrene ($M_w = 1.2M_n = 1.8 \times 10^6$) in Aroclors 1248. Points: measurements by Ashare (1967, 1968). Curves: calculated by Carreau (1972) from constitutive equation number 6 (model B) in Tables 6.4(1) and 6.4(2) using constants [10.3(1)] chosen to give the best fit to all the data shown here.

over ranges of up to four decades in the independent variables, the theory can be seen to be remarkably successful. Data taken by Ashare (1967, 1968) were used.

(3)–(4) The theory was then subjected to much more stringent tests by comparing its predictions with data obtained in shear flow using a step-function shear rate (Macdonald, 1968). Following a sudden startup of shear flow, the transient behavior of N_1 and σ was determined [Figs. 10.3(3),(4)].

(5)–(6) Following sudden cessation of shear flow, the relaxation of N_1 and σ was determined [Figs. 10.3(5), (6)]. It is seen that with the possible exception of N_1 relaxation [Fig. 10.3(5)], the agreement is good.

In order to measure correctly the N_1 transients for viscous liquids in a Weissenberg rheogoniometer, it is important to use a gap of suitable angle and radius and it may be necessary to stiffen the apparatus (Meissner, 1972). Macdonald (1968, p. 71) conducted response-time tests using Newtonian liquids which indicated that the response times for sudden changes of axial

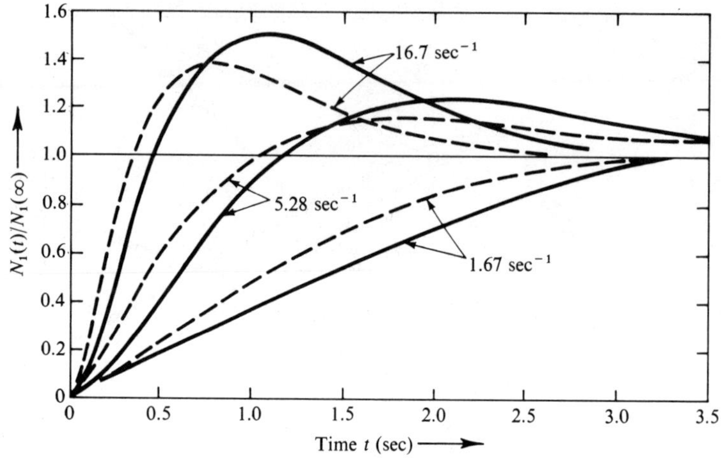

Figure 10.3(3) Variation with time of primary normal stress difference N_1 during shear flow at constant shear rate starting at $t = 0$ for the polymer solution of 10.3(1) and 10.3(2). Full curves: measurements by Macdonald (1968). Dashed curves: calculated by Carreau (1972) for model B with constants chosen to fit data of Fig. 10.3(2).

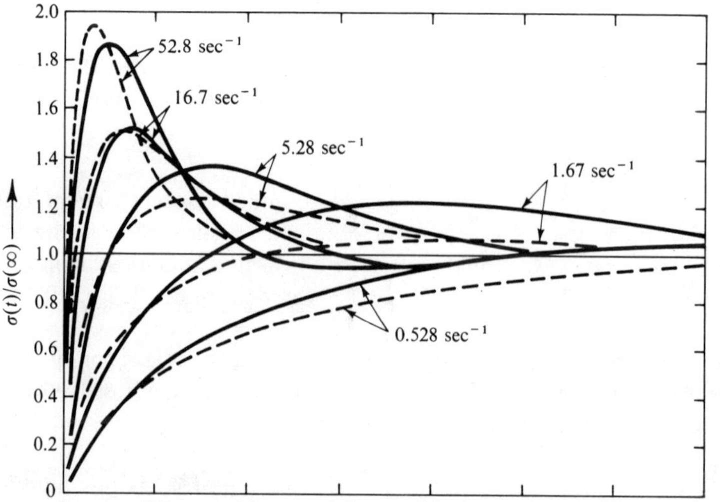

Figure 10.3(4) Similar to Fig. 10.3(3), with shear stress σ in place of N_1.

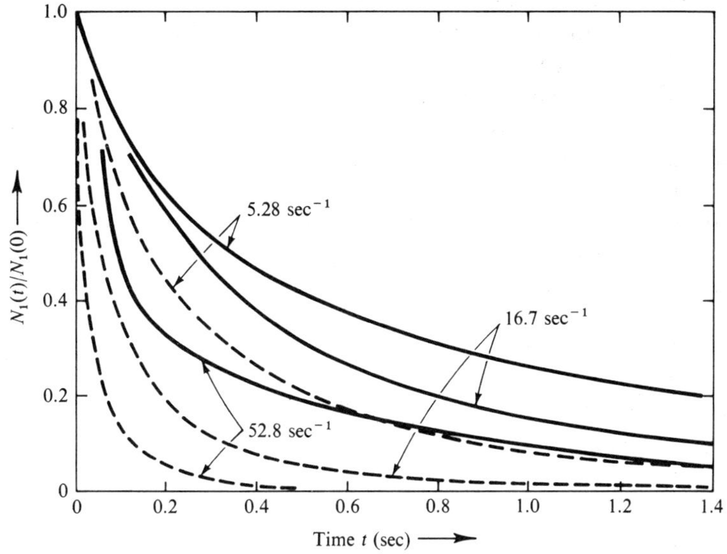

Figure 10.3(5) Variation with time of primary normal stress difference N_1 following cessation, at $t = 0$, of steady shear flow. Other details as in Fig. 10.3(3).

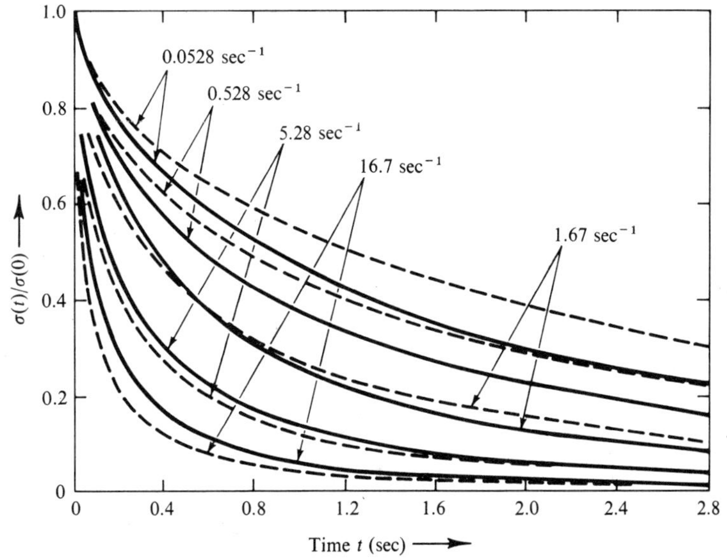

Figure 10.3(6) Similar to Fig. 10.3(5), with shear stress σ in place of N_1.

thrust were some ten times shorter than times involved in the growth and decrease of axial thrust observed after the start of shearing with viscoelastic liquids. This gives reason for confidence in the reliability of his N_1 transient data. Ki Chang (private communication) has recently measured transients with similar viscoelastic liquids, and finds a big change in the time-dependence of axial thrust when the servo system (supplied by the manufacturers) is replaced by a cantilever spring system of the type used by Meissner. This suggests that further investigation is required, even for polymer solutions having viscosities of the order 10^3 poise.

Van Es and Christensen (1973) have recently proposed an experimental test to decide whether a given liquid has a constitutive equation of the form given in line 5 of Table 6.4(1), and report that certain polymer solutions used by Huppler *et al.* (1967) do not satisfy the test. It should, perhaps, be emphasized that this test depends on the fact that the memory function is allowed to depend (in any way) on strain-rate invariants evaluated at time t' [in the notation of Table 6.4(1)], not at other times. The test does not therefore apply to other constitutive equations [such as the Carreau B equation in line 6 of Table 6.4(1)], as one can readily verify; these equations allow the memory function to involve a functional of strain-rate invariants over the time interval (t', t).

10.4 Measurements of N_1 and N_2

A recent review of measurements and theory of N_1 and N_2, the primary and secondary normal stress differences in unidirectional shear flow, has been given by Pipkin and Tanner (1973). Here, we select a few points for special mention. Shortage of space prevents us from giving a comprehensive discussion of the experimental data.

The logarithmic variation of pressure with radius, given by 9.4(17) for the cone-and-plate system, has been verified experimentally by various workers [e.g., Markovitz and Brown (1964), Adams and Lodge (1964)]. In suitable circumstances, the deviations of experimental data points from a straight line, when pressure is plotted as a function of log r, is quite insignificant; using five data points obtained for one polyisobutylene liquid at a shear rate of 6.7 $\sec^{-1}$, and fitting a straight line through the points by the method of least squares, Lipson and Lodge (1968) found that the standard deviation of the slope of the line was as little as 0.27% of the slope.

These results give strong support to the assumptions used to derive 9.4(17), namely, that the extra stress tensor at a particle P can be expressed as an isotropic, one-particle function of the metric tensor history at P [no particular form for this function is used in the derivation of 9.4(17)]. Stark

(1968) has shown, however, that this support is not conclusive; by adding to the integrand of the constitutive equation in Table 6.4(2) terms of the form

$$\gamma^{-1} : \nabla_t \left(a_1 \frac{\partial^2 \gamma'^{-1}}{\partial t'^2} + a_2 \gamma^{-1} \cdot \frac{\partial^2 \gamma'}{\partial t'^2} \cdot \gamma^{-1} \right) \tag{1}$$

where a_1 and a_2 are constants and ∇_t denotes the covariant derivative formed using the body metric tensor at time t, Stark showed that a term

$$-\mu_0 \dot{s}^2 (3a_1 + 4a_2)/r^2 \tag{2}$$

must be added to the logarithmic term in the equation 9.4(17) for the pressure distribution. Here, $\dot{s}$ denotes shear rate, and μ_0 a constant equal to $\int_0^\infty \mu(t)\, dt$, where $\mu(t - t')$ is the memory function.

The presence of the covariant derivative in (1) means that $\mathbf{\Pi}$ is no longer expressible as a *one*-particle function of the history of γ. The calculations verify that, in general, multiparticle functions will give departures from the observed logarithmic pressure-distribution relation; in special cases, however [e.g., using (1) with $3a_1 + 4a_2 = 0$, $a_1 \neq 0$], *the logarithmic relation can be obtained with a multiparticle constitutive function.* Stark also showed that in this case, the effects of the terms (1) would show up in the pressures generated by unidirectional shear flows in other systems (parallel plates and concentric cylinders).

There is, fortunately, no evidence available which suggests that we should consider anything other than one-particle constitutive functions for polymeric liquids. The terms (1) considered by Stark were arbitrarily chosen in an attempt to find a possible explanation of inconsistencies between values of N_1 and N_2 which Adams and Lodge (1964) derived from pressure measurements in cone-and-plate and parallel-plate systems. A more plausible explanation of these inconsistencies, in terms of a systematic error p_{H} introduced into pressure measurements by the use of small holes filled with the viscoelastic liquid under test, was made by Broadbent *et al.* (1968). It is perhaps worth noting that terms of the type (1), since they involve only the body metric tensor and absolute scalars, are automatically admissible in a constitutive equation.

The importance of the hole pressure error p_{H} has been the subject of recent investigations too numerous to review here and can now be regarded as well established for viscoelastic liquids. In slow flow past a hole (or slit), streamlines dip into and out of the hole; if the liquid is Newtonian, symmetry considerations show that there is no effect on the pressure at the base of the hole; if the liquid is viscoelastic, however, the curvature of the shear surfaces combined with nonzero N_1 and N_2 gives rise to a substantial change in the pressure at the base of the hole (Tanner and Pipkin, 1969). Crudely speaking, the effect of a positive N_1 (which is, in a sense, equivalent to a tension along the shear lines) is to pull the liquid out of the hole, so that p_H is negative.

Calculations of p_H have been made for a "second-order liquid" in two-dimensional flow past a slit (Tanner and Pipkin, 1969; Kearsley, 1970) and for general viscoelastic liquids flowing past slits or holes of circular cross section (Higashitani and Pritchard, 1972; Higashitani, 1973). For holes of circular cross section, Higashitani derived the equation

(3) $$p_H \equiv (-p_{22})_{\text{pert}} - (-p_{22})_{\text{unp}} = \int_0^{\sigma_1} \frac{N_2 - N_1}{3\sigma}\, d\sigma,$$

where the integration over shear stress σ is taken up to the value σ_1 appropriate to the local unidirectional shear flow unperturbed by the presence of the hole; $N_2 - N_1$ is to be regarded as a function of σ alone. This result represents an estimate of the effects of hole-induced curvature of "shear" surfaces; the calculation involves neglect of certain terms, but does not require assumptions to be made about the form of the constitutive equation.

(4) Higashitani (1973) found good agreement between the predictions of (3) and values of p_H measured, for aqueous solutions of a polyacrylamide, in a slit-cross-section apparatus containing flush-mounted, and hole-mounted, pressure transducers [Fig. 10.4(4)]; values of N_1 and σ were obtained from measurements in a Weissenberg rheogoniometer; N_2 was assumed to be negligible compared to N_1. Han and Kim (1973) have obtained different results for p_H for similar polymer solutions; they did not, however, change the flow direction to test for effects of slight nonparallelism of the channel walls, and the values of p_H quoted were less than the manufacturer's stated errors for the pressure transducers used. Higashitani used more sensitive transducers.

It now appears that on taking p_H into account wherever appropriate, measured values for N_2 are all negative. Values for N_1 are all positive. In particular, negative values for N_2 have recently been obtained by Christiansen and Miller (1971), Christiansen and Leppard (1974), and van Es (1972), without use of holes.

Pressure distributions for liquids in the cone-and-plate system, measured by Christiansen and Miller using flush-mounted pressure transducers, are qualitatively similar to the corresponding distributions obtained by Hall (1968) in the shear of polymeric solids in the rubberlike state; a flowing polymeric liquid and a sheared rubberlike solid both generate pressures which decrease from a maximum at $r = 0$ to a positive value at $r = R$, where R is the radius of the sample.

Since there is reason to believe that p_H varies linearly with N_1 and N_2, measurements of p_H might be put to practical use in obtaining continuous

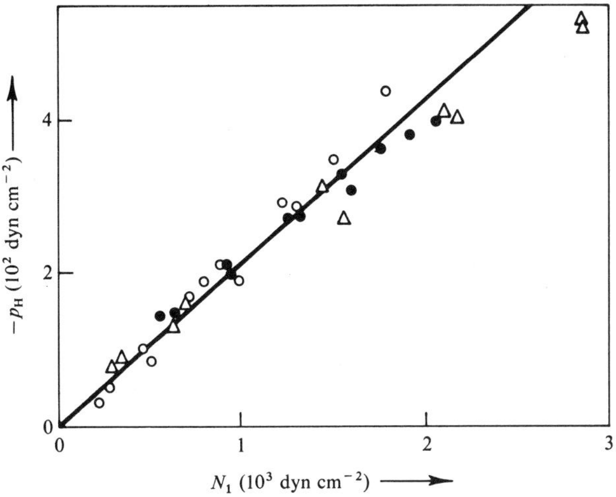

Figure 10.4(4) Relation between hole pressure error p_H and primary normal stress difference N_1. Points represent values of p_H measured using flush-mounted and hole-mounted pressure transducers in the slit-cross-section flow-tube apparatus of Higashitani (1973). The line represents the values of p_H calculated (for holes of circular cross section) from Higashitani's equation 10.4(3) with $N_2 = 0$, using values of N_1 and σ obtained from measurements in a Weissenberg rheogoniometer. Three concentrations of Separan AP30 in water are used: 0.6% (open circles); 1.5% (triangles); 3% (filled circles); for some points, the flow direction along the tube is reversed.

information (different from that obtainable, say, from viscosity measurements) from polymer solutions or molten polymers flowing through pipes (Kearsley, 1970; Lodge, 1973b).

11 RELATIONS BETWEEN BODY- AND SPACE-TENSOR FORMALISMS

11.1 Convected components

In continuum mechanics, a so-called "convected coordinate system" has been used by Hencky (1925) to describe strain rates; by Brillouin (1925) to simplify the stress–strain relation for a perfectly elastic solid [see 6.3(4)]; by Deuker (1941) to express the stress equations of motion; by Gleyzal (1949), Green and Zerna (1968), and Green and Adkins (1960) to formulate and solve problems in finite-strain elasticity theory; and by Oldroyd (1950a) to formulate admissible constitutive equations for viscoelastic materials. To establish the relation between our present body tensor formalism and these convected coordinate system, or convected component, formalisms, it will be sufficient to consider Oldroyd's (1950a) paper, which is of fundamental and far-reaching importance in polymer rheology. The body tensor formalism can be regarded as a "coordinate-free convected component formalism" and represents but a small step further along the road taken by Oldroyd (Lodge, 1951).

According to the definitions given in Chapters 2 and 4, *a tensor is not the same thing as its components.* In order to define and use body tensor fields and to grasp the distinction between them and space tensor fields, *it is essential to avoid using the common convention which* (for example) *uses the phrase "the tensor g_{ij}" and thus confuses a tensor with its components.* We use instead the phrase "the tensor $\boldsymbol{g}$ whose components in a coordinate system S are g_{ij}." It is, of course, widely recognized that this is what the phrase really means, but no definition of the tensor $\boldsymbol{g}$ is usually offered; the tensor components g_{ij} alone are defined. In applications (such as general relativity theory and differential geometry) *where one and only one geometric manifold is involved,* the confusion sometimes engendered by the common convention can perhaps be tolerated; in rheology, on the other hand, where two distinct geometric manifolds (body and space) are involved, the common convention wastes more time than it saves and should, we believe, be abandoned.

Since Oldroyd (1950a) uses the common convention (on p. 524, for example, he writes "where g_{ik} is the metric tensor") without defining the term tensor as distinct from its components, it appears that, in the first instance at least, he is using *components of space tensor fields referred to a space coordinate system* (to use our terminology). What he then calls "convected components" *are components* (of the space metric tensor field, for example) *referred to members of a one-parameter family of space coordinate systems chosen in a special manner*, as will now be defined.

(1) **Definition** Any given body coordinate system B and motion of the body together define a one-parameter family $\{\hat{S}(B, t)\}$ of space coordinate systems in which the time t is the parameter and for each value of t, the space coordinate system $\hat{S}(B, t)$ is congruent to B at time t. The set $\{\hat{S}(B, t)\}$ is called a *convected family of space coordinate systems.* Using $\hat{x}^i$ to denote the coordinates of a place Q referred to the convected coordinate system $\hat{S}(B, t)$, we have $\hat{S}(B, t) \overset{t}{\equiv} B$ and hence

$$\text{(2)} \qquad \hat{x}^i = \xi^i \qquad (\text{Q occupied by P at time } t; \quad B : \text{P} \to \xi),$$

from the definition 2.2(23) of congruent coordinate systems.

It should be noted that *a convected family of space coordinate systems is to be distinguished from the body coordinate system that generates it*: The convected family depends on the motion; the body coordinate system does not. *The "convected coordinates" $\hat{x}^i$ and the body coordinates ξ^i are thus different in definition although* [from (2)] *equal in value.*

For any given value of t, $\hat{S}(B, t)$ is a space coordinate system and therefore possesses fields of covariant and contravariant base vectors which we shall denote by $\hat{\boldsymbol{b}}^i$ and $\hat{\boldsymbol{b}}_i$, where

$$\text{(3)} \qquad \hat{\boldsymbol{b}}^i = \hat{\boldsymbol{b}}^i(t) = \boldsymbol{b}^i(\hat{S}(B, t), \text{Q}) \qquad \text{and} \qquad \hat{\boldsymbol{b}}_i = \hat{\boldsymbol{b}}_i(t) = \boldsymbol{b}_i(\hat{S}(B, t), \text{Q}),$$

so that from the definitions 2.3(35) and 2.4(21), we have

$$\text{(4)} \qquad \hat{\boldsymbol{b}}^i\hat{S} = \delta^i, \qquad \hat{\boldsymbol{b}}_i\hat{S} = \delta_i \qquad (i = 1, 2, 3).$$

It follows from 2.3(22), 2.4(7), and 5.4(5) that the "convected base vectors" $\hat{\boldsymbol{b}}^i$ and $\hat{\boldsymbol{b}}_i$ are simply related to the body base vectors $\boldsymbol{\beta}^i(B, \mathrm{P})$ and $\boldsymbol{\beta}_i(B, \mathrm{P})$ for the body coordinate system B that generates the convected family:

$$\text{(5)} \qquad \hat{\boldsymbol{b}}^i(t) = \mathbb{T}(t)\boldsymbol{\beta}^i, \qquad \hat{\boldsymbol{b}}_i(t) = \mathbb{T}(t)\boldsymbol{\beta}_i.$$

Recalling the fact that the body base vectors are independent of time, it is evident that the time dependence of the convected base vectors can be attributed entirely to the time dependence of the transfer operator $\mathbb{T}(t)$.

(6) **Problem** Prove that the hydrodynamic derivatives of the convected base vectors are given by the equations

$$D\hat{\boldsymbol{b}}_i/Dt = \hat{\boldsymbol{b}}_i \cdot \nabla\boldsymbol{v}, \qquad D\hat{\boldsymbol{b}}^i/Dt = -(\nabla\boldsymbol{v}) \cdot \hat{\boldsymbol{b}}^i.$$

If we use $\hat{g}_{ij}(\hat{x}, t)$ to denote the convected components of the space metric tensor field $\boldsymbol{g}(\mathrm{Q})$ (i.e., the components of $\boldsymbol{g}$ referred to the space coordinate system $\hat{S}$), then we have [as an application of 2.5(18)(b)]

$$\text{(7)} \qquad \boldsymbol{g}(\mathrm{Q}) = \hat{g}_{ij}(\hat{x}, t)\hat{\boldsymbol{b}}^i(t)\hat{\boldsymbol{b}}^j(t), \qquad \boldsymbol{g}^{-1}(\mathrm{Q}) = \hat{g}^{ij}(\hat{x}, t)\hat{\boldsymbol{b}}_i(t)\hat{\boldsymbol{b}}_j(t).$$

Using the transfer operator $\mathbb{T}^{-1}(t)$ on these equations, we obtain the equations

$$\text{(8)} \qquad \boldsymbol{\gamma}(\mathrm{P}, t) = \hat{g}_{ij}(\hat{x}, t)\boldsymbol{\beta}^i\boldsymbol{\beta}^j, \qquad \boldsymbol{\gamma}^{-1}(\mathrm{P}, t) = \hat{g}^{ij}(\hat{x}, t)\boldsymbol{\beta}_i\boldsymbol{\beta}_j;$$

on comparing these with 2.6(16)–(17), it follows that

$$\text{(9)} \qquad \hat{g}_{ij}(\hat{x}, t) = \gamma_{ij}(\xi, t), \qquad \hat{g}^{ij}(\hat{x}, t) = \gamma^{ij}(\xi, t).$$

These are particular examples of the following general result.

(10) **Theorem** The convected components $\hat{F}^{i\cdots}_{r\ldots}(\hat{x}, t)$ of an arbitrary space tensor field $\boldsymbol{F}(\mathrm{Q})$ are equal to the corresponding components $\Phi^{i\cdots}_{r\ldots}(\xi, t)$ of the body tensor field $\boldsymbol{\Phi}(\mathrm{P}, t) = \mathbb{T}^{-1}(t)\boldsymbol{F}(\mathrm{Q})$ referred to the body coordinate system B that is used to generate the convected family of space coordinate systems.

Proof We have $\boldsymbol{F}(\mathrm{Q}) = \hat{F}^{i\cdots}_{r\ldots}(\hat{x}, t)\hat{\boldsymbol{b}}_i(t) \cdots \hat{\boldsymbol{b}}^r(t) \cdots$, where the base vectors are to be arranged in the correct order; hence $\boldsymbol{\Phi}(\mathrm{P}, t) = \mathbb{T}^{-1}(t)\boldsymbol{F}(\mathrm{Q}) = \hat{F}^{i\cdots}_{r\ldots}(\hat{x}, t)\boldsymbol{\beta}_i \cdots \boldsymbol{\beta}^r \cdots = \Phi^{i\cdots}_{r\ldots}(\xi, t)\boldsymbol{\beta}_i \cdots \boldsymbol{\beta}^r \cdots$. Hence $\hat{F}^{i\cdots}_{r\ldots} = \Phi^{i\cdots}_{r\ldots}$, which proves the theorem.

We can also use Cartesian vectors for the tangent and normal convected base vectors; the Cartesian equations corresponding to (7) are evidently

$$\text{(11)} \qquad \mathbf{U} = \hat{g}_{ij}(\hat{x}, t)\hat{\mathbf{b}}^i(t)\hat{\mathbf{b}}^j(t) = \hat{g}^{ij}(\hat{x}, t)\hat{\mathbf{b}}_i(t)\hat{\mathbf{b}}_j(t),$$

from 4.5(13)(i).

Finally, we may also define the convected components $\hat{p}^{ij}$ of stress, where

$$\text{(12)} \qquad \boldsymbol{p}(\mathrm{Q}, t) = \hat{p}^{ij}(\hat{x}, t)\hat{\boldsymbol{b}}_i(t)\hat{\boldsymbol{b}}_j(t), \qquad \hat{p}^{ij}(\hat{x}, t) = \pi^{ij}(\xi, t).$$

11.2 Embedded vectors

In the book "Elastic Liquids" (Lodge, 1964), the rheological behavior of a material undergoing arbitrary *homogeneous* deformations is described in terms of "embedded vectors" $\mathbf{e}_i(t)$; a given vector function of time is said to be *embedded* in the deforming material if it is always represented by a material straight line joining a given pair of particles. The embedded vectors "join" particles, which are not restricted to be neighboring, and are introduced via the familiar laws of elementary vector analysis. *Embedded vectors should therefore be regarded as Cartesian space vectors* according to the definition given in Section 4.2. Vectors associated with any two places belong to the same linear space, and no distinction is made between covariant vectors and contravariant vectors. Before we derive the relation between the embedded vector formalism and the body tensor formalism, it is helpful to note the equivalence of various *alternative definitions of homogeneous deformation*:

(1) **Problem** Prove that the following statements are equivalent:

(i) $\boldsymbol{\nabla}_t \boldsymbol{\gamma}(\mathrm{P}, t') = \mathbf{0}$;

(ii) $\boldsymbol{\nabla}_{t'} \boldsymbol{\gamma}(\mathrm{P}, t) = \mathbf{0}$;

(iii) there exists a $B : \mathrm{P} \to \xi$ such that $\gamma_{ij}(\xi, t')$ and $\gamma_{ij}(\xi, t)$ are independent of ξ;

(iv) there exists a rectangular Cartesian $C : \mathrm{Q} \to y$ such that $y'^i = b^i{}_j y^j + a^i$, where $b^i{}_j$ and a^i are independent of y^i and y'^i;

(v) $\boldsymbol{\nabla} \boldsymbol{C}_t(\mathrm{Q}, t') = \mathbf{0}$;

(vi) the components of the Cauchy strain tensor $\boldsymbol{C}_t(\mathrm{Q}, t')$ referred to any rectangular Cartesian $C : \mathrm{Q} \to y$ are independent of y.

The relation between the embedded vector and body tensor formalisms is embodied in the following problem.

(2) **Problem** If $\mathbf{e}_i$ $(i = 1, 2, 3)$ denote any three linearly independent embedded vectors, prove that there exists a body coordinate system $B : \mathrm{P} \to \xi$

such that for any two states related by a homogeneous deformation, we have $\gamma_{ij} = \mathbf{e}_i \cdot \mathbf{e}_j$, $\gamma^{ij} = \mathbf{e}^i \cdot \mathbf{e}^j$, $\overrightarrow{P_0 P} = \xi^i \mathbf{e}_i$, and $\mathbf{f} = \pi^{ij} n_i \mathbf{e}_j$, where the ξ^i are the coordinate differences for any two particles P_0 and P, the $\mathbf{e}^i$ are reciprocal to the $\mathbf{e}_i$, and $\mathbf{f}$ is the traction acting across a plane of unit normal $\mathbf{n} = n_i \mathbf{e}^i$.

From our experience with body tensors, we can now see why the embedded vector formalism has advantages for the description of rheological behavior of viscoelastic materials subjected to flow histories which are homogeneous but otherwise arbitrary: The embedded vector formalism shares the advantages of the body tensor formalism (referred in this situation to a nonrectangular Cartesian body coordinate system), and so, for example, the formulation of admissible constitutive equations using π^{ij} and γ_{ij} is straightforward and the admissible time derivatives are simply time derivatives of π^{ij} and γ_{ij}. The restriction to flow histories which are homogeneous enables one to define π^{ij} and γ_{ij} in terms of Cartesian vectors without using the somewhat more formidable apparatus of body tensor fields and general tensor analysis. A principal aim of this book is thus to generalize the elementary treatment given in "Elastic Liquids" (Lodge, 1964) in order to remove the restriction to homogeneous flow histories.

11.3 Objectivity condition for space-tensor constitutive equations

Most of the textbooks and much of the published literature in rheology and continuum mechanics use vectors and tensors which are Cartesian, as defined (for example) in Section 4.2, when judged according to the following properties:

(1) *Characteristic properties of Cartesian vectors*

- (i) A given Cartesian vector has one, and only one, magnitude.
- (ii) There is a Cartesian vector $\mathbf{r} = \overrightarrow{QQ'}$ for *any* two given places Q and Q′.
- (iii) All Cartesian vectors associated with a given geometric manifold belong to one linear space; in particular, vectors associated with nonneighboring places can be added, and no distinction can be drawn between covariant vectors and contravariant vectors.

For example, in the book "Theoretical Elasticity" by Green and Zerna (1968), the various boldface symbols (used, e.g., on pp. 18, 19) representing vectors are derived from a "position vector $\mathbf{R}$" $= \overrightarrow{OP}$, where O and P are places not restricted to be neighboring. Although the tangent base vectors $\mathbf{g}_i = \partial \mathbf{R}/\partial \theta^i$ associated with an arbitrary system of curvilinear coordinates θ^i

are called "covariant" base vectors and the reciprocal set $\mathbf{g}^r$ is called "contravariant" base vectors [see (1.9.6), (1.9.7), and (1.9.12)], they do, in fact, belong to the same linear space, as one can see, for example, from the equation $\mathbf{v} = v^r \mathbf{g}_r = v_s \mathbf{g}^s$ (1.9.22). Green and Zerna [1968, p. 21] write†: "The three elements v^r are sometimes called the *contravariant components* of the vector $\mathbf{v}$, and the three elements v_r are called the *covariant components* of $\mathbf{v}$.... The word 'component' is frequently omitted and we refer to v_r and v^r as covariant and contravariant vectors respectively, but we emphasize that from this point of view they are not different vectors but two aspects of the single space vector $\mathbf{v}$." Equation (1.9.22) of Green and Zerna is the same as our Cartesian vector equation 4.5(7) when the appropriate changes of notation are made (i.e., for $\mathbf{g}_i$, $\mathbf{g}^i$, and θ, put $\mathbf{b}_i$, $\mathbf{b}^i$, and x). For these reasons, it follows that *the boldface symbols of Green and Zerna represent Cartesian space vectors*, not general space vectors as defined in Sections 2.3 and 2.4. Similar conclusions can be drawn (Lodge, 1972, Section 6) for the vectors in the books of Sedov (1966), Green and Adkins (1960), Eringen (1962), Fredrickson (1964), Coleman *et al.* (1966), and Truesdell and Noll (1965).

We have already remarked in Section 11.1 that we do *not* follow the common convention, quoted from Green and Zerna: "The word 'component' is frequently omitted and we refer to v_r and v^r as covariant and contravariant vectors, respectively...." It is perhaps this convention which is mainly responsible for the remarkable fact that it is only recently (Lodge, 1951, 1972) that body tensor fields have been defined and used in continuum mechanics. We have shown, in Chapters 2 and 4 especially, that *it is essential to regard a vector and its components as two different entities if one wishes to recognize the differences among Cartesian space vector fields, general space vector fields, and body vector fields.*

Following Oldroyd's convected-component formulation of general constitutive equations for viscoelastic materials (Oldroyd, 1950a), various papers appeared in which the fundamental kinematic variables were Cartesian tensors describing the deformation gradient $\mathbf{F}(y, t, t')$ (Noll, 1955, 1958; Coleman *et al.*, 1966) or the displacement gradient in combination with gradients of velocity, acceleration, etc. (Rivlin and Ericksen, 1955). Using Cartesian base vectors $\mathbf{b}_i(C) = \mathbf{b}^i(C)$ for a rectangular Cartesian space coordinate system $C : \mathrm{Q} \to y$, the deformation gradient tensor $\mathbf{F}$ can be defined by the equations

$$(2) \qquad \mathbf{F}(y, t, t') = (\partial y'^i / \partial y^j) \mathbf{b}_i(C) \mathbf{b}^j(C); \qquad y'^i = \chi^i(y, t, t'),$$

where the partial derivatives are formed using the functions $\chi^i(y, t, t')$, whose values y'^i are the coordinates of the place Q′ occupied at time t' by the particle P that is at place Q (of coordinates y^i) at time t. Using the definition of

† A. E. Green and W. Zerna, "Theoretical Elasticity," 2nd. ed., p. 21. © 1968, Oxford University Press; by permission of The Clarendon Press, Oxford.

∇ for Cartesian tensor fields given in 8.2(2), we see that

$$\mathbf{F}_t(t') \equiv \mathbf{F}(y, t, t') = \widetilde{\nabla \mathbf{y}'}, \qquad \mathbf{y}' = y'^i \mathbf{b}_i(C), \tag{3}$$

where $\mathbf{y}' = \overrightarrow{\mathrm{OQ}'}$ and O denotes the origin of C. The definition (2), requiring the transpose in (3), is used in order to agree with that of Coleman *et al.* (1966).

In the context of formulating admissible constitutive equations for viscoelastic materials using different tools, the Cartesian deformation-gradient formalism has been compared with the convected-component formalism by Walters (1965) and with the body tensor formalism by Lodge and Stark (1972). For brevity, we shall simply state the main features of the latter comparison for isotropic, incompressible materials whose constitutive equations can be expressed in either of the following forms:

$$\mathbf{\Pi}(\mathrm{P}, t) = \underset{-\infty}{\overset{t}{\mathscr{F}}} \{\gamma(\mathrm{P}, t')\} \qquad \text{(body tensors)}; \tag{4}$$

$$\mathbf{P}(\mathbf{y}, t) = \underset{-\infty}{\overset{t}{\mathscr{H}}} \{\mathbf{F}(\mathbf{y}, t, t')\} \qquad \text{(Cartesian space tensors)}. \tag{5†}$$

Equation (5) expresses the Cartesian extra-stress tensor $\mathbf{P}(\mathbf{y}, t) \equiv \mathbb{T}(t)\mathbf{\Pi}(\mathrm{P}, t)$ as a function $\mathscr{H}$ of the deformation gradient history. The argument $\mathbf{y}$ denotes the position vector of the place that particle P occupies at time t.

An important difference between the two formalisms is the following. *Every body tensor equation of the form* (4) (in which $\mathscr{F}$, being a t-isotropic, one-particle function, can involve the unit tensor $\boldsymbol{\delta}$ but no other tensor) *is admissible as a constitutive equation. Not every Cartesian tensor equation of the form* (5) *is admissible as a constitutive equation*, but only those that satisfy the "material objectivity" condition [see (8)]. It is instructive to compare this condition with two others, as expressed in the following equations:

$$\Lambda \mathscr{F}\{\gamma'\}\tilde{\Lambda} = \mathscr{F}\{\tilde{\Lambda}^{-1}\gamma'\Lambda^{-1}\} \qquad \begin{pmatrix}\text{body tensor condition; arbitrary} \\ \Lambda\text{, independent of time}\end{pmatrix} \tag{6}$$

$$L\mathscr{H}\{\mathsf{F}\}\tilde{L} = \mathscr{H}\{L\mathsf{F}\tilde{L}\} \qquad \begin{pmatrix}\text{Cartesian tensor condition; arbi-} \\ \text{trary orthogonal } L\text{, independent} \\ \text{of time}\end{pmatrix} \tag{7}$$

$$R(t)\mathscr{H}\{\mathsf{F}\}\tilde{R}(t) = \mathscr{H}\{R(t')\mathsf{F}\tilde{R}(t)\} \qquad \begin{pmatrix}\text{material objectivity condition; ar-} \\ \text{bitrary orthogonal } R(t')\text{, depend-} \\ \text{ent on time}\end{pmatrix} \tag{8}$$

Equation (6) follows from the matrix representations $\Pi = \mathscr{F}\{\gamma'\}$ and $\bar{\Pi} = \mathscr{F}\{\bar{\gamma}'\}$ of (4) in two arbitrary body coordinate systems B and $\bar{B}$, on using the transformation laws $\bar{\Pi} = \Lambda\Pi\tilde{\Lambda}$ and $\gamma' = \tilde{\Lambda}\bar{\gamma}'\Lambda$, which come from 2.5(8), where

† Equations (5) and (8) describe a "simple fluid."

$\Lambda \equiv \Lambda_P(B, \bar{B}) = [\partial\bar{\xi}^r/\partial\xi^c]$. At any given particle P, the transformation matrix Λ can be assigned an arbitrary value. Similarly, (7) follows from the matrix representations $\mathsf{P} = \mathscr{H}\{\mathsf{F}\}$ and $\bar{\mathsf{P}} = \mathscr{H}\{\bar{\mathsf{F}}\}$ of (5) in two arbitrary rectangular Cartesian coordinate systems C and $\bar{C}$, on using the transformation laws $\bar{\mathsf{P}} = L\mathsf{P}\tilde{L}$ and $\bar{\mathsf{F}} = L\mathsf{F}\tilde{L}$ [see 4.3(2)], where $L \equiv L(C, \bar{C})$ is an arbitrary, constant, orthogonal matrix. Finally, Lodge and Stark (1972) have shown that the material objectivity condition (8) can readily be derived from the body tensor condition (6) and that (4) and (5) [with $\mathscr{H}$ satisfying (8)] are equivalent and alternative formulations of the same constitutive equation; when referred to an arbitrary body coordinate system B and rectangular space coordinate system C, the two functions are related by the matrix equation

$$\text{(9)} \qquad \gamma^{1/2}\mathscr{F}\{\gamma'\}\gamma^{1/2} = \mathscr{H}\{(\gamma^{-1/2}\gamma'\gamma^{-1/2})^{1/2}\} \qquad (|\gamma| = |\gamma'|),$$

where γ and γ' denote the values at times t and t' of an arbitrary, positive-definite (symmetric) matrix function of time which satisfies the constant-volume condition $|\gamma| = |\gamma'|$. If $\mathscr{H}\{\mathsf{F}\}$ is given, where F is the deformation gradient matrix (so that $|\mathsf{F}| = 1$, for constant volume), then (9) enables one to determine $\mathscr{F}$, for an arbitrary body coordinate system B. If $\mathscr{F}\{\gamma'\}$ is given, so that $\mathscr{F}\{\gamma'\}$ is known for an arbitrary B, then $\mathscr{H}\{\mathsf{F}\}$ for an arbitrary rectangular Cartesian C can be determined from the equation

$$\text{(10)} \qquad \mathscr{H}\{\mathsf{F}\} = \mathscr{F}\{\tilde{\mathsf{F}}\mathsf{F}\} \qquad (B \stackrel{t}{\equiv} C;\ \ \gamma = I),$$

where F denotes the deformation gradient matrix $\mathbf{F}C$, which satisfies the constant-volume condition $|\mathsf{F}| = 1$. In (9), $\gamma^{1/2}$ denotes the square root of a positive-definite matrix, as defined, say, by 1.3(19) and is not the representative of a body tensor. The square root of a mixed tensor has been defined in 5.7(7); γ represents a covariant tensor $\boldsymbol{\gamma}$, which cannot possess a square root because the contraction of two covariant tensors is not an admissible tensor operation.

It is easy to show from (8) and 1.3(23) that (5) can be rewritten in the form $\mathbf{P} = \mathscr{H}\{(\tilde{\mathbf{F}}\cdot\mathbf{F})^{1/2}\} \equiv \mathscr{G}\{\tilde{\mathbf{F}}\cdot\mathbf{F}\}$, say. But it follows from 5.4(57) with $S \equiv C$ that the Cartesian Cauchy strain tensor $\mathbf{C}$ satisfies the equation

$$\text{(11)} \qquad \mathbf{C}_t(\mathrm{Q}, t') = \tilde{\mathbf{F}}\cdot\mathbf{F} \qquad [\mathbf{F} = \mathbf{F}(\mathbf{y}, t, t') \equiv \mathbf{F}_t(t')]$$

and hence (5) is equivalent to the equation

$$\text{(12)} \qquad \mathbf{P}(\mathrm{Q}, t) = \mathop{\mathscr{G}}_{t'=-\infty}^{t}\{\mathbf{C}_t(\mathrm{Q}, t')\},$$

where the functional $\mathscr{G}$ satisfies the material objectivity condition

$$\text{(13)} \qquad R(t)\mathscr{G}[\mathsf{C}_t(\mathrm{Q}, t')]\tilde{R}(t) = \mathscr{G}[R(t)\mathsf{C}_t(\mathrm{Q}, t')\tilde{R}(t)] \qquad [\tilde{R}(t) = R^{-1}(t)],$$

which follows from (8) and (11).

Since $R(t')$ is absent from this equation [but is present in (8)] and since this equation is similar in form to the Cartesian tensor condition

$$(14)\qquad L\mathcal{G}\{\mathbf{C}_t(\mathrm{Q}, t')\}\tilde{L} = \mathcal{G}\{L\mathbf{C}_t(\mathrm{Q}, t')\tilde{L}\}, \qquad (\tilde{L} = L^{-1}, \text{ independent of } t)$$

it is just possible that the reader may be tempted to think that every Cartesian tensor equation in which the only tensors are $\mathbf{P}(\mathrm{Q}, t)$, $\mathbf{C}_t(\mathrm{Q}, t')$, and $\mathbf{U}$ (and these are everywhere evaluated at the same particle) is an admissible constitutive equation, in conflict with the statement made after Eq. (5). The following counterexample suffices to prove that *the Cartesian tensor condition* (14) *is not sufficient to ensure that* (12) *is admissible as a constitutive equation*:

$$(15)\qquad \mathbf{P}(\mathrm{Q}, t) = \int_{-\infty}^{t} m(t - t')\frac{D}{Dt}\mathbf{B}_t(\mathrm{Q}, t')\,dt' \qquad \text{(not admissible).}$$

Here, $\mathbf{B}_t(\mathrm{Q}, t') \equiv \mathbf{C}_t^{-1}(\mathrm{Q}, t')$ denotes the Cartesian Finger strain tensor, and the use of the (Cartesian) hydrodynamic derivative operator

$$(16)\qquad \frac{D}{Dt} = \frac{\partial}{\partial t} + \mathbf{v}(\mathrm{Q}, t)\cdot\nabla$$

ensures that Eq. (15) is a "one-particle" equation. It is evidently a Cartesian tensor equation, whose representative in any C satisfies (14). To show, however, that it is inadmissible as a constitutive equation, it is perhaps simplest to obtain the corresponding body tensor equation obtained by transfer to the body at time t. From 5.4(14)(ii), 5.4(47), and the extra-stress analog of 5.6(14), namely

$$(17)\qquad \mathbf{P}(\mathrm{Q}, t) = \mathbb{T}(t)\boldsymbol{\Pi}(\mathrm{P}, t) \qquad (\mathrm{P} \overset{t}{\leftrightarrow} \mathrm{Q}),$$

it follows that (15) leads to the equation

$$(18)\qquad \boldsymbol{\Pi}(\mathrm{P}, t) = \int_{-\infty}^{t} m(t - t')\mathbb{T}^{-1}(t)\frac{D}{Dt}\mathbf{B}_t(\mathrm{Q}, t')\,dt') \qquad \text{(not admissible).}$$

It follows from 5.4(64), 5.4(60), and (16), after some straightforward reduction, that

$$(19)\qquad \mathbb{T}(t)\frac{\partial\boldsymbol{\pi}(\mathrm{P}, t)}{\partial t} = \left\{\frac{D}{Dt}\mathbf{p}(\mathrm{Q}, t) + \mathbf{w}\cdot\mathbf{p} - \mathbf{p}\cdot\mathbf{w}\right\} - \frac{1}{2}(\mathbf{A}^1\cdot\mathbf{p} + \mathbf{p}\cdot\mathbf{A}^1),$$

where

$$(20)\qquad 2\mathbf{w} \equiv \mathrm{Curl}\ \mathbf{v} = \nabla\mathbf{v} - \widetilde{\nabla\mathbf{v}}, \qquad \mathbf{A}^1 = \nabla\mathbf{v} + \widetilde{\nabla\mathbf{v}},$$

so that

$$(21)\qquad \nabla\mathbf{v} = \tfrac{1}{2}\mathbf{A}^1 + \mathbf{w}, \qquad \widetilde{\nabla\mathbf{v}} = \tfrac{1}{2}\mathbf{A}^1 - \mathbf{w}.$$

w is called the (Cartesian) *vorticity tensor*; on transfer to the body, it gives a unique absolute, covariant, second-rank *body vorticity tensor* $\boldsymbol{\omega}$:

$$\boldsymbol{\omega}(\mathrm{P}, t) = \mathbb{T}^{-1}(t)\mathbf{w}(\mathrm{Q}, t) \qquad (\mathrm{P} \overset{t}{\leftrightarrow} \mathrm{Q}). \tag{22}$$

Equation (19) is valid for any absolute, contravariant, second-rank body tensor field $\boldsymbol{\pi}(\mathrm{P}, t)$, where $\mathbf{p}(\mathrm{Q}, t) = \mathbb{T}(t)\boldsymbol{\pi}(\mathrm{P}, t)$ is the corresponding Cartesian space tensor field, and, from 5.4(60) and 5.4(59), leads to the following equation for the transfer of the hydrodynamic derivative of a second-rank Cartesian tensor field $\mathbf{p}(\mathrm{Q}, t)$:

$$\mathbb{T}^{-1}(t)\frac{D}{Dt}\mathbf{p}(\mathrm{Q}, t) = \left\{\frac{\partial}{\partial t}\boldsymbol{\pi}(\mathrm{P}, t) - \boldsymbol{\gamma}^{-1}\cdot\boldsymbol{\omega}\cdot\boldsymbol{\pi} + \boldsymbol{\pi}\cdot\boldsymbol{\omega}\cdot\boldsymbol{\gamma}^{-1}\right\} + \frac{1}{2}(\boldsymbol{\gamma}^{-1}\cdot\dot{\boldsymbol{\gamma}}\cdot\boldsymbol{\pi} + \boldsymbol{\pi}\cdot\dot{\boldsymbol{\gamma}}\cdot\boldsymbol{\gamma}^{-1}). \tag{23}$$

This involves the contravariant tensor $\boldsymbol{\pi}$; a similar equation for the covariant tensor $\boldsymbol{\pi}^0 = \mathbb{T}^{-1}(t)\mathbf{p}$, obtainable from 5.4(66) with $\boldsymbol{\pi}^0 = \boldsymbol{\gamma}\cdot\boldsymbol{\pi}\cdot\boldsymbol{\gamma}$, is readily found to be expressible as follows:

$$\mathbb{T}^{-1}(t)\frac{D}{Dt}\mathbf{p}(\mathrm{Q}, t) = \left\{\frac{\partial}{\partial t}\boldsymbol{\pi}^0(\mathrm{P}, t) - \boldsymbol{\omega}\cdot\boldsymbol{\gamma}^{-1}\cdot\boldsymbol{\pi}^0 + \boldsymbol{\pi}^0\cdot\boldsymbol{\gamma}^{-1}\cdot\boldsymbol{\omega}\right\} - \frac{1}{2}(\dot{\boldsymbol{\gamma}}\cdot\boldsymbol{\gamma}^{-1}\cdot\boldsymbol{\pi}^0 + \boldsymbol{\pi}^0\cdot\boldsymbol{\gamma}^{-1}\cdot\dot{\boldsymbol{\gamma}}). \tag{24}$$

The different sign for the $\dot{\boldsymbol{\gamma}}$ terms is the essential difference between (23) and (24).

We may apply (23) with $\mathbf{p}$ and $\boldsymbol{\pi}$ replaced by $\mathbf{B}_t(\mathrm{Q}, t')$ and $\boldsymbol{\gamma}'^{-1} = \boldsymbol{\gamma}^{-1}(\mathrm{P}, t')$, since these are related by the operator $\mathbb{T}(t)$, according to 5.4(14)(ii). Since $\partial\boldsymbol{\gamma}'^{-1}/\partial t = 0$, the result is

$$\mathbb{T}^{-1}(t)\frac{D}{Dt}\mathbf{B}_t(\mathrm{Q}, t') = \{\boldsymbol{\gamma}'^{-1}\cdot\boldsymbol{\omega}\cdot\boldsymbol{\gamma}^{-1} - \boldsymbol{\gamma}^{-1}\cdot\boldsymbol{\omega}\cdot\boldsymbol{\gamma}'^{-1}\} + \frac{1}{2}(\boldsymbol{\gamma}^{-1}\cdot\dot{\boldsymbol{\gamma}}\cdot\boldsymbol{\gamma}'^{-1} + \boldsymbol{\gamma}'^{-1}\cdot\dot{\boldsymbol{\gamma}}\cdot\boldsymbol{\gamma}^{-1}). \tag{25}$$

When this expression is substituted in (18), it is seen that we obtain an equation which involves not only the body stress tensor $\boldsymbol{\pi}$ and metric tensor $\boldsymbol{\gamma}$, but also the body vorticity tensor $\boldsymbol{\omega}$. *Since the body vorticity tensor is present, the equation is inadmissible as a constitutive equation* because the body vorticity tensor introduces a dependence on motion of the body relative to axes fixed in space, as the following simple example shows.

(26) **Problem** A body rotates rigidly at an angular speed $\dot{\phi}$ about an axis fixed in space. Show that the body vorticity tensor is $\boldsymbol{\omega} = \dot{\phi}(\boldsymbol{\beta}^2\boldsymbol{\beta}^3 - \boldsymbol{\beta}^3\boldsymbol{\beta}^2)$, where $\boldsymbol{\beta}^i$ is an orthonormal set of covariant body vectors with $\boldsymbol{\beta}^1$ along the axis of rotation.

11.4 Formulation of constitutive equations: historical note

Zaremba (1903), having shown that certain equations proposed by Natanson were inadmissible as constitutive equations, proposed certain equations [6.5(1)(iv)], involving the "corotational" (or "Jaumann") time derivative of the Cartesian stress tensor, which are admissible. Zaremba (1903, pp. 619, 620) also remarks that the use of "the variables introduced in hydrodynamics by Lagrange" enables one to avoid formulating inadmissible constitutive equations, but neither justifies this statement nor explains what it means.

According to Batchelor (1967, p. 71), a change from Eulerian variables to Lagrangian variables in hydrodynamics means (for example) a change of independent variables from y^i, t to a^i, t *without changing the dependent variables* v^i in a simultaneous set of partial differential equations—the Navier–Stokes equations, if the fluid is viscous. Here, the v^i denote fluid velocity components referred to a rectangular Cartesian space coordinate system $C: \mathrm{Q} \to y$; the y^i denote the coordinates of the place Q which a typical fluid particle P occupies at time t; and the a^i are the values at P of any ordered set of three variables ("Lagrange variables") used to furnish time-independent labels for the fluid particles. With this interpretation of "Lagrangian variables," Zaremba's statement is incorrect. If, in addition, one changes the *dependent* variables from Cartesian components to convected components (or body tensor components), as Oldroyd (1950a) showed, then it is easy to formulate constitutive equations that are admissible, and to avoid formulating inadmissible equations. The following alternative ways of representing the Cartesian stress tensor field $\mathbf{p}(\mathrm{P}, t)$ make these distinctions clear:

(1) $\mathbf{p}(\mathrm{P}, t) = p^{ij}(y, t)\mathbf{b}_i(C)\mathbf{b}_j(C)$ $\left(\begin{array}{l}\text{Cartesian components,}\\ \text{Eulerian variables}\end{array}\right)$;

(2) $\mathbf{p}(\mathrm{P}, t) = p_{\mathrm{L}}^{ij}(a, t)\mathbf{b}_i(C)\mathbf{b}_j(C)$ $\left(\begin{array}{l}\text{Cartesian components,}\\ \text{Lagrangian variables}\end{array}\right)$;

(3) $\mathbf{p}(\mathrm{P}, t) = \hat{p}^{ij}(\hat{x}, t)\hat{\mathbf{b}}_i(t)\hat{\mathbf{b}}_j(t)$ (convected components);

(4) $\boldsymbol{\pi}(\mathrm{P}, t) = \pi^{ij}(\xi, t)\boldsymbol{\beta}_i(B, \mathrm{P})\boldsymbol{\beta}_j(B, \mathrm{P})$ $\left(\begin{array}{l}\text{body tensor components;}\\ \mathbf{p} = \mathbb{T}(t)\pi\end{array}\right)$;

(5) $a^i = \hat{x}^i = \xi^i, \quad \hat{p}^{ij}(\hat{x}, t) = \pi^{ij}(\xi, t)$ $(B \overset{t}{\equiv} \hat{S})$;

(6) $p^{ij}(y, t) = p_{\mathrm{L}}^{ij}(a, t) \neq \hat{p}^{ij}(\hat{x}, t)$ (for $t_0 \leq t \leq t_1$).

The subscript L (for Lagrangian) is used to emphasize that p^{ij} and p_{L}^{ij} denote different functions (whose values at P are the same).

Expressing 11.3(15) in terms of Lagrangian variables does not help one to recognize its inadmissibility; expressing it in convected components (or body tensor components) does, because of the presence of the convected vorticity components $\hat{w}_{ij}$ [$= \omega_{ij}$, as in 11.3(25) referred to B].

The advantages of using convected components in certain aspects of continuum mechanics had previously been noted by Hencky (1925), Brillouin (1925), Deuker (1941), and Gleyzal (1949), but it was Oldroyd (1950a) who recognized the advantages in formulating constitutive equations for materials in general (when motion relative to space-fixed axes is irrelevant) and for viscoelastic materials (especially when subjected to finite-strain histories) in particular.

Appreciation of Oldroyd's (1950a) paper was not immediate and universal, if one judges by various papers published years later using space tensor components and deformation gradient variables. In the context of Oldroyd's contribution, the extra contribution, to be found in later papers, to the problem of formulating admissible constitutive equations in general can, we believe, be fairly summarized in the following statement: By using space tensor fields referred to space coordinate systems and by using the deformation gradient and velocity gradient, acceleration gradient, etc. as the basic kinematic variables, one introduces an unwanted dependence on motion relative to axes fixed in space; this dependence must be eliminated by the material objectivity condition or its equivalent. Noll (1972, p. 1) has recently noted the extra complications attendant upon the use of the space-tensor/deformation-gradient formalism.

We take issue with the historical account given by Coleman *et al.* (1966, p. 86), who state, "The first to formulate a properly invariant general theory applicable to finite deformations, however, were Green and Rivlin in 1957.... Noll employed ideas similar to those used by Green and Rivlin .... Noll introduced ... a general notion of fluidity, namely the concept of a simple fluid on which the present book is based." Oldroyd's (1950a) paper contains a properly invariant general theory applicable to finite deformations, which includes all Noll's "simple fluids," as Lodge and Stark (1972) have proved. It is sometimes argued that because Oldroyd described constitutive equations as a set of integrodifferential equations, his theory is somehow less general than that of Noll, who refers instead to constitutive functionals; this argument is invalid, however, because Oldroyd (1950a, p. 526, ¶ 3) also states that, ". . . the equations of state are essentially relationships between the six independent functions of the time $\gamma_{jl}(\xi, t')$ and the six independent functions $\pi_{jl}(\xi, t')$." Oldroyd's theory is thus of generality at least equal to that of Noll's theory; whether Oldroyd's theory is in fact of greater generality depends on whether Oldroyd's equations equivalent to 6.1(1) can always (in principle) be solved to give the stress explicitly as a functional of metric tensor history, and this is not yet known.

Rivlin (1966) refers to Oldroyd's 1950 arguments as "somewhat heuristic"; Pipkin (1964a, p. 1035) describes them as "semi-intuitive." We do not see the justification for these comments; nor has any been offered in print. All workers in this field use the assumption that "rheological properties should not depend on the orientation or motion of a material element relative to axes fixed in space." This assumption is not a precise statement, but it can be made precise in various ways, e.g., (i) by postulating that all constitutive equations having the form of body tensor equations, in which the only tensor variables are $\boldsymbol{\pi}$ and $\boldsymbol{\gamma}$, shall be admissible; (ii) by using Oldroyd's corresponding statement equivalent to (i) expressed in terms of components of $\boldsymbol{\pi}$ and $\boldsymbol{\gamma}$ referred to any one body coordinate system; or (iii) by postulating a Cartesian tensor relation involving a Cartesian stress tensor and a Cartesian deformation gradient tensor, subject to a certain restriction of invariance of form with respect to certain rigid rotations (leading to a material objectivity condition). It is self-evident (and requires no analytic proof) that the use of convected coordinate systems and convected components $\hat{p}_{ij}$ and $\hat{g}_{ij}$ automatically gives equations which are consistent with the preceding assumption. By formulating their basic assumptions in terms of the deformation gradient (or velocity gradient, etc.), and by using Cartesian space tensors, other authors start further back and, in consequence, require a more elaborate analysis to produce constitutive equations that are admissible and in a conveniently usable form. It is, perhaps, the simplicity and brevity of Oldroyd's treatment which has misled people about its rigor.

APPENDIX A

EQUATIONS IN CYLINDRICAL POLAR COORDINATES

(1) S: $\quad x^i$: $\phi, z, r \qquad [C : Q \to y; \quad z = y^2, \quad r^2 = (y^1)^2 + (y^3)^2,$
$\tan \phi = y^3/y^1];$

(2) $$h_i: \quad r, 1, 1;$$

(3) $\sqrt{|g|} = r; \qquad G^1_{13} = G^1_{31} = r^{-1}, \qquad G^3_{11} = -r, \qquad \text{other} \quad G^i_{jk} = 0.$

Stress equations of motion:

(4) $$\frac{\partial \widehat{rr}}{\partial r} + \frac{1}{r}\frac{\partial \widehat{r\phi}}{\partial \phi} + \frac{\partial \widehat{rz}}{\partial z} + \frac{\widehat{rr} - \widehat{\phi\phi}}{r} = \rho(a_r - X_r),$$

(5) $$\frac{\partial \widehat{r\phi}}{\partial r} + \frac{1}{r}\frac{\partial \widehat{\phi\phi}}{\partial \phi} + \frac{\partial \widehat{\phi z}}{\partial z} + 2\frac{\widehat{r\phi}}{r} = \rho(a_\phi - X_\phi),$$

(6) $$\frac{\partial \widehat{rz}}{\partial r} + \frac{1}{r}\frac{\partial \widehat{\phi z}}{\partial \phi} + \frac{\partial \widehat{zz}}{\partial z} + \frac{\widehat{rz}}{r} = \rho(a_z - X_z).$$

$$\widehat{\phi\phi} = r^2 p^{11}, \qquad \widehat{\phi z} = rp^{12}, \qquad \widehat{\phi r} = rp^{13}, \tag{7}$$

$$\widehat{zz} = p^{22}, \qquad \widehat{zr} = p^{23}, \qquad \widehat{rr} = p^{33}.$$

Acceleration:

$$a_\phi = ra^1 = r\left(\frac{\partial v^1}{\partial t} + v^1 \frac{\partial v^1}{\partial \phi} + v^2 \frac{\partial v^1}{\partial z} + v^3 \frac{\partial v^1}{\partial r}\right) + 2v^1 v^3, \tag{8}$$

$$a_z = a^2 = \frac{\partial v^2}{\partial t} + v^1 \frac{\partial v^2}{\partial \phi} + v^2 \frac{\partial v^2}{\partial z} + v^3 \frac{\partial v^2}{\partial r}, \tag{9}$$

$$a_r = a^3 = \frac{\partial v^3}{\partial t} + v^1 \frac{\partial v^3}{\partial \phi} + v^2 \frac{\partial v^3}{\partial z} + v^3 \frac{\partial v^3}{\partial r} - r(v^1)^2. \tag{10}$$

Velocity:

$$\frac{v_\phi}{r} = v^1 = \dot{\phi}(x, t); \qquad v_z = v^2 = \dot{z}(x, t), \qquad v_r = v^3 = \dot{r}(x, t). \tag{11}$$

$$(\text{Div } \boldsymbol{v})S = \frac{1}{r}\frac{\partial v_\phi}{\partial \phi} + \frac{\partial v_z}{\partial z} + \frac{1}{r}\frac{\partial}{\partial r}(rv_r). \tag{12}$$

APPENDIX B
EQUATIONS IN SPHERICAL POLAR COORDINATE SYSTEMS

(1) $$S: \quad x^i: \ \phi, \theta, r \quad \begin{pmatrix} C: Q \to y; & y^1 = r \sin\theta \cos\phi, \\ & y^2 = r \sin\theta \sin\phi, \\ & y^3 = r \cos\theta \end{pmatrix};$$

(2) $$h_i: \ r \sin\theta, r, 1;$$

(3) $$\sqrt{|g|} = r^2 \sin\theta, \qquad G^1_{21} = G^1_{12} = \cot\theta, \qquad G^1_{31} = G^1_{13} = r^{-1},$$
$$G^2_{11} = -\sin\theta\cos\theta, \qquad G^2_{32} = G^2_{23} = r^{-1}, \qquad G^3_{11} = -r\sin^2\theta,$$
$$G^3_{22} = -r; \qquad \text{all other } G^i_{jk} \text{ are zero.}$$

Stress equations of motion:

(4) $$\frac{\partial \widehat{r\phi}}{\partial r} + \frac{1}{r}\frac{\partial \widehat{\theta\phi}}{\partial \theta} + \frac{1}{r\sin\theta}\frac{\partial \widehat{\phi\phi}}{\partial \phi} + \frac{1}{r}(3\widehat{r\phi} + 2\widehat{\theta\phi}\cot\theta) = \rho(a_\phi - X_\phi);$$

(5) $$\frac{\partial \widehat{r\theta}}{\partial r} + \frac{1}{r}\frac{\partial \widehat{\theta\theta}}{\partial \theta} + \frac{1}{r\sin\theta}\frac{\partial \widehat{\theta\phi}}{\partial \phi} + \frac{1}{r}[(\widehat{\theta\theta} - \widehat{\phi\phi})\cot\theta + 3\widehat{r\theta}] = \rho(a_\theta - X_\theta);$$

(6) $$\frac{\partial \widehat{rr}}{\partial r} + \frac{1}{r}\frac{\partial \widehat{r\theta}}{\partial \theta} + \frac{1}{r \sin \theta}\frac{\partial \widehat{r\phi}}{\partial \phi} + \frac{1}{r}(2\widehat{rr} - \widehat{\theta\theta} - \widehat{\phi\phi} + \widehat{r\theta} \cot \theta) = \rho(a_r - X_r).$$

(7) $$\widehat{\phi\phi} = r^2 p^{11} \sin^2 \theta, \qquad \widehat{\theta\theta} = r^2 p^{22}, \qquad \widehat{rr} = p^{33},$$

$$\widehat{r\theta} = rp^{32}, \qquad \widehat{r\phi} = rp^{13} \sin \theta, \qquad \widehat{\phi\theta} = r^2 p^{12} \sin \theta.$$

Acceleration:

(8) $$\frac{a_\phi}{r \sin \theta} = a^1 = \frac{\partial v^1}{\partial t} + v^1 \frac{\partial v^1}{\partial \phi} + v^2 \frac{\partial v^1}{\partial \theta} + v^3 \frac{\partial v^1}{\partial r} + 2v^1 v^2 \cot \theta + \frac{2}{r} v^1 v^3;$$

(9) $$r^{-1} a_\theta = a^2 = \frac{\partial v^2}{\partial t} + v^1 \frac{\partial v^2}{\partial \phi} + v^2 \frac{\partial v^2}{\partial \theta} + v^3 \frac{\partial v^3}{\partial r} - (v^1)^2 \sin \theta \cos \theta + \frac{2}{r} v^2 v^3;$$

(10) $$a_r = a^3 = \frac{\partial v^3}{\partial t} + v^1 \frac{\partial v^3}{\partial \phi} + v^2 \frac{\partial v^3}{\partial \theta} + v^3 \frac{\partial v^3}{\partial r} - r(v^1 \sin \theta)^2 - r(v^2)^2.$$

Velocity:

(11) $$\frac{v_\phi}{r \sin \theta} = v^1 = \dot{\phi}(x, t), \qquad \frac{v_\theta}{r} = v^2 = \dot{\theta}(x, t), \qquad v_r = v^3 = \dot{r}(x, t);$$

(12) $$(\text{Div } v)S = \frac{1}{r \sin \theta}\left\{\frac{\partial v_\phi}{\partial \phi} + \frac{\partial}{\partial \theta}(v_\theta \sin \theta)\right\} + \frac{1}{r^2}\frac{\partial}{\partial r}(r^2 v_r).$$

APPENDIX C
EQUATIONS IN ORTHOGONAL COORDINATE SYSTEMS

(1) Orthogonal $S: \mathrm{Q} \to x; \quad g = \mathrm{diag}[h_1{}^2, h_2{}^2, h_3{}^2]; \quad |g|^{1/2} = h_1 h_2 h_3.$

(2) $$G^1_{1k} = G^1_{k1} = \frac{1}{h_1}\frac{\partial h_1}{\partial x^k} \quad (k = 1, 2, 3);$$

$$G^1_{22} = -\frac{h_2}{h_1{}^2}\frac{\partial h_2}{\partial x^1}; \qquad G^1_{33} = -\frac{h_3}{h_1{}^2}\frac{\partial h_3}{\partial x^1}; \qquad G^1_{23} = G^1_{32} = 0.$$

(3) Physical components of stress: $\widehat{p_{ki}} = \widehat{p^{ki}} = h_{\hat{k}} h_{\hat{\imath}} p^{\hat{k}\hat{\imath}}.$

(4) Physical components of extra stress: $\widehat{P_{ki}} = \widehat{p_{ki}} + p\delta_{ki}.$

Stress equations of motion:

(5) $$-\frac{\partial p}{\partial x^i} + \frac{h_{\hat{\imath}}}{h_1 h_2 h_3}\frac{\partial}{\partial x^k}\left(\frac{h_1 h_2 h_3}{h_k}\widehat{P_{k\hat{\imath}}}\right) + \frac{1}{h_k}\left(\widehat{P_{k\hat{\imath}}}\frac{\partial h_{\hat{\imath}}}{\partial x^k} - \widehat{P_{kk}}\frac{\partial h_k}{\partial x^i}\right)$$

$$= \rho h_{\hat{\imath}}(\widehat{a_{\hat{\imath}}} - \widehat{X_{\hat{\imath}}}).$$

Acceleration:

(6) $$\widehat{a_i}/h_i = a^i = \left(\frac{\partial}{\partial t} + v^k \frac{\partial}{\partial x^k}\right) v^i(x, t) + G^i_{jk} v^j v^k; \qquad v^i = \widehat{v_i}/h_i.$$

(7) $$(\text{Div } \boldsymbol{v})S = \frac{1}{h_1 h_2 h_3} \frac{\partial}{\partial x^i} \left(\frac{h_1 h_2 h_3}{h_i} \widehat{v_i}\right).$$

Divergence of stress

(8)

$$\widehat{p^{k1}{}_{,k}} = \frac{1}{h_k} \frac{\partial \widehat{p^{k1}}}{\partial x^k} + \frac{\widehat{p^{11}} - \widehat{p^{22}}}{\rho_{13}} + \frac{\widehat{p^{11}} - \widehat{p^{33}}}{\rho_{12}} + \widehat{p^{21}}\left(\frac{2}{\rho_{23}} + \frac{1}{\rho_{21}}\right) + \widehat{p^{31}}\left(\frac{2}{\rho_{32}} + \frac{1}{\rho_{31}}\right);$$

$$\rho_{31} = h_2 h_3 \Big/ \frac{\partial h_2}{\partial x^3}$$

= the principal radius of curvature for the surface x^3 = const associated with line of curvature given by the intersection with x^1 = const etc.

The expressions for $\widehat{p^{k2}{}_{,k}}$ and $\widehat{p^{k3}{}_{,k}}$ are of similar form.

APPENDIX D
SUMMARY OF DEFINITIONS FOR UNIDIRECTIONAL SHEAR FLOW

(1) Stress components p^{ij}, referred to a rectangular Cartesian coordinate system, for a unidirectional shear flow of velocity components $v^1 = \dot{s}y^2$, $v^2 = v^3 = 0$, are shown in Fig. D(1). For curvilinear shear flow, $Oy^1y^2y^3$ can be regarded as a local rectangular Cartesian coordinate system and $p^{ij} = \widehat{p^{ij}}$ as the physical components of stress in an orthogonal space coordinate system $S: Q \to x$ having x^2 surfaces as shear surfaces and x^1 lines as shear lines. The following notation is used:

(2) $\sigma \equiv p^{21}, \qquad N_1 \equiv p^{11} - p^{22}, \qquad N_2 \equiv p^{22} - p^{33};$

$p^{11} > 0$ for a tensile component.

(3) $$\sigma/\dot{s} = \eta,$$

the viscosity, when σ and $\dot{s}$ are constant.

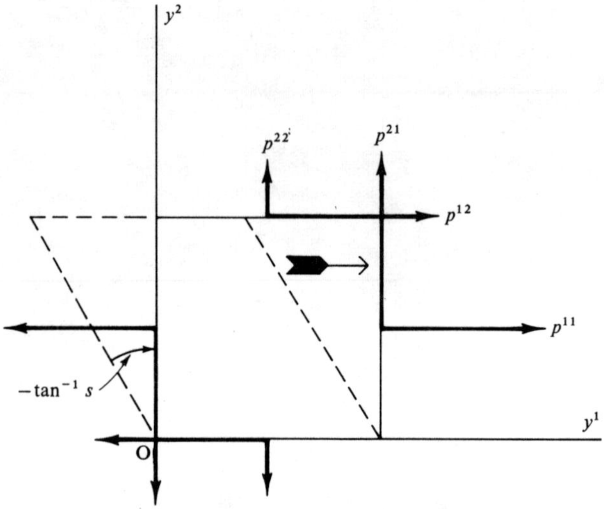

Figure D(1) Stress components in unidirectional shear flow; p^{33} is not shown; $p^{13} = p^{23} = 0$.

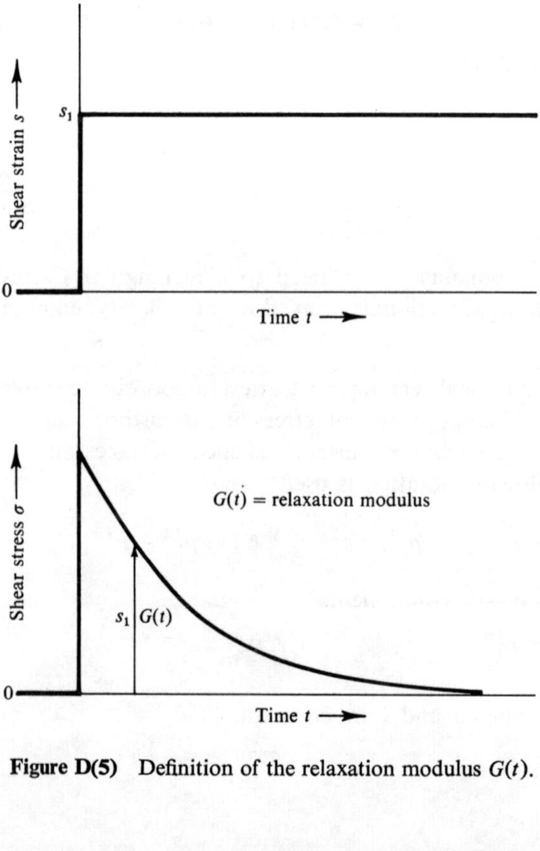

Figure D(5) Definition of the relaxation modulus $G(t)$.

Small-amplitude oscillatory shear

(4)
$$\dot{s} = \alpha\omega \cos \omega t, \qquad \sigma = \alpha\omega(\eta' \cos \omega t + \eta'' \sin \omega t),$$
$$\eta' = \omega^{-1}G'' = \text{dynamic viscosity},$$
$$G' = \omega\eta'' = \text{dynamic modulus},$$
$$\eta^* = \eta' - i\eta'' = \text{complex viscosity},$$
$$\delta \equiv \tan^{-1}(\eta'/\eta'') = \text{loss angle},$$
$$G^* = G' + iG'' = \text{complex modulus},$$
$$G^{*-1} = J' - iJ'' = \text{complex compliance}.$$

(5) *Step-function shear strain.* See Fig. D(5).

(6) *Step-function shear stress.* See Fig. D(6).

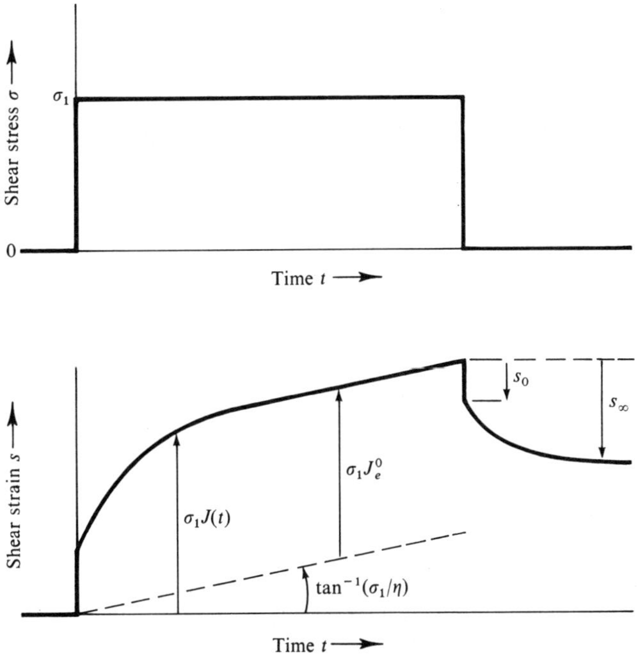

Figure D(6) Definition of compliance $J(t)$, steady-state compliance J_e^0, instantaneous constrained shear recovery s_0, and ultimate constrained shear recovery s_∞.

APPENDIX E

SUMMARY OF $\mathbb{T}$ OPERATIONS FOR COVARIANT STRAIN TENSORS

The table appearing on the following page is similar to that given by Armstrong and Bird (1973). Approximations in columns I and V are as follows: * small $\partial u_i/\partial y^j$; † small v_i; ‡ small $\partial u_i/\partial y^j$ and small v_i.

$$\mathbf{C}_{t'}(Q', t) = \frac{\partial x^i}{\partial x'^r}\frac{\partial x^j}{\partial x'^s}\, g_{ij}(x)(\mathbf{b}^r\mathbf{b}^s)_{S,\,Q'}.$$

$$\dot{\mathbf{C}}_{t'}(Q', t) = \frac{\partial x^i}{\partial x'^r}\frac{\partial x^j}{\partial x'^s}\, A_{ij}^{(1)}(x, t)(\mathbf{b}^r\mathbf{b}^s)_{S,\,Q'}.$$

$$\ddot{\mathbf{C}}_{t'}(Q', t) = \frac{\partial x^i}{\partial x'^r}\frac{\partial x^j}{\partial x'^s}\, A_{ij}^{(2)}(x, t)(\mathbf{b}^r\mathbf{b}^s)_{S,\,Q'}.$$

$$\mathbf{A}^2(Q, t) = (\mathbf{b}^i\mathbf{b}^j + \mathbf{b}^j\mathbf{b}^i)_{S,\,Q}\left\{\frac{1}{2}\left(\frac{\partial}{\partial t} + v^k\frac{\partial}{\partial x^k}\right)A_{ij}^{(1)}(x, t) + A_{ik}^{(1)}\frac{\partial v^k}{\partial x^j}\right\}.$$

$$\mathbf{u} = u_i(y, t, t')\mathbf{b}^i(C) = \mathbf{r}(P, t) - \mathbf{r}(P, t'). \qquad \mathbf{v} = v_i(y, t)\mathbf{b}^i(C) = \partial\mathbf{r}(P, t)/\partial t.$$

Particle P occupies places Q, Q′ at times t, t'. Space coordinate systems: $S: Q \to x$ (arbitrary); $C: Q \to y$ (rectangular Cartesian).

I (≡ V)	II		III		IV	V
(as in column V)	$\stackrel{*}{\doteqdot} \mathbf{C}_{t'}(Q', t) - \mathbf{U}$	$\stackrel{t'}{\longleftrightarrow}$	$\gamma(P, t) - \gamma(P, t')$	$\stackrel{t}{\longleftrightarrow}$	$\mathbf{U} - \mathbf{C}_t(Q, t')$	$\stackrel{*}{\doteqdot} (\mathbf{b}^i\mathbf{b}^j + \mathbf{b}^j\mathbf{b}^i)_C \dfrac{\partial}{\partial y^i} u_j(y, t, t')$
	$\Big\downarrow \frac{\partial}{\partial t}$		$\Big\downarrow \frac{\partial}{\partial t}$		$\Big\downarrow \frac{\mathscr{D}}{\mathscr{D}t}$	$\dagger \Big\downarrow \frac{\partial}{\partial t}$
$\mathbf{A}^1(Q, t)$	$\stackrel{*}{\doteqdot} \dot{\mathbf{C}}_{t'}(Q', t)$	$\stackrel{t'}{\longleftrightarrow}$	$\dot{\gamma}(P, t)$	$\stackrel{t}{\longleftrightarrow}$	$\mathbf{A}^1(Q, t)$	$= (\mathbf{b}^i\mathbf{b}^j + \mathbf{b}^j\mathbf{b}^i)_C \dfrac{\partial}{\partial y^i} v_j(y, t)$
	$\Big\downarrow \frac{\partial}{\partial t}$		$\Big\downarrow \frac{\partial}{\partial t}$		$\Big\downarrow \frac{\mathscr{D}}{\mathscr{D}t}$	$\Big\downarrow \frac{\partial}{\partial t}$
$\dfrac{\partial}{\partial t} \mathbf{A}^1(Q, t)$	$\stackrel{\ddagger}{\doteqdot} \ddot{\mathbf{C}}_{t'}(Q', t)$	$\stackrel{t'}{\longleftrightarrow}$	$\ddot{\gamma}(P, t)$	$\stackrel{t}{\longleftrightarrow}$	$\mathbf{A}^2(Q, t)$	$\stackrel{\ddagger}{\doteqdot} \dfrac{\partial}{\partial t} \mathbf{A}^1(Q, t)$

APPENDIX F
CALCULATIONS FOR VISCOELASTIC LIQUIDS*

Rotating sphere	Giesekus (1965a), Thomas and Walters (1964a), Walters and Waters (1964), Walters and Savins (1965), Thomas and Walters (1966).
Plate and rotating cone, secondary flow	Bhatnagar and Rathna (1962), Mohan Rao (1962), Giesekus (1965a), Adams and Lodge (1964), Walters and Waters (1968), Griffiths and Walters (1970), Kocherov *et al.* (1973).
Cylinder and rotating disk	Bhatnagar (1963), Griffiths *et al.* (1969), Kramer and Johnson (1972), Hill (1972).
Flow through pipe of noncircular cross section	Green and Rivlin (1956), Schechter (1961), Rivlin and Langlois (1963), Pipkin (1965), Young and Wheeler (1964), Jones (1964, 1965), Rivlin (1964), Giesekus (1965b), Wheeler and Wissler (1965, 1966), Jones and Jones (1966), Jones (1967), Semjonow (1967), Mitsuishi and Aoyasi (1969), Arai and Toyoda (1970), Dodson *et al.* (1974).
Flow through a rotating pipe	Jones and Lewis (1968), Vidyanidhi and Sithapathi (1970), Gunn *et al.* (1972).

* This Appendix includes a few experimental papers, and has been prepared with the generous assistance of Professor K. Walters.

Flow through a curved pipe	Jones (1960), Thomas and Walters (1963, 1965), Jones and Walters (1968), Barnes and Walters (1969), Walters (1972).
Flow through a corrugated pipe	Bhatnagar and Mathur (1966), Dodson *et al.* (1971).
Pulsatile flow	Pipkin (1964b,c), Jones and Walters (1967), Walters and Townsend (1970), Barnes *et al.* (1971), Townsend (1973).
Boundary layer flows	Rajeswari and Rathna (1962), Beard and Walters (1964), Davies (1966).
Converging flows	Langlois and Rivlin (1959), Collins and Schowalter (1963), Kaloni (1965a,b), Giesekus (1968, 1969), Duda and Vrentas (1973).
Journal bearings	Horowitz and Steidler (1960), Tanner (1963), Pearson (1967), Fix and Paslay (1967), Ehrlich and Slattery (1968), Reiner *et al.* (1969), Davies and Walters (1973b).
Squeezing flow	Scott (1931), Tanner (1965), Kuzma (1967), Parlato (1969), Kramer (1972), Leider and Bird (1973).
Flow past submerged bodies	Leslie and Tanner (1961), Caswell and Schwarz (1962).
Free surface near rotating rod	Kaye (1973), Joseph and Fosdick (1973), Joseph *et al.* (1973).
Stability of flow	Jain (1957), Thomas and Walters (1964b), Datta (1964), Chan Man Fong (1965), Chan Man Fong and Walters (1965), Rubin and Elata (1966), Beard *et al.* (1966), Goddard and Miller (1967), Lockett and Rivlin (1968), Jones and Walters (1968), Denn and Roisman (1969), Ginn and Denn (1969), Feinberg and Schowalter (1969), Matovich and Pearson (1969), Denn *et al.* (1971), Mook (1971), Chang and Lodge (1971), Giesekus (1972), Hayes and Hutton (1972), Smith and Rivlin (1972), Chang *et al.* (1972), Chang (1973).

REFERENCES

Abbott, T. N. G., and Walters, K. (1970). *J. Fluid Mech.* **40**, 205.

Adams, N., and Lodge, A. S. (1964). *Phil. Trans. Roy. Soc. London Ser. A* **256**, 149.

Arai, I., and Toyoda, H. (1970). *Proc. Int. Congr. Rheol., 5th* (S. Onogi, ed.), **4**, p. 461. Univ. of Tokyo Press, Tokyo, and Univ. Park Press, Manchester, U. K.

Armstrong, R. C. (1973). Ph. D. Thesis, Univ. of Wisconsin, Madison.

Armstrong, R. C., and Bird, R. B. (1973). *J. Chem. Phys.* **58**, 2715.

Ashare, E. (1967). Ph. D. Thesis, Univ. of Wisconsin, Madison.

Ashare, E. (1968). *Trans. Soc. Rheol.* **12**, 535.

Astarita, G. (1972). stated at *Int. Congr. Rheol., 6th. Lyon.*

Bancroft, D. M., and Kaye, A. (1970). *Rheol. Acta* **9**, 595.

Barnes, H. A., and Walters, K. (1969). *Proc. Roy. Soc. Ser. A* **314**, 85.

Barnes, H. A., Townsend, P., and Walters, K. (1971). *Rheol. Acta* **10**, 517.

Batchelor, G. K. (1967). "An Introduction to Fluid Dynamics," p. 71. Cambridge Univ. Press, London and New York.

Beard, D. W., and Walters, K. (1964). *Proc. Cambridge Phil. Soc.* **60**, 667.

Beard, D. W., Davies, M. H., and Walters, K. (1966). *J. Fluid Mech.* **24**, 321.

Bernstein, B., and Fosdick, R. L. (1970). *Rheol. Acta* **9**, 186.

Bernstein, B., Kearsley, E. A., and Zapas, L. J. (1963). *Trans. Soc. Rheol.* **7**, 391.

Bhatnagar, R. K. (1963). *Proc. Indian Acad. Sci. A* **58**, 279.

Bhatnagar, R. K., and Mathur, M. N. (1966). *J. Indian Inst. Sci.* **48**, 1.

Bhatnagar, P. L., and Rathna, S. L. (1962). *Quart. J. Mech. Appl. Math.* **16**, 329.
Bird, R. B. (1972). *Z. Angew. Math. Phys.* **23**, 157.
Bird, R. B., and Armstrong, R. C. (1972). *J. Chem. Phys.* **56**, 3680.
Bird, R. B., and Carreau, P. J. (1968). *Chem. Eng. Sci.* **23**, 427.
Bird, R. B., and Harris, E. K. (1968). *AIChE J.* **14**, 758.
Bishop, R. L., and Goldberg, S. I. (1968). "Tensor Analysis on Manifolds." Macmillan, New York.
Blyler, L. L., and Kurtz, S. J. (1967). *J. Appl. Polym. Sci.* **11**, 127.
Bogue, D. C., and Doughty, J. O. (1966). *Ind. Eng. Chem. Fundam.* **5**, 243.
Boltzmann, L. (1874). *Sitzungsber Math. Naturwiss. Kl. Kaiserl Akad. Wiss. Wien* **70** (2), 275.
Boltzmann, L. (1876). *Ann. Phys. (Leipzig)* **7**, 624.
Booij, H. C. (1966). *Rheol. Acta* **5**, 215.
Booij, H. C. (1970). Ph. D. Thesis, Univ. of Leiden.
Brillouin, L. (1925). *Ann. Phys. (Paris)* [10] **3**, 251.
Brillouin, L. (1938). "Les Tenseurs en Mécanique et en Elasticité," p. 240, Eq. X.89. 1st French ed., Masson, 1938. [Reprinted by Dover, New York, 1946.] [Transl. by R. O. Brennan, Academic Press, New York, 1964].
Brindley, G., and Broadbent, J. M. (1973). *Rheol. Acta* **12**, 48.
Broadbent, J. M., and Lodge, A. S. (1971). *Rheol. Acta* **10**, 557.
Broadbent, J. M., Kaye, A., Lodge, A. S., and Vale, D. G. (1968). *Nature (London)* **217**, 55.
Bronson, R. (1969). "Matrix Methods: An Introduction." Academic Press, New York.
Carreau, P. J. (1972). *Trans. Soc. Rheol.* **16**, 99.
Carroll, L. (1872). Through the looking glass. "The Complete Works of Lewis Carroll," p. 196. Nonesuch Press, London, 1939.
Caswell, B. (1967). *Arch. Ration. Mech. Anal.* **26**, 385.
Caswell, B., and Schwarz, W. H. (1962). *J. Fluid Mech.* **13**, 417.
Chacon, R. V. S., and Rivlin, R. S. (1964). *Z. Angew. Math. Phys.* **15**, 444.
Chan Man Fong, C. F. (1965). *Rheol. Acta* **4**, 37.
Chan Man Fong, C. F., and Walters, K. (1965). *J. Mec.* **4**, 439.
Chang, H. (1973). Ph. D. Thesis, Univ. of Wisconsin, Madison.
Chang, H., and Lodge, A. S. (1971). *Rheol. Acta* **10**, 448.
Chang, H., and Lodge, A. S. (1972). *Rheol. Acta* **11**, 127.
Chang, H., Lodge, A. S., and McLeod, J. B. (1972). Math. Res. Center Rep. TSR 1246. Univ. of Wisconsin, Madison.
Chen, I.-J., and Bogue, D. C. (1972). *Trans. Soc. Rheol.* **16**, 59.
Christensen, R. M. (1971). "Theory of Viscoelasticity: An Introduction." Academic Press, New York.
Christiansen, E. B., and Leppard, W. R. (1974). *Trans. Soc. Rheol.* **18**, 65.
Christiansen, E. B., and Miller, M. J. (1971). *Trans. Soc. Rheol.* **15**, 189.
Coleman, B. D. (1961). *Arch. Ration. Mech. Anal.* **9**, 273.
Coleman, B. D. (1962). *Trans. Soc. Rheol.* **6**, 293.
Coleman, B. D. (1964). *Arch. Ration. Mech. Anal.* **17**, 230.
Coleman, B. D. (1968). *Proc. Roy. Soc. Ser. A* **306**, 449.
Coleman, B. D., and Markovitz, H. (1964). *J. Appl. Phys.* **35**, 1.
Coleman, B. D., and Noll, W. (1960). *Arch. Ration. Mech. Anal.* **6**, 355.
Coleman, B. D., and Noll, W. (1961). *Rev. Mod. Phys.* **33**, 239.
Coleman, B. D., Markovitz, H., and Noll, W. (1966). "Viscometric Flows of Non-Newtonian Fluids." Springer-Verlag, Berlin and New York.
Collins, M., and Schowalter, W. R. (1963) *AIChE J.* **9**, 98.

Curtiss, C. F. (1956). *J. Chem. Phys.* **24**, 225.
Dahler, J. S., and Scriven, L. E. (1961). *Nature* (*London*) **192**, 36.
Datta, S. K. (1964). *Phys. Fluids* **7**, 1915.
Davies, J. M., and Walters, K. (1972). *In* "The Mathematics Student" (M. K. Singal, ed.), P. L. Bhatnagar anniversary vol. **40**. Indian Math. Soc. Univ. of Delhi.
Davies, J. M., and Walters, K. (1973a). *J. Phys. D: Appl. Phys.* **6**, 2259.
Davies, J. M., and Walters, K. (1973b). *In* "The Rheology of Lubricants" (T. C. Davenport, ed.), p. 65. Wiley, New York.
Davies, M. H. (1966). *Z. Angew. Math. Phys.* **17**, 189.
Denn, M. M., and Roisman, J. (1969). *AIChE J.* **15**, 454.
Denn, M. M., Sun, Z. S., and Ruston, B. D. (1971). *Trans. Soc. Rheol.* **15**, 415.
Deuker, E. A. (1941). *Deut. Math.* **5**, 546.
Dirckes, A. C., and Schowalter, W. R. (1966). *Ind. Eng. Chem. Fundam.* **5**, 263.
Dodson, A. G., Townsend, P., and Walters, K. (1971). *Rheol. Acta* **10**, 508.
Dodson, A. G., Townsend, P., and Walters, K. (1974). "Computers and Fluids," Vol. 2. Pergamon, Oxford. To be published.
Duda, J. L., and Vrentas, J. S. (1973). *Trans. Soc. Rheol.* **17**, 89.
Eilenberg, S., and Steenrod, N. (1952). "Foundations of Algebraic Topology," p. 141, Theorem 9.5. Princeton Univ. Press, Princeton, New Jersey.
Eisenhart, L. P. (1909). "A Treatise on the Differential Geometry of Curves and Surfaces." Ginn, Boston, Massachusetts.
Eisenhart, L. P. (1926). "Riemannian Geometry." Princeton Univ. Press, Princeton, New Jersey.
Eisenschitz, R., Rabinowitsch, B., and Weissenberg, K. (1929). *Mitt. Deut. Mat. Prüf. Anst. Sond.* **9**, 91.
Ehrlich, R., and Slattery, J. C. (1968). *Ind. Eng. Chem. Fundam.* **7**, 239.
Ericksen, J. L. (1960). *In* "Viscoelasticity: Phenomenological Aspects" (J. T. Bergen, ed.), p. 77. Academic Press, New York.
Eringen, A. C. (1962). "Non-Linear Theory of Continuous Media." McGraw-Hill, New York.
Feinberg, M. R., and Schowalter, W. R. (1969). *Proc. Int. Congr. Rheol. 5th* (S. Onogi, ed.) **1**, p. 201. Univ. of Tokyo Press, Tokyo, and Univ. Park Press, Manchester, England.
Ferry, J. D. (1970). "Viscoelastic Properties of Polymers," 2nd ed. Wiley, New York.
Finkbeiner, D. T., II (1960). "Introduction to Matrices and Linear Transformations," p. 196. Freeman, San Francisco, California.
Fix, G. J., and Paslay, P. R. (1967). *J. Appl. Mech.* **34**, 579.
Flory, P. J. (1953). "Principles of Polymer Chemistry." Cornell Univ. Press, Ithaca, New York.
Frazer, R. A., Duncan, W. J., and Collar, A. R. (1952). "Elementary Matrices." Cambridge Univ. Press, London and New York.
Fredrickson, A. G. (1962). *Chem. Eng. Sci.* **17**, 155.
Fredrickson, A. G. (1964). "Rheology: Theory and Applications." Prentice-Hall, Englewood Cliffs, New Jersey.
Fromm, H. (1933). *Ing. Arch.* **4**, 432.
Fromm, H. (1948). *Z. Angew. Math. Mech.* **28**, 43.
Garner, F. H., Nissan, A. H., and Wood, G. F. (1950). *Phil. Trans. Roy. Soc. London Ser. A* **243**, 37.
Gaskins, F. H., and Philippoff, W. (1959). *Trans. Soc. Rheol.* **3**, 181.
Giesekus, H. (1962). *Rheol. Acta* **2**, 50.

Giesekus, H. (1965a). *Proc. Int. Congr. Rheol. 4th, 1963* (E. H. Lee, ed.) **1**, p. 249. Wiley (Interscience), New York.
Giesekus, H. (1965b). *Rheol. Acta* **4**, 299.
Giesekus, H. (1968). *Rheol. Acta* **7**, 127.
Giesekus, H. (1969). *Rheol. Acta* **8**, 411.
Giesekus, H. (1972). *Progr. Heat Mass Transfer* **5**, 187.
Ginn, R. F., and Denn, M. M. (1969). *AIChE J.* **15**, 450.
Ginn, R. F., and Metzner, A. B. (1969). *Trans. Soc. Rheol.* **13**, 429.
Gleyzal, A. (1949). *Quart. Appl. Math.* **6**, 429.
Goddard, J. D., and Miller, C. (1966). *Rheol. Acta* **5**, 177.
Goddard, J. D., and Miller, C. (1967). Tech. Rep. No. 066673-8-T. Univ. of Michigan, Ann Arbor.
Green, A. E., and Adkins, J. E. (1960). "Large Elastic Deformations and Non-Linear Continuum Mechanics." Oxford Univ. Press (Clarendon), London and New York.
Green, A. E., and Rivlin, R. S. (1956). *Quart. Appl. Math.* **14**, 299.
Green, A. E., and Rivlin, R. S. (1957). *Arch. Ration. Mech. Anal.* **1**, 1.
Green, A. E., and Rivlin, R. S. (1960). *Arch. Ration. Mech. Anal.* **4**, 387.
Green, A. E., and Zerna, W. (1968). "Theoretical Elasticity," 2nd ed. Oxford Univ. Press (Clarendon), London and New York.
Green, A. E., Rivlin, R. S., and Shield, R. T. (1952). *Proc. Roy. Soc. Ser. A* **211**, 128.
Green, A. E., Rivlin, R. S., and Spencer, A. J. M. (1959). *Arch. Ration. Mech. Anal.* **3**, 82.
Green, M. S., and Tobolsky, A. V. (1946). *J. Chem. Phys.* **14**, 80.
Greensmith, H. W., and Rivlin, R. S. (1953). *Phil. Trans. Roy. Soc. London Ser. A.* **245**, 399.
Griffiths, D. F., and Walters, K. (1970). *J. Fluid Mech.* **42**, 379.
Griffiths, D. F., Jones, D. T., and Walters, K. (1969). *J. Fluid Mech.* **36**, 161.
Gunn, R. W., Mena, B., and Walters, K. (1972). *Proc. Int. Congr. Rheol. 6th Lyon, Rheol. Acta.* To be published.
Hall, M. M. (1968). CoA Note Mat. No. 17. The College of Aeronaut., Cranfield, Bletchley, England.
Han, C. D., and Kim, K. U. (1973). *Trans. Soc. Rheol.* **17**, 129.
Harris, J. (1961). Ph. D. Thesis, Univ. of Wales, Swansea.
Harris, J. (1963). *Brit. J. Appl. Phys.* **14**, 275.
Harris, J. (1965). *Proc. Int. Congr. Rheol., 4th, 1963* (E. H. Lee, ed.) **3**, p. 417. Wiley (Interscience), New York.
Hayes, J. W., and Hutton, J. F. (1972). *Progr. Heat Mass Transfer* **5**, 195.
Hayes, J. W., and Tanner, R. I. (1965). *Proc. Int. Congr. Rheol. 4th, 1963* (E. H. Lee, ed.) **3**, p. 389. Wiley (Interscience), New York.
Hencky, H. (1925). *Z. Angew. Math. Mech.* **5**, 144.
Higashitani, K. (1973). Ph. D. Thesis, Univ. of Wisconsin, Madison.
Higashitani, K., and Pritchard, W. G. (1972). *Trans. Soc. Rheol.* **16**, 687.
Hill, C. T. (1972). *Trans. Soc. Rheol.* **16**, 213.
Horowitz, H. H., and Steidler, F. E. (1960). *Trans. Amer. Soc. Lub. Engr.* **3**, 124.
Huilgol, R. R. (1969). *Trans. Soc. Rheol.* **13**, 513.
Huppler, J. D. (1965). *Trans. Soc. Rheol.* **9**, 273.
Huppler, J. D., Macdonald, I. F., Ashare, E., Spriggs, T. W., and Bird, R. B. (1967). *Trans. Soc. Rheol.* **11**, 181.
Hutton, J. F. (1963). *Nature (London)* **200**, 646.
Jackson, R. (1967). Ph. D. Thesis, Univ. of Manchester, U. K.
Jackson, R., and Kaye, A. (1966). *Brit. J. Appl. Phys.* **17**, 1355.
Jain, M. K. (1957). *J. Sci. Eng. Res.* **1**, 195.

James, H. M. (1947). *J. Chem. Phys.* **15**, 651.
James, H. M., and Guth, E. (1947). *J. Chem. Phys.* **15**, 669.
Janeschitz-Kriegl, H. (1965). *Proc. Int. Congr. Rheol., 4th, 1963* (E. H. Lee, ed.), **1**, p. 143. Wiley (Interscience), New York.
Janeschitz-Kriegl, H. (1969). *Fortschr. Hochpolym. Forsch.* **6**, 170.
Jeffreys, H. (1931). "Cartesian Tensors." Cambridge Univ. Press, London and New York.
Jones, D. T., and Walters, K. (1968). *AIChE J.* **14**, 658.
Jones, J. R. (1960). *Quart. J. Mech. Appl. Math.* **13**, 428.
Jones, J. R. (1964). *J. Mec.* **3**, 79.
Jones, J. R. (1965). *J. Mec.* **4**, 121.
Jones, J. R., and Jones, R. S. (1966). *J. Mec.* **5**, 375.
Jones, J. R., and Lewis, M. K. (1968). *Rheol. Acta* **7**, 307.
Jones, J. R., and Walters, T. S. (1967). *Rheol. Acta* **6**, 240.
Jones, J. R., and Walters, T. S. (1968). *Rheol. Acta* **7**, 290.
Jones, R. S. (1967). *J. Mec.* **6**, 443.
Joseph, D. D., and Fosdick, R. L. (1973). *Arch. Ration. Mech. Anal.* **49**, 321.
Joseph, D. D., Beavers, G. S., and Fosdick R. L. (1973). *Arch. Ration. Mech. Anal.* **49**, 381.
Kaelble, D. H. (1969). *J. Appl. Poly. Sci.* **13**, 2547.
Kaloni, P. N. (1965a). *J. Phys. Soc. Japan* **20**, 132.
Kaloni, P. N. (1965b). *J. Phys. Soc. Japan* **20**, 610.
Kataoka, T., and Ueda, S. (1969). *J. Poly. Sci. Part A-2* **7**, 475.
Kaye, A. (1962). CoA Note No. 134. The College of Aeronaut., Cranfield, Bletchley, England.
Kaye, A. (1966). *Brit. J. Appl. Phys.* **17**, 803.
Kaye, A. (1973). *Rheol. Acta* **13**, 206.
Kaye, A., and Saunders, D. W. (1964). *J. Sci. Instrum.* **41**, 139.
Kaye, A., Lodge, A. S., and Vale, D. G. (1968). *Rheol. Acta* **7**, 368.
Kearsley, E. A. (1970). *Trans. Soc. Rheol.* **14**, 419.
Kocherov, V. L., Lukach, Yu. L., Spryagin, E. A., and Vinogradov, G. V. (1973). *Polym. Eng. Sci.* **13**, 194.
Kohlrausch, F. (1866). *Ann. Phys. (Leipzig)* **128**, 207, 399.
Kotaka, T., Kurata, M., and Tamura, M. (1959). *J. Appl. Phys.* **30**, 1705.
Kramer, J. M. (1972). Rep. No. RRC-15. Rheol. Res. Center, Univ. of Wisconsin, Madison.
Kramer, J. M., and Johnson, M. W., Jr. (1972). *Trans. Soc. Rheol.* **16**, 197.
Krieger, I. M., and Maron, S. H. (1954). *J. Appl. Phys.* **25**, 72.
Kuzma, D. C. (1967). *Appl. Sci. Res. Sect. A.* **18**, 15.
Landau, L. D., and Lifschitz, E. M. (1959). "Fluid Mechanics," p. 64. Addison-Wesley, Reading, Massachusetts.
Lang, S. (1962). "Differential Manifolds." Wiley (Interscience), New York.
Langlois, W. E., and Rivlin, R. S. (1959). Tech. Rep. No. 3. Div. of Appl. Math., Brown Univ., Providence, Rhode Island.
Leider, P., and Bird, R. B. (1973). Rep. No. RRC-22. Rheol. Res. Center, Univ. of Wisconsin, Madison.
Leonov, A. I. (1964). *J. Appl. Mech. Tech. Phys. (USSR)* **4**, 78.
Leslie, F. M., and Tanner, R. I. (1961). *Quart. J. Mech. Appl. Math.* **14**, 36.
Lipson, J. M., and Lodge, A. S. (1968). *Rheol. Acta* **7**, 364.
Livingston, P. M., and Curtiss, C. F. (1959). *J. Chem. Phys.* **31**, 1643.
Lockett, F. J., and Rivlin, R. S. (1968). *J. Mec.* **7**, 475.
Lodge, A. S. (1951). *Proc. Cambridge Phil. Soc.* **47**, 575.

Lodge, A. S. (1954). *Proc. Int. Congr. Rheol. 2nd, 1953* (V. G. W. Harrison, ed.), p. 229. Butterworth, London.
Lodge, A. S. (1956). *Trans. Faraday Soc.* **52**, 120.
Lodge, A. S. (1958). *In* "Rheology of Elastomers" (P. Mason and N. Wookey, eds.), p. 70. Pergamon, Oxford.
Lodge, A. S. (1960). *Kolloid-Z.* **171**, 46.
Lodge, A. S. (1964). "Elastic Liquids." Academic Press, New York.
Lodge, A. S. (1968). *Rheol. Acta* **7**, 379.
Lodge, A. S. (1970). *Proc. Int. Congr. Rheol., 5th* (S. Onogi, ed.), **4**, p. 169. Univ. of Tokyo Press, Tokyo and Univ. Park Press, Manchester, England.
Lodge, A. S. (1972). *Rheol. Acta* **11**, 106.
Lodge, A. S. (1973a). *In* "The Karl Weissenberg 80th Birthday Celebration Essays" (J. Harris, ed.). East African Literature Bureau, Ngong Rd., P.O. Box 30022, Nairobi, Kenya.
Lodge, A. S. (1973b). U.S. Patent No. 3,777,549.
Lodge, A. S., and Meissner, J. (1972). *Rheol. Acta* **11**, 351.
Lodge, A. S., and Meissner, J. (1973). *Rheol. Acta* **12**, 41.
Lodge, A. S., and Stark, J. H. (1972). *Rheol. Acta* **11**, 119.
Lodge, A. S., and Wu, Y.-J. (1971). *Rheol. Acta* **10**, 539.
Lodge, A. S., and Wu, Y.-J. (1972). Math. Res. Center, Rep. No. TSR 1250. Univ. of Wisconsin, Madison.
Love, A. E. H. (1927). "The Mathematical Theory of Elasticity," 4th ed. Macmillan, New York. [Reprinted by Dover, New York, 1944].
McConnell, A. J. (1957). "Applications of Tensor Analysis." Dover, New York.
Macdonald, I. F. (1968). Ph. D. Thesis, Univ. of Wisconsin, Madison.
Macosko, C. W., and Starita, J. M. (1971). *SPE J.* **27**, 38.
Markovitz, H., and Brown, D. R. (1964). *Int. Symp. Second-Order Effects in Elasticity, Plasticity, and Fluid Dynamics, Haifa, 1962*, p. 585. Macmillan, New York.
Markovitz, H., and Coleman, B. D. (1964). *Phys. Fluids* **7**, 833.
Marsh, B. D., and Pearson, J. R. A. (1968). *Rheol. Acta* **7**, 326.
Matovich, M. A., and Pearson, J. R. A. (1969). *Ind. Eng. Chem. Fundam.* **8**, 512.
Maxwell, B. (1967). *Polym. Eng. Sci.* **7**, 45.
Maxwell, B., and Chartoff, R. P. (1965). *Trans. Soc. Rheol.* **9**, 41.
Maxwell, B., and Scalora, A. J. (1959). *Mod. Plast.* **37**, No. 2, 107, 202.
Meissner, J. (1971). *Rheol. Acta* **10**, 320.
Meissner, J. (1972). *J. Appl. Polym. Sci.* **16**, 2877.
Meissner, J. (1973). Private communication.
Meister, B. J. (1971). *Trans. Soc. Rheol.* **15**, 63.
Metzner, A. B., Houghton, W. T., Sailor, R. A., and White, J. L. (1961). *Trans. Soc. Rheol.* **5**, 133.
Middleman, S., and Gavis, J. (1961). *Phys. Fluids* **4**, 355.
Miller, M., Ferry, J. D., Schremp, F. W., and Eldridge, J. E. (1951). *J. Phys. Colloid Chem.* **55**, 1387.
Miller, M. J., and Christiansen, E. B. (1972). *AIChE J.* **18**, 600.
Mitsuishi, N., and Aoyasi, Y. (1969). *Chem. Eng. Sci.* **24**, 309.
Mohan Rao, D. K. (1962). *Proc. Indian Acad. Sci.* *A***56**, 198.
Mook, D. T. (1971). *Phys. Fluids* **14**, 2769.
Mooney, M. (1940). *J. Appl. Phys.* **11**, 582.
Nevanlinna, F., and Nevanlinna, R. (1959). "Absolute Analysis," p. 204. Springer-Verlag, Berlin and New York.
Noble, B., and Sewell, M. J. (1972). *J. Inst. Math. Appl.* **9**, No. 2, 123.

Noll, W. (1955). *J. Ration. Mech. Anal.* **4**, 13.
Noll, W. (1958). *Arch. Ration. Mech. Anal.* **2**, 197.
Noll, W. (1962). *Arch. Ration. Mech. Anal.* **11**, 97.
Noll, W. (1972). *Arch. Ration. Mech. Anal.* **48**, 1.
Noll, W. (1973). *Arch. Ration. Mech. Anal.* **52**, 62.
Olabisi, O., and Williams, M. C. (1972). *Trans. Soc. Rheol.* **16**, 727.
Oldroyd, J. G. (1950a). *Proc. Roy. Soc. Ser. A* **200**, 523.
Oldroyd, J. G. (1950b). *Proc. Roy. Soc. Ser. A* **202**, 345.
Oldroyd, J. G. (1958). *Proc. Roy. Soc. Ser. A* **245**, 278.
Oldroyd, J. G. (1965). *Proc. Roy. Soc. Ser. A* **283**. 115.
Oldroyd, J. G. (1970). *Proc. Roy. Soc. Ser. A* **316**, 1.
Osaki, K., Tamura, M., Kurata, M., and Kotaka, T. (1963). *Zairyo* **12**, 239.
Padden, F. J., and DeWitt, T. W. (1954). *J. Appl. Phys.* **25**, 1086.
Parlato, P. (1969). M. S. Thesis, Univ. of Delaware, Newark.
Pearson, J. R. A. (1967). *In* "Nonlinear Partial Differential Equations" (W. F. Ames, ed.), p. 73. Academic Press, New York.
Penrose, R. (1968). *In* "Structure of Space–Time," Battelle Rencontres, 1967 Lectures in Math. and Phys. (C. M. De Witt and J. A. Wheeler, eds.), p. 135. Benjamin, New York.
Pipkin, A. C. (1964a). *Rev. Mod. Phys.* **36**, 1034.
Pipkin, A. C. (1964b). *Arch. Ration. Mech. Anal.* **15**, 1.
Pipkin, A. C. (1964c). *Phys. Fluids* **7**, 1143.
Pipkin, A. C. (1965). *Proc. Int. Congr. Rheol., 4th 1963* (E. H. Lee, ed.), **1**, p. 213. Wiley (Interscience), New York.
Pipkin, A. C. (1966). *In* "Modern Developments in the Mechanics of Continua" (S. Eskinazy, ed.), p. 89. Academic Press, New York.
Pipkin, A. C. (1968). *Trans. Soc. Rheol.* **12**, 397.
Pipkin, A. C., and Owen, D. R. (1967). *Phys. Fluids* **10**, 836.
Pipkin, A. C., and Tanner, R. I., (1973). *In* "Mechanics Today (1972)" (S. Nemat-Nasser, ed.). Pergamon Press, Oxford.
Prager, W. (1961). "Introduction to the Mechanics of Continua." Ginn, Boston, Massachusetts.
Pritchard, W. G. (1971). *Phil. Trans. Roy. Soc. London Ser. A* **270**, 507.
Rabinowitsch, B. (1929). *Z. Phys. Chem. Abt. A* **145**, 1.
Rajeswari, G, K., and Rathna, S. L. (1962). *Z. Angew. Math. Phys.* **13**, 43.
Reiner, M. (1945). *Amer. J. Math.* **67**, 350.
Reiner, M., Hanin, M., and Harney, A. (1969). *Isr. J. Technol.* **7**, 273.
Riesz, F., and Sz.-Nagy, B. (1955). "Functional Analysis." Ungar, New York.
Rinehart, R. F. (1955). *Amer. Math. Monthly* **62**, 395.
Rivlin, R. S. (1948). *Phil. Trans. Roy. Soc. London Ser. A* **241**, 47.
Rivlin, R. S. (1956). *J. Ration. Mech. Anal.* **5**, 179.
Rivlin, R. S. (1962). *Z. Angew. Math. Phys.* **13**, 589.
Rivlin, R. S. (1964). *Int. Symp. Second-Order Effects in Elasticity, Plasticity, and Fluid Dynamics, Haifa, 1962*, p. 668. Macmillan, New York.
Rivlin, R. S. (1966). *J. Franklin Inst.* **282**, 338.
Rivlin, R. S., and Ericksen, J. L. (1955). *J. Ration. Mech. Anal.* **4**, 323.
Rivlin, R. S., and Langlois, W. E. (1963). *Rend. Mat.* **22**, 169.
Roberts, J. E. (1952). Rep. ADE 13/52. Ministry of Supply, U. K.
Roberts, J. E. (1954). *Proc. Int. Congr. Rheol., 2nd, 1953* (V. G. W. Harrison, ed.), p. 91. Butterworth, London.

Robison, G. B. (1969). "An Introduction to Mathematical Logic," §1.1, p. 1. Prentice-Hall, Englewood Cliffs, New Jersey.
Rosenfeld, L. (1965). "Theory of Electrons," 2nd ed., p. 14. Dover, New York.
Rouse, P. E., Jr. (1953). *J. Chem. Phys.* **21**, 1272.
Rubin, H., and Elata, C. (1966). *Phys. Fluids* **9**, 1929.
Rudin, W. (1964). "Principles of Mathematical Analysis," 2nd ed. McGraw-Hill, New York.
Russell, R. J. (1946). Ph. D. Thesis, Univ. of London, U.K.
Sakiadis, B. L. (1962). *AIChE J.* **8**, 317.
Schechter, R. S. (1961). *AIChE J.* **7**, 445.
Scott, J. R. (1931). *Trans. Inst. Rubber Ind.* **7**, 169.
Sedov, L. I. (1966). "Foundations of Non-Linear Mechanics of Continua," translated from the Russian ed. of 1962. Pergamon, Oxford.
Semjonow, V. (1967). *Rheol. Acta* **6**, 171.
Shen, K.-S. (1971). Ph. D. Thesis, Univ. of Missouri, Columbia; Order No. 72-18, 192, Univ. Microfilms, Ann Harbor, Michigan.
Simmons, J. M. (1966). *J. Sci. Instrum.* **43**, 887.
Simmons, J. M. (1968). *Rheol. Acta* **7**, 184.
Sips, R. (1951). *J. Polym. Sci.* **7**, 191, eq. (70).
Smith, M. M., and Rivlin, R. S. (1972). *J. Mec.* **11**, 70.
Spriggs, T. W. (1965). *Chem. Eng. Sci.* **20**, 931.
Spriggs, T. W., and Bird, R. B. (1964). *Ind. Eng. Chem. Fundam.* **4**, 182.
Spriggs, T. W., Huppler, J. D., and Bird, R. B. (1966). *Trans. Soc. Rheol.* **10**, 191.
Stark, J. H. (1968). Ph. D. Thesis, Univ. of Manchester, U.K.
Synge, J. L., and Schild, A. (1949). "Tensor Calculus." Univ. of Toronto Press, Toronto.
Tadmor, Z., and Bird, R. B. (1974). *Polym. Eng. Sci.* **14**, 124.
Tanner, R. I. (1963). *Aust. J. Appl. Sci.* **14**, 129.
Tanner, R. I. (1965). *Trans. Amer. Soc. Lub. Engr.* **8**, 179.
Tanner, R. I. (1969). *AIChE J.* **15**, 177.
Tanner, R. I. (1970). *Trans. Soc. Rheol.* **14**, 483.
Tanner, R. I., and Kuo, Y. (1972). *Proc. Int. Congr. Rheol. 6th, Lyon, Abstracts* **3**, p. 197.
Tanner, R. I., and Pipkin, A. C. (1969). *Trans. Soc. Rheol.* **13**, 471.
Tanner, R. I., and Simmons, J. M. (1967). *Chem. Eng. Sci.* **22**, 1803.
Thomas, R. H., and Walters, K. (1963). *J. Fluid Mech.* **16**, 228.
Thomas, R. H., and Walters, K. (1964a). *Quart. J. Mech. Appl. Math.* **17**, 39.
Thomas, R. H., and Walters, K. (1964b). *J. Fluid Mech.* **19**, 557.
Thomas, R. H., and Walters, K. (1965). *J. Fluid Mech.* **21**, 173.
Thomas, R. H., and Walters, K. (1966). *Rheol. Acta* **5**, 23.
Townsend, P. (1973). *Rheol. Acta* **12**, 13.
Treloar, L. R. G. (1958). "The Physics of Rubber Elasticity," 2nd ed. Oxford Univ. Press, (Clarendon), London and New York.
Treloar, L. R. G. (1973). *Rep. Progr. Phys.* **36**, 755.
Truesdell, C., and Noll, W. (1965). *In* "Handbuch der Physik" (S. Flügge, ed.), Vol. III/3. Springer-Verlag, Berlin and New York.
van Es, H. E. (1972). *Proc. Int. Congr. Rheol. 6th Lyon. Rheol. Acta.* To be published.
van Es, H. E., and Christensen, R. M. (1973). *Trans. Soc. Rheol.* **17**, 325
Vidyanidhi, V., and Sithapathi, A. (1970). *J. Phys. Soc. Japan*, **29**, 215.
Wales, J. L. S., and Philippoff, W. (1973). *Rheol. Acta* **12**, 25.
Wall, F. T., (1942). *J. Chem. Phys.* **10**, 485.

Walters, K. (1960). *Quart. J. Mech. Appl. Math.* **13**, 444.
Walters, K. (1962). *Quart. J. Mech. Appl. Math.* **15**, 63.
Walters, K. (1965). *Nature (London)* **207**, 826.
Walters, K. (1970a). *Z. Angew. Math. Phys.* **21**, 592.
Walters, K. (1970b). *J. Fluid Mech.* **40**, 191.
Walters, K. (1972). *Progr. Heat Mass Transfer* **5**, 217.
Walters, K., and Savins, J. S. (1965). *Trans. Soc. Rheol.* **9**, 407.
Walters, K., and Townsend, P. (1970). *Proc. Int. Congr. Rheol. 5th* (S. Onogi, ed.), **4**, p. 471. Univ. of Tokyo Press, Tokyo, and Univ. Park Press, Manchester, U. K.
Walters, K., and Waters, N. D. (1964). *Rheol. Acta* **3**, 312.
Walters, K., and Waters, N. D. (1968). *In* "Polymer Systems: Deformation and Flow" (R. E. Wetton and R. W. Whorlow, ed.), p. 211. Macmillan, New York.
Walters, K., and Jones, T. E. R. (1968). *Proc. Int. Congr. Rheol. 5th* (S. Onogi, ed.), **4**, p. 337. Univ. of Tokyo Press, Tokyo, and Univ. Park Press, Manchester, U. K.
Wang, C.-C. (1965). *Arch. Ration. Mech. Anal.* **20**, 329.
Ward, A. F. H., and Jenkins, G. M. (1958). *Rheol. Acta* **1**, 110.
Ward, I. M. (1971). "Mechanical Properties of Solid Polymers." Wiley (Interscience), New York.
Warner, F. W. (1971), "Foundations of Differentiable Manifolds and Lie Groups." Scott Foresman, Glenview, Illinois.
Weatherburn, C. E. (1939). "Differential Geometry of Three Dimensions." Cambridge Univ. Press, London and New York.
Weissenberg, K. (1931). *Abh. Preuss. Akad. Wiss. Phys.-Math. Kl.* **No. 2**.
Weissenberg, K. (1935). *Arch. Sci. Phys. Natur.* [5] **140**, 44, 130.
Weissenberg, K. (1947). *Nature (London)* **159**, 310.
Weyl, H. (1922). "Space–Time–Matter," translated by H. L. Brose. Methuen, London.
Weyl, H. (1939). "The Classical Groups." Princeton Univ. Press, Princeton, New Jersey. [2nd ed. 1946].
Wheeler, J. A., and Wissler, E. H. (1965). *AIChE J.* **11**, 207.
Wheeler, J. A., and Wissler, E. H. (1966). *Trans. Soc. Rheol.* **10**, 353.
White, J. L., and Metzner, A. B. (1963). *J. Polym. Sci.* **7**, 1867.
Williams, M. C. (1965). *Chem. Eng. Sci.* **20**, 693.
Wineman, A. S., and Pipkin, A. C. (1966). *Acta Mech.* **II/1**, 104.
Yamamoto, M. (1971). *Trans. Soc. Rheol.* **15**, 783.
Yamamoto, M. (1972). "Buttai no Henkeigaku." Seibundō Shinkōsha, Tokyo.
Yin, W.-L., and Pipkin, A. C. (1970). *Arch. Ration. Mech. Anal.* **37**, 111.
Young, D. M., and Wheeler, M. F. (1964). *In* "Nonlinear Problems in Engineering" (W. F. Ames, ed.), Academic Press, New York.
Zaremba, S. (1903). *Bull. Int. Acad. Polon. Sci. Lett. Cl. Sci. Math. Natur.* pp, 85, 380, 403, 594, 614.
Zaremba, S. (1937). *Memor. Sci. Math.* **No. 82**.
Zimm, B. H. (1956). *J. Chem. Phys.* **24**, 269.

SOLUTIONS TO PROBLEMS

Problem 1.2(4) $e_1, \ldots, e_n$ are linearly independent (l.i.), so the space is at least n dimensional. Suppose, if possible, that $x_1, \ldots, x_{n+1}$ are l.i. Then $x_1, \ldots, x_n$ are l.i., $x_i = a_{ij} e_j$ $(i, j = 1, 2, \ldots, n)$, and $|a_{ij}| \neq 0$, because, if $|a_{ij}| = 0$, there exist numbers b_i such that $b_i x_i = \mathbf{0}$. Hence these equations can be solved to give $e_i = c_{ij} x_j$, say, and these e_i can be substituted in the equation $x_{n+1} = a_{n+1,j} e_j$ to show that $x_1, \ldots, x_{n+1}$ are not l.i., which is impossible. Hence the dimension cannot be greater than n. Q.E.D.

Problem 1.2(6) For any numbers a and b, $a^2\|x\|^2 + 2ab\langle x, y\rangle + b^2\|y\|^2 = \|ax + by\|^2 \geq 0$, and hence $\langle x, y\rangle \leq \|x\|\,\|y\|$. Therefore $(\|x\| + \|y\|)^2 - \|x + y\|^2 = 2(\|x\|\,\|y\| - \langle x, y\rangle) \geq 0$, which proves the inequality. If the equality holds, then $\langle x, y\rangle = \|x\|\,\|y\|$, so that there exist real numbers a and b, not both zero, such that $\|ax + by\| = 0$; but this is possible only if $ax + by = \mathbf{0}$, by 1.2(5), which proves that x and y are linearly dependent, as stated.

Problem 1.3(2) An elementary proof is given by Lodge (1964, p. 288).

Problem 1.3(3) (i) follows on taking the determinant of both sides of 1.3(1). Using 1.3(1) with A and a^{cr} replaced by a and α_{cr}, respectively, we have $A^{-1} = |A|^{-1}\tilde{a}$ and $a^{-1} = |a|^{-1}\tilde{\alpha}$, and hence $\tilde{\alpha} = |a|a^{-1} = |a|(|A|\tilde{A}^{-1})^{-1} = |A|^{n-1}|A|^{-1}\tilde{A}$. Q.E.D.

Problem 1.3(4) Given any $y = [y_r]$, we have $\tilde{y}A^{-1}y = \tilde{x}Ax$, where $x = A^{-1}y$, and so $\tilde{y}A^{-1}y \geq 0$, as stated. The equality sign holds if, and only if, $x = 0$, and therefore if, and only if, $y = 0$. Q.E.D.

Problem 1.3(8) On writing out the matrix equation $\tilde{R}R = I$, we obtain

(1) $\quad l_i^2 + m_i^2 + n_i^2 = 1 \quad (i = 1, 2, 3); \qquad l_2 l_3 + m_2 m_3 + n_2 n_3 = 0,$

and two similar equations; these six equations form a necessary and sufficient set of conditions that (l_i, m_i, n_i) for $i = 1, 2, 3$ shall be the actual direction cosines of three mutually orthogonal directions in space. From 1.3(6), it follows that $R\tilde{R} = I$ and hence the further stated result follows.

Problem 1.3(15) If possible, let $\lambda = a + ib$ be a complex eigenvalue of A, and let $x = u + iv$ be the corresponding complex eigenvector (where a, b, u, and v are real). On left-multiplication of 1.3(10), i.e., $Ax = \lambda x$, by $\tilde{u} - i\tilde{v}$, we see that $\tilde{u}Au + \tilde{v}Av + i(\tilde{u}Av - \tilde{v}Au) = (a + ib)(\tilde{u}u + \tilde{v}v)$, since $\tilde{u}v = \tilde{v}u$. Because $\tilde{A} = A$, we have $\tilde{u}Av = \tilde{v}Au$, and hence $b = 0$. The eigenvector equation $Ax = \lambda x$ is thus equivalent to three scalar equations, homogeneous in the elements of x, with real coefficients and vanishing determinant; there exists, therefore, a real nonzero solution for x. Q.E.D.

Problem 1.3(16) Multiplying 1.3(10) from the left by $\tilde{x}^{(2)}$, subtracting the corresponding equation obtained by interchanging 1 and 2, and using the fact that $\tilde{A} = A$, it follows that $(\lambda_1 - \lambda_2)\tilde{x}^{(1)}x^{(2)} = 0$, from which the required result follows.

Problem 1.3(17) The required equation can be rewritten in the form $AR = R\,\mathrm{diag}[\lambda_1, \lambda_2, \lambda_3]$—a 3×3 matrix equation which is equivalent to three matrix equations, of eigenvector form, in which the columns of R play the role of eigenvectors. We therefore show that an R can be constructed from the eigenvectors of A, normalized to make R orthogonal.

Case (i) λ_1, λ_2, and λ_3 all different. It follows from 1.3(16) that there exist three mutually orthogonal eigenvectors $x^{(i)}$ $(i = 1, 2, 3)$ and for each the arbitrary multiplicative constant can be chosen so that $\tilde{x}^{(1)}x^{(1)} = 1$, etc. Writing $x^{(i)} = [l_i, m_i, n_i]$, it is easy to verify that the matrix R, whose ith column is $x^{(i)}$, is orthogonal, and that the eigenvector equations $Ax^{(1)} = \lambda_1 x^{(1)}$, etc., can be written together in the form $AR = R\,\mathrm{diag}[\lambda_1, \lambda_2, \lambda_3]$. Q.E.D.

Case (ii) $\lambda_1 = \lambda_2 \neq \lambda_3$. As in case (i), we can write $x^{(i)} = [l_i, m_i, n_i]$ for $i = 2$ and 3, where (l_2, m_2, n_2) and (l_3, m_3, n_3) are actual direction cosines of two perpendicular directions. We can now define (l_1, m_1, n_1) as the actual direction cosines of a direction perpendicular to each of these two; it is easy to verify algebraically that this is always possible, by showing that the appropriate equations (1) always have a solution for l_1, m_1, and n_1. We can again define an orthogonal R as in case (i), but we do not know whether the first column of R is an eigenvector of A or not. We can, however, write (for some numbers a, b, c)

$$A\begin{bmatrix} l_1 \\ m_1 \\ n_1 \end{bmatrix} = \begin{bmatrix} al_1 \\ bm_1 \\ cn_1 \end{bmatrix}, \qquad AR = \begin{bmatrix} al_1 & \lambda_2 l_2 & \lambda_3 l_3 \\ bm_1 & \lambda_2 m_2 & \lambda_3 m_3 \\ cn_1 & \lambda_2 n_2 & \lambda_3 n_3 \end{bmatrix};$$

multiplying the second of these equations from the left by $\tilde{R}$, we have

$$\tilde{R}AR = \begin{bmatrix} al_1{}^2 + bm_1{}^2 + cn_1{}^2 & 0 & 0 \\ al_1 l_2 + bm_1 m_2 + cn_1 n_2 & \lambda_2 & 0 \\ al_1 l_3 + bm_1 m_3 + cn_1 n_3 & 0 & \lambda_3 \end{bmatrix}.$$

Since $\tilde{A} = A$, it follows that the right-hand matrix must be symmetric,

$$(2) \qquad al_1 l_2 + bm_1 m_2 + cn_1 n_2 = al_1 l_3 + bm_1 m_3 + cn_1 n_3 = 0,$$

and $\tilde{R}AR = \text{diag}[\theta, \lambda_2, \lambda_3]$, where $\theta = al_1{}^2 + bm_1{}^2 + cn_1{}^2$. Thus $\tilde{R}AR$ is diagonal, as stated, and it remains only to prove that $\theta = \lambda_1$. Since $\tilde{R}(A - \lambda I)R = \tilde{R}AR - \lambda I$, it follows that $|\tilde{R}(A - \lambda I)R| = |A - \lambda I| = (\theta - \lambda)(\lambda_2 - \lambda)(\lambda_3 - \lambda)$. Hence $\theta = \lambda_1$. Q.E.D.

Case (iii) $\lambda_1 = \lambda_2 = \lambda_3$. It follows from 1.3(12) and 1.3(13) that $A_{\text{I}}{}^2 = 3A_{\text{II}}$, and hence that $(A_{22} - A_{33})^2 + (A_{33} - A_{11})^2 + (A_{11} - A_{22})^2 + 6(A_{23}^2 + A_{31}^2 + A_{12}^2) = 0$. Hence each term must vanish separately, and so $A = A_{11}I$. Hence $\lambda_1 = \lambda_2 = \lambda_3 = A_{11}$, and the stated result is valid with $R = I$.

Problem 1.3(18) Let x denote an arbitrary, real, $n \times 1$ column matrix. Writing $y = Rx$, we have $Q \equiv \tilde{x}\tilde{R}ARx = \tilde{y}Ay \geq 0$, since A is positive definite. Thus $Q \geq 0$, and $Q = 0$ if, and only if, $y = 0$, i.e., if and only if $x = R^{-1}y = 0$. (R^{-1} exists, because R is orthogonal.) Thus Q is positive definite.

Problem 1.3(20) The underlying idea is that A, A^{-1}, and $A^{1/2}$, have the same eigenvectors and are diagonalized by the same orthogonal matrix R. Since A is symmetric, we can write, from 1.3(17), $A = R\,\text{diag}[\lambda_1, \lambda_2, \lambda_3]\tilde{R}$, for some orthogonal R. Hence $A^{-1} = R\,\text{diag}[\lambda_1^{-1}, \lambda_2^{-1}, \lambda_3^{-1}]\tilde{R}$ (A^{-1} exists because A is positive definite), and $A^{1/2} = R\,\text{diag}[\lambda_1^{1/2}, \lambda_2^{1/2}, \lambda_3^{1/2}]\tilde{R}$. From 1.3(4), it follows that A^{-1} is positive definite, and therefore $(A^{-1})^{1/2}$ exists, and is defined by the equation $(A^{-1})^{1/2} = R\,\text{diag}[\lambda_1^{-1/2}, \lambda_2^{-1/2}, \lambda_3^{-1/2}]\tilde{R}$, which follows from 1.3(19). The right-hand side of this equation is evidently equal to $(A^{1/2})^{-1}$, from the previous equation, which proves the result.

Problem 1.3(21) Since A and B are symmetric and have the same eigenvectors, they are diagonalized by the same orthogonal matrix R, i.e., $A = R\,\text{diag}[\lambda_1,$

$\lambda_2, \lambda_3]\tilde{R}$ and $B = R \operatorname{diag}[\mu_1, \mu_2, \mu_3]\tilde{R}$, where λ_i and μ_i denote the eigenvalues of A and B, respectively. Hence $AB = R \operatorname{diag}[\lambda_1\mu_1, \lambda_2\mu_2, \lambda_3\mu_3]\tilde{R} = BA$, and the other results with $A^{1/2}$ and $B^{1/2}$ can be proved in a similar way.

Problem 1.3(22) Let x be an arbitrary, real, $n \times 1$ column matrix, and F a given $n \times n$ nonsingular matrix. Then $Q \equiv \tilde{x}\tilde{F}Fx = \tilde{y}y \geq 0$, where $y = Fx$, and $Q = 0$ if, and only if, $y = 0$. But F^{-1} exists, so $x = F^{-1}y$, and $y = 0$ if, and only if, $x = 0$. Hence $\tilde{F}F$ is positive definite, and similarly $F\tilde{F}$ is positive definite. Q.E.D.

Problem 1.3(34) Let A and B be positive-definite, $n \times n$ matrices, and let y be an $n \times 1$ column matrix. Then for any y, we have $\tilde{y}Ay \geq 0$ and $\tilde{y}By \geq 0$, and hence $\tilde{y}(A + B)y \geq 0$. Moreover, if $\tilde{y}(A + B)y = 0$, then either (i) $y = 0$, or (ii) $\tilde{y}Ay = \tilde{y}By = 0$, or (iii) one of $\tilde{y}Ay$ and $\tilde{y}By$ is negative. But (iii) is impossible, since A and B are positive definite, and (ii) is possible only if $y = 0$, for the same reason. Hence $y = 0$, and so $A + B$ is positive definite. Q.E.D.

Problem 1.3(35) Since A is arbitrary, we can first take $A = I$; hence M is orthogonal, and so (for any A) we have $AM = MA$. Taking $A = \operatorname{diag}[a_1, a_2, a_3]$, $M = [M_{rc}]$, and writing out the equation $AM = MA$, we see that $M_{23} = M_{32} = 0$, etc., showing that $M = \operatorname{diag}[M_{11}, M_{22}, M_{33}]$. On adding a nonzero element $A_{12} = A_{21} = x$ to the previously chosen A [and x and a_i can evidently be chosen, according to 1.3(2), to make A positive definite] and using again the equation $AM = MA$, we find that $M_{11} = M_{22}$. Similarly, $M_{22} = M_{33}$. Hence $M = M_{11}I$. Q.E.D.

Problem 1.3(36) By 1.3(17), since A is positive definite, and therefore symmetric according to our convention, there exists an orthogonal matrix R such that $A = R \operatorname{diag}[\lambda_1, \lambda_2, \lambda_3]\tilde{R}$, where the λ_i are the eigenvalues of A. Hence, for any column matrix x, we have $\tilde{x}Ax = \lambda_i y_i^2$, where $\tilde{R}x = [y_r]$. Hence the quadratic form $\lambda_i y_i^2$ must be positive definite, and therefore $\lambda_i > 0$ $(i = 1, 2, 3)$ [from 1.3(2)]. Q.E.D.

Problem 1.3(37) $e^{1jk}e_{1rs} = 1$ if $(jkrs) = (2323)$ or (3232), -1 if $(jkrs) = (2332)$ or (3223), and 0 otherwise; these statements can be expressed in the form $e^{1jk}e_{1rs} = (\delta_r^j\delta_s^k - \delta_s^j\delta_r^k)(\delta_2^j + \delta_3^j)$. Adding the two similar equations for $e^{2jk}e_{2rs}$ and $e^{3jk}e_{3rs}$ and using the result $\delta_1^j + \delta_2^j + \delta_3^j = 1$, we obtain the stated result.

Problem 2.2(21) Let P (of coordinates ξ^i) lie on $\sigma(\xi) = c$, and let P_1 (of coordinates $\xi^i + d\xi^i$ lie on $\sigma(\xi) = c + dc$. Then $\sigma_i d\xi^i = dc$, where $\sigma_i \equiv \partial\sigma/\partial\xi^i$, which gives a constraint $\sigma_i\,\delta(d\xi^i) = 0$ for variations of $d\xi^i$ when P_1 moves on its surface while P remains fixed. The required separation dh is that value of $ds = PP_1$ for which $\delta(ds)^2 = 2\gamma_{ij}\,d\xi^i\,\delta(d\xi^j) = 0$ [using 2.2(12) and 2.2(15)] for arbitrary $\delta(d\xi^j)$ subject to the preceding constraint. Using a Lagrange multiplier θ in connection with this constraint, we have $(\gamma_{ij}\, d\xi^i - \theta\sigma_j)\, \delta(d\xi^j) = 0$. We may choose θ to make the coefficient of $(d\xi^1)$ zero; $\delta(d\xi^2)$ and $\delta(d\xi^3)$ can then be given arbitrary values, and so their coefficients, too, must vanish. Hence $\gamma_{ij}\, d\xi^i = \theta\sigma_j$, and so $dc = \sigma_i\, d\xi^i = \theta\gamma^{ik}\sigma_i\sigma_k$, which gives θ. Substituting these values for $d\xi^i$ and θ in the equation $(dh)^2 = \gamma_{ij}\, d\xi^i\, d\xi^j$, we obtain the required result.

Problem 2.3(7) On squaring 2.3(6), we see that the stated result would be false if and only if $\tilde{\Lambda}_P(B, \bar{B})\tilde{\Lambda}_P(\bar{B}, \bar{\bar{B}}) = \tilde{\Lambda}_P(\bar{B}, \bar{\bar{B}})\tilde{\Lambda}_P(B, \bar{B})$ for all B, $\bar{B}$, and $\bar{\bar{B}}$. But when B, $\bar{B}$, and $\bar{\bar{B}}$ are chosen so that (for example) $\bar{\xi}^1 = \xi^1 + b\xi^2$, $\bar{\xi}^2 = \xi^2$, $\bar{\xi}^3 = \xi^3$, $\bar{\bar{\xi}}^1 = \bar{\xi}^1$, $\bar{\bar{\xi}}^2 = a\bar{\xi}^1 + \bar{\xi}^2$, $\bar{\bar{\xi}}^3 = \bar{\xi}^3$, it is easy to verify that the matrices $\tilde{\Lambda}_P(B, \bar{B})$ and $\tilde{\Lambda}_P(\bar{B}, \bar{\bar{B}})$ do not commute. This proves the stated result.

Problem 2.3(15) From 2.3(11), we require to find $\bar{B}$, given B and θ, such that $\theta = \Lambda\theta$, where $\Lambda \equiv \Lambda_P(B, \bar{B})$, i.e., we require a nonsingular matrix Λ (not equal to I) that has a given column matrix $[\theta^r]$ as an eigenvector belonging to the eigenvalue 1. This can always be done; a simple example is

$$\Lambda = \begin{bmatrix} 1 & x\theta^3 & -x\theta^2 \\ \theta^3 & 1 & -\theta^1 \\ 0 & 0 & 1 \end{bmatrix},$$

where x is chosen so that $|\Lambda| = 1 - x(\theta^3)^2 \neq 0$. Since we are working at one particle P, we can "integrate 2.3(9) locally," treating Λ as constant, to obtain $\bar{\xi}^i = \Lambda^i{}_j\xi^j + b^i$, which define $\bar{B}$ locally at P, given B.

Problem 2.3(16) Given $\bar{B}$ and $\bar{\theta}$, we define $\boldsymbol{\varphi}$ by the equation $\boldsymbol{\varphi}B = \Lambda^{-2}(B, \bar{B})\bar{\theta}$, for all B. Then, for any $\bar{\bar{B}}$, we have $\boldsymbol{\varphi}\bar{\bar{B}} = \bar{\bar{\theta}} = \Lambda^{-2}(\bar{\bar{B}}, \bar{B})\bar{\theta}$. If we now define $\boldsymbol{\varphi}'$ by the equation $\boldsymbol{\varphi}'B = \Lambda^{-2}(B, \bar{\bar{B}})\bar{\bar{\theta}} = \Lambda^{-2}(B, \bar{\bar{B}})\Lambda^{-2}(\bar{\bar{B}}, \bar{B})\bar{\theta}$, this can equal $\boldsymbol{\varphi}B$ (for all $\bar{\theta}$) if and only if $\Lambda^{-2}(B, \bar{\bar{B}})\Lambda^{-2}(\bar{\bar{B}}, \bar{B}) = \Lambda^{-2}(B, \bar{B})$; taking the reciprocal transpose, this requirement conflicts with 2.3(7) (in which $\bar{B}$ can be taken as given, and B and $\bar{\bar{B}}$ arbitrary). Hence $\boldsymbol{\varphi} \neq \boldsymbol{\varphi}'$, and so $\boldsymbol{\varphi}$ depends on the choice of $\bar{B}$. Q.E.D.

Problem 2.3(40) The motion is described by $x^i = f^i(\xi, t)$ in $S: \mathrm{Q} \to x$ and by $\bar{x}^i = \bar{f}^i(\xi, t)$ in $\bar{S}: \mathrm{Q} \to \bar{x}$; the coordinate transformation $S \to \bar{S}$ is described by $x^i \to \bar{x}^i = \bar{x}^i(x)$. Thus $\bar{f}^i(\xi, t) = \bar{x}^i(f(\xi, t))$, and hence

$$\bar{v}^i = \frac{\partial \bar{f}^i(\xi, t)}{\partial t} = \frac{\partial \bar{x}^i}{\partial x^j}\frac{\partial f^j(\xi, t)}{\partial t} = \frac{\partial \bar{x}^i}{\partial x^j} v^j, \qquad \text{or} \qquad \bar{v} = L_Q(S, \bar{S})v \quad \text{Q.E.D.}$$

Problem 2.4(3) Define $\boldsymbol{\psi}'$ by the equation $\boldsymbol{\psi}'B = \tilde{\Lambda}_P(B, \bar{\bar{B}})\bar{\bar{\psi}}$ for all B; substituting for $\bar{\bar{\psi}}$, we have $\boldsymbol{\psi}'B = \tilde{\Lambda}_P(B, \bar{\bar{B}})\tilde{\Lambda}_P(\bar{\bar{B}}, \bar{B})\bar{\psi} = \tilde{\Lambda}_P(B, \bar{B})\bar{\psi}$, by 2.3(6). Hence, by 2.4(2), we have $\boldsymbol{\psi}'B = \boldsymbol{\psi}B$, for all B, and so $\boldsymbol{\psi}' = \boldsymbol{\psi}$. Q.E.D.

Problem 2.4(4) The equation $(a_i \boldsymbol{\psi}^i)B = \tilde{\Lambda}_P(B, \bar{B})(a_i \psi^i)$, valid for all B, is of the form 2.4(2) and therefore defines a covariant vector $a_i \boldsymbol{\psi}^i$ at P. The operations of addition and multiplication by numbers, so defined, can easily be seen to satisfy the axioms 1.2(1)(i, ii) with the zero element $\mathbf{0}$ defined by $\mathbf{0}B = 0$. Q.E.D.

Problem 2.4(8) $\boldsymbol{\psi}(\mathrm{P}) - \psi_i \boldsymbol{\beta}^i(B, \mathrm{P})$ is a covariant body vector at P, whose representative matrix in B is $(\boldsymbol{\psi} - \psi_i \boldsymbol{\beta}^i)B = \psi - \psi_i \delta^i = \psi - \psi = 0$, and is therefore the zero vector. Q.E.D.

Problem 2.4(9) θ and ψ are equal if, and only if, $\Lambda^{-1}\theta = \tilde{\Lambda}\bar{\theta}$ [from 2.3(12) and 2.4(2)], i.e., if and only if $\Lambda\tilde{\Lambda}\bar{\theta} = \bar{\theta}$. It is obviously possible, given any $\bar{B}$ and $\bar{\theta}$, to choose a B such that Λ [$\equiv \Lambda_P(B, \bar{B})$] is not orthogonal and $\Lambda\tilde{\Lambda}$ does not have $\bar{\theta}$ as an eigenvector belonging to the eigenvalue 1.

Problem 2.4(12) On differentiating the scalar transformation law $\bar{\sigma}(\bar{\xi}) = \sigma(\xi)$, we have $\partial\bar{\sigma}/\partial\bar{\xi}^i = (\partial\sigma/\partial\xi^j)(\partial\xi^j/\partial\bar{\xi}^i)$ which, from 2.3(3), is of the form 2.4(2), namely, $(\operatorname{grad} s)B = [\partial\sigma/\partial\xi^r] = \tilde{\Lambda}_P(B, \bar{B})[\partial\bar{\sigma}/\partial\bar{\xi}^r]$. Q.E.D.

Problem 2.4(17) $\boldsymbol{\beta}^1(B, \mathrm{P})$ is a covariant body vector such that $\boldsymbol{\beta}^1 B = \delta^1$, from 2.4(6). From 2.4(15) with $\sigma(\xi) \equiv \xi^1$, a normal to the ξ^1 surface at P is a covariant body vector at P whose representative matrix in B is of the form $\lambda[\partial\xi^1/\partial\xi^r] = \lambda\delta^1$, where λ is some scalar. Hence $\boldsymbol{\beta}^1$ is a normal at P to $\xi^1 =$ const; similarly, $\boldsymbol{\beta}^i$ is a normal at P to $\xi^i =$ const, for $i = 2, 3$. Q.E.D.

Problem 2.4(26) (i) $\boldsymbol{\beta}_i \cdot \boldsymbol{\beta}^j$ is an absolute scalar whose value in B is $\delta_i^k \delta_k^j = \delta_i^j$, from 2.3(22) and 2.4(7). Q.E.D. (ii) and (iii) follow at once on applying (i) to 2.3(24) and 2.4(8). The other stated results are just the corresponding results for space vectors, and can be proved similarly or regarded as proved, because the results are evidently valid whatever manifold is used.

Problem 2.4(27) Let $\boldsymbol{b}_i$ $(i = 1, 2, 3)$ be any three linearly independent contravariant space vectors at P. Let $[b_i{}^r] = \boldsymbol{b}_i S$ $(i = 1, 2, 3)$ be the representative matrices in any S. Define a square matrix $A = [b_r{}^c]$. Then A is non-singular, because the only solution of the equation $x^i \boldsymbol{b}_i = \mathbf{0}$, and therefore of $x^i b_i{}^r = 0$ $(r = 1, 2, 3)$, must be $x^i = 0$, since the $\boldsymbol{b}_i$ are linearly independent. If a reciprocal set $\boldsymbol{b}^k$ (of covariant vectors at P) exists, then $\delta_i^k = \boldsymbol{b}_i \cdot \boldsymbol{b}^k = b_i{}^r b_r^{*k}$, where $[b_r^{*k}] = \boldsymbol{b}^k S$ $(k = 1, 2, 3)$. Hence it is necessary that $A^* \equiv [b_r^{*c}]$ is the reciprocal of A. Thus A^* exists and is unique. Given $\boldsymbol{b}_i$ and any S, we can thus determine A and A^*, and define $\boldsymbol{b}^k$ as covariant vectors at P such that $\boldsymbol{b}^k S = [b_r^{*k}]$. Thus a unique reciprocal set $\boldsymbol{b}^k$ exists. A similar proof applies if we start with covariant (instead of contravariant) vectors, and if we use body (instead of space) vectors. Q.E.D.

Problem 2.5(4) For any B, we have

$$\begin{aligned}\mathbf{\Theta}'B &= \Lambda^{-1}(B, \bar{\bar{B}})\bar{\bar{\Theta}}\tilde{\Lambda}^{-1}(B, \bar{\bar{B}})\\ &= \Lambda^{-1}(B, \bar{\bar{B}})\Lambda^{-1}(\bar{\bar{B}}, \bar{B})\bar{\Theta}\tilde{\Lambda}^{-1}(\bar{\bar{B}}, \bar{B})\tilde{\Lambda}^{-1}(B, \bar{\bar{B}}) \quad \text{[from 2.5(5)]}\\ &= \Lambda^{-1}(B, \bar{B})\bar{\Theta}\tilde{\Lambda}^{-1}(B, \bar{B}) \quad \text{[from 2.3(5) with } \bar{B} \text{ and } \bar{\bar{B}} \text{ interchanged]}.\end{aligned}$$

Hence $\mathbf{\Theta}'B = \mathbf{\Theta}B$, for all B, and so $\mathbf{\Theta}' = \mathbf{\Theta}$. Q.E.D.

Problem 2.5(6) Writing $\boldsymbol{\beta}_i(B, \mathrm{P}) = \boldsymbol{\beta}_i$, we have $\boldsymbol{\beta}_i \boldsymbol{\beta}_j B = \delta_i \tilde{\delta}_j$ (a 3×3 matrix having one in row i and column j, and zero everywhere else). Given any contravariant, second-rank tensor $\mathbf{\Theta}$ at P, we can write $\mathbf{\Theta}B = [\Theta^{rc}] = \Theta^{ij}\delta_i \tilde{\delta}_j$, where $\Theta^{rc} = \Theta^{rc}(\xi)$ are the components of $\mathbf{\Theta}$ in B. Hence $\mathbf{\Theta} - \Theta^{ij}\boldsymbol{\beta}_i \boldsymbol{\beta}_j$ is a contravariant tensor at P, whose representative matrix in one coordinate system (namely B) is zero, and is therefore the zero tensor.

Q.E.D.

Problem 2.5(11) Suppose, if possible, that, given a second-rank tensor $\mathbf{\Theta}$, we can write $\mathbf{\Theta} = \mathbf{\Theta}_{\mathrm{S}} + \mathbf{\Theta}_{\mathrm{A}}$, where $\tilde{\mathbf{\Theta}}_{\mathrm{S}} = \mathbf{\Theta}_{\mathrm{S}}$ and $\tilde{\mathbf{\Theta}}_{\mathrm{A}} = -\mathbf{\Theta}_{\mathrm{A}}$. Then $\tilde{\mathbf{\Theta}} = \mathbf{\Theta}_{\mathrm{S}} - \mathbf{\Theta}_{\mathrm{A}}$, and so, on adding and subtracting these equations, we see that

$$\text{(3)} \qquad \mathbf{\Theta}_{\mathrm{S}} = \tfrac{1}{2}(\mathbf{\Theta} + \tilde{\mathbf{\Theta}}), \qquad \mathbf{\Theta}_{\mathrm{A}} = \tfrac{1}{2}(\mathbf{\Theta} - \tilde{\mathbf{\Theta}}).$$

The required decomposition is thus both possible and unique. Q.E.D.

Problem 2.5(12) (i) is evidently a particular case of (ii). To prove (ii), given any three linearly independent, contravariant vectors $\boldsymbol{\theta}_i$ at P, we have shown in 2.4(27) that there exists a unique reciprocal set, $\boldsymbol{\psi}^i$ say, of covariant vectors at P, satisfying the defining equation (iii), and that the 3×3 matrices $A = [\theta^c_{(r)}]$ and $A^* = [\psi^{(c)}_r]$ (where B is any coordinate system, $\theta_i = [\theta^r_{(i)}] = \boldsymbol{\theta}_i B$, and $\psi^k = [\psi^{(k)}_r] = \boldsymbol{\psi}^k B$) are reciprocal to one another. From the mixed tensor analog of 2.5(2), we have $(\boldsymbol{\theta}_i \boldsymbol{\psi}^i)B = \theta_i \psi^i = [\theta^r_{(i)}\psi^{(i)}_c] = I$, because $AA^* = I$ implies that $A^*A = I$. Hence $\boldsymbol{\theta}_i \boldsymbol{\psi}^i$ is a mixed tensor whose representative matrix in any coordinate system is I; therefore $\boldsymbol{\theta}_i \boldsymbol{\psi}^i = \boldsymbol{\delta}$. Q.E.D. The result $\boldsymbol{\psi}^i \boldsymbol{\theta}_i = \tilde{\boldsymbol{\delta}}$ is proved in a similar fashion.

Problem 2.5(13) The result follows at once from the definitions of second-rank tensors given in 2.5(8) and the fact that the operations transpose and taking the reciprocal, when performed on square matrices, commute.

Problem 2.5(16) (ii) By definition of the dot product, we have $\mathbf{\Theta} \cdot \boldsymbol{\psi}B = \Theta\psi$ and $\mathbf{\Theta} \cdot \boldsymbol{\psi}\bar{B} = \bar{\Theta}\bar{\psi}$. From 2.5(8), we have $\Theta\psi = \Lambda^{-1}\bar{\Theta}\tilde{\Lambda}^{-1}\tilde{\Lambda}\bar{\psi} = \Lambda^{-1}\bar{\Theta}\bar{\psi}$, which is the transformation law for a contravariant vector. Q.E.D. The remaining results can all be proved in a similar manner, on using the appropriate transformation laws chosen from 2.5(8).

Problem 2.5(17) By definition of the reciprocal of a contravariant tensor $\mathbf{\Phi}$, we have $\mathbf{\Phi} \cdot \mathbf{\Phi}^{-1}B = \Phi\Phi^{-1} = I$, and, from 2.5(14) and 2.5(15), it follows that

$\mathbf{\Phi} \cdot \mathbf{\Phi}^{-1}$ is a right-covariant mixed tensor. Hence $\mathbf{\Phi} \cdot \mathbf{\Phi}^{-1} = \boldsymbol{\delta}$, as stated. The other result with order of factors interchanged can be proved in a similar manner. If $\mathbf{\Psi}$ is another right inverse of $\mathbf{\Phi}$, then $\mathbf{\Phi} \cdot \mathbf{\Psi} = \boldsymbol{\delta}$, and so $\mathbf{\Phi} \cdot \mathbf{\Psi} B = \Phi\Psi = I$; thus $\Psi = \Phi^{-1}$, and, since $\mathbf{\Psi}$ must be a right-covariant mixed tensor, like $\mathbf{\Phi}^{-1}$, it follows that $(\mathbf{\Psi} - \mathbf{\Phi}^{-1})B = \Psi - \Phi^{-1} = 0$. Hence $\mathbf{\Psi} = \mathbf{\Phi}^{-1}$, showing that the reciprocal is unique. The corresponding results for the cases in which $\mathbf{\Phi}$ is covariant, or right-covariant mixed, are readily proved in a similar manner. Finally, if $\mathbf{\Phi}$ is covariant (or right-covariant mixed), then $\mathbf{\Phi} \cdot \boldsymbol{\delta}$ is covariant (or right-covariant mixed), and for any B, we have $\mathbf{\Phi} \cdot \boldsymbol{\delta} B = \Phi \cdot I = \Phi = \mathbf{\Phi} B$; hence $\mathbf{\Phi} \cdot \boldsymbol{\delta} = \mathbf{\Phi}$. Q.E.D.

Problem 2.6(31) Differentiating the equation $\boldsymbol{\gamma}^{-1} \cdot \boldsymbol{\gamma} = \boldsymbol{\delta}$, we have $(\partial \boldsymbol{\gamma}^{-1}/\partial t) \cdot \boldsymbol{\gamma} + \boldsymbol{\gamma}^{-1} \cdot \dot{\boldsymbol{\gamma}} = \mathbf{0}$. Contracting with $\boldsymbol{\gamma}^{-1}$ from the right, we obtain (i). Differentiating (i), and using (i) again to simplify the result, we obtain (ii).

Problem 2.6(32) Differentiating the equation $\boldsymbol{\gamma} \cdot \boldsymbol{\gamma}^{-1} = \tilde{\boldsymbol{\delta}}$ a total of n times, using Leibniz' theorem, the stated result is obtained immediately.

Problem 2.6(33) From 1.3(33) (with $A = \gamma = \boldsymbol{\gamma} B$, for an arbitrary B), we have $\mathrm{Tr}(\gamma^{-1}\dot{\gamma}) = 0$. Differentiating this equation, we have

$$\mathrm{Tr}(\gamma^{-1}\ddot{\gamma}) = -\mathrm{Tr}(\partial\gamma^{-1}/\partial t\dot{\gamma}) = \mathrm{Tr}(\gamma^{-1}\dot{\gamma}\gamma^{-1}\dot{\gamma})$$

[from 2.6(31)(i)]. This is evidently the matrix representation, in an arbitrary B, of the tensor equation stated, which is therefore proved.

Problem 2.7(3) Consider $\langle \boldsymbol{\theta}_1, \boldsymbol{\theta}_2 \rangle \equiv \boldsymbol{\theta}_1 \cdot \boldsymbol{\gamma} \cdot \boldsymbol{\theta}_2 = \tilde{\theta}_1 \gamma \theta_2$ in any B. This is symmetric in $\boldsymbol{\theta}_1$ and $\boldsymbol{\theta}_2$, because $\boldsymbol{\gamma}$ is symmetric; is linear in $\boldsymbol{\theta}_1$ and $\boldsymbol{\theta}_2$; and $\langle \boldsymbol{\theta}, \boldsymbol{\theta} \rangle > 0$ when $\theta \neq 0$, because $\boldsymbol{\gamma}$ is positive definite. Finally, $\langle \mathbf{0}, \mathbf{0} \rangle = 0$, so that all the requirements 1.2(5) for an inner product are satisfied. Q.E.D.

Problem 2.7(4) From 2.7(1) and 2.6(16), we have

$$\|\boldsymbol{\beta}_k\| = (\boldsymbol{\beta}_{\bar{k}} \cdot \boldsymbol{\gamma} \cdot \boldsymbol{\beta}_{\bar{k}})^{1/2} = (\boldsymbol{\beta}_{\bar{k}} \cdot \gamma_{ij} \boldsymbol{\beta}^i \boldsymbol{\beta}^j \cdot \boldsymbol{\beta}_{\bar{k}})^{1/2} = (\gamma_{ij} \delta^i_{\bar{k}} \delta^j_{\bar{k}})^{1/2} = (\gamma_{\bar{k}\bar{k}})^{1/2}. \quad \text{Q.E.D.}$$

The other results can be proved in a similar manner.

Problem 2.7(6) From the definition 2.4(15), the unit normal $\mathbf{v}$ must satisfy the equation $\mathbf{v}B = \lambda[\sigma_r]$, where $\sigma_r \equiv \partial\sigma/\partial\xi^r$ and λ is a scalar chosen to make $\|\mathbf{v}\| = 1$. From 2.7(1), we have $\|\mathbf{v}\|^2 = \mathbf{v} \cdot \boldsymbol{\gamma}^{-1} \cdot \mathbf{v} = \lambda^2 \sigma_i \gamma^{ij} \sigma_j$. Hence $\lambda = \pm(\gamma^{ij}\sigma_i\sigma_j)^{1/2}$, and the stated result (i) follows. (ii) is just the space manifold analog of (i).

Problem 2.7(9) Surfaces $\xi^1 = \text{const}$ and $\xi^2 = \text{const}$ of $B : \mathrm{P} \to \xi$ are orthogonal at P if $\boldsymbol{\beta}^1(B, \mathrm{P})$ and $\boldsymbol{\beta}^2(B, \mathrm{P})$ are orthogonal, i.e., if $0 = \boldsymbol{\beta}^1 \cdot \boldsymbol{\gamma}^{-1}(\mathrm{P}, t) \cdot \boldsymbol{\beta}^2 = \delta^1_i \gamma^{ij}(\xi, t)\delta^2_j = \gamma^{12}(\xi, t)$. Similarly, for the other two pairs of surfaces to be orthogonal, it is necessary and sufficient that $\gamma^{23} = \gamma^{31} = 0$. Q.E.D.

Problem 2.7(11) From 2.7(7)(i), the angle χ between contravariant vectors $\boldsymbol{\theta}$ and $\boldsymbol{\gamma}^{-1} \cdot \boldsymbol{\psi}$ is given by the equation $\|\boldsymbol{\theta}\| \, \|\boldsymbol{\gamma}^{-1} \cdot \boldsymbol{\psi}\| \cos \chi = \boldsymbol{\theta} \cdot \boldsymbol{\gamma} \cdot \boldsymbol{\gamma}^{-1} \cdot \boldsymbol{\psi} = \boldsymbol{\theta} \cdot \boldsymbol{\psi}$. From 2.7(7)(ii), the angle χ' between covariant vectors $\boldsymbol{\psi}$ and $\boldsymbol{\gamma} \cdot \boldsymbol{\theta}$ is given by the equation $\|\boldsymbol{\psi}\| \, \|\boldsymbol{\gamma} \cdot \boldsymbol{\theta}\| \cos \chi' = \boldsymbol{\psi} \cdot \boldsymbol{\gamma}^{-1} \cdot \boldsymbol{\gamma} \cdot \boldsymbol{\theta} = \boldsymbol{\psi} \cdot \boldsymbol{\theta} = \boldsymbol{\theta} \cdot \boldsymbol{\psi}$. But $\|\boldsymbol{\gamma}^{-1} \cdot \boldsymbol{\psi}\|^2 = (\boldsymbol{\gamma}^{-1} \cdot \boldsymbol{\psi}) \cdot \boldsymbol{\gamma} \cdot (\boldsymbol{\gamma}^{-1} \cdot \boldsymbol{\psi}) = \boldsymbol{\psi} \cdot \boldsymbol{\gamma}^{-1} \cdot \boldsymbol{\psi} = \|\boldsymbol{\psi}\|^2$. Similarly,

$$\|\boldsymbol{\gamma} \cdot \boldsymbol{\theta}\| = \|\boldsymbol{\theta}\|.$$

Hence $\cos \chi = \cos \chi'$. Q.E.D.

Problem 2.7(13) We have

$$\begin{aligned}(\|\boldsymbol{v}_1\| \, \|\boldsymbol{v}_2\| \sin \chi)^2 &= \|\boldsymbol{v}_1\|^2 \|\boldsymbol{v}_2\|^2 - (\|\boldsymbol{v}_1\| \, \|\boldsymbol{v}_2\| \cos \chi)^2 \\ &= (g_{ri} v_1^r v_1^i)(g_{sj} v_2^s v_2^j) - (g_{rs} v_1^r v_2^s)(g_{ij} v_1^i v_2^j) \\ &= (g_{ri} g_{sj} - g_{rs} g_{ij}) v_1^i v_2^j v_1^r v_2^s . \quad \text{Q.E.D.}\end{aligned}$$

Problem 2.7(14) From 2.4(12) and 2.4(15), it follows that grad σ is normal to $\sigma = c$ and grad σ' is normal to $\sigma' = c'$. Since these are covariant vectors, it follows from 2.7(7)(ii) that the condition that they be orthogonal is $(\text{grad}\ \sigma) \cdot \boldsymbol{\gamma}^{-1} \cdot (\text{grad}\ \sigma') = 0$, whose representative in an arbitrary $B : P \to \xi$ is $\gamma^{ij} \sigma_i \sigma_j = 0$. Q.E.D.

Problem 2.7(16) Suppose $x^i \boldsymbol{\alpha}_i = 0$ for some scalars x^i. Then we have $0 = x^i \boldsymbol{\alpha}_i \cdot \boldsymbol{\gamma} \cdot \boldsymbol{\alpha}_k = x^i \, \delta_{ik} = x^k$ for $k = 1, 2, 3$; therefore the $\boldsymbol{\alpha}_i$ are linearly independent, as stated in (i). We have $\boldsymbol{\alpha}^i \cdot \boldsymbol{\alpha}_j = \boldsymbol{\alpha}_i \cdot \boldsymbol{\gamma} \cdot \boldsymbol{\alpha}_j = \delta_{ij}$, which proves (ii). From (i), the $\boldsymbol{\alpha}_i$ are linearly independent and form a basis for contravariant vectors at P; therefore, by 2.5(7) and 1.2(9)(iii), their outer products $\boldsymbol{\alpha}_i \boldsymbol{\alpha}_j$ form a basis for second-rank contravariant tensors at P. We may therefore write $\boldsymbol{\gamma}^{-1} = x^{ij} \boldsymbol{\alpha}_i \boldsymbol{\alpha}_j$, for some scalars x^{ij} to be determined. Therefore $\boldsymbol{\alpha}^k \cdot \boldsymbol{\gamma}^{-1} \cdot \boldsymbol{\alpha}^m = x^{ij} \delta_i^k \, \delta_j^m = x^{km}$, by (ii). But $\boldsymbol{\gamma}^{-1} \cdot \boldsymbol{\alpha}^k = \boldsymbol{\alpha}_k$ (from the definition of $\boldsymbol{\alpha}^k$), and so $\boldsymbol{\alpha}^m \cdot \boldsymbol{\gamma}^{-1} \cdot \boldsymbol{\alpha}^k = \boldsymbol{\alpha}^m \cdot \boldsymbol{\alpha}_k = \delta_k^m = x^{km}$. Hence $\boldsymbol{\gamma}^{-1} = \delta^{km} \boldsymbol{\alpha}_k \boldsymbol{\alpha}_m = \boldsymbol{\alpha}_k \boldsymbol{\alpha}_k$. Q.E.D. Similarly, one can prove that $\boldsymbol{\gamma} = \boldsymbol{\alpha}^i \boldsymbol{\alpha}^i$ (using the fact that $\boldsymbol{\alpha}^i \boldsymbol{\alpha}^j$ form a basis for covariant, second-rank tensors) and $\boldsymbol{\delta} = \boldsymbol{\alpha}_i \boldsymbol{\alpha}^i$ (using the fact that $\boldsymbol{\alpha}_i \boldsymbol{\alpha}^j$ form a basis for right-covariant mixed tensors).

In view of the fundamental importance of 2.7(16), we give a *direct proof of the statement that $\boldsymbol{\alpha}_i \boldsymbol{\alpha}_j$ form a basis for contravariant, second-rank tensors when $\boldsymbol{\alpha}_i$ are three linearly independent, contravariant vectors.* We work at the same particle P throughout. In any $\bar{B}$, let $\boldsymbol{\alpha}_i \bar{B} = \bar{\alpha}_i$. We first seek a transformation $\bar{B} \to B$ such that $\boldsymbol{\alpha}_i B = \delta_i$: we have $\delta_i = \Lambda^{-1}(B, \bar{B}) \bar{\alpha}_i$, and so $\Lambda \delta_i = \bar{\alpha}_i$; therefore $\Lambda_k^r \delta_i^k = \bar{\alpha}_{(i)}^r$, i.e., $\Lambda_i^r = \bar{\alpha}_i^r$. It is always possible to find a B to satisfy this equation, because the only restriction on Λ (in order that it describe a coordinate transformation at P) is that it be nonsingular; but the matrix $[\bar{\alpha}_{(c)}^r]$ is nonsingular, because the $\boldsymbol{\alpha}_i$ are linearly independent [see the proof for Problem 2.4(27)]. Hence there exists a coordinate system B such that

$$(4) \qquad \boldsymbol{\alpha}_i B = \delta_i$$

provided only that $\boldsymbol{\alpha}_i$ are three linearly independent vectors. Now, for any given second-rank, contravariant tensor $\boldsymbol{\Theta}$, whose components in B are Θ^{ij}, we have $\boldsymbol{\Theta}B = [\Theta^{rc}] = \Theta^{rc}\delta_r\tilde{\delta}_c$. But from 2.5(2), we have $\Theta^{rc}\boldsymbol{\alpha}_r\boldsymbol{\alpha}_c B = \Theta^{rc}\delta_r\tilde{\delta}_c$. Hence $\boldsymbol{\Theta} - \Theta^{rc}\boldsymbol{\alpha}_r\boldsymbol{\alpha}_c$ is a contravariant, second-rank tensor, whose representative matrix in one coordinate system is the zero matrix, and is therefore the zero tensor. Since $\boldsymbol{\Theta}$ is arbitrary, this proves that $\boldsymbol{\alpha}_r\boldsymbol{\alpha}_c$ form a basis for second-rank contravariant tensors at P, as stated. The corresponding results for covariant tensors and mixed tensors can be proved in similar ways.

Problem 2.7(23) Since B is orthogonal at t, we have $\gamma^{ij} = \gamma^{\hat\imath\hat\imath}\delta_{\hat\imath\hat\jmath}$, and hence, from 2.7(18), $\widehat{\gamma^{ij}} = (\gamma^{\hat\imath\hat\imath}\gamma^{\hat\jmath\hat\jmath})^{-1/2}\gamma^{\hat\imath\hat\jmath} = \delta_{ij}$; similarly, $\widehat{\gamma_{ij}} = \delta_{ij}$, which proves (i). From the definition of $\boldsymbol{\pi}^0$, we have $\pi_{ij}^{(0)} = \gamma_{ik}\pi^{km}\gamma_{mj} = \gamma_{\hat\imath\hat\imath}\pi^{\hat\imath\hat\jmath}\gamma_{\hat\jmath\hat\jmath}$, since B is orthogonal; using 2.7(19) for $\widehat{\pi_{ij}^{(0)}}$ and $\widehat{\pi_{ij}}$, and the fact that for orthogonal B we have $\gamma^{\hat\imath\hat\imath} = (\gamma_{\hat\imath\hat\imath})^{-1}$, we obtain (ii). From the definition of $\boldsymbol{\pi}^0$, we have $\boldsymbol{\pi} = \boldsymbol{\gamma}^{-1} \cdot \boldsymbol{\pi}^0 \cdot \boldsymbol{\gamma}^{-1}$; differentiating this, we have $\dot{\boldsymbol{\pi}} = \boldsymbol{\gamma}^{-1} \cdot \dot{\boldsymbol{\pi}}^0 \cdot \boldsymbol{\gamma}^{-1} + \dot{\boldsymbol{\gamma}}^{-1} \cdot \boldsymbol{\pi}^0 \cdot \boldsymbol{\gamma}^{-1} + \boldsymbol{\gamma}^{-1} \cdot \boldsymbol{\pi}^0 \cdot \dot{\boldsymbol{\gamma}}^{-1}$; using 2.6(31)(i) to express $\dot{\boldsymbol{\gamma}}^{-1}$ in terms of $\dot{\boldsymbol{\gamma}}$, and then taking physical components [using the fact that, as shown in (i), $\boldsymbol{\gamma}$ and $\boldsymbol{\gamma}^{-1}$ have physical components δ_{ij}], we immediately obtain (iii).

Problem 2.8(15) Using 2.8(2)(ii), we have $\boldsymbol{\mu}^2 = (\lambda_i\boldsymbol{\theta}_i\boldsymbol{\psi}^i) \cdot (\lambda_j\boldsymbol{\theta}_j\boldsymbol{\psi}^j) = \lambda_i{}^2\boldsymbol{\theta}_i\boldsymbol{\psi}^i$, by 2.8(2)(i), which proves the required result for the case $n = 2$; the same method obviously proves the result for any positive integer n. When all λ_i are different from zero, $|\boldsymbol{\mu}| = \lambda_1\lambda_2\lambda_3 \neq 0$, and so $\boldsymbol{\mu}^{-1}$ exists. Using 2.8(2)(i), we have $\boldsymbol{\mu} \cdot (\lambda_i^{-1}\boldsymbol{\theta}_i\boldsymbol{\psi}^i) = \boldsymbol{\theta}_i\boldsymbol{\psi}^i = \boldsymbol{\delta}$, by 2.5(12)(ii), since $\boldsymbol{\theta}_i$ and $\boldsymbol{\psi}^i$, according to 2.8(2)(i), are reciprocal sets of vectors. This proves that $\boldsymbol{\mu}^{-1} = \lambda_i^{-1}\boldsymbol{\theta}_i\boldsymbol{\psi}^i$, and hence, by the result already proved, the required result is valid for all integers n.

Problem 2.8(16) From 2.8(2)(ii), we have $\boldsymbol{\mu} = \lambda_1\boldsymbol{\theta}_i\boldsymbol{\psi}^i = \lambda_1\boldsymbol{\delta}$, by 2.5(12)(ii) and 2.8(2)(i). Q.E.D. Further, for any contravariant vector $\boldsymbol{\varphi}$, we have $\boldsymbol{\mu} \cdot \boldsymbol{\varphi} = \lambda_1\boldsymbol{\delta} \cdot \boldsymbol{\varphi} = \lambda_1\boldsymbol{\varphi}$, showing that $\boldsymbol{\varphi}$ is a right eigenvector of $\boldsymbol{\mu}$; similarly, every covariant vector is a left eigenvector of $\boldsymbol{\mu}$. Q.E.D.

Problem 2.8(19) The right-covariant mixed tensor $\boldsymbol{\mu} = \boldsymbol{\gamma}^{-1} \cdot \boldsymbol{\Psi}$ satisfies the conditions of 2.8(2) and therefore has three right-eigenvectors $\boldsymbol{\theta}_i$, say, such that $\boldsymbol{\mu} \cdot \boldsymbol{\theta}_i = \lambda_{\hat\imath}\boldsymbol{\theta}_{\hat\imath}$ $(i = 1, 2, 3)$, where the λ_i are the principal values. Contracting from the left with $\boldsymbol{\gamma}$, we obtain the required result (i). We have $\boldsymbol{\theta}_2 \cdot \boldsymbol{\Psi} \cdot \boldsymbol{\theta}_1 = \lambda_1\boldsymbol{\theta}_2 \cdot \boldsymbol{\gamma} \cdot \boldsymbol{\theta}_1$ and $\boldsymbol{\theta}_1 \cdot \boldsymbol{\Psi} \cdot \boldsymbol{\theta}_2 = \lambda_2\boldsymbol{\theta}_1 \cdot \boldsymbol{\gamma} \cdot \boldsymbol{\theta}_2$; but $\boldsymbol{\Psi}$ and $\boldsymbol{\gamma}$ are symmetric, and so, assuming that $\lambda_1 \neq \lambda_2$, we have $\boldsymbol{\theta}_1 \cdot \boldsymbol{\gamma} \cdot \boldsymbol{\theta}_2 = 0$. Similarly, assuming that the

λ_i are all different, we have $\boldsymbol{\theta}_i \cdot \boldsymbol{\gamma} \cdot \boldsymbol{\theta}_j = x_i \delta_{ij}$, for some scalars x_i. But each $\boldsymbol{\theta}_i$ is, as yet, undetermined to the extent of an arbitrary scalar multiple; so each x_i can be chosen so that the required equation (ii) is satisfied. (iii) and (iv) can be proved in a similar manner. The right eigenvectors $\boldsymbol{\theta}_i$ of $\boldsymbol{\mu} = \boldsymbol{\gamma}^{-1} \cdot \boldsymbol{\Psi}$ have a reciprocal set $\boldsymbol{\psi}^i$ [the left eigenvectors, given by 2.8(1) and 2.8(2)] and, by (ii) (just proved), a reciprocal set $\boldsymbol{\gamma} \cdot \boldsymbol{\theta}_i$; since a given set of linearly independent vectors has a unique reciprocal set, according to 2.4(27), it follows that $\boldsymbol{\psi}^i = \boldsymbol{\gamma} \cdot \boldsymbol{\theta}_i = \boldsymbol{\theta}_i \cdot \boldsymbol{\gamma}$, since $\tilde{\boldsymbol{\gamma}} = \boldsymbol{\gamma}$. From 2.8(2)(ii), we then have $\boldsymbol{\mu} = \boldsymbol{\gamma}^{-1} \cdot \boldsymbol{\Psi} = \lambda_i \boldsymbol{\theta}_i \boldsymbol{\psi}^i = \lambda_i \boldsymbol{\theta}_i \boldsymbol{\theta}_i \cdot \boldsymbol{\gamma}$; contracting with $\boldsymbol{\gamma}^{-1}$ from the right, we obtain the required result 2.8(20) (left). The second result 2.8(20) can be proved in a similar manner.

Problem 2.8(22) From the definition 2.8(21), the contravariant principal axes $\boldsymbol{\theta}_i$ of strain for states t and t' are right eigenvectors of $\boldsymbol{\mu} = \boldsymbol{\gamma}^{-1} \cdot \boldsymbol{\gamma}'$, and hence also of $\boldsymbol{\Psi} = \boldsymbol{\gamma}'$, by 2.8(19), and therefore can be chosen to satisfy the equations

(5) $$(\boldsymbol{\gamma}' - e_i^2 \boldsymbol{\gamma}) \cdot \boldsymbol{\theta}_i = \mathbf{0}, \qquad \boldsymbol{\theta}_i \cdot \boldsymbol{\gamma} \cdot \boldsymbol{\theta}_j = \delta_{ij},$$

where e_i^2 are the principal values. Hence $\boldsymbol{\theta}_i$, the principal axes of strain, which have been chosen to be orthonormal at t, also satisfy the equations

(6) $$\boldsymbol{\theta}_j \cdot \boldsymbol{\gamma}' \cdot \boldsymbol{\theta}_i = e_i^2 \delta_{ij},$$

[obtained from (5) by contracting with $\boldsymbol{\theta}_j$ from the left), and are therefore orthogonal (though not orthonormal) at t'. Since $\boldsymbol{\theta}_i$ are contravariant vectors at P, they are tangents to three material lines through P (at all times); from these properties of $\boldsymbol{\theta}_i$, it follows that these three material lines are orthogonal at time t and orthogonal at time t'. Q.E.D.

Problem 2.8(23) We are given that $\boldsymbol{\gamma}^{-1} \cdot \boldsymbol{\Psi} \cdot \boldsymbol{\gamma}^{-1} = \lambda_l \boldsymbol{\theta}_l \boldsymbol{\theta}_l$, where $\boldsymbol{\theta}_i \cdot \boldsymbol{\gamma} \cdot \boldsymbol{\theta}_j = \delta_{ij}$. It follows that $\boldsymbol{\Psi} \cdot \boldsymbol{\theta}_k = \lambda_i \boldsymbol{\gamma} \cdot \boldsymbol{\theta}_i \boldsymbol{\theta}_i \cdot \boldsymbol{\gamma} \cdot \boldsymbol{\theta}_k = \lambda_{\bar{k}} \boldsymbol{\gamma} \cdot \boldsymbol{\theta}_{\bar{k}}$. Q.E.D. The corresponding result for a given covariant tensor can be proved in a similar manner.

Problem 2.8(24) The $\dot{s}_i(\text{P}, t)$ are the roots in $\dot{s}$ of the equation $|\boldsymbol{\gamma}^{-1} \cdot \dot{\boldsymbol{\gamma}} - \dot{s}\boldsymbol{\delta}| = 0$. The $e_i(t, t')$ are the roots in λ of the equation

$$0 = |\boldsymbol{\gamma}^{-1} \cdot \boldsymbol{\gamma}' - \lambda^2 \boldsymbol{\delta}| = |\boldsymbol{\gamma}^{-1} \cdot (\boldsymbol{\gamma}' - \boldsymbol{\gamma})/(t' - t) - (\lambda^2 - 1)(t' - t)^{-1} \boldsymbol{\delta}|;$$

on taking the limit $t' \to t$, the required result follows.

Problem 2.8(25) From the definition 2.7(7)(i) of angle, we have

$$\|\boldsymbol{d\xi}_1\| \, \|\boldsymbol{d\xi}_2\| \cos \chi = \boldsymbol{d\xi}_1 \cdot \boldsymbol{\gamma} \cdot \boldsymbol{d\xi}_2,$$

where χ denotes the angle at any time t between the material line elements $\boldsymbol{d\xi}_1$ and $\boldsymbol{d\xi}_2$. The time derivative of this equation is

$$-\|\boldsymbol{d\xi}_1\| \, \|\boldsymbol{d\xi}_2\| \dot{\chi} \sin \chi + \cos \chi \, \partial(\|\boldsymbol{d\xi}_1\| \, \|\boldsymbol{d\xi}_2\|)/\partial t = \boldsymbol{d\xi}_1 \cdot \dot{\boldsymbol{\gamma}} \cdot \boldsymbol{d\xi}_2.$$

If we now choose $\boldsymbol{d\xi}_1$ and $\boldsymbol{d\xi}_2$ to be tangential at time t to contravariant principal axes of strain rate, we have (at t) $\cos\chi = 0$ and

$$\dot{\boldsymbol{\gamma}} \cdot \boldsymbol{d\xi}_2 = \dot{s}_2 \boldsymbol{\gamma} \cdot \boldsymbol{d\xi}_2, \tag{7}$$

which follows from the definition 2.8(18), namely $(\boldsymbol{\gamma}^{-1} \cdot \dot{\boldsymbol{\gamma}} - \dot{s}_2 \boldsymbol{\delta}) \cdot \boldsymbol{d\xi}_2 = \boldsymbol{0}$, on contracting from the left with $\boldsymbol{\gamma}$. Hence, at time t, we have $\|\boldsymbol{d\xi}_1\| \; \|\boldsymbol{d\xi}_2\| \dot{\chi} = -\dot{s}_2 \, \boldsymbol{d\xi}_1 \cdot \boldsymbol{\gamma} \cdot \boldsymbol{d\xi}_2 = 0$, and so $\dot{\chi} = 0$. Q.E.D.

Problem 2.8(27) The main axes of the strain ellipsoid at time t [which was a sphere 2.8(26)(i) at t'] are defined by those values $\boldsymbol{d\xi}_i$, say, of the contravariant vector $\boldsymbol{d\xi}$ for which the quantity $(ds)^2 = \boldsymbol{d\xi} \cdot \boldsymbol{\gamma} \cdot \boldsymbol{d\xi}$ has a stationary value for arbitrary variations $\delta(\boldsymbol{d\xi})$ subject to the constraint $0 = \delta(ds')^2 = 2\, \boldsymbol{d\xi} \cdot \boldsymbol{\gamma}' \cdot \delta(\boldsymbol{d\xi})$. Introducing a Lagrange multiplier x in conjunction with this constraint, the condition for a stationary value can be written in the form $\boldsymbol{d\xi} \cdot (\boldsymbol{\gamma} - x\boldsymbol{\gamma}') \cdot \delta(\boldsymbol{d\xi}) = 0$; when written in component form, the usual Lagrange multiplier method enables one to choose x to make the coefficient of, say, $\delta(d\xi^1)$ equal to zero; the remaining two quantities $\delta(d\xi^2)$ and $\delta(d\xi^3)$ can then be given arbitrary values and so their coefficients, too, must vanish; hence the main axes are the roots in $\boldsymbol{d\xi}$ of the equation $\boldsymbol{d\xi} \cdot (\boldsymbol{\gamma}' - x^{-1}\boldsymbol{\gamma}) = \boldsymbol{0}$ and therefore coincide with the principal axes of strain for the states t and t'. Q.E.D. Moreover, from 2.8(21), we have $(\boldsymbol{\gamma}' - e_i^2 \boldsymbol{\gamma}) \cdot \boldsymbol{d\xi}_i = \boldsymbol{0}$, and therefore $\boldsymbol{d\xi}_i \cdot \boldsymbol{\gamma}' \cdot \boldsymbol{d\xi}_i = e_i^2 \, \boldsymbol{d\xi}_i \cdot \boldsymbol{\gamma} \cdot \boldsymbol{d\xi}_i$, showing that lengths of main axes change by factors e_i. Q.E.D.

Problem 2.8(28) From (7), it follows that $\dot{s}_2$ and, similarly, $\dot{s}_1$ and $\dot{s}_3$ are the roots in $\dot{s}$ of $|\dot{\boldsymbol{\gamma}} - \dot{s}\boldsymbol{\gamma}| = 0$. Q.E.D.

Problem 3.2(21) From the given conditions, both the ξ^2 surfaces and the ξ^3 surfaces are specified; only the ξ^1 surfaces are at our disposal. Let us first choose these surfaces in any manner, and seek the condition that a surface $\sigma(\xi) = c$ shall have lines of intersection with ξ^2 surfaces which are orthogonal to the ξ^1 lines. Any tangent to such a line of intersection will have components $(d\xi^1, 0, d\xi^3)$ that satisfy the single equation $\sigma_1 \, d\xi^1 + \sigma_3 \, d\xi^3 = 0$, where $\sigma_i \equiv \partial\sigma/\partial\xi^i$. A tangent to a ξ^1 curve will have components of the form $(d\bar{\xi}_1, 0, 0)$, and, for the two tangents to be orthogonal, it is necessary and sufficient that $0 = \gamma_{ij} \, d\bar{\xi}^i d\xi^j = d\bar{\xi}^1(\gamma_{11} \, d\xi^1 + \gamma_{13} \, d\xi^3)$. Hence a surface $\sigma(\xi) = c$ will exist if, and only if, $\gamma_{13}\sigma_1 = \gamma_{11}\sigma_3$; but this partial differential equation always has solutions $\sigma(\xi) = c$, and so a one-parameter family of surfaces exists, having the required properties. Finally, we can choose $\xi^1 \equiv \sigma$, which proves the required result.

Problem 3.3(13) Using the identity $\cos\zeta = (1 + \tan^2 \zeta)^{-1/2}$, the required results (i) follow immediately from 3.2(13), 3.2(15)(iii), and 3.3(12). The results (ii) and (iii) simply involve rotation, through an angle ζ, of the axes

$(\boldsymbol{\alpha}_1, \boldsymbol{\alpha}_3)$ to get the axes $(\boldsymbol{\alpha}_4, \boldsymbol{\alpha}_5)$, as shown in Fig. 3.3(6): Since the $\boldsymbol{\beta}_i$ form a basis for contravariant vectors at P, we may write $\boldsymbol{\alpha}_4 = x^i\boldsymbol{\beta}_i$ $(i = 1, 2, 3)$. Then $\boldsymbol{\beta}^2 \cdot \boldsymbol{\alpha}_4 = x^2 = 0$, since $\boldsymbol{\alpha}_4$ is tangential to a shearing surface. From 3.2(15), we have, for some scalar c, since $\boldsymbol{\alpha}_4$ is, by definition, tangential to a shear line,

$$\begin{aligned}\boldsymbol{\alpha}_4 &= c(d\xi^1\, \boldsymbol{\beta}_1 + d\xi^3\, \boldsymbol{\beta}_3) = c'(\gamma_{33}\,\dot{\gamma}_{12}\boldsymbol{\beta}_1 + \gamma_{11}\,\dot{\gamma}_{32}\,\boldsymbol{\beta}_3)\\ &= c'[(\gamma_{33})^{1/2}\dot{\gamma}_{12}\,\boldsymbol{\alpha}_1 + (\gamma_{11})^{1/2}\dot{\gamma}_{32}\,\boldsymbol{\alpha}_3]\\ &= c_1(\cos\zeta\,\boldsymbol{\alpha}_1 + \sin\zeta\,\boldsymbol{\alpha}_3) = c_2(\dot{s}_1\boldsymbol{\alpha}_1 + \dot{s}_3\,\boldsymbol{\alpha}_3);\end{aligned}$$

we have here used 3.3(7) and 3.3(12), as well as 3.2(15). Finally, to determine the scalar coefficient c_2, we use the fact that $\boldsymbol{\alpha}_4$ is a unit vector: $\boldsymbol{\alpha}_4 \cdot \boldsymbol{\gamma} \cdot \boldsymbol{\alpha}_4 = 1 = c_2{}^2(\dot{s}_1{}^2\|\boldsymbol{\alpha}_1\|^2 + \dot{s}_3{}^2\|\boldsymbol{\alpha}_3\|^2) = c_2{}^2\dot{s}^2$, so $c_2 = \pm\dot{s}^{-1}$. The choice of sign is made in accordance with Fig. 3.3(6). Then (iii) can be proved since $\boldsymbol{\alpha}_5 \cdot \boldsymbol{\gamma} \cdot \boldsymbol{\alpha}_4 = 0$.

Problem 3.3(30) At any given instant t, we can choose a shear flow basis $\boldsymbol{\alpha}^i$ so that $\boldsymbol{\alpha}_1$ is tangential to a shear line; then 3.3(26) becomes $\dot{\boldsymbol{\gamma}} = \dot{s}(\boldsymbol{\alpha}^1\boldsymbol{\alpha}^2 + \boldsymbol{\alpha}^2\boldsymbol{\alpha}^1)$, and when this is substituted in the principal axis equation $(\dot{\boldsymbol{\gamma}} - \dot{s}\boldsymbol{\gamma}) \cdot \boldsymbol{\theta} = \mathbf{0}$, with $\boldsymbol{\theta} = x^i\boldsymbol{\alpha}_i$, it is found that the equation is satisfied if $x^3 = 0$ and $x^1 = x^2$. Thus the principal strain-rate axis $\boldsymbol{\theta}$ is tangential to the surface containing $\boldsymbol{\alpha}_1$ and $\boldsymbol{\alpha}_2$ and is equally inclined to these vectors. Q.E.D.

Problem 3.3(31) In an arbitrary shear flow, the mixed tensor $\boldsymbol{\mu} \equiv \boldsymbol{\gamma}^{-1} \cdot \dot{\boldsymbol{\gamma}}$ has principal values 0, $\dot{s}$, and $-\dot{s}$. In an arbitrary coordinate system, the matrix μ, representing the tensor $\boldsymbol{\mu}$, satisfies the Cayley–Hamilton theorem [1.3(24)], which, in this case, reduces to the equation $\mu^3 = \dot{s}^2\mu$, since $\mu_{\mathrm{I}} = 0$ [from 1.3(12) with $A = \mu$, $\lambda_1 = 0$, $\lambda_2 = \dot{s}$, and $\lambda_3 = -\dot{s}$], $\mu_{\mathrm{II}} = -\dot{s}^2$ [from 1.3(13)], and $\mu_{\mathrm{III}} = 0$ [from 1.3(14)]. Since the coordinate system is arbitrary, it follows that the matrix equation given here implies the validity of the corresponding tensor equation $\boldsymbol{\mu}^3 = \dot{s}^2\boldsymbol{\mu}$; on left-contraction with $\boldsymbol{\gamma}$, the stated result is obtained.

Problem 3.3(32) Taking $\dot{s}_3 = 0$ and writing $\dot{s}_1 = \dot{s}$ in 3.3(26) and 3.3(27), we obtain the following equations for an arbitrary unidirectional shear flow:

(8) $$\dot{\boldsymbol{\gamma}} = \dot{s}(\boldsymbol{\alpha}^1\boldsymbol{\alpha}^2 + \boldsymbol{\alpha}^2\boldsymbol{\alpha}^1);$$

(9) $$\ddot{\boldsymbol{\gamma}} = \ddot{s}(\boldsymbol{\alpha}^1\boldsymbol{\alpha}^2 + \boldsymbol{\alpha}^2\boldsymbol{\alpha}^1) + 2\dot{s}^2\boldsymbol{\alpha}^2\boldsymbol{\alpha}^2;$$

(10) $$\dot{\boldsymbol{\gamma}} \cdot \boldsymbol{\gamma}^{-1} \cdot \dot{\boldsymbol{\gamma}} = \dot{s}^2(\boldsymbol{\alpha}^1\boldsymbol{\alpha}^1 + \boldsymbol{\alpha}^2\boldsymbol{\alpha}^2);$$

(11) $$\boldsymbol{\gamma} = \boldsymbol{\alpha}^i\boldsymbol{\alpha}^i \qquad \text{[from 2.7(16)]};$$

(12) $$\dddot{\boldsymbol{\gamma}} = \dddot{s}(\boldsymbol{\alpha}^1\boldsymbol{\alpha}^2 + \boldsymbol{\alpha}^2\boldsymbol{\alpha}^1) + 6\dot{s}\ddot{s}\boldsymbol{\alpha}^2\boldsymbol{\alpha}^2 \qquad \left[\begin{array}{l}\text{from the derivative of}\\ \text{(9), using 3.3(14)}\end{array}\right].$$

Equations (8)–(10) can be solved for $\boldsymbol{\alpha}^1\boldsymbol{\alpha}^1$ (by eliminating $\boldsymbol{\alpha}^2\boldsymbol{\alpha}^2$ and $\boldsymbol{\alpha}^1\boldsymbol{\alpha}^2 + \boldsymbol{\alpha}^2\boldsymbol{\alpha}^1$), yielding the required equation 3.3(32)(i). These equations, together with (11), can be solved to give $\boldsymbol{\alpha}^3\boldsymbol{\alpha}^3$, yielding 3.3(32)(iii). The solution involves division by $\dot{s}$, and is therefore valid at any instant when $\dot{s} \neq 0$. The corresponding results 3.3(32)(ii, iv) for contravariant vectors are obtained in a similar manner from the following equations:

(13) $$\frac{\partial \gamma^{-1}}{\partial t} = -\dot{s}(\boldsymbol{\alpha}_1\boldsymbol{\alpha}_2 + \boldsymbol{\alpha}_2\boldsymbol{\alpha}_1) \qquad \text{[from 3.3(28)]};$$

(14) $$\frac{\partial^2 \gamma^{-1}}{\partial t^2} = -\ddot{s}(\boldsymbol{\alpha}_1\boldsymbol{\alpha}_2 + \boldsymbol{\alpha}_2\boldsymbol{\alpha}_1) + 2\dot{s}^2\boldsymbol{\alpha}_1\boldsymbol{\alpha}_1 \qquad \text{[from 3.3(29)]};$$

(15) $$\frac{\partial \gamma^{-1}}{\partial t}\cdot\gamma\cdot\frac{\partial \gamma^{-1}}{\partial t} = \dot{s}^2(\boldsymbol{\alpha}_1\boldsymbol{\alpha}_1 + \boldsymbol{\alpha}_2\boldsymbol{\alpha}_2) \qquad \text{[from (13) and 2.7(16)]};$$

(16) $$\gamma^{-1} = \boldsymbol{\alpha}_i\boldsymbol{\alpha}_i \qquad \text{[from 2.7(16)]};$$

(17) $$\frac{\partial^3 \gamma^{-1}}{\partial t^3} = -\dddot{s}(\boldsymbol{\alpha}_1\boldsymbol{\alpha}_2 + \boldsymbol{\alpha}_2\boldsymbol{\alpha}_1) + 6\dot{s}\ddot{s}\boldsymbol{\alpha}_1\boldsymbol{\alpha}_1 \qquad \begin{bmatrix}\text{from derivative of}\\ \text{(14), using 3.3(14)}\end{bmatrix}.$$

Equations (12) and (17), not used for this problem, are included here for use with the following problem.

Problem 3.3(37) On eliminating $\boldsymbol{\alpha}^1\boldsymbol{\alpha}^2 + \boldsymbol{\alpha}^2\boldsymbol{\alpha}^1$ and $\boldsymbol{\alpha}^2\boldsymbol{\alpha}^2$ from (8), (9), and (12), the required result is immediately obtained. It is perhaps worth noting that on transferring this equation to space at time t, using the definition 5.4(40) for the mth rate-of-strain covariant space tensor $A^m = A^m(\mathrm{Q}, t)$, we obtain the equation

(18) $$\dot{s}^2 A^3 = (\dot{s}\dddot{s} - 3\ddot{s}^2)A^1 + 3\dot{s}\ddot{s}A^2 \qquad \begin{bmatrix}\text{unidirectional shear flow}\\ \text{of shear rate } \dot{s}\end{bmatrix}.$$

Further, using (13), (14), and (17), it is easy to show that in 3.3(37), γ can be replaced by γ^{-1}, and hence that in (18), A^m can be replaced by A_m.

Problem 3.3(38) On putting $\dot{s}_3 = 0$, and writing $\dot{s}_1 = \dot{s}$ in 3.3(20) and 3.3(17), we have

(19) $$\gamma' - \gamma = s(\boldsymbol{\alpha}^1\boldsymbol{\alpha}^2 + \boldsymbol{\alpha}^2\boldsymbol{\alpha}^1) + s^2\boldsymbol{\alpha}^2\boldsymbol{\alpha}^2 \qquad \text{(unidirectional shear flow)}.$$

Eliminating $\boldsymbol{\alpha}^1\boldsymbol{\alpha}^2 + \boldsymbol{\alpha}^2\boldsymbol{\alpha}^1$ and $\boldsymbol{\alpha}^2\boldsymbol{\alpha}^2$ between this equation, (8), and (9), we obtain the required result for the covariant strain tensor. From 3.3(21), we obtain the equation

(20) $$\gamma'^{-1} - \gamma^{-1} = -s(\boldsymbol{\alpha}_1\boldsymbol{\alpha}_2 + \boldsymbol{\alpha}_2\boldsymbol{\alpha}_1) + s^2\boldsymbol{\alpha}_1\boldsymbol{\alpha}_1.$$

Eliminating $\boldsymbol{\alpha}_1\boldsymbol{\alpha}_2 + \boldsymbol{\alpha}_2\boldsymbol{\alpha}_1$ and $\boldsymbol{\alpha}_1\boldsymbol{\alpha}_1$ from (13), (14), and (20), we obtain the required equation for the contravariant strain tensor.

Problem 3.9(19) Suppose, if possible, that there exists a flow which is a shear flow and is also shear free. The principal strain rates must be of the form $\dot{\kappa}_1 = \dot{s}$, $\dot{\kappa}_2 = -\dot{s}$, and $\dot{\kappa}_3 = 0$, where $\dot{s}$ denotes the shear rate, and the principal strain rate axes, $\boldsymbol{\varepsilon}_i$ say, must be tangential to the ξ^i curves of a body coordinate system B that is always orthogonal; from 3.9(10), we thus have $\dot{\boldsymbol{\gamma}} = \dot{s}(\boldsymbol{\varepsilon}^1\boldsymbol{\varepsilon}^1 - \boldsymbol{\varepsilon}^2\boldsymbol{\varepsilon}^2)$. From 3.3(30), however, $\boldsymbol{\varepsilon}^1$ and $\boldsymbol{\varepsilon}^2$ must lie in the plane of the shear-flow base vectors $\boldsymbol{\alpha}^1$ and $\boldsymbol{\alpha}^2$ and make angles of $\pm\pi/4$ with them. Hence $\boldsymbol{\varepsilon}^1 = (\boldsymbol{\alpha}^1 + \boldsymbol{\alpha}^2)/\sqrt{2}$, $\boldsymbol{\varepsilon}^2 = (\boldsymbol{\alpha}^2 - \boldsymbol{\alpha}^1)/\sqrt{2}$, and therefore $\dot{\boldsymbol{\gamma}} = \dot{s}(\boldsymbol{\alpha}^2\boldsymbol{\alpha}^2 - \boldsymbol{\alpha}^1\boldsymbol{\alpha}^1)$, where $\boldsymbol{\alpha}^2$ is normal to a shearing surface and $\boldsymbol{\gamma} \cdot \boldsymbol{\alpha}^1$ is tangential to a shear line at t. But this is consistent with (8) if, and only if, $\dot{s} = 0$, i.e., if, and only if, the motion is rigid. Q.E.D.

Problem 4.5(9) From the correspondence $\boldsymbol{v} \rightarrow \mathbf{v}$, we have

$$\boldsymbol{v}(\mathrm{Q})C = \mathbf{v}(\mathrm{Q})C = v^i(x)\mathbf{b}_i(S, \mathrm{Q})C = v^i(x)\boldsymbol{b}_i(S, \mathrm{Q})C,$$

by (8) and (7). Hence $\boldsymbol{v} - v^i\boldsymbol{b}_i$ is a contravariant vector at Q, whose representative matrix in one coordinate system C is zero, and is therefore the zero vector; hence the $v^i(x)$ are the components in S of the contravariant vector $\boldsymbol{v}$ which corresponds to the Cartesian vector $\mathbf{v}$. Q.E.D. The corresponding result for the covariant vector $\boldsymbol{u}$ can be proved in a similar manner by using the equation

$$\boldsymbol{b}^i(S, \mathrm{Q})C = \mathbf{b}^i(S, \mathrm{Q})C; \tag{21}$$

this can be deduced from 4.5(7) and the fact that the square matrices $[b_c^{(r)}]$ and $[b^r_{(c)}]$, whose elements are components (in any appropriate coordinate system) of reciprocal sets of vectors, are reciprocal to one another.

Problem 4.5(10) $\mathbf{b}_i(S, \mathrm{Q})\mathbf{b}^i(S, \mathrm{Q})C = \boldsymbol{b}_i(S, \mathrm{Q})\boldsymbol{b}^i(S, \mathrm{Q})C = IC = I = \mathbf{U}C$; $\mathbf{b}_i(S, \mathrm{Q})\mathbf{b}^i(S, \mathrm{Q}) - \mathbf{U}$ is a second-rank Cartesian tensor, which has zero as its matrix representative in one coordinate system, and is therefore the zero Cartesian tensor. Q.E.D. The other, similar result can be obtained from this one on taking the transpose, since the transpose of a Cartesian tensor is a Cartesian tensor.

Problem 4.5(11) From 4.5(6), we have

$$(ds)^2 = d\mathbf{r} \cdot d\mathbf{r} = dx^i\, dx^j\, \mathbf{b}_i(S, \mathrm{Q}) \cdot \mathbf{b}_j(S, \mathrm{Q}) = g_{ij}(x)\, dx^i\, dx^j,$$

for all values of dx^i; since the coefficients of $dx^i\, dx^j$ are both symmetric, they must be equal. Q.E.D. The other result can be deduced from this one by using the reciprocal matrix argument employed in the previous proof.

Problem 4.5(13) The proofs are similar to those already given. We have $\boldsymbol{p}(\mathrm{Q})C = \mathbf{p}(\mathrm{Q})C = p^{ij}(x)\mathbf{b}_i(S, \mathrm{Q})\mathbf{b}_j(S, \mathrm{Q})C = p^{ij}(x)\boldsymbol{b}_i(S, \mathrm{Q})\boldsymbol{b}_j(S, \mathrm{Q})C$, and so the contravariant tensors at the beginning and end are equal, which proves one of the required results. The others can be proved in a similar manner.

Problem 4.5(14) Writing $\boldsymbol{b}_i(S, \text{Q}) = \boldsymbol{b}_i$, we have

$$0 = (\boldsymbol{g}^{-1} - g^{ij}\boldsymbol{b}_i\,\boldsymbol{b}_j)C = (\mathbf{U} - g^{ij}\mathbf{b}_i\,\mathbf{b}_j)C,$$

since $\boldsymbol{g}^{-1}C = \boldsymbol{g}C = \mathbf{U}C = I$. Hence $\mathbf{U} = g^{ij}\mathbf{b}_i\,\mathbf{b}_j$. Q.E.D. The other result is proved similarly. Contraction with $\mathbf{b}^k$ from the right gives the result $\mathbf{b}^k = g^{ik}\mathbf{b}_i = g^{ki}\mathbf{b}_i$. Q.E.D.

Problem 4.5(15) From Definition 4.5(3) applied to any two coordinate systems $S: \text{Q} \to x$ and $\bar{S}: \text{Q} \to \bar{x}$, we have

$$\mathbf{b}_i(S, \text{Q}) = \frac{\partial \mathbf{r}}{\partial x^i} = \frac{\partial \mathbf{r}}{\partial \bar{x}^k}\frac{\partial \bar{x}^k}{\partial x^i} = \mathbf{b}_k(\bar{S}, \text{Q})\frac{\partial \bar{x}^k}{\partial x^i}. \tag{22}$$

On taking $\bar{S} \equiv C: \text{Q} \to y$, the required result is obtained. The second result can be obtained from it as follows: writing $\mathbf{b}^j(S, \text{Q}) = a_i^j\,\mathbf{b}^i(\bar{S}, \text{Q})$, we have

$$\delta_k^j = \mathbf{b}^j(S, \text{Q}) \cdot \mathbf{b}_k(S, \text{Q}) = a_i^j\,\mathbf{b}^i(\bar{S}, \text{Q}) \cdot \mathbf{b}_m(\bar{S}, \text{Q})(\partial \bar{x}^m/\partial x^k) = a_i^j(\partial \bar{x}^i/\partial x^k).$$

Hence $a_i^j = \partial x^j/\partial \bar{x}^i$. Q.E.D.

Problem 4.5(16) From 4.2(7) and 4.5(1), we have $\mathbf{r}(\text{Q})C = [y^r - y_0^r]$. From 4.5(3), we have $\mathbf{b}_i(C) = \partial \mathbf{r}/\partial y^i$. Hence $\mathbf{b}_i(C)C = \partial(\mathbf{r}C)/\partial y^i = \delta_i$. Therefore $[\mathbf{r} - (y^i - y_0^i)\mathbf{b}_i(C)]C = [y^r - y_0^r] - (y^i - y_0^i)\delta_i = 0$. Thus

$$\mathbf{r}(\text{Q}) - (y^i - y_0^i)\mathbf{b}_i(C)$$

is a Cartesian vector, whose representative matrix in one coordinate system is zero, and is therefore the zero vector. Q.E.D.

Problem 5.2(12) From 5.2(9), we have $\boldsymbol{v}_1 \cdot \boldsymbol{v}_2 \times \boldsymbol{v}_3 = v_{(1)}^i e_{ijk} v_{(2)}^j v_{(3)}^k = |v_{(j)}^i|$, from 1.3(25) and 1.3(26), which proves (iv). A single cyclic permutation of factors is equivalent to two interchanges of adjacent factors and therefore to two interchanges of columns in the determinant $|v_{(j)}^i|$, and has no effect on the value, proving (ii). There are two interchanges of factors involved in (i), and it is therefore valid for the same reason. If the $\boldsymbol{v}_i$ are linearly independent, then the determinant is nonzero, and so $s = (\boldsymbol{v}_1 \times \boldsymbol{v}_2 \cdot \boldsymbol{v}_3)^{-1}$ exists; by 5.2(9), s^{-1} is of weight $3n - 1$, and so $\boldsymbol{u}^i$ is of weight $2n - 1 - (3n - 1) = -n$, as stated. Furthermore [still taking $\boldsymbol{u}^i$ as being defined by 5.2(13)], it is clear that $\boldsymbol{u}^1 \cdot \boldsymbol{v}_1 = 1$, etc., and $\boldsymbol{u}^1 \cdot \boldsymbol{v}_2 = 0$, etc., because each equals a determinant with two equal columns; thus $\boldsymbol{u}^i \cdot \boldsymbol{v}_j = \delta_j^i$, showing that the $\boldsymbol{u}^i$ are reciprocal to the $\boldsymbol{v}_j$, as stated; since the reciprocal set is unique, the proof of (v) is complete.

Problem 5.3(2) Let us define a function $\boldsymbol{I}: \mathscr{S} \to A_{3\times 1}$ such that $\boldsymbol{I}S = I$ (for all S). Let $\boldsymbol{\theta}$ be an arbitrary contravariant vector. Then $I\theta = \theta$ (where $\boldsymbol{\theta}S = \theta$), and the right-hand side is the S representative of a contravariant vector. Hence $\boldsymbol{I}$ is a right-covariant mixed tensor. Q.E.D.

Problem 5.4(39) On differentiating the transformation law $X^i = (\partial x^i/\partial \bar{x}^k)\bar{X}^k$ for an absolute, contravariant vector field $\boldsymbol{X}(\mathrm{Q})$ and contracting with Y^i (expressed in terms of $\bar{Y}^s$ on the right-hand side), we obtain the equation

$$Y^j \frac{\partial X^i}{\partial x^j} = \frac{\partial^2 x^i}{\partial \bar{x}^k\, \partial \bar{x}^s} \bar{X}^k \bar{Y}^s + \frac{\partial x^i}{\partial \bar{x}^k} \frac{\partial \bar{X}^k}{\partial \bar{x}^s} \bar{Y}^s.$$

The first term on the right-hand side is of "nontensor form," but is symmetric in X and Y; when we subtract the equation obtained by interchanging X and Y, this term contributes nothing, and we obtain the required result.

Problem 5.4(45) Result (ii) follows from the definition 5.4(14)(i) and the fact 5.4(22) that $\mathbb{T}(t)$ commutes with $\partial/\partial t'$ when $t \neq t'$. Result (iii) follows from (ii) on letting $t' \to t$ and using the definition 5.4(40). To prove (iv), we use 5.4(31)(iii) with $F \equiv C$ and $\Phi \equiv \gamma'$, with the result

$$\dot{\boldsymbol{C}}_t(\mathrm{Q}, t')S = \left[\frac{\partial \xi^u}{\partial x^r} \frac{\partial \xi^v}{\partial x^c} \dot{\gamma}_{uv}(\xi, t')\right] \qquad \text{(for any } S \text{ and } B). \tag{23}$$

We may now choose $B \stackrel{t'}{\equiv} S$, so that $\xi^u = x'^u$, and $\dot{\gamma}_{uv}(\xi, t') = A^{(1)}_{uv}(x', t')$, from 5.4(4) (with t' for t) and 5.4(40) (with $m = 1$, and t' for t). The stated result is thus proved.

Problem 5.4(55) From 5.4(31)(iii) applied to contravariant tensors $F \equiv \dot{\boldsymbol{B}}$ and $\Phi \equiv \partial\gamma'^{-1}/\partial t'$, and the definition 5.4(54), we have

$$\dot{\boldsymbol{B}}_t(\mathrm{Q}, t')S = \left[\frac{\partial x^r}{\partial \xi^i} \frac{\partial x^c}{\partial \xi^j} \frac{\partial \gamma^{ij}(\xi, t')}{\partial t'}\right] \qquad \text{(for any } S \text{ and } B). \tag{24}$$

We may now choose $B \stackrel{t'}{\equiv} S$, so that $\xi^i = x'^i$ and $\partial\gamma^{ij}(\xi, t')/\partial t' = A^{ij}_{(1)}(x', t')$, by 5.4(9) and 5.4(46). This proves (i), for any S. If we transfer 2.6(31)(i) to space at time t, using the definitions 5.4(40) and 5.4(46), we obtain the equation

$$\boldsymbol{A}_1 = -\boldsymbol{g}^{-1} \cdot \boldsymbol{A}^1 \cdot \boldsymbol{g}^{-1}. \tag{25}$$

Using this equation at time t' referred to any S, we obtain (ii). Finally, if we transfer the identity

$$\int_{t''}^{t} \frac{\partial \gamma(\mathrm{P}, t')}{\partial t'}\, dt' = \gamma(\mathrm{P}, t) - \gamma(\mathrm{P}, t'') \tag{26}$$

to space at time t, using 5.4(47), 5.4(45)(i), and 5.4(14), we have

$$\int_{t''}^{t} \dot{\boldsymbol{C}}_t(\mathrm{Q}, t')\, dt' = \boldsymbol{g}(\mathrm{Q}) - \boldsymbol{C}_t(\mathrm{Q}, t''). \tag{27}$$

The representative of this equation in any S, with 5.4(45)(iv), gives (iii).

Problem 5.4(56) On integration by parts, using the definition (ii) for G, we obtain the identity

$$(28)\quad \int_{t''}^{t} \mu(t-t')\gamma'^{-1}\,dt' = \gamma^{-1}G(0) - \gamma''^{-1}G(t-t'') - \int_{t''}^{t} G(t-t')\frac{\partial\gamma'^{-1}}{\partial t'}\,dt'.$$

On transferring this equation to space at time t, using 5.4(47), 5.4(14)(ii), and 5.4(54), we obtain the required equation (i).

Problem 5.4(57) From 5.4(14)(i) and 5.4(31)(iii) (with $F \equiv C$ and $\Phi \equiv \gamma'$), we have

$$(29)\qquad C_t(Q, t')S = \left[\frac{\partial \xi^u}{\partial x^r}\frac{\partial \xi^v}{\partial x^c}\gamma_{uv}(\xi, t')\right] \qquad \text{(for any } S \text{ and } B).$$

We may now take $B \overset{t'}{\equiv} S$, so that $\xi^u = x'^u$ and $\gamma_{uv}(\xi, t') = g_{uv}(x')$, which proves (i). Result (ii) can be proved in a similar manner, using 5.4(14)(ii) and 5.4(31)(iii) with $F \equiv B$ and $\Phi \equiv \gamma'^{-1}$.

Problem 5.5(10) In any B, let the given surface have an equation $\sigma(\xi) = c$. Since $\boldsymbol{d\xi}'$ and $\boldsymbol{d\xi}''$ are tangential to the surface, we have $\sigma_i\, d\xi'^i = 0$ and $\sigma_i\, d\xi''^i = 0$, where $\sigma_i = (\partial\sigma/\partial\xi^i)_{\mathrm{P}}$. The solution of these two equations for σ_i is of the form $\sigma_i = ae_{ijk}\, d\xi'^j\, d\xi''^k$, and so, from the definition of $\mathbf{v}$, we have $\mathbf{v} = b$ grad σ, so that the present definition of $\mathbf{v}$ agrees with the definition of the normal to a material surface given in 2.4(15), proving (ii). Further, since $\mathbf{v}$ is a unit vector (at time t), its components must satisfy 2.7(6)(i), which contains no reference to P′ or P″; hence $\mathbf{v}$ is independent of the choice of P′ and P″, proving (i). Since dA is an absolute scalar, its value is independent of the choice of coordinate system used to evaluate it; we may therefore choose a coordinate system B such that $d\xi'^i = d\xi^1\, \delta_1^i$ and $d\xi''^j = d\xi^2\, \delta_2^j$, as stated in (iv). Then from the definition of $\mathbf{v}$, we have $(\mathbf{v}\, dA)B = [v_r\, dA]$, where

$$v_i\, dA = |\gamma|^{1/2} e_{ijk}(d\xi^1\,\delta_1^j)(d\xi^2\,\delta_2^k) = |\gamma|^{1/2} e_{i12}\, d\xi^1\, d\xi^2$$
$$= |\gamma|^{1/2}\,\delta_i^3\, d\xi^1\, d\xi^2,$$

from 1.3(25). But $1 = \|\mathbf{v}\| = \gamma^{ij}v_i v_j = \gamma^{33}{v_3}^2$, since $v_1 = v_2 = 0$. The required result (iv) follows at once, with a suitable choice of signs. Finally, using 2.7(13) (body tensor analog), we have

$$(\mathrm{PP'\ PP''} \sin \angle \mathrm{P'PP''})^2 = (\gamma_{ri}\gamma_{sj} - \gamma_{rs}\gamma_{ij})\, d\xi^1\,\delta_1^i\, d\xi^2\,\delta_2^j\, d\xi^1\,\delta_1^r\, d\xi^2\,\delta_2^s$$
$$= (\gamma_{12}^2 - \gamma_{11}\gamma_{22})(d\xi^1\, d\xi^2)^2 = |\gamma|\,\gamma^{33}(d\xi^1\, d\xi^2)^2;$$

using (iv), this gives (iii), when the sign is chosen appropriately.

Problem 5.6(7) It follows from 5.6(6)(i) that $(\boldsymbol{\pi}\cdot\boldsymbol{\gamma} - \sigma_1\boldsymbol{\delta})\cdot \boldsymbol{d\xi}_1 = \mathbf{0}$, where $\boldsymbol{d\xi}_1$ $(= \boldsymbol{\gamma}^{-1}\cdot\boldsymbol{\psi}^1)$ is a contravariant principal axis of stress associated with the principal value σ_1. Since it is given that $\boldsymbol{d\xi}_1$ is always tangential to the same material line at P, we can choose the normalization of $\boldsymbol{d\xi}_1$ so that $\boldsymbol{d\xi}_1$ is independent of time. Since $\partial\sigma_1/\partial t = 0$, it follows that $\partial(\boldsymbol{\pi}\cdot\boldsymbol{\gamma})/\partial t \cdot \boldsymbol{d\xi}_1 = \mathbf{0}$, with two similar equations for the other two principal axes. Since the principal axes are represented by linearly independent vectors, it follows that $\partial(\boldsymbol{\pi}\cdot\boldsymbol{\gamma})/\partial t = \mathbf{0}$, as stated.

Problem 5.6(8) From 5.6(6)(i), the principal values σ_i of stress are the roots in σ of the equation

$$(30) \qquad |\pi\gamma - \sigma I| = 0 \qquad \text{(in any } B\text{)}.$$

From 1.3(12) and 1.3(14) (with $A = \pi\gamma$), it follows that $\sigma_1\sigma_2\sigma_3 = |\pi\gamma| = |\boldsymbol{\pi}|\,|\boldsymbol{\gamma}|$, and $\sigma_1 + \sigma_2 + \sigma_3 = \mathrm{tr}(\pi\gamma) = \boldsymbol{\pi}:\boldsymbol{\gamma}$. Q.E.D.

Problem 5.6(12) With the notation of 5.6(11), we have $\boldsymbol{\pi}:\mathbf{v}^{\hat{\imath}}\mathbf{v}^{\hat{\imath}} = -p$, for $i = 1, 2, \ldots, 6$, and hence $(\boldsymbol{\pi} + p\boldsymbol{\gamma}^{-1}):\mathbf{v}^{\hat{\imath}}\mathbf{v}^{\hat{\imath}} = 0$, since $\boldsymbol{\gamma}^{-1}:\mathbf{v}^{\hat{\imath}}\mathbf{v}^{\hat{\imath}} = 1$, because the $\mathbf{v}^i$ are unit vectors. From the proof of 5.6(11), using 5.6(9), it follows that $(\boldsymbol{\pi} + p\boldsymbol{\gamma}^{-1})B = 0$, for some B, and therefore $\boldsymbol{\pi} + p\boldsymbol{\gamma}^{-1} = \mathbf{0}$. Q.E.D. For any plane, the normal component of traction is $\mathbf{vv}:\boldsymbol{\pi} = -p\mathbf{vv}:\boldsymbol{\gamma}^{-1} = -p$ Q.E.D. Finally, the contravariant traction vector for any plane is $\boldsymbol{\pi}\cdot\mathbf{v} = -p\boldsymbol{\gamma}^{-1}\cdot\mathbf{v}$, and so the covariant traction vector is $\boldsymbol{\gamma}\cdot\boldsymbol{\pi}\cdot\mathbf{v} = -p\mathbf{v}$, which is in the direction of the normal $\mathbf{v}$; there is therefore no tangential component of traction on any plane. Q.E.D.

Problem 5.6(20) A rigid rotation of the body can be described by the equation $\mathbf{r} = \mathbf{R}(t)\cdot\mathbf{r}_0$, where $\mathbf{R}$ is an orthogonal tensor, and $\mathbf{r}_0$ and $\mathbf{r}$ are the position vectors at times t_0 and t of a given particle, the origin particle being supposed fixed in space. Choosing $B \overset{t_0}{\equiv} C$, we have $\xi^i = y_0^i$, and hence, since $y^i = R^i_j y^j_0$, we have $[\partial y^r/\partial\xi^c] = R(t) = \mathbf{R}(t)C$. Since $\boldsymbol{\pi} \overset{t}{\rightarrow} \mathbf{p}$, we have $\mathbf{p}C = p = R\pi\tilde{R}$, where π is independent of t, by 5.6(19), since the stress is constant. If $\pi = cI$, then $p = cI = \text{const}$; but if the stress is not isotropic, then $\pi \neq cI$, and it is evident that p varies with t because R does: For example, for rotation about the Oy^3 axis, we have $R^1_1 = R^2_2 = \cos\theta(t)$, $R^1_2 = -R^2_1 = \sin\theta(t)$, and $R^3_i = R^i_3 = \delta_{3i}$; if, say, the only nonzero π^{ij} are $\pi^{12} = \pi^{21}$, then it is easy to show that $p^{11} = \pi^{21}\sin 2\theta(t)$, showing that $\mathbf{p}$ varies with t.

Problem 5.6(22) From 2.2(21) and 2.7(6)(i), we have

$$(31) \qquad v_k = \pm\sigma_k\, dh/dc.$$

Hence $f_n = \boldsymbol{\pi}:\mathbf{vv} = \pi^{jk}\sigma_j\sigma_k(dh/dc)^2$; since π^{jk}, σ_j, and dc are all independent of t, it follows that $f_n/(dh)^2$ is independent of t. Q.E.D.

Problem 5.6(23) Let $\overrightarrow{\mathrm{PP}'} = \boldsymbol{d\xi} = \boldsymbol{\gamma}^{-1}\cdot\mathbf{v}x$, where x is a scalar to be determined; this equation expresses the fact that $\overrightarrow{\mathrm{PP}'}$ is normal to a plane element of unit normal $\mathbf{v}$. Then $\varepsilon^2/f_n = (\mathrm{PP'})^2 = \boldsymbol{\gamma}:\boldsymbol{d\xi}\,\boldsymbol{d\xi} = \boldsymbol{\gamma}^{-1}:\mathbf{vv}x^2 = x^2$. Hence the equation (with $\boldsymbol{d\xi}$ as a variable vector) of the given polar diagram is $\boldsymbol{d\xi}\cdot\boldsymbol{\gamma}\cdot\boldsymbol{\pi}\cdot\boldsymbol{\gamma}\cdot\boldsymbol{d\xi} = \varepsilon^2\boldsymbol{\pi}:\mathbf{vv}/f_n = \varepsilon^2$. Since ε and $\boldsymbol{\gamma}\cdot\boldsymbol{\pi}\cdot\boldsymbol{\gamma}$ are independent of t, it follows that the surface is always composed of the same particles. Q.E.D.

Problem 5.7(45) Writing $\boldsymbol{d\gamma} \equiv \boldsymbol{\gamma}' - \boldsymbol{\gamma} = \boldsymbol{\gamma}\cdot\boldsymbol{\lambda}'$ [$= O(\varepsilon)$], the constant-volume condition gives $0 = d|\boldsymbol{\gamma}| = \boldsymbol{\gamma}^{-1}:\boldsymbol{d\gamma} + O(\varepsilon^2)$, from 1.3(33). Hence $\boldsymbol{\gamma}^{-1}:\boldsymbol{d\gamma} = \boldsymbol{\gamma}^{-1}:(\boldsymbol{\gamma}\cdot\boldsymbol{\lambda}') = \boldsymbol{\delta}:\boldsymbol{\lambda}' = \lambda_{\mathrm{I}}' = O(\varepsilon^2)$. Q.E.D.

Problem 6.3(5) For arbitrary infinitesimal strain $\boldsymbol{d\gamma}$, at constant volume and constant temperature, it follows from 6.3(1) and 6.3(2) that $dA = \boldsymbol{\pi}:\boldsymbol{d\gamma}/2\rho = \boldsymbol{\Pi}:\boldsymbol{d\gamma}/2\rho$, since $\boldsymbol{\gamma}^{-1}:\boldsymbol{d\gamma} = 0$ because the volume is constant. From 6.2(1), we then have $dA = (ckTN_0/2\rho)\boldsymbol{\gamma}_0^{-1}:\boldsymbol{d\gamma}$, which integrates at once to give the stated result.

Problem 6.3(6) We have the result $(\partial\varphi/\partial\varepsilon_{ij})\,d\varepsilon_{ij} = d\varphi$; thus the quantities $\partial\varphi/\partial\varepsilon_{ij}$, when contracted twice with the components $d\varepsilon_{ij}$ of an arbitrary covariant tensor of weight n, give an absolute scalar $d\varphi$. By a quotient theorem, it follows that $\partial\varphi/\partial\varepsilon_{ij}$ are components of a contravariant second-rank tensor of weight $-n$. Q.E.D.

Problem 6.3(13) Let $\boldsymbol{\varphi}$ be a covariant principal axis of strain belonging to the principal strain $e_i(t, t_0) = x$, say. Then $\boldsymbol{\gamma}_0^{-1}\cdot\boldsymbol{\varphi} = x\boldsymbol{\gamma}^{-1}\cdot\boldsymbol{\varphi}$, and so

$$\boldsymbol{\gamma}^{-1}\cdot\boldsymbol{\gamma}_0\cdot\boldsymbol{\gamma}^{-1}\cdot\boldsymbol{\varphi} = x^{-1}\boldsymbol{\gamma}^{-1}\cdot\boldsymbol{\varphi}.$$

From 6.3(11), we now see that

$$\boldsymbol{\pi}\cdot\boldsymbol{\varphi} = 2\rho\{(I_2A_2 + I_3A_3) + A_1x - A_2x^{-1}\}\boldsymbol{\gamma}^{-1}\cdot\boldsymbol{\varphi}. \quad \text{Q.E.D.}$$

Problem 6.3(15) We have $J_2 = \mathrm{tr}(\boldsymbol{\gamma}\cdot\boldsymbol{\gamma}_0^{-1})^2$, and so $dJ_2 = 2(\boldsymbol{\gamma}_0^{-1}\cdot\boldsymbol{\gamma}\cdot\boldsymbol{\gamma}_0^{-1}):\boldsymbol{d\gamma}$. With $dJ_1 = dI_1$ and $dJ_3 = dI_3$, given by 6.3(10), the stated result can be obtained, by a similar method to that used to get 6.3(11), from 6.3(1) and 6.3(2).

Problem 7.1(15) The required equation is obtained at once from 7.1(11) on contracting from the left and from the right with $\boldsymbol{\gamma}$, and making use of 2.6(31).

Problem 7.1(16) σ, N_1, and N_2 are defined for unidirectional shear flow. For such a flow, using 3.3(28), we have, for a Newtonian liquid, the result $\boldsymbol{\pi} + p\boldsymbol{\gamma}^{-1} = -\eta_0\,\partial\boldsymbol{\gamma}^{-1}/\partial t = \dot{s}\eta_0(\boldsymbol{\alpha}_2\boldsymbol{\alpha}_1 + \boldsymbol{\alpha}_1\boldsymbol{\alpha}_2)$. Hence $\sigma \equiv \boldsymbol{\alpha}^1\boldsymbol{\alpha}^2:\boldsymbol{\pi} = \dot{s}\eta_0$, $N_1 \equiv (\boldsymbol{\alpha}^1\boldsymbol{\alpha}^1 - \boldsymbol{\alpha}^2\boldsymbol{\alpha}^2):\boldsymbol{\pi} = 0$, and $N_2 \equiv (\boldsymbol{\alpha}^2\boldsymbol{\alpha}^2 - \boldsymbol{\alpha}^3\boldsymbol{\alpha}^3):\boldsymbol{\pi} = 0$. Q.E.D.

Problem 7.1(17) For unidirectional shear flow, we have, from 6.2(2) and 3.3(21), $\boldsymbol{\pi} = -p\boldsymbol{\gamma}^{-1} + \int_{-\infty}^{t}\mu(t-t')[s^2\boldsymbol{\alpha}_1\boldsymbol{\alpha}_1 - s(\boldsymbol{\alpha}_1\boldsymbol{\alpha}_2 + \boldsymbol{\alpha}_2\boldsymbol{\alpha}_1)]\,dt'$, where $s = \int_t^{t'}\dot{s}''\,dt''$. Using the definitions 7.1(3) and remembering that $\boldsymbol{\alpha}_i$ and $\boldsymbol{\alpha}^i$ depend on t but not on t', the stated results follow at once.

Problem 8.1(13) Contracting $\boldsymbol{W}$ with an arbitrary, absolute, covariant vector field $\boldsymbol{u}(\mathrm{Q})$, we have $\boldsymbol{u} \cdot \boldsymbol{W} = \boldsymbol{v}$, say, where $\boldsymbol{v}(\mathrm{Q})$ is a relative, contravariant vector of weight 1. We can therefore apply 8.1(7), which shows that we have a relative scalar (of weight 1) $s \equiv \partial v^i/\partial x^i = (\partial u_j/\partial x^i)W^{ji} + (\partial W^{ji}/\partial x^i)u_j$. Hence $(\partial W^{ji}/\partial x^i)u_j = s - \frac{1}{2}W^{ji}(\partial u_j/\partial x^i - \partial u_i/\partial x^j)$, since $W^{ji} = -W^{ij}$. From 8.1(5), it follows that the right-hand side is a relative scalar of weight 1; but the u_j are components of an arbitrary covariant vector field, and hence, from a quotient theorem, it follows that $\partial W^{ji}/\partial x^i$ are the components of a relative contravariant vector field of weight 1. Q.E.D.

Problem 8.1(14) The stated results follow at once from 8.1(8) and 8.1(9), when we use the results $\sqrt{|g|} = h_1 h_2 h_3$, $\widehat{v^i} = v^{\hat{\imath}}/\sqrt{g^{\hat{\imath}\hat{\imath}}} = h_{\hat{\imath}} v^{\hat{\imath}}$, and $g^{ij} = h_{\hat{\imath}}^{-2}\delta_{\hat{\imath}j}$.

Problem 8.2(18) From 1.3(33) (with $A = g$ and $t = x^i$), we have

$$2|g|^{-1/2}\,\partial|g|^{1/2}/\partial x^i = 2\partial\big(\log_e \sqrt{|g|}\big)/\partial x^i = \partial(\log_e|g|)/\partial x^i = \operatorname{tr}(g^{-1}\,\partial g/\partial x^i)$$
$$= g^{jk}\,\partial g_{kj}/\partial x^i = g^{jk}(G_{kji} + G_{jki}) = G^j_{ji} + G^k_{ki} = 2G^k_{ki}. \quad \text{Q.E.D.}$$

We have here made use of 8.2(17) and 8.2(16).

Problem 8.2(19) From 8.2(17) with $i = k$, we have $G_{\hat{k}\hat{k}j} = h_{\hat{k}}\,\partial h_{\hat{k}}/\partial x^j$. Hence, from 8.2(16), we have $G^{\hat{k}}_{\hat{k}j} = h_{\hat{k}}^{-1}\partial h_{\hat{k}}/\partial x^j$; this gives the first row [and hence also the first column, because of 8.2(10)] of the square matrix $[G^1_{rc}]$. The other elements are obtained directly from 8.2(16) on substituting $g_{ij} = h_{\hat{\imath}}^2\delta_{\hat{\imath}j}$.

Problem 8.2(29) We have $P^{ij}{}_{,k} = p^{ij}{}_{,k} + (g^{ij}p)_{,k} = p^{ij}{}_{,k} + g^{ij}\,\partial p/\partial x^k$, since $\nabla \boldsymbol{g} = \mathbf{0}$ and p is an absolute scalar. Q.E.D.

Problem 8.6(3) A linear relation between $\boldsymbol{v}(\mathrm{Q})$ and $\boldsymbol{v}(\mathrm{Q}|\mathrm{Q}')$ must be of the form $\boldsymbol{v}(\mathrm{Q}|\mathrm{Q}') = \boldsymbol{H} \cdot \boldsymbol{v}(\mathrm{Q})$, where $\boldsymbol{H}$ is a linear operator (independent of the $\boldsymbol{v}$'s). Writing $\boldsymbol{v}(\mathrm{Q}) = v^i(x)\boldsymbol{b}_i(S, \mathrm{Q})$ and $\boldsymbol{v}(\mathrm{Q}|\mathrm{Q}') = v^j(x|x')\boldsymbol{b}_j(S, \mathrm{Q})$, substituting in this equation, and contracting with $\boldsymbol{b}^k(S, \mathrm{Q}')$, we have $v^i(x|x') = H^i_j(x|x')v^j(x)$, where $H^i_j(x|x') = \boldsymbol{b}^i(S, \mathrm{Q}') \cdot \boldsymbol{H} \cdot \boldsymbol{b}_j(S, \mathrm{Q})$. The assumption that the mapping is also linear in $\boldsymbol{dx}$ is equivalent to a requirement of differentiability of $\boldsymbol{H}$, so that we have $H^i_j(x|x') = K^i_j(x) - G^i_{jk}(x)\,dx^k + O(dx)^2$. Finally, the requirement that the mapping reduce to the identity when $\mathrm{Q} \equiv \mathrm{Q}'$ can be met if and only if $K^i_j = \delta^i_j$. These results together yield 8.6(2), as required.

Problem 8.6(4) With the preceding notation, since the operator $\boldsymbol{H}$ is independent of coordinate systems, we have [on using the space analog of 2.4(28)] the result

$$\bar{H}^i_j(\bar{x}|\bar{x}') = \boldsymbol{b}^i(\bar{S}, \mathrm{Q}') \cdot \boldsymbol{H} \cdot \boldsymbol{b}_j(\bar{S}, \mathrm{Q}) = \left(\frac{\partial \bar{x}^i}{\partial x^k}\right)_{\mathrm{Q}'} H^k_m(x|x')\left(\frac{\partial x^m}{\partial \bar{x}^j}\right)_{\mathrm{Q}},$$

and hence

$$\delta^i_j - G^i_{jk}(\bar{x})\, d\bar{x}^k = \left(\frac{\partial \bar{x}^i}{\partial x^k} + \frac{\partial^2 \bar{x}^i}{\partial x^k\, \partial x^s}\right)_Q (\delta^k_m - G^k_{mp}(x)\, dx^p)\left(\frac{\partial x^m}{\partial \bar{x}^j}\right)_Q,$$

to first order in dx. Expressing $d\bar{x}^k$ in terms of dx^s, we see that the terms independent of dx^s cancel, while the terms linear in dx^s give the required result after multiplication by $\partial\bar{x}^j/\partial x^r$ with summation over j; the letters used for indices are different in this analysis and in the stated result. Finally, in the stated result 8.6(3), if $G^m_{pq} = G^m_{qp}$, the left-hand side is symmetric in p and q and therefore so also must the right-hand side be; expressing this statement as an equation, interchanging dummy indices r and s, and multiplying by appropriate "inverse factors" $\partial x^q/\partial \bar{x}^k$, etc., it is easy to see that $\bar{G}^i_{rs} = \bar{G}^i_{sr}$. Q.E.D.

Problem 8.6(5) Since $\boldsymbol{dx} \to \boldsymbol{dX}$, we have $dX^i = dx^i - G^i_{jk}\, dx^j\, \delta x^k$; since $\boldsymbol{\delta x} \to \boldsymbol{\delta X}$, we have $\delta X^i = \delta x^i - G^i_{jk}\, \delta x^j\, dx^k$. The difference of coordinates of Q_1 and Q_2 is equal to $(\delta x^i + dX^i) - (dx^i + \delta X^i) = G^i_{jk}(-dx^j\, \delta x^k + \delta x^j\, dx^k) = 0$ for all dx^j and δx^k if and only if $G^i_{jk} = G^i_{kj}$. Q.E.D.

Problem 8.6(7) We have $\boldsymbol{v}(Q) \xrightarrow{G(Q)} \boldsymbol{v}_1(Q|Q_1) \xrightarrow{G(Q_1)} \boldsymbol{V}_1(Q')$; applying 8.6(2) twice, we have $V^i_{(1)}(x') = (\delta^i_j - G^i_{jk}(x_1)\, d_2 x^k) v^j_{(1)}(x|x_1)$ and $v^j_{(1)}(x|x_1) = (\delta^j_m - G^j_{mr}(x)\, d_1 x^r) v^m(x)$. Eliminating $v^j_{(1)}(x|x_1)$, writing $G^i_{jk}(x_1) = G^i_{jk}(x + d_1 \boldsymbol{x}) = G^i_{jk}(x) + (\partial G^i_{jk}/\partial x^r)\, d_1 x^r$, and ordering the result, we have

(32)

$$V^i_{(1)}(x')$$
$$= \left\{\delta^i_m - G^i_{mk}(d_1 x^k + d_2 x^k) + \left(G^i_{jk} G^j_{mr} - \frac{\partial G^i_{mk}}{\partial x^r}\right) d_1 x^r\, d_2 x^k + \cdots\right\}_Q v^m(x).$$

This equation is obtained by going from Q to Q′ via Q_1; the corresponding equation for $V^i_{(2)}(x')$ is obtained from this by interchanging indices 1 and 2 on the right-hand side, corresponding to the route from Q to Q′ via Q_2. On forming the difference $V^i_{(1)} - V^i_{(2)}$ from these equations, it is seen that the terms of zeroth and first order in $d_1 x$ and $d_2 x$ cancel and that the terms of second order are expressible in the form

$$(V^i_{(1)} - V^i_{(2)})_{Q'} = (R^i_{mkr} v^m\, d_1 x^r\, d_2 x^k)_Q + O(dx)^3, \tag{33}$$

where (with appropriate changes of indices) R^i_{mkr} is given in 8.6(9). It follows that $V_1(Q') = V_2(Q')$ if, and only if, $R^i_{mkr} = 0$, as stated. The tensor character of $\boldsymbol{R}$ is far from obvious in 8.6(9); however, on the right-hand side of (33), we can replace Q by Q′ [because the resulting error is of order $(dx)^3$, and therefore negligible], and we then see that R^i_{mkr}, when contracted with arbitrary vectors, gives a vector; hence the R^i_{mkr} are components of a fourth-rank tensor, by a quotient theorem. Q.E.D. It is obvious from 8.6(9) that R^i_{jsk} is antisymmetric in s and k, as stated in 8.6(10) and that when $G^i_{jk} = G^i_{kj}$, the two groups of terms in R^i_{jsk} involve cyclic interchanges of indices j, s, and k and hence that on adding the three equations obtained by cyclic interchanges of these indices the various terms cancel one another, proving 8.6(11).

Problem 8.6(15) On putting $\bar{G}^i_{rs} = 0$ in 8.6(4), we see that $G^m_{qp} = G^m_{pq}$; the condition is therefore necessary. Suppose now that we are given the values of $G^m_{qp} = G^m_{pq}$ at some point Q_0, of coordinates x^i_0 in S. Consider the local coordinate transformation $S \to \bar{S}$ $(x^i \to \bar{x}^i)$ defined by the equations

$$\bar{x}^i - \bar{x}^i_0 = a^i_j\,(x^j - x^j_0) + b^i_{jk}\,(x^j - x^j_0)(x^k - x^k_0) + O(x - x_0)^3,$$

where a^i_j and b^i_{jk} can depend on x_0 but not on x. Then $(\partial\bar{x}^i/\partial x^j)_{Q_0} = a^i_j$ and $(\partial^2\bar{x}^i/\partial x^j\,\partial x^k)_{Q_0} = b^i_{jk}$. Choose any constants a^i_j such that $|a^i_j| \neq 0$; then $|\partial\bar{x}^i/\partial x^j| = |a^i_j| + O(\bar{x} - x)$ and is therefore nonzero in a neighborhood of Q_0. The transformation then is a coordinate transformation in a neighborhood of Q_0. We may now choose $b^i_{jk} = a^i_m G^m_{jk}$, and then we have, from 8.6(4), $\bar{G}^i_{rs} = 0$, after multiplication by $(\partial x^q/\partial\bar{x}^m)_{Q_0}$ and $(\partial x^p/\partial\bar{x}^n)_{Q_0}$. Q.E.D.

Problem 8.6(16) In a flat space, the vector obtained at Q by parallel transfer of a given vector at Q_0 is independent of the path taken, and therefore gives rise to a uniquely defined vector field. For any two neighboring places Q and Q′, the vectors $\boldsymbol{v}(Q)$ and $\boldsymbol{v}(Q')$ of this field satisfy the equation $\boldsymbol{v}(Q) = \boldsymbol{v}(Q'|Q)$, where $\boldsymbol{v}(Q') \xrightarrow{G(Q')} \boldsymbol{v}(Q'|Q)$, by the definition of parallel transfer. Hence, from the definition 8.6(12) of the covariant derivative operator $\boldsymbol{\nabla}$, it follows that $\boldsymbol{\nabla v}(Q) = \boldsymbol{0}$. Since Q is an arbitrary place, it follows that $\boldsymbol{\nabla v} = \boldsymbol{0}$ everywhere, so the field is homogeneous, according to the definition 8.2(28). Q.E.D.

Problem 8.6(17) For any two neighboring places Q and Q′, we have $\boldsymbol{v}(Q) \xrightarrow{G(Q)} \boldsymbol{v}(Q|Q') = \boldsymbol{H}\cdot\boldsymbol{v}(Q)$, where $\boldsymbol{H}$ is a linear operator with a unique inverse, since the transformation is one-to-one, by definition. If $\boldsymbol{v}_i(Q)$ are any three linearly independent vectors at Q and θ^i are any three scalars, then $\boldsymbol{H}\cdot(\theta^i\boldsymbol{v}_i(Q)) = \theta^i\boldsymbol{H}\cdot\boldsymbol{v}_i(Q) = \theta^i\boldsymbol{v}_i(Q|Q')$, where the $\boldsymbol{v}_i(Q|Q')$ are the corresponding vectors at Q′ obtained by parallel transfer. Suppose, if possible, that $\theta^i\boldsymbol{v}_i(Q|Q') = \boldsymbol{0}$; then $\boldsymbol{H}\cdot(\theta^i\boldsymbol{v}_i(Q)) = \boldsymbol{0}$, and therefore $\theta^i\boldsymbol{v}_i(Q) = \boldsymbol{0}$ [because $\boldsymbol{H}^{-1}(\boldsymbol{0}) = \boldsymbol{0}$, since $\boldsymbol{H}$ is one-to-one]. But the $\boldsymbol{v}_i(Q)$ are linearly independent, by hypothesis; therefore $\theta^i = 0$, and so the $\boldsymbol{v}_i(Q|Q')$ are linearly independent. Thus parallel transfer along any given curve cannot destroy linear independence. Q.E.D.

Problem 8.6(18) From 8.2(24)(iii), if $\boldsymbol{B}(Q)$ is a homogeneous, absolute, covariant vector field (i.e., such that $\boldsymbol{\nabla B} = \boldsymbol{0}$), its components in any S satisfy the equations $\partial B_j/\partial x^k = G^r_{jk} B_r = \partial B_k/\partial x^j$ if $G^r_{jk} = G^r_{kj}$. Hence Curl $\boldsymbol{B} = \boldsymbol{0}$, from 8.1(5). Q.E.D.

Problem 8.7(1) Let χ be the angle between any two given contravariant vectors $\boldsymbol{v}(\mathrm{Q})$ and $\boldsymbol{v}'(\mathrm{Q})$; then for any scalars a and b, we have, from 2.7(1) and 2.7(7), $\|a\boldsymbol{v} + b\boldsymbol{v}'\|^2 = a\|\boldsymbol{v}\|^2 + b\|\boldsymbol{v}'\|^2 + 2ab\|\boldsymbol{v}\|\ \|\boldsymbol{v}'\| \cos\chi$. By hypothesis, $\boldsymbol{v}$ and $\boldsymbol{v}'$ are unchanged on parallel transfer; therefore $\cos\chi$ is unchanged on parallel transfer. Q.E.D. Further, on expressing the equation $\|\boldsymbol{v}(\mathrm{Q})\|^2 = \|\boldsymbol{v}(\mathrm{Q}\,|\,\mathrm{Q}')\|^2$ in components in an arbitrary S, we have

$$
\begin{aligned}
(g_{ij} v^i v^j)_{\mathrm{Q}} &= g_{ij}(x') v^i(x\,|\,x') v^j(x\,|\,x') \\
&= \left\{ \left(g_{ij} + \frac{\partial g_{ij}}{\partial x^k}\, dx^k \right) (v^i - G^i_{rs} v^r\, dx^s)(v^j - G^j_{mn} v^m\, dx^n) \right\}_{\mathrm{Q}} \\
&= (g_{ij} v^i v^j)_{\mathrm{Q}} + \left\{ \left(\frac{\partial g_{ij}}{\partial x^k} - G_{jik} - G_{ijk} \right) v^i v^j\, dx^k \right\}_{\mathrm{Q}} + O(dx)^2.
\end{aligned}
$$

The first term cancels; the term linear in dx^k is therefore zero, and since v^i and dx^k are arbitrary and the coefficient of $v^i v^j\, dx^k$ is symmetric in i and j, it follows that this coefficient must vanish, which proves 8.7(2). If $\boldsymbol{G}$ is symmetric, then $G_{ijk} = G_{ikj}$ and, on subtracting 8.7(2) from the sum of two similar equations obtained by successive cyclic interchanges of indices i, j, an k, one obtains 8.7(3). Following the procedure used previously for a contravariant vector and using 8.6(14) for an arbitrary covariant vector $\boldsymbol{u}(\mathrm{Q})$, it is easy to show that

(34)

$$\|\boldsymbol{u}(\mathrm{Q}\,|\,\mathrm{Q}')\|^2 - \|\boldsymbol{u}(\mathrm{Q})\|^2 = \left(\frac{\partial g^{rt}}{\partial x^s} + G^r_{is} g^{it} + G^t_{is} g^{ir} \right)_{\mathrm{Q}} u_r(x) u_t(x)\, dx^s + O(dx)^2.$$

Since $g_{ij} g^{jt} = \delta^t_j$, we have

$$g^{jt}(\partial g_{ij}/\partial x^s) = -g_{ij}(\partial g^{jt}/\partial x^s). \tag{35}$$

Using this equation with 8.7(2), it is easy to show that

$$-\partial g^{rt}/\partial x^s = G^r_{is} g^{it} + G^t_{is} g^{ir}. \tag{36}$$

Hence, from (34), we have $\|\boldsymbol{u}(\mathrm{Q}\,|\,\mathrm{Q}')\| = \|\boldsymbol{u}(\mathrm{Q})\|$ [to $O(dx)$], showing that the magnitude of an absolute covariant vector is unchanged on parallel transfer Q.E.D. The constancy of angle between covariant vectors during parallel transfer can now be proved in a similar manner to the proof given for contravariant vectors.

Problem 8.7(4) If S exists having normal base vectors $\boldsymbol{b}^i(S, \mathrm{Q}) = \boldsymbol{B}^i(\mathrm{Q})$, then $\boldsymbol{B}^i(\mathrm{Q})S = \delta^i$, and therefore Curl $\boldsymbol{B}^i = \boldsymbol{0}$, from 8.1(6); the condition Curl $\boldsymbol{B}^i = \boldsymbol{0}$

is thus necessary for the existence of such an S. Let $\bar{B}_j^{(i)}(\bar{x})$ denote the components of $\boldsymbol{B}^i(\mathrm{Q})$ in an arbitrary $\bar{S}$; if an S exists with the required property, then its covariant base vectors $\boldsymbol{b}^i(S, \mathrm{Q})$ have components δ^i_j in S and components $\delta^i_k\, \partial x^k/\partial \bar{x}^j = \partial x^i/\partial \bar{x}^j$ in $\bar{S}$. We require, therefore, to find a coordinate transformation $\bar{x} \to x$ such that $\partial x^i/\partial \bar{x}^j = B_j^{(i)}(\bar{x})$. This is a set of partial differential equations for the unknown functions $x^i(\bar{x})$, the functions $B_j^{(i)}(\bar{x})$ being given; this is of standard form, and a necessary and sufficient condition of integrability is $\partial B_j^{(i)}/\partial \bar{x}^k = \partial B_k^{(i)}/\partial \bar{x}^j$, i.e., Curl $\boldsymbol{B}^i = \mathbf{0}$. Hence the required functions $x^i(\bar{x})$ exist and they define a coordinate transformation $\bar{x} \to x$ since the Jacobian $|\partial x^i/\partial \bar{x}^j| = |B_j^{(i)}|$ vanishes nowhere, because the $\boldsymbol{B}^i$ are linearly independent. Q.E.D.

Problem 8.7(5) If a Cartesian coordinate system S exists, then in S, we have $\partial g_{ij}/\partial x^k = 0$ everywhere, by definition; hence $G^i_{jk} = 0$, from 8.7(2) and 8.7(3), and so $R^i_{jsk} = 0$ and $\boldsymbol{R} = \mathbf{0}$, from 8.6(9); the condition $\boldsymbol{R} = \mathbf{0}$ is therefore necessary for the existence of a Cartesian coordinate system. To show that the condition is also sufficient, let us take any three linearly independent, absolute covariant vectors $\boldsymbol{B}^i(\mathrm{Q}_0)$ at any given point Q_0. By parallel transfer along arbitrary paths (since the space is flat) we can generate three absolute, covariant vector fields $\boldsymbol{B}^i(\mathrm{Q})$ defined over the whole space. From the covariant vector analog of 8.6(16), these fields are homogeneous, and therefore, by 8.6(18), satisfy the conditions Curl $\boldsymbol{B}^i = \mathbf{0}$. Moreover, from 8.6(18), the vectors $\boldsymbol{B}^i(\mathrm{Q})$ are linearly independent, and therefore satisfy all the conditions of 8.7(4). It follows that a coordinate system S (for the whole space) exists having $\boldsymbol{B}^i$ as its normal base vectors. Referred to S, the components of $\boldsymbol{B}^i(\mathrm{Q})$ are therefore $B_j^{(i)}(x) = \delta^i_j$, and the condition $\nabla \boldsymbol{B}^i = \mathbf{0}$ (for homogeneous $\boldsymbol{B}^i$) takes the form $0 = G^r_{jk}\delta^i_r = G^i_{jk}$. It follows from 8.7(2) and 8.7(3) that S is Cartesian. Q.E.D.

Problem 8.7(6) From the definition, we have $\gamma^{nhk} = \boldsymbol{B}_k \cdot \nabla \boldsymbol{B}^n \cdot \boldsymbol{B}_h = \boldsymbol{B}_h \cdot \widetilde{\nabla \boldsymbol{B}^n} \cdot \boldsymbol{B}_k$. Hence $\gamma^{nhk} - \gamma^{nkh} = \boldsymbol{B}_k \cdot (\nabla \boldsymbol{B}^n - \widetilde{\nabla \boldsymbol{B}^n}) \cdot \boldsymbol{B}_h = \boldsymbol{B}_k \cdot \mathrm{Curl}\, \boldsymbol{B}^n \cdot \boldsymbol{B}_h$, since $\nabla \boldsymbol{B}^n - \widetilde{\nabla \boldsymbol{B}^n}$ and Curl $\boldsymbol{B}^n$ are both absolute, second-rank, covariant tensors whose components at any place Q are equal in a coordinate system chosen so that $G^i_{jk} = 0$ at Q; according to 8.6(15), such a coordinate system exists when $\boldsymbol{G}$ is symmetric; the tensors are therefore equal so (i) is proved. Since the $\boldsymbol{B}^i$ are linearly independent, so also are the $\boldsymbol{B}_i$; hence $\gamma^{nhk} = \gamma^{nkh}$ if and only if Curl $\boldsymbol{B}^n = \mathbf{0}$, and therefore if and only if surfaces exist which are everywhere normal to the vectors $\boldsymbol{B}^n$, by 8.7(4). Q.E.D.

Problem 8.8(11) Since $\boldsymbol{R}(\gamma(\mathrm{P}, t)) = \mathbf{0}$, there exists a body coordinate system $B : \mathrm{P} \to \xi$ that is Cartesian at time t [from 8.7(5)] and therefore has $\Gamma^i_{jk}(\xi, t) = 0$. Covariant differentiation at time t therefore reduces to partial differentiation with respect to ξ^i (in B), and the homogeneous deformation condition

$\nabla_t \gamma(\mathrm{P}, t') = \mathbf{0}$ reduces to $\partial \gamma_{ij}(\xi, t')/\partial \xi^k = 0$, which implies that $\partial \gamma^{ij}(\xi, t')/\partial \xi^k = 0$. It then follows that $\partial \theta^{ij}(\xi)/\partial \xi^k = 0$. Further, since the sum of positive multiples of positive-definite tensors is a positive-definite tensor [from the tensor analog of 1.3(34)], it follows that $\boldsymbol{\theta}$ is positive definite (and symmetric). Hence [from the tensor analog of 1.3(4)] $\boldsymbol{\theta}^{-1}$ exists and is positive definite (and symmetric); also, $\boldsymbol{\theta}^{-1}$ is covariant since $\boldsymbol{\theta}$ is contravariant. This result implies that $\partial \theta_{ij}/\partial \xi^k = 0$, where $\theta_{ij}\theta^{jm} = \delta_j^m$, and therefore $\boldsymbol{R}(\boldsymbol{\theta}^{-1}) = \mathbf{0}$. By 8.8(7), it follows that $\boldsymbol{\theta}^{-1}$ is the metric tensor field in some state t^*. Moreover, since B is Cartesian and $\partial \theta_{ij}/\partial \xi^k = 0$, it follows that $\nabla_t \boldsymbol{\theta}^{-1} = \mathbf{0}$, and therefore the deformation $t \to t^*$ is homogeneous, and therefore continuous. Q.E.D.

Problem 10.2(20) From 7.4(4) and 10.2(3), we have

$$\boldsymbol{\Pi}^+ = (\widehat{p_+^{ii}} + p_+)\boldsymbol{\alpha}_i^+\boldsymbol{\alpha}_i^+ = \boldsymbol{\Pi}^- = (\widehat{p_-^{ii}} + p_-)\boldsymbol{\alpha}_i^-\boldsymbol{\alpha}_i^-, \tag{37}$$

where $\boldsymbol{\alpha}_i^+ = (\boldsymbol{\gamma}^{-1} \cdot \boldsymbol{\alpha}^i)^+$ is a shear-free flow basis, orthonormal at t^+, etc. From 3.9(5) and 3.9(7), remembering that $\boldsymbol{\beta}_\iota$ is constant, it follows that $\boldsymbol{\alpha}_i^+ = (\gamma_{\bar{i}\bar{i}}^+/\gamma_{\bar{i}\bar{i}}^-)^{-1/2}\boldsymbol{\alpha}_i^- = \boldsymbol{\alpha}_i^-/e_i$, where $e_1 = \lambda$, $e_2 = e_3 = \lambda^{-1/2}$. Hence $\widehat{p_+^{\bar{i}\bar{i}}} + p_+ = e_i^2(\widehat{p_-^{\bar{i}\bar{i}}} + p_-)$ $(i = 1, 2, 3)$, from which we find that

$$\widehat{p_-^{11}} + p_- = \frac{x^+ - x^-/\lambda}{\lambda^2 - \lambda^{-1}} \quad \text{and} \quad \widehat{p_-^{22}} + p_- = \frac{x^+ - \lambda^2 x^-}{\lambda^2 - \lambda^{-1}}, \tag{38}$$

where $x \equiv \widehat{p^{11}} - \widehat{p^{22}}$. From 7.4(4) and 7.4(7) with $e_2 = e_3$ (the flow history having always been one of simple elongation, at constant volume) it follows that $\widehat{p^{22}} = \widehat{p^{33}}$ $(= \widehat{p_{22}} = \widehat{p_{33}})$. From (37) and 10.2(18), we have $(\mu_0^*)^3 = |\boldsymbol{\Pi}^- \cdot \boldsymbol{\gamma}^-| = |(\widehat{p_-^{ii}} + p_-)\boldsymbol{\alpha}_i^-\boldsymbol{\alpha}_-^i| = (\widehat{p_-^{11}} + p_-)(\widehat{p_-^{22}} + p_-)^2$; substituting from (38), we obtain the required result for μ_0^*. Further, since $(\mu_0^*)^3 = |\boldsymbol{\Pi}^- \cdot \boldsymbol{\gamma}^-|$, it follows that μ_0^* is uniquely determined by the flow history in the interval $(-\infty, t^-)$ and is therefore independent of λ, which is a principal elongation ratio for the strain $t^- \to t^+$. Q.E.D.

Problem 11.1(6) We use any given space coordinate $S: \mathrm{Q} \to x$, with base vectors $\boldsymbol{b}_i(S, \mathrm{Q})$, and a convected family $\hat{S}(B, t)$ with base vectors $\hat{\boldsymbol{b}}_i(\mathrm{Q}, t)$. Since, at any one t, $\hat{S}(B, t)$ is a space coordinate system $\hat{S}: \mathrm{Q} \to \hat{x}$, we have, from the space analog of 2.4(28), with 2.2(4) and 11.1(2),

$$\hat{\boldsymbol{b}}_i(\mathrm{Q}, t) = \boldsymbol{b}_j(S, \mathrm{Q})\frac{\partial x^j}{\partial \hat{x}^i} = \boldsymbol{b}_j(S, \mathrm{Q})\frac{\partial f^j(\xi, t)}{\partial \xi^i}. \tag{39}$$

Hence, from 2.2(8), the derivative of (39) gives $\partial \hat{\boldsymbol{b}}_i/\partial t = \boldsymbol{b}_j \partial v^j/\partial \xi^i$, and, taking now $S \equiv \hat{S}$ at time t, so that $v^j = \hat{v}^j$, we see that $\partial \hat{\boldsymbol{b}}_i/\partial t = \boldsymbol{b}_j\, \partial \hat{v}^j/\partial \hat{x}^i =$

$\hat{\boldsymbol{b}}_j(\hat{v}^j{}_{,i} - \hat{G}^j_{si}\hat{v}^s) = \hat{\boldsymbol{b}}_i \cdot \nabla \boldsymbol{v} - \boldsymbol{v} \cdot \nabla \hat{\boldsymbol{b}}_i$, from 8.2(24)(ii, viii). Using 8.2(23), we obtain the first of the required equations. The second may be obtained in a similar manner or, alternatively, by taking the hydrodynamic derivative of the equation $\hat{\boldsymbol{b}}_i \cdot \hat{\boldsymbol{b}}^j = \delta^j_i$, remembering that D/Dt has properties of a differential operator.

Problem 11.2(1) *Proof that (iii)⇒(i).* Since B is Cartesian at t, we $\nabla_t \gamma' B = [\partial\gamma_{ij}(\xi, t')/\partial\xi^k] = 0$, since B is also Cartesian at t'; hence (iii)⇒(i).

Proof that (i)⇒(iii). Since $\boldsymbol{R}(\gamma) = \boldsymbol{0}$, there exists a B which is Cartesian at t, and so $0 = \nabla_t \gamma' B = [\partial\gamma_{ij}(\xi, t')/\partial\xi^k]$; hence B is also Cartesian at t', proving that (i)⇒(iii). Interchanging t and t', it follows that (ii)⇒(iii), since (i)⇒(iii). Hence (ii) ≡ (iii), so (i), (ii), and (iii) are equivalent. From 5.4(14) and 8.3(3), it follows that (v) ≡ (i), so (i), (ii), (iii), and (v) are equivalent. Further, (v) and (vi) are clearly equivalent, because Cartesian space coordinate systems certainly exist, and in them $G^i_{jk} = 0$ and covariant differentiation reduces to partial differentiation with respect to y^i. Thus we have proved that all but (iv) are equivalent. Given (iv), it follows that $\partial y'^i/\partial y^j$, and therefore [from 5.4(47) with $S \equiv C$] $C_t(\mathrm{Q}, t')C$, is independent of y; thus (iv)⇒(vi). All that remains, therefore, is to prove that (vi)⇒(iv), for then (i)–(vi) will have been shown to be equivalent, as stated.

Proof that (vi)⇒(iv). Applying the polar decomposition theorem 1.3(23) to the nonsingular matrix $F = [\partial y'^r/\partial y^c]$, we see that $F = R[C_{rc}]^{1/2}$, where the C_{rc} are the components of $C_t(\mathrm{Q}, t')$ in a rectangular Cartesian coordinate system, and R is an orthogonal matrix. We require to prove, therefore, that, if $\tilde{F}F = [C_{rc}]$ is independent of y [as (vi) states], then so also are R and F [as (iv) states]; this result is very close to 4.1(2), which states that a constant orthogonal transformation $y \to y'$ is necessarily linear. Since $t' \to t$ is a deformation, the Jacobian $|\partial y'^i/\partial y^j|$ is nonzero, and so the transformation $y'^i \to y^i$ can be reinterpreted as a coordinate transformation. For any two neighboring particles, we have $(ds')^2 = \delta_{ij}\, dy'^i\, dy'^j = C_{ij}\, dy^i\, dy^j$. Since $[C_{ij}]$ is positive definite (and symmetric), there exists an orthogonal matrix Q such that $[C_{rc}] = \tilde{Q}\,\mathrm{diag}[x_1, x_2, x_3]Q$, for some positive numbers x_i (from [1.3(17)]; since $[C_{rc}]$ is independent of y, so also are Q and x_i. Writing $dY^i = x_i^{1/2} Q^i{}_j\, dy^j$, we have $\delta_{ij}\, dy'^i\, dy'^j = \delta_{ij}\, dY^i\, dY^j$ for all y. Hence, by 4.1(2), the transformation $y'^i \to Y^i$ is linear, and therefore the transformation $y'^i \to y^i$ is linear. Q.E.D.

Problem 11.2(2) We consider homogeneous deformations only. It is easy to verify that, for any given embedded basis $\mathbf{e}_i = \overrightarrow{\mathrm{P}_0\mathrm{P}_i}$, where P_0, P_1, P_2, and P_3 are any four given noncoplanar particles, there exists a state t that can be reached by homogeneous deformation and in which the $\mathbf{e}_i$ are orthonormal. Let us choose a rectangular Cartesian coordinate system $C\colon \mathrm{Q} \to y$ with base vectors $\mathbf{b}_i(C) = \mathbf{e}_i(t)$, and let us choose a body coordinate system $B : \mathrm{P} \to \xi$ such that $B \overset{t}{\equiv} C$. Then $\boldsymbol{\beta}_i(B, \mathrm{P}) \overset{t}{\to} \mathbf{b}_i(C) = \mathbf{e}_i(t)$. Also, in any other state t', we have, from 5.4(31)(iii), $\mathbb{T}(t')\boldsymbol{\beta}_i = (\partial y'^k/\partial \xi^j)\beta^j_{(i)}\mathbf{b}_k(C) = (\partial y'^k/\partial y^i)\mathbf{e}_k(t)$. From 11.2(1)(iv), we have $y'^i = b^i{}_j(t, t')y^j + a^i$, where $b^i{}_j \doteq \partial y'^i/\partial y^j$. If P_k is at $y'^i_{(k)}$ at t' and at $y^i_{(k)}$ at t, then $\mathbf{e}_k(t') \equiv (\overrightarrow{\mathrm{P}_0\mathrm{P}_k})_{t'} = (y'^i_{(k)} - y'^i_{(0)})\mathbf{b}_i(C) = b^i{}_k(t, t')\mathbf{e}_i(t)$, since $\mathbf{e}_k(t) = (y^i_{(k)} - y^i_{(0)})\mathbf{b}_i(C)$. Hence

$$\mathbb{T}(t')\boldsymbol{\beta}_i(B, \mathrm{P}) = \mathbf{e}_i(t') \tag{40}$$

(for all states t' reached by homogeneous deformation). Since the relation of vectors to their reciprocal set is invariant under $\mathbb{T}$, it follows that

$$\mathbb{T}(t')\boldsymbol{\beta}^i(B, \mathrm{P}) = \mathbf{e}^i(t'). \tag{41}$$

If ξ^i denotes the differences of body coordinates for P and P_0 in B, then $\xi^i = y^i - y^i_{(0)}$, since $B \overset{t}{\equiv} C$; y^i denotes the coordinates of the place occupied by P at t. Hence $(\overrightarrow{\mathrm{P}_0\mathrm{P}})_{t'} = (y'^i - y'^i_{(0)})\mathbf{b}_i(C) = b^i{}_j(y^j - y^j_{(0)})\mathbf{b}_i(C) = \xi^i\mathbf{e}_i(t')$, from the preceding results. Q.E.D. Further, since $\mathbb{T}(t')\boldsymbol{\gamma}(\mathrm{P}, t') = \mathbf{U}$, we have, using (40), $\gamma_{ij}(\xi, t') = \boldsymbol{\beta}_i(B, \mathrm{P}) \cdot \boldsymbol{\gamma}(\mathrm{P}, t') \cdot \boldsymbol{\beta}_j(B, \mathrm{P}) = \mathbf{e}_i(t') \cdot \mathbf{U} \cdot \mathbf{e}_j(t') = \mathbf{e}_i(t') \cdot \mathbf{e}_j(t')$. Q.E.D. Similarly, $\gamma^{ij}(\xi, t') = \mathbf{e}^i(t') \cdot \mathbf{e}^j(t')$. We can express 5.6(2) in the form $\boldsymbol{d\varphi}/dA = \pi^{ij}\boldsymbol{\beta}_i\,\boldsymbol{\beta}_j \cdot \boldsymbol{\nu}$; on transfer to space at time t, since $\boldsymbol{d\varphi}/dA \to \mathbf{f}$ and $\boldsymbol{\nu} \to \mathbf{n}$, we obtain the result $\mathbf{f} = \pi^{ij}\mathbf{e}_i\,\mathbf{e}_j \cdot \mathbf{n}$. From (40), it is clear that an equation of the same form will result if the corresponding body tensor equation at any other time t' is transferred to space at time t'. Q.E.D.

Problem 11.3(26) Using an orthonormal set of Cartesian base vectors $\mathbf{b}_i$ having $\mathbf{b}_1$ along the axis of rotation, the velocity $\mathbf{v}$ of a particle of position vector $\mathbf{r} = y^i\mathbf{b}_i$ (with respect to an origin on the axis of rotation) is given by the equation $\mathbf{v} = \dot{\phi}\mathbf{b}_1 \times \mathbf{r} = \dot{\phi}(y^2\mathbf{b}_3 - y^3\mathbf{b}_2)$. Hence $\nabla\mathbf{v} = \mathbf{b}_i\,\partial\mathbf{v}/\partial y^i = \dot{\phi}(\mathbf{b}_2\,\mathbf{b}_3 - \mathbf{b}_3\,\mathbf{b}_2)$ and so $\mathbf{w} \equiv \frac{1}{2}(\nabla\mathbf{v} - \widetilde{\nabla\mathbf{v}}) = \dot{\phi}(\mathbf{b}_2\,\mathbf{b}_3 - \mathbf{b}_3\,\mathbf{b}_2)$. Hence the covariant body vorticity tensor $\boldsymbol{\omega} = \mathbb{T}^{-1}(t)\mathbf{w} = \dot{\phi}(\boldsymbol{\beta}^2\boldsymbol{\beta}^3 - \boldsymbol{\beta}^3\boldsymbol{\beta}^2)$, where the $\boldsymbol{\beta}^i$ are orthonormal covariant body vectors with $\boldsymbol{\beta}^1$ along the axis of rotation. Q.E.D.

AUTHOR INDEX

Numbers in italics refer to the pages on which the complete references are listed.

SUBJECT INDEX

H

I

J

K

L

M

T

A 4
B 5
C 6
D 7
E 8
F 9
G 0
H 1
I 2
J 3